Plant Invasions and Global Climate Change

Sachchidanand Tripathi •
Rahul Bhadouria • Priyanka Srivastava •
Rishikesh Singh • Daizy R. Batish
Editors

Plant Invasions and Global Climate Change

Springer

Editors
Sachchidanand Tripathi
Department of Botany
Deen Dayal Upadhyaya College
University of Delhi
Dehli, India

Rahul Bhadouria
Department of Environmental Studies
Delhi College of Arts and Commerce, University of Delhi
Delhi, India

Priyanka Srivastava
Department of Botany
Sri Guru Tegh Bahadur Khalsa College
University of Delhi
Delhi, India

Rishikesh Singh
School of Earth & Environmental Sciences
Amity University
Mohali, Punjab, India

Daizy R. Batish
Department of Botany
Panjab University
Chandigarh, India

ISBN 978-981-99-5909-9 ISBN 978-981-99-5910-5 (eBook)
https://doi.org/10.1007/978-981-99-5910-5

This Springer imprint is published by the registered company Springer Nature Singapore Pte Ltd.
The registered company address is: 152 Beach Road, #21-01/04 Gateway East, Singapore 189721, Singapore

Paper in this product is recyclable.

Foreword

The world flora has been greatly impacted by the introduction, invasion, and colonization of many exotic weeds from time to time. While a vast majority of such exotics have become naturalized, living in harmony with native flora, yet several, particularly the recently introduced alien weeds have become invasive, posing great challenges to agriculture, flora, and even human health, globally. As these invasive alien species act synergistically with various components of changing climate, an understanding of interaction between species invasion and climate change would be a prime factor for forecasting future shifts in biodiversity and ecosystem management. This book, *Plant Invasions and Global Climate Change* just does that and fulfils the long felt need to bridge the knowledge gap between the processes facilitating plant invasion in ecosystems under changing climate regimes and further helps to effectively prevent, control, and even eradicate invasive species. Various experts in the area have contributed to diverse aspects covering, from plant invasions in different ecosystems under changing climate to forecasting invasions, management and control of weed invasions and even policy interventions for prevention and control of invasions. I hope and trust the book will be of immense use to students and teachers of ecology, scientists, agriculturists, foresters, environmental scientists, and even policy makers. I congratulate all the editors of the book—Drs. Sachchidanand Tripathi, Rahul Bhadouria, Priyanka Srivastava, Rishikesh Singh, and Daizy R. Batish for bringing out this useful contribution. I have my best wishes for this important publication.

J.S.S. Academy of Science (Deemed University)
Mysuru, Karnataka, India

INSA
New Delhi, India

CSIR
Bengaluru, Karnataka, India

Karnataka State Environmental Appraisal Committee
Bengaluru, Karnataka, India

R. Raghavendra Rao

10th June, 2023

Preface

Emergence of novel ecosystems in response to human-induced changes, both biotic and abiotic has posed threat of biotic homogenization. In current scenario, plant invasions are expansive and significant component of anthropogenic global climate change. Moreover, the invasion may be supplemental to the other components of climate change. Invasive alien species (IAS) under current scenario have been suggested as a major threat to biodiversity. Temperature variations may further compromise the adaptability of native species, thereby stressing them and decreasing the resistance potential of natural communities to invasion. It is also predicted that increasing disturbances or extreme events such as fires, floods, cyclones, storms, heatwaves, and droughts will be the direct consequences of changing climate that may facilitate IAS spread and establishment. Rising carbon dioxide levels may not only impact the native plants but also to the ecosystem as a whole in terms of increasing availability of resources and changes in fire regime, thereby providing novel opportunity of spread to invasive species.

A comprehensive understanding of interaction between species invasion and climate change will be supplemental in forecasting future shifts in biodiversity. Recent studies suggested some of the mechanisms that could trigger in promoting plant invasions. Further, different predictive models indicate a plausible increase in the abundance and impact of IAS which may have direct implications for future research and target oriented policy and decision-making. However, these predictions become more complicated considering the complexity of interactions between the impacts of changing climate with other components of global change (changes in land use, nitrogen deposition, etc.) that are affecting the distribution of native plant species, ecosystem dynamics as well as non-native/invasive species. Bioclimatic models have been proved to be useful in predicting the impacts of climate change on biodiversity and future distribution of species. Overall, invasive species are not only affecting the terrestrial and aquatic ecosystems but also to those areas which were erstwhile considered pristine like mountainous ecosystem. And the consequences may further be increased with the changing climate scenario.

This book is having a global approach and targeting the students, teachers, researchers, environmental engineers, and policy makers currently working in this area to augment the state-of-the-art knowledge. The book provides an ensemble of the researches/knowledge related to the challenges, impacts, and precautionary measures for tackling plant invasions under the climate change perspectives in different regions/ecosystems of the world. The book has 17 chapters, which have been further divided into five distinct themes, viz., (1) Plant invasion and climate change: Background, science and mechanistic approach; (2) Plant invasion in different ecosystems: Case studies; (3) Plant invasion: Assessment, mapping, and forecasting; (4). Plant invasion: Management and control; and (5) Plant invasion and policy interventions. Overall, 65 authors from 11 countries (Argentina, Brazil, Canada, China, India, Pakistan, Portugal, Spain, Tunisia, Türkiye, Uruguay) have contributed their chapters in the book.

Chapter 1, entitled 'Plant Invasion and Climate Change: A Global Overview' by Aditi Sharma et al. from India provides a global overview on plant invasion under the changing climate scenario. The background of invasion, the invasion process, important hypotheses, potential effects, and future of plant invasion in the context of global climate change have been covered in this chapter as a comprehensive framework for understanding plant invasion.

Chapter 2, entitled 'Impacts of Plant Invasions on Ecosystem Functionality: A Perspective for Ecosystem Health and Ecosystem Services' by Adrián Lázaro-Lobo et al. from Spain focuses on the effects of plant invasions on ecosystem function and examined how climate change may be incremental to the effects of invasive plants on ecosystem functions and ecosystem services. Additionally, the authors have offered several suggestions for future research, particularly to focus towards individual ecosystem process and services and to assess how climate change influences the impact of invasive plants on a variety of ecosystem processes and services.

Chapter 3, entitled 'Menace of Plant Invasion: A View from Ecological Lens' by Abhishek Raj et al. from India examined the ecology of plant invasion and how it affects the environment and natural resources. This chapter also discusses the mapping, detection, and monitoring of invasive plants using geospatial techniques. It also explores how climate change affects invasive species and carbon (C) dynamics.

Chapter 4, entitled 'Role of Extreme Climate Events in Amplification of Plant Invasion' by Sundari Devi Laishram and Rashmi Shakya from India highlighted the importance of various climatic events that are intensifying as a result of global warming, ecological responses, and their contributions to the spread of exotic and invasive plant species.

Chapter 5, entitled 'Plant Invasion as Gleaned from Parasara's Vrkshayurveda' by D.A. Patil from India explored 34 exotic plant species belonging to 32 genera and 21 families of angiosperms. With regard to plant invasion, several taxa are highlighted, and economic and socio-religious developments are also explored in this chapter.

Chapter 6, entitled 'Water, Wind, and Fire: Extreme Climate Events Enhance the Spread of Invasive Plants in Sensitive North American Ecosystems' by Jennifer Grenz and David R. Clements from Canada examined how invasive plants react to significant weather extremes in the setting of North America. The authors explored site-specific elements that affect vegetation responses and offered insights into the site-specific heterogeneity of vegetation trajectories to climatic events. In light of the rising frequency and scale of climate-related events, they recommended monitoring, mitigation, and proactive management strategies for future research and restoration planning.

Chapter 7, entitled 'Understanding Eco-geographical Relationship in Invaded Ranges by *Acacia longifolia* (Andrews) Willd.—An Intercontinental Case Study on *Acacia* Invasions' by Jorge Luis P. Oliveira-Costa et al. belonging to four/five countries analysed invasive species dynamics on a global scale (in terms of invasiveness/invasibility), concentrating on regions invaded by *Acacia longifolia* with various natural and socio-ecological traits.

Chapter 8, entitled 'Invasive Plants in India—Their Adaptability, Impact, and Response to Changing Climate' by Sonia Rathee et al. from India addressed the invasive flora of India, climatic appropriateness and invasion hotspots, introduction methods, the significance of climate change in plant invasion, adaptations of invasive plants to changing climate, and socio-economic and socio-ecological implications of plant invasion in India.

Chapter 9, entitled 'Plant Invasion in an Aquatic Ecosystem: A New Frontier Under Climate Change' by Reema Mishra et al. from India provided a scientific description on complicated relationships between plant invasion in the aquatic system and climate change.

Chapter 10, 'Plant Invasion and Soil Processes: A Mechanistic Understanding' by Talat Afreen et al. from India outlined the effects of invasive plants on the soil nutrient profile, microbial activity, and phyto-diversity.

Chapter 11, entitled 'Plant Invasion Dynamics in Mountain Ecosystems Under Changing Climate Scenario' by Mushtaq Ahmad Dar et al. from India outlined the dynamics of plant invasion, gave instances of a few significant invasive species found in the Indian Himalayan Region, and made recommendations for the containment of invasive plant species.

Chapter 12, entitled 'The Role of Epigenetics on Plant Invasions Under Climate Change Scenario' by Mehmet Arslan et al. from Turkiye demonstrated how epigenetic pathways can let alien species become invasive in recently established locations.

Chapter 13, entitled 'Comparative Assessment of Machine Learning Algorithms for Habitat Suitability of *Tribulus terrestris* (Linn): An Economically Important Weed' by Manish Mathur and Preet Mathur from India has undertaken a comparative account to evaluate the relative performance of several methods (both regression and machine learning based) for the evaluation of *Tribulus terrestris* habitat suitability inside arid and semi-arid environments of the Indian sub-continent.

Chapter 14, entitled 'The Role of Halophytic Plant Invasions for the Conservation and Restoration of Degraded Agricultural Lands' by Rida Zainab et al. from Pakistan elaborated several strategies in order to emphasize the dynamics of invasive ecosystems and included diverse halophytic plant restoration adaptations to the effects of climate change.

Chapter 15, entitled 'Plant Invasion and Climate Change: An Overview on History, Impacts, and Management Practices' by Rituraj Singh et al. from India provided scientific evidence of history of invasion ecology and key characteristics of invasive species. The authors of this chapter also explored the relationship between invasive species and global warming, the effects of invasive species on ecological, social, and economic systems, as well as the main management strategies now being used.

The penultimate Chap. 16, entitled 'Biochar: A Tool for Combatting Both Invasive Species and Climate Change' by Leeladarshini Sujeeun and Sean C. Thomas from Canada addressed the ecological implications of invasive species, how allelopathy facilitates plant invasions, and how biochar may be used to decrease the impact of invasive plants and climate change.

The ultimate Chap. 17, entitled 'An Action Plan to Prevent and Manage Alien Plant Invasions in India' by Achyut Kumar Banerjee and K.V. Sankaran from China and India outlined the issues in India in regulating the entry and spread of invasive alien plant species, as well as the current rules and capacities in place. In addition, the authors provided a national strategy and action plan for overcoming the challenges associated with managing IAS in the nation, and how to effectively put the recommendations made in India's 5th National Report to the Convention on Biological Diversity into practice.

Overall, the book provides state-of-the-art knowledge on plant invasions in changing climate scenario. Further, it also emphasizes on the theoretical and practical aspects to evaluate the threat level posed by plant invasives along with their management and potential directions for future studies.

New Delhi, India — Sachchidanand Tripathi
New Delhi, India — Rahul Bhadouria
New Delhi, India — Priyanka Srivastava
Mohali, India — Rishikesh Singh
Chandigarh, India — Daizy R. Batish

Contents

Editors and Contributors

About the Editors

Sachchidanand Tripathi Ph.D. is an associate professor, Department of Botany, Deen Dayal Upadhyaya College, University of Delhi, Delhi, India. He obtained his doctoral degree from the Department of Botany, Banaras Hindu University, India. The areas of his interest are plant ecology, soil ecology, ecophysiology, and urban ecology. He has published more than 50 publications (including research publications, books and book chapters, conference proceedings) with reputed international journals and publishers.

Rahul Bhadouria Ph.D. is an assistant professor at the Department of Environmental Studies, Delhi College of Arts and Commerce, University of Delhi, New Delhi, India. He obtained his doctoral degree from the Department of Botany, Banaras Hindu University, Varanasi, India. He has published more than 22 papers, 22 book chapters, and 8 edited books in internationally reputed journals/publishers. His current research areas are management of soil C dynamics to mitigate climate change, a perspective on tree seedling survival and growth attributes in tropical dry forests under the realm of climate change, plant community assembly, functional diversity and soil attributes along the forest-savanna-grassland continuum in India, recovery of degraded mountains in central Himalayas and urban ecology.

Priyanka Srivastava Ph.D. is an assistant professor at Sri Guru Tegh Bahadur Khalsa College (University of Delhi, New Delhi, India). She obtained her doctoral degree from the Department of Botany, Banaras Hindu University, Varanasi, India. She has published several research/review papers in reputed journals of ecology and environmental science. Her current research areas are 'Effect of invasive plant/soil organism/nutrient on the ecological properties of forest ecosystem', 'soil microbes in ecosystem functioning in a tropical dry forest under the realm of climate change', and 'Ecosystem analysis for biodiversity conservation, management and its sustainable development'.

Rishikesh Singh Ph.D. is working as Assistant Professor at School of Earth & Environment Sciences, Amity University, Mohali, Punjab, India. He has worked as National Post-doctoral Fellow (NPDF) at Panjab University, Chandigarh, India. Dr. Singh obtained his doctoral degree from the Institute of Environment and Sustainable Development, Banaras Hindu University, Varanasi. He is an environmental scientist with his research interests in soil carbon dynamics, land-use change and management, waste management, environmental contaminants, biochar, and carbon sequestration. He has published several research and review articles and is a reviewer of several international journals of Elsevier, Wiley, Taylor and Francis, Frontiers, PLoS, and Springer Nature groups. He has published six books in leading international publishers such as Elsevier, Springer Nature, and Wiley. Dr. Singh has more than 70 Scopus publications with >1100 citations and a Scopus h-index 21.

Daizy R. Batish Ph.D. is Professor at the Department of Botany, Panjab University, Chandigarh, India. Prof. Batish obtained her doctoral degree from the Department of Botany, Panjab University, Chandigarh. She is a well-known researcher in the field of Invasion Ecology and has more than 30 years of teaching and research experience. She has worked in different dimensions of invasion ecology such as allelopathy, phytotoxicity, species interactions, phytochemicals extraction and utilization, and essential oils. She has published several research and review articles and is working as a reviewer of several international journals of Elsevier, Wiley, Taylor and Francis, and Springer Nature groups. She has published seven books related to different dimensions of invasion ecology. Prof. Batish has more than 170 Scopus publications with >6800 citations and a Scopus h-index >46.

Contributors

Zainul Abideen Dr. Muhammad Ajmal Khan Institute of Sustainable Halophyte Utilization, University of Karachi, Karachi, Pakistan

Talat Afreen PG Centre of Botany, Gaya College (A Constituent Unit of Magadh University), Bodhgaya, Bihar, India

Preeti Agarwal Department of Botany, Gargi College, University of Delhi, Delhi, India

Mustaqeem Ahmad Laboratory for the Conservation of Endangered Species (LaCONES), CSIR-CCMB, Hyderabad, Telangana, India

Álvaro Alonso Biological Invasions Research Group, Department of Life Sciences, University of Alcalá, Alcalá de Henares, Madrid, Spain

Mehmet Arslan Erciyes University, Talas/Kayseri, Türkiye

Achyut Kumar Banerjee State Key Laboratory of Biocontrol and Guangdong Provincial Key Laboratory of Plant Resources, School of Life Sciences, Sun Yat-sen University, Guangzhou, China

Arnab Banerjee Department of Environmental Science, Sant Gahira Guru Vishwavidyalaya, Sarguja, Ambikapur, Chhattisgarh, India

Daizy R. Batish Department of Botany, Panjab University, Chandigarh, India

Rahul Bhadouria Department of Environmental Studies, Delhi College of Arts and Commerce, University of Delhi, New Delhi, India

Pilar Castro-Díez Biological Invasions Research Group, Department of Life Sciences, University of Alcalá, Alcalá de Henares, Madrid, Spain

Rigzin Chuskit Department of Botany, Panjab University, Chandigarh, India

David R. Clements Department of Biology, Trinity Western University, Langley, BC, Canada

Mushtaq Ahmad Dar Department of Botany, Panjab University, Chandigarh, India

Rajkumari S. Devi Department of Botany, Deen Dayal Upadhyaya College (University of Delhi), New Delhi, India

S. Dinesha Pt. Deendayal Upadhyay College of Horticulture and Forestry, Dr. Rajendra Prasad Central Agriculture University, Pusa, Samastipur, Bihar, India

César Fagundez Centro Universitario Regional del Este (CURE), Universidad de La Republica (UdelaR), Rocha, Uruguay

Romina D. Fernández Institute of Regional Ecology, National University of Tucumán-CONICET, Tucumán, Argentina

Rui Ferreira de Figueiredo Department of Geography and Tourism, CEGOT, Faculty of Letters and Humanities, University of Coimbra, Coimbra, Portugal

Elena Granda Biological Invasions Research Group, Department of Life Sciences, University of Alcalá, Alcalá de Henares, Madrid, Spain

Jennifer Grenz Department of Forest Resources Management, University of British Columbia, Vancouver, BC, Canada

Maria Hasnain Department of Biotechnology, Lahore College for Women University, Lahore, Pakistan

Ramesh Kumar Jha Pt. Deendayal Upadhyay College of Horticulture and Forestry, Dr. Rajendra Prasad Central Agriculture University, Pusa, Samastipur, Bihar, India

Manoj Kumar Jhariya Department of Farm Forestry, Sant Gahira Guru Vishwavidyalaya, Sarguja, Ambikapur, Chhattisgarh, India

Amarpreet Kaur Department of Botany, Panjab University, Chandigarh, India

Pritam Kaur Department of Botany, Gargi College, University of Delhi, Delhi, India

Shalinder Kaur Department of Botany, Panjab University, Chandigarh, India

Hammad Afzal Kayani Department of Biosciences, The Shaheed Zulfikar Ali Bhutto Institute of Science and Technology, Karachi, Pakistan

Ravinder K. Kohli Department of Botany, Panjab University, Chandigarh, India
Amity University, Mohali, Punjab, India

Sweety Kumari PG Centre of Botany, Gaya College (A Constituent Unit of Magadh University), Bodhgaya, Bihar, India

Krishna Kumar Pt. Deendayal Upadhyay College of Horticulture and Forestry, Dr. Rajendra Prasad Central Agriculture University, Pusa, Samastipur, Bihar, India

Sundari Devi Laishram Department of Botany, Miranda House, University of Delhi, Delhi, India

Adrián Lázaro-Lobo Biological Invasions Research Group, Department of Life Sciences, University of Alcalá, Alcalá de Henares, Madrid, Spain
Department of Biology and Geology, King Juan Carlos University, Móstoles, Spain

Manish Mathur ICAR-Central Arid Zone Research Institute, Jodhpur, India

Preet Mathur Jodhpur Institute of Engineering and Technology, Jodhpur, India

Taher Mechergui Faculty of Sciences of Bizerte (FSB), Jarzouna, Tunisia
Sylvo-Pastoral Resources Laboratory of Tabarka, Tabarka, Tunisia

Reema Mishra Department of Botany, Gargi College, University of Delhi, Delhi, India

Lia Montti Instituto de Investigaciones Marinas y Costeras (IIMyC), FCEyN, Universidad Nacional de Mar del Plata-CONICET, Mar del Plata, Argentina
Instituto de Geología de Costas y del Cuaternario (IGCyC), FCEyN, Universidad Nacional de Mar del Plata-CIC, Mar del Plata, Argentina

Jorge Luis P. Oliveira-Costa Centre of Studies in Geography and Spatial Planning (CEGOT), University of Coimbra, Coimbra, Portugal

D. A. Patil Post-Graduate Department of Botany, S.S.V.P. Sanstha's L.K. Dr. P.R. Ghogrey Science College, Dhule, Maharashtra, India

Vânia Regina Pivello Department of Ecology, Institute of Biosciences, University of São Paulo (USP), São Paulo, Brazil

Abhishek Raj Pt. Deendayal Upadhyay College of Horticulture and Forestry, Dr. Rajendra Prasad Central Agriculture University, Pusa, Samastipur, Bihar, India

Prakash Rajak Ecosystem Analysis Laboratory, Department of Botany, Banaras Hindu University, Varanasi, India

Sonia Rathee Department of Botany, Panjab University, Chandigarh, India

Alberto Romero-Blanco Biological Invasions Research Group, Department of Life Sciences, University of Alcalá, Alcalá de Henares, Madrid, Spain

Asunción Saldaña-López Biological Invasions Research Group, Department of Life Sciences, University of Alcalá, Alcalá de Henares, Madrid, Spain

K. V. Sankaran Kerala Forest Research Institute, Peechi, Thrissur, India

Rashmi Shakya Department of Botany, Miranda House, University of Delhi, Delhi, India

Aditi Sharma Department of Botany, Panjab University, Chandigarh, India

Padma Sharma Department of Environment Studies, Panjab University, Chandigarh, India

Ashutosh Kumar Singh Centre of Science and Society, University of Allahabad, Prayagraj, India

Garvita Singh Department of Botany, Gargi College, University of Delhi, Delhi, India

Harminder Pal Singh Department of Environment Studies, Panjab University, Chandigarh, India

Hema Singh Ecosystem Analysis Laboratory, Department of Botany, Banaras Hindu University, Varanasi, India

Rishikesh Singh Department of Botany, Panjab University, Chandigarh, India
School of Earth & Environment Sciences, Amity University, Mohali, Punjab, India

Ritturaj Singh Ecosystem Analysis Laboratory, Department of Botany, Banaras Hindu University, Varanasi, India

Swati Singh Central Pollution Control Board, Delhi, India

Renu Soni Department of Botany, Gargi College, University of Delhi, Delhi, India

Leeladarshini Sujeeun Institute of Forestry and Conservation, University of Toronto, Toronto, ON, Canada

Sean C. Thomas Institute of Forestry and Conservation, University of Toronto, Toronto, ON, Canada

Sachchidanand Tripathi Department of Botany, Deen Dayal Upadhyaya College (University of Delhi), New Delhi, India

Ahmet Uludağ Çanakkale Onsekiz Mart University, Çanakkale, Türkiye

İlhan Üremiş Hatay Mustafa Kemal University, Antakya, Hatay, Türkiye

Rida Zainab Department of Biotechnology, Lahore College for Women University, Lahore, Pakistan

Part I

Plant Invasion and Climate Change: Background, Science and Mechanistic Approach

Plant Invasion and Climate Change: A Global Overview

1

Aditi Sharma, Amarpreet Kaur, Shalinder Kaur, Ravinder K. Kohli, and Daizy R. Batish

Abstract

The phenomenon of plant invasion is a consequence of invading plants' exceptional range expansion into new geographic areas. Even though older naturalists were aware of the problem of plant invasion, research on the subject has intensified mainly in the last two decades. The attributes of migrated alien plants, as well as the biotic and abiotic aspects of the introduced environment—which may be investigated with the aid of numerous hypotheses—are what lead to successful plant invasions. After going through an introduction-naturalization-invasion continuum, these species dominate the invaded ecosystem, homogenize the floristic composition, jeopardize rare and unique species, disturb ecosystem stability, and incur high social and financial losses. In the future, it is anticipated that the range of these species will increase significantly, in part due to the expansion of global trade, agriculture, and other human activities, and somewhat due to anthropogenically induced climate change. Most of the invasive plant species respond positively to various consequences of climate change, viz. rising temperatures, augmented nitrogen accumulation, enhanced CO_2 levels, erratic precipitation regimes, etc. With the growing fierceness of the recognized invaders and the continuous appearance of novel invaders, the threats and difficulties pertaining to the invasive aliens are continuously increasing. Furthermore, biological invasions and climate change may act concomitantly and magnify each other's effect, which makes it important to study both phenomena

A. Sharma · A. Kaur · S. Kaur · D. R. Batish (✉)
Department of Botany, Panjab University, Chandigarh, India
e-mail: daizybatish@pu.ac.in

R. K. Kohli
Department of Botany, Panjab University, Chandigarh, India

Amity University, Mohali, Punjab, India

S. Tripathi et al. (eds.), *Plant Invasions and Global Climate Change*,
https://doi.org/10.1007/978-981-99-5910-5_1

collectively to devise a better approach to mitigate their effects. This chapter offers a general framework for understanding plant invasion, including the fundamental background, the process of invasion, key hypotheses, consequences, and future of plant invasion in the global climate change scenario.

Keywords

Climate change · Ecological impacts · Invasion hypotheses · Invasive species · Socio-economic impacts

1.1 Introduction

The dissolution of biogeographic boundaries and the improvement of global trade, transportation, and tourism have increased the cross-border migration of non-indigenous plant species, giving rise to the global environmental challenge of plant invasion (Bonnamour et al. 2021; Byrne et al. 2022). The term *plant invasion* refers to the unusual range expansion of species into new geographical areas, whereas the term *invasive plant species* refers to the tiny percentage of migratory plant species introduced purposely or accidentally outside their natural range which are able to acclimate the novel environments, establish self-sustainable populations, and have a negative ecological and socio-economic influence on the introduced habitat (Kaur et al. 2019; Shackleton et al. 2019). Richardson et al. (2000) provided a general overview of the plant invasion as a complex multistage process (Fig. 1.1), the major steps of which are explained hereunder:

Introduction: A plant or its propagule must be transported via any agency across the primary intercontinental and/or intracontinental geographic barriers to begin the invasion process. It is mostly mediated by humans, but other elements may also be responsible. At this stage, the species could be described as "alien", "exotic", "non-native", "non-indigenous", or "introduced".

Acclimatization: The primary obstacle an alien plant species faces when it is introduced in a novel habitat is the environmental barriers composed of biotic and abiotic components. An invasive alien species must learn to adapt to such environmental variations to survive in the new habitat.

Naturalization: After establishment, a species must get past any obstacles preventing it from reproducing continuously. At this stage of invasion, a species is classified as "casual" or "naturalized". Casuals are characterized as imported species that can persist and occasionally breed but are unable to create populations that can replace themselves. As a result, they are dependent on frequent introductions for their survival within novel non-native bounds. Contrarily, plants capable of reproducing on their own, independently, and for numerous generations are called as naturalized.

Invasion: The term "invasive" is referred to a naturalized organism, which produces enormous off-springs through vegetative and/or generative mechanisms and

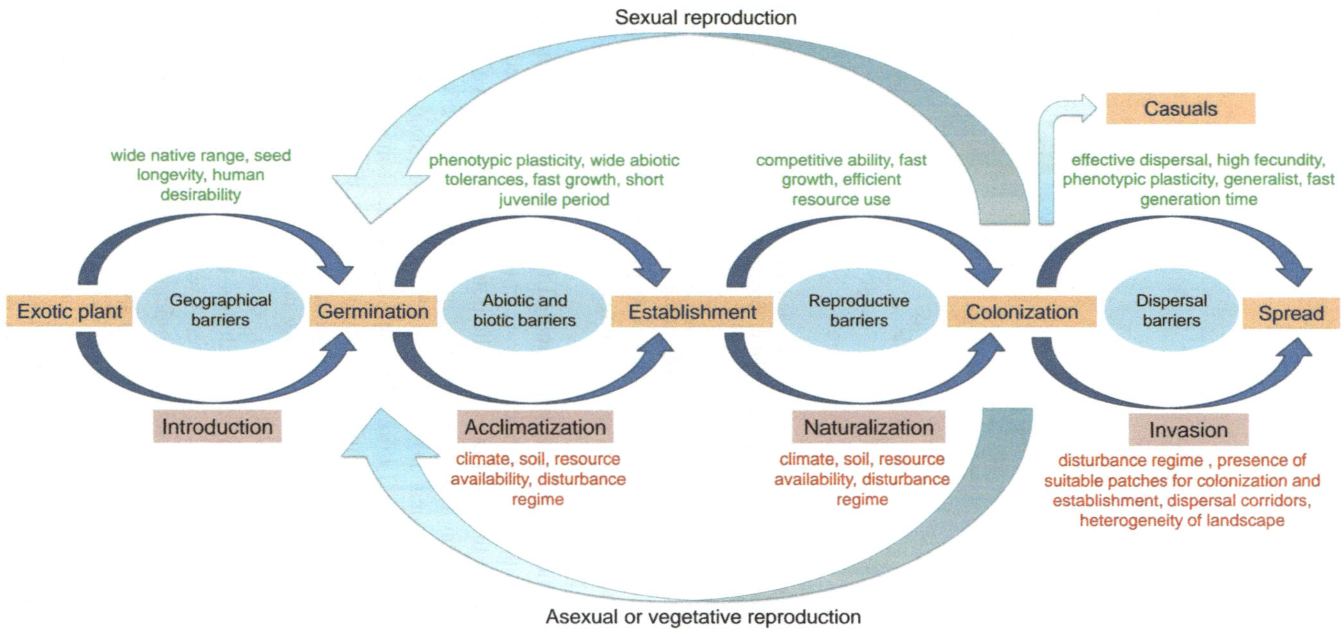

Fig. 1.1 Scheme of the plant invasion process summarizing significant stages, barriers, and plant and habitat attributes facilitating the process. (After Richardson et al. 2000; Theoharides and Dukes 2007)

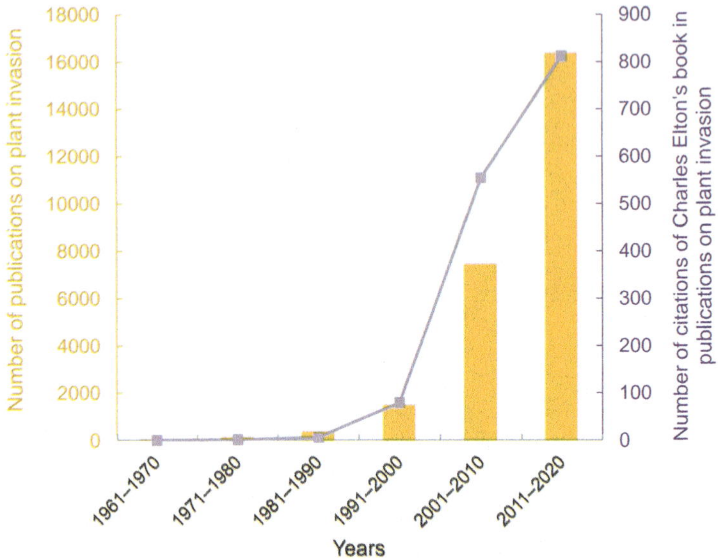

Fig. 1.2 Rate of publications on plant invasion and citations of Charles Elton's book "The Ecology of Invasions by Animals and Plants" (Elton 1958) in those publications over the years

overcomes local/regional dispersal obstacles, thereby spreading far from the parent plants. At this stage, the species begins to interfere with the native vegetation of the introduced region, leading to severe consequences.

Since ancient times, naturalists have studied the phenomenon of invasion, and various nineteenth and twentieth century scientists have mentioned such terms and/or descriptions in their writings. Yet the notion did not perceive much significance until Charles Elton, a British naturalist from the twentieth century, described the precise idea of biological invasion in his book "The Ecology of Invasions by Animals and Plants" (Elton 1958). Elton introduced the concept of invasion and emphasized the unnatural distribution of invasive species, their impacts on biodiversity, and the reasons for their spread. Even today, the hypotheses put forward by him with limited experimental evidence are duly considered and proven from time to time (Richardson and Pyšek 2008). Both, the number of publications on plant invasion and the acknowledgement of Elton's book in those publications have readily increased over time (Fig. 1.2). Later it was also established that biological invasion is the second most important ecological disturbance that endangers global biodiversity (Bellard et al. 2016) and is the primary factor responsible for island ecosystems' loss of species (Tershy et al. 2002).

Lately, invasion dynamics are changing unprecedently due to global climate change. Invasive plant species respond positively to various components of climate change, viz. global warming, augmented nitrogen accumulation, enhanced CO_2 levels, erratic precipitation regimes, etc. (Gao et al. 2018; Johnson and Hartley

2018; Howell et al. 2020; Ren et al. 2022). With the growing aggressiveness of the established invasive species and the ongoing appearance of novel invaders, the threats and difficulties pertaining to plant invasion are continuously increasing. Furthermore, biological invasions and climate change may act concomitantly and magnify each other's effects (Sage 2020), which makes it important to study both phenomena collectively to devise a better approach to mitigate their effects.

The present chapter discusses the background and concepts of plant invasion. This discussion aims to improve understanding of the phenomenon of plant invasion by outlining the key factors responsible for the establishment of an alien plant species in a new geographic range. The subsequent presentation of the global status of invasive plant species along with their negative effects at ecological and socio-economic levels and their response toward global climate change emphasizes the current and futuristic issues pertaining to plant invasion.

1.2 Factors Affecting the Success of Invasive Alien Plants

The introduction-naturalization-invasion continuum relies heavily on the functional traits of the introduced species as well as those of the invaded habitat (Roiloa et al. 2020; Ibáñez et al. 2021). Additionally, it is acknowledged that invasion is governed by a variety of factors and is not dependent on a single theory or hypothesis (Dai et al. 2020). For example, three factors, including climatic conditions, habitat resistance, and vigour of the invasive species, are used in the Invasion Factor Framework presented by Young et al. (2022) to elucidate the establishment of invasive plant species. Likewise, the combined outcome of species introduction and plantation record, changes in the introduced regions, and dispersal passages, all contribute to the species richness of alien taxa in the natural forest ecosystems (Wagner et al. 2021). Further, Liao et al. (2021) stated that different functional characteristics account for different aspects of the process. For instance, the characteristics linked to population growth and evolutionary adaptation may determine the breadth of invasion, whereas traits associated with relative competitiveness define the severity of invasion impact (Ni et al. 2021). Fuentes-Lillo et al. (2021), on the contrary, believe that the anthropogenic influences may outweigh the abiotic factors in being the most significant driver of the distribution of alien plant species in certain cases (Fuentes-Lillo et al. 2021). Similar frameworks and hypotheses have been put forth by researchers from time to time explaining the mechanisms underlying the invasion process, and some of those express opposing viewpoints (Enders et al. 2018).

1.2.1 Hypotheses Ascertaining the Influence of Habitat Characteristics on Invasion

Community ecologists have long recognized the importance of habitat characteristics in the success of alien plant invasion. It is becoming more widely acknowledged that community ecology's ideas and experimental methods could significantly advance our knowledge of plant invasions and ability to control them (Huston 2004). Certain well-accepted hypotheses that are proven to enhance the invasibility of an ecosystem are explained hereunder:

Disturbance: The hypothesis states that compared to the ecosystems have not been disturbed, perturbed ecosystems are more likely to be the target of alien species invasions (Elton 1958; Hobbs and Huenneke 1992). Disturbance facilitates greater seedling recruitment for many invasive species, thus, having an essential role in their success (Pearson et al. 2022). Guo et al. (2022) demonstrated that even among different habitat types, disturbance-related factors affected the invasibility of a species more than phylogenetic and native plant diversity.

Empty Niche or opportunity windows: Invaders are drawn to establish and reproduce when there are resources or unfilled niches available (MacArthur 1970). Holzmueller and Jose (2011) reported that the patches of *Imperata cylindrica* (L.) P.Beauv. quickly increase in size and density to fill vacant niches that emerged after a disturbance, such as a fire or a hurricane.

Fluctuating resource: Any natural or anthropogenic disturbance increases or decreases resource availability, and hence, impacts the vegetation patterns causing dominance of invaders (Davis et al. 2000). Ibáñez et al. (2021) observed that fluctuation of resources is strongly linked with the performance of alien invasive plants, particularly in case of decreasing water availability and/or increasing light and nutrient availability.

Diversity-invasibility or biotic resistance: The hypothesis states that biodiverse communities are more capable of fending off the invasion by alien species in comparison to ecosystems with lesser diversity (Elton 1958). The shreds of experimental evidence collected at small geographical scales support the hypothesis contending that biologically diverse communities are fiercely competitive (Ernst et al. 2022). Li et al. (2022a) demonstrated a negative association between resident species diversity and grassland invasion by *Ambrosia artemisiifolia* L., which remained constant even after the nutrient addition. Likewise, in various community types and ecoregions of the United States, Beaury et al. (2020) showed a negative association between native richness and alien occurrence.

Biotic acceptance: In contrary to the biotic resistance hypothesis, certain researchers state that the presence of rich and diverse native populations supports invasion by non-native species (Stohlgren et al. 2006). In the riparian forests of the Warta River Valley (Poland), Dyderski et al. (2015) discovered an affirmative association between the richness of alien and indigenous woody perennials, which indicates that diverse ecosystems readily attract exotic invaders.

Enemy release: According to this hypothesis, the lack of native foes (pests, diseases, and predators) in the invaded zone encourages the unrestrained spread of the alien species (Elton 1958; Keane and Crawley 2002). Native habitats have greater control over plant populations via natural enemies than non-native habitats (Lucero et al. 2019). For instance, the establishment of *Ambrosia trifida* L. in the alien range was influenced by the release from both above- and below-ground enemies, which used to attack the weed at different life stages (Zhao et al. 2020).

Specialist–generalist: Ecosystems that have specialized local pests/predators and generalist local mutualists are more vulnerable to invasion (Callaway et al. 2004). A study by Eschtruth and Battles (2009) examined the role of the white-tailed deer (*Odocoileus virginianus* Zimmermann), a generalist herbivore, in the invasion of three exotic plant species (*Microstegium vimineum* (Trin.) A.Camus, *Alliaria petiolata* (M.Bieb.) Cavara and Grande, and *Berberis thunbergii* DC.) in American forest ecosystems, and the findings of the study showed that herbivory patterns exhibited by the deer can hasten the spread of exotic plants.

Island susceptibility: Compared to continental ecosystems, islands are additionally vulnerable to the onslaught and effects of invaders (Jeschke 2008). Gimeno et al. (2006) examined the comparative susceptibility of islands to the invasion of *Oxalis pes-caprae* L. in comparison to the neighbouring mainland regions of Spain and found that the islands occupy a larger share of habitats preferred by the weed.

1.2.2 Hypotheses Ascertaining the Influence of Plant Characteristics on Invasion

Ecologists have also long sought to predict which species are likely to invade new habitats, and recently, quantitative studies have been employed to do so. In this context, the suggested hypothesis and investigations, despite being limited to a small number of taxa, provide valuable insights into the establishment and dissemination of invaders (Kolar and Lodge 2001). Some of these hypotheses are listed hereunder:

Ideal weed: Certain characteristics of an invasive plant species determine the chances of its successful establishment in the introduced habitat (Elton 1958; Rejmánek and Richardson 1996). Functional traits such as the ability to germinate under diverse conditions, fast nutrient acquisition, high growth and reproduction rate, quick life cycle, etc. play a significant role during the introduction phase of the invasion process, making it easier for the introduced plants to survive and colonize new ranges (Dai et al. 2020; Montesinos 2022).

Limiting similarity: The likelihood of invasion by a species will increase as the disparity between native and foreign species grows (MacArthur and Levins 1967). The concept of limiting similarity is built on the idea that antagonism within species would be the highest amongst phylogenetically closer species and for species to coexist, they need to be functionally distinct (Price and Pärtel 2013).

Darwin's naturalization conundrum: Charles Darwin presented two opposing hypotheses related to plant invasion: the "pre-adaptation hypothesis" stating that pre-adapted traits in an exotic species would be crucial for environmental filtering and its survival in a particular habitat and the "naturalization hypothesis" stating that trait disparities in an exotic species allow it to successfully establish via niche differentiation and competitive exclusion (Park et al. 2020). A recent study by Omer et al. (2022) demonstrates that the correlation between phylogenetic remoteness to the indigenous vegetation and the successful establishment of a non-native plant species shifts from one step of the invasion process to the next one, thus proving both hypotheses.

Evolution of increased competitive ability: When herbivory is reduced as a result of the dearth of usual foes in the novel habitat, invaders choose higher growth rates and improved competitiveness over defense (Blossey and Nötzold 1995). According to Feng et al. (2011), *Ageratina adenophora* (Spreng.) R.M.King & H.Rob. populations from non-native ranges (China and India) allocate more nitrogen to photosynthesis and less to cell walls than native populations, indicating a shift away from defense and towards growth and development. Likewise, invasive plants change the composition of secondary metabolites to produce fewer compounds that are used to protect them from herbivores and more chemicals that are used to help them adapt to their abiotic environment (Xiao et al. 2020).

Phenotypic plasticity: Invasive species can function superior in a variety of new localities by altering their phenotypic traits in response to environmental conditions (Williams et al. 1995). According to Rathee et al. (2021), phenotypic alterations in reproductive traits assisted *Parthenium hysterophorus* L. to invade and spread well in mountainous ecosystems.

Propagule pressure: An invasive species has a competitive advantage over native species if it can produce long-lasting, viable seeds (Lockwood et al. 2013). For example, high propagule pressure considerably increases the dry weight and dominance index of *Solidago canadensis* L. (Liu et al. 2022).

Invasional meltdown: Ecosystem disruption caused by invaders allows other alien species to establish themselves (Simberloff and von Holle 1999; Sax et al. 2007). Fruit preferences and foraging strategies of an invasive fruit-eating mammal, *Macaca fascicularis* Raffles enhance the seed dispersal of invasive plants in remnant forests of Mauritius (Reinegger et al. 2022). Likewise, in China, *A. philoxeroides* (Mart.) Griseb. has been noted to act as a wintertime insulator for an alien mosquitofish (*Gambusia affinis* S. F. Baird & Girard), which allows the spread of the fish as farther as the plant is expanding with climate change (Xiong et al. 2019).

Novel weapon: Phytochemicals (known as allelochemicals) generated by an invading plant species mediate novel interactions among plants and between plants and microbes (known as allelopathy), changing how the ecosystem functions (Callaway and Ridenour 2004). Invasive plants such as *P. hysterophorus*, *Verbesina encelioides* (Cav.) A.Gray, *Calyptocarpus vialis* Less., etc. exhibit phytotoxicity against various crop and weed species by releasing toxic chemical compounds via leachation, root exudation, and residual decomposition (Lal et al. 2021; Mehal et al. 2023a; Kaur et al. 2022a). The latest investigation indicated that allelopathy is present in 72% of the 524 invasive species studied, suggesting it to be a ubiquitous mechanism of invasion (Kalisz et al. 2021).

Community ecology: Invasive alien species with evolved phenologies become easily acclimated to the non-native ranges, particularly under climate change scenarios (Wolkovich and Cleland 2011). *P. hysterophorus*' varied phenology in response to shifting temperature and humidity conditions, as described by Kaur et al. (2017), explains the weed's ability to adapt and invade a variety of non-native habitats.

The invasion aspect of a species may be affected by one of these or several additional elements that have not been taken into account by these hypotheses. Although we have a solid grasp of the principles underlying successful plant invasions, yet there is much more to investigate and learn considering the complexity of ecological components and functions and the constantly increasing frequency of invasions.

1.3 Statistics of Invasive Alien Plants

It is anticipated that about one-sixth of the earth's landmass, which also constitutes 16% of the world's biodiversity hotspots, is vulnerable to invasion (Early et al. 2016). In the world's 843 continental and island locations, 13,168 naturalized vascular plant species have been reported that account for nearly 3.9% of the global extant flora (van Kleunen et al. 2015). According to recent estimates, non-native species currently make up more than one-fourth of island floras (Brock and Daehler 2022). The scientists also claimed that while the Pacific Islands had the highest accretion rate of naturalized flora, North America had the maximum naturalized flora (van Kleunen et al. 2015). This study was supported by a second investigation on the naturalized alien flora of the world, which identified California, North America as having the most diverse naturalized alien flora with 1753 species of alien plants (Pyšek et al. 2017). Likewise, more than 2677 naturalized exotics are recorded from various countries in South America (Zenni et al. 2022). South Africa, with 1139 species, constitutes the maximum number of naturalized non-natives among African countries (Richardson et al. 2022). In Europe, the maximum naturalized flora is reported from England (1379 species), followed by Sweden (874 species), Scotland (861 species), Wales (835 species), France (716 species), the European part of Russia (649 species), Ukraine (626 species), and Norway (595 species), showing that northern Europe is the most heavily invaded (Pyšek et al. 2022). An inventory of global plant invaders is presented in Table 1.1 (Global Invasive Species Database 2023).

A comparatively limited number of families and genera contain the majority of global invaders (Mack et al. 2000). Asteraceae, which includes 1343 species, has contributed the most to the world's naturalized flora, trailed by 1267 species of Poaceae and 1189 species of Fabaceae (1189 species) (Pyšek et al. 2017). Global representative genera of naturalized alien plants are *Solanum*, *Euphorbia*, and *Carex* with 112, 108, and 106 species, respectively (Pyšek et al. 2017). It has also been determined that transportation and naive possessions contribute to the majority of

Table 1.1 List of the global invasive alien plant species as provided by Global Invasive Species Database (2023)

Family	Plant species
Acanthaceae	*Acanthus mollis*; *Asystasia gangetica*; *Hygrophila polysperma*; *Ruellia brevifolia*; *Thunbergia grandiflora*
Aceraceae	*Acer ginnaal*, *A. platanoides*
Agavaceae	*Agave americana*, *A. sisalana*; *Furcraea foetida*; *Phormium tenax*
Aizoaceae	*Carpobrotus edulis*
Araliaceae	*Hedera helix*
Alismataceae	*Sagittaria platyphylla*, *S. sagittifolia*
Amaranthaceae	*Alternanthera philoxeroides*, *A. sessilis*
Anacardiaceae	*Cotinus coggygria*; *Rhus longipes*; *Schinus terebinthifolius*
Annonaceae	*Annona glabra*, *A. squamosa*
Apiaceae	*Heracleum mantegazzianum*
Apocynaceae	*Funtumia elastica*; *Thevetia peruviana*; *Vinca major*
Araceae	*Epipremnum pinnatum*; *Pistia stratiotes*; *Syngonium podophyllum*; *Zantedeschia aethiopica*
Araliaceae	*Schefflera actinophylla*
Arecaceae	*Archontophoenix cunninghamiana*; *Elaeis guineensis*; *Livistona chinensis*; *Phoenix canariensis*; *Trachycarpus fortunei*
Asclepiadaceae	*Cryptostegia grandiflora*, *C. madagascariensis*; *Cynanchum rossicum*
Asteraceae	*Ageratina adenophora*, *A. riparia*; *Ageratum conyzoides*; *Ambrosia artemisiifolia*; *Austroeupatorium inulifolium*; *Bellis perennis*; *Bidens pilosa*; *Carduus nutans*; *Centaurea biebersteinii*, *C. diffusa*, *C. melitensis*, *C. solstitialis*; *Chromolaena odorata*; *Chrysanthemoides monilifera*; *Cirsium arvense*, *C. vulgare*; *Conyza floribunda*; *Cynara cardunculus*; *Delairea odorata*; *Dyssodia tenuiloba*; *Elephantopus mollis*; *Erigeron karvinskianus*; *Eupatorium cannabinum*; *Euryops multifidus*; *Gymnocoronis spilanthoides*; *Hieracium aurantiacum*, *H. floribundum*, *H. pilosella*; *Hypochaeris radicata*; *Launaea intybacea*; *Mikania micrantha*; *Nypa fruticans*; *Onopordum acanthium*; *Parthenium hysterophorus*; *Pluchea carolinensis*, *P. indica*; *Senecio angulatus*, *S. inaequidens*, *S. jacobaea*, *S. squalidus*, *S. viscosus*, *S. vulgaris*; *Sonchus asper*, *S. oleraceus*; *Sphagneticola trilobata*; *Taraxacum officinale*; *Tithonia diversifolia*; *Tussilago farfara*; *Xanthium spinosum*
Balsaminaceae	*Impatiens glandulifera*, *I. walleriana*
Basellaceae	*Anredera cordifolia*
Begoniaceae	*Begonia cucullata*
Berberidaceae	*Berberis buxifolia*, *B. darwinii*, *B. thunbergii*
Betulaceae	*Alnus glutinosa*
Bignoniaceae	*Macfadyena unguis-cati*; *Spathodea campanulata*; *Tabebuia heterophylla*; *Tecoma capensis*, *T. stans*
Boraginaceae	*Cynoglossum officinale*; *Heliotropium curassavicum*
Brassicaceae	*Alliaria petiolata*; *Brassica elongata*, *B. tournefortii*; *Camelina sativa*; *Cardamine flexuosa*, *C. glacialis*; *Lepidium latifolium*, *L. virginicum*
Buddlejaceae	*Buddleja davidii*, *B. madagascariensis*
Butomaceae	*Butomus umbellatus*
Cabombaceae	*Cabomba caroliniana*

(continued)

Table 1.1 (continued)

Family	Plant species
Cactaceae	*Acanthocereus tetragonus*; *Opuntia cochenillifera*, *O. ficus-indica*, *O. monacantha*, *O. stricta*
Cannaceae	*Canna indica*
Caprifoliaceae	*Lonicera japonica*, *L. maackii*
Caryophyllaceae	*Cerastium fontanum*; *Sagina procumbens*; *Stellaria alsine*, *S. media*
Casuarinaceae	*Casuarina equisetifolia*
Cecropiaceae	*Cecropia peltata*, *C. schreberiana*
Celastraceae	*Celastrus orbiculatus*; *Euonymus alata*, *E. fortunei*
Ceratophyllaceae	*Ceratophyllum demersum*
Chenopodiaceae	*Salsola tragus*
Chrysobalanaceae	*Chrysobalanus icaco*
Clusiaceae	*Hypericum perforatum*
Combretaceae	*Terminalia catappa*
Commelinaceae	*Commelina benghalensis*; *Tradescantia fluminensis*, *T. spathacea*
Convolvulaceae	*Ipomoea aquatic*, *I. cairica*, *I. setosa* ssp. *pavonii*; *Merremia peltata*, *M. tuberosa*
Crassulaceae	*Crassula helmsii*; *Kalanchoe pinnata*
Cucurbitaceae	*Coccinia grandis*; *Sechium edule*
Cyperaceae	*Cyperus rotundus; Oxycaryum cubense*
Dioscoreaceae	*Dioscorea bulbifera*, *D. oppositifolia*
Elaeagnaceae	*Elaeagnus angustifolia*, *E. pungens*, *E. umbellata*
Ericaceae	*Calluna vulgaris*; *Rhododendron ponticum*
Euphorbiaceae	*Aleurites moluccana*; *Antidesma bunius*; *Euphorbia esula*; *Jatropha gossypiifolia*; *Ricinus communis*; *Triadica sebifera*
Fabaceae	*Abrus precatorius*; *Acacia concinna*, *A. confuse*, *A. farnesiana*, *A. longifolia*, *A. mangium*, *A. mearnsii*, *A. melanoxylon*, *A. nilotica*, *A. pycnantha*, *A. retinodes*, *A. saligna*; *Adenanthera pavonina*; *Albizia julibrissin*, *A. lebbeck*; *Caesalpinia decapetala*; *Coronilla varia*; *Cytisus scoparius*, *C. striatus*; *Dalbergia sissoo*; *Dichrostachys cinerea*; *Dipogon lignosus*; *Falcataria moluccana*; *Flemingia strobilifera*; *Genista monspessulana*; *Haematoxylum campechianum*; *Lespedeza cuneata*; *Leucaena leucocephala*; *Lotus corniculatus*; *Melilotus alba*; *Mimosa diplotricha*, *M. pigra*, *M. pudica*; *Prosopis glandulosa*, *P. juliflora*; *Psoralea pinnata*; *Pueraria montana* var. *lobata*; *Robinia pseudoacacia*; *Samanea saman*; *Senegalia catechu*; *Sesbania punicea*; *Trifolium dubium*, *T. repens*; *Ulex europaeus*; *Vachellia drepanolobium*; *Wisteria floribunda*, *W. sinensis*
Flacourtiaceae	*Flacourtia indica*
Geraniaceae	*Erodium cicutarium*
Goodeniaceae	*Scaevola sericea*
Gunneraceae	*Gunnera manicata*, *G. tinctoria*
Haloragaceae	*Myriophyllum aquaticum*, *M. heterophyllum*, *M. spicatum*
Hydrocharitaceae	*Egeria densa*; *Elodea canadensis*; *Halophila stipulacea*; *Hydrilla verticillata*; *Hydrocharis morsus-ranae*; *Lagarosiphon major*; *Vallisneria nana*, *V. spiralis*

(continued)

Table 1.1 (continued)

Family	Plant species
Iridaceae	*Iris pseudacorus*
Juncaceae	*Juncus tenuis*; *Luzula campestris*
Lamiaceae	*Ocimum gratissimum*
Lardizabalaceae	*Akebia quinata*
Lauraceae	*Cinnamomum camphora*, *C. verum*; *Litsea glutinosa*
Lemnaceae	*Landoltia punctata*
Lentibulariaceae	*Utricularia gibba*
Liliaceae	*Agapanthus praecox*; *Asparagus densiflorus*, *A. officinalis*; *Sansevieria hyacinthoides*, *S. trifasciata*
Limnocharitaceae	*Limnocharis flava*
Lythraceae	*Cuphea ignea*; *Lythrum salicaria*; *Trapa natans*
Malpighiaceae	*Hiptage benghalensis*
Malvaceae	*Abelmoschus moschatus*
Melastomataceae	*Clidemia hirta*; *Melastoma candidum*; *Miconia calvescens*; *Tibouchina urvilleana*
Meliaceae	*Cedrela odorata*; *Melia azedarach*
Menyanthaceae	*Nymphoides peltata*
Moraceae	*Castilla elastica*; *Ficus microcarpus*, *F. rubiginosa*; *Morus alba*
Myricaceae	*Morella faya*
Myrsinaceae	*Ardisia acuminata*, *A. crenata*, *A. elliptica*
Myrtaceae	*Eugenia uniflora*; *Kunzea ericoides*; *Melaleuca quinquenervia*; *Pimenta dioica*; *Psidium cattleianum*, *P. guajava*; *Rhodomyrtus tomentosa*; *Syzygium cumini*, *S. jambos*; *Waterhousea floribunda*
Najadaceae	*Najas minor*
Nymphaeaceae	*Nymphaea odorata*
Oleaceae	*Fraxinus floribunda*; *Ligustrum lucidum*, *L. robustum*, *L. sinense*, *L. vulgare*; *Olea europaea*
Onagraceae	*Fuchsia boliviana*, *F. magellanica*; *Ludwigia peruviana*
Orchidaceae	*Oeceoclades maculata*
Oxalidaceae	*Oxalis corniculata*, *O. latifolia*, *O. pes-caprae*
Passifloraceae	*Passiflora edulis*, *P. foetida*, *P. maliformis*, *P. suberosa*, *P. tarminiana*
Piperaceae	*Piper aduncum*
Pittosporaceae	*Pittosporum tenuifolium*, *P. undulatum*, *P. viridiflorum*
Plantaginaceae	*Veronica serpyllifolia* ssp. *humifusa*
Poaceae	*Aegilops triuncialis*; *Agrostis capillaries*, *A. gigantean*; *Ammophila arenaria*; *Andropogon virginicus*; *Arundo donax*; *Bambusa vulgaris*; *Bothriochloa pertusa*; *Bromus inermis*, *B. rubens*, *B. tectorum*; *Cenchrus ciliaris*, *C. clandestinus*, *C. echinatus*, *C. macrourus*, *C. polystachios*, *C. setaceus*; *Cortaderia jubata*, *C. selloana*; *Cynodon dactylon*; *Dactylis glomerata*; *Glyceria maxima*; *Heteropogon contortus*; *Holcus lanatus*; *Imperata cylindrica*; *Ischaemum polystachyum*; *Melinis minutiflora*; *Microstegium vimineum*; *Miscanthus sinensis*; *Nassella neesiana*, *N. tenuissima*; *Neyraudia reynaudiana*; *Oplismenus undulatifolius*; *Panicum repens*; *Paspalum scrobiculatum*, *P. urvillei*, *P. vaginatum*; *Phalaris arundinacea*; *Phragmites australis*; *Phyllostachys flexuosa*; *Poa*

(continued)

Table 1.1 (continued)

Family	Plant species
	annua, P. pratensis; *Rottboellia cochinchinensis*; *Sacciolepis indica*; *Schismus barbatus*; *Setaria verticillata*; *Sorghum halepense*; *Spartina alterniflora, S. anglica, S. densiflora*; *Sporobolus africanus*; *Urochloa maxima, U. mutica*; *Vulpia bromoides*; *Zizania latifolia*
Polygalaceae	*Polygala paniculata*
Polygonaceae	*Persicaria perfoliata*; *Polygonum cuspidatum*; *Rumex acetosella, R. crispus, R. obtusifolius*
Pontederiaceae	*Antigonon leptopus*; *Eichhornia crassipes*
Portulacaceae	*Montia fontana*
Potamogetonaceae	*Potamogeton crispus, P. perfoliatus*
Proteaceae	*Grevillea robusta*
Ranunculaceae	*Clematis terniflora, C. vitalba*; *Ranunculus ficaria*
Rhamnaceae	*Colubrina asiatica*; *Frangula alnus*; *Rhamnus alaternus, R. cathartica*; *Ziziphus mauritiana*
Rhizophoraceae	*Rhizophora mangle*
Rosaceae	*Duchesnea indica*; *Eriobotrya japonica*; *Fragaria vesca*; *Prunus campanulata*; *Pyrus calleryana*; *Rosa bracteata, R. multiflora*; *Rubus discolor, R. ellipticus, R. moluccanus, R. niveus, R. pinnatus, R. rosifolius*; *Spiraea japonica*
Rubiaceae	*Cinchona pubescens*; *Paederia foetida*; *Spermacoce verticillata*
Rutaceae	*Triphasia trifolia*
Salicaceae	*Populus alba*; *Salix babylonica, S. cinerea, S. humboldtiana*
Sapindaceae	*Cardiospermum grandiflorum*; *Cupaniopsis anacardioides*
Saururaceae	*Houttuynia cordata*
Scrophulariaceae	*Bacopa monnieri*; *Limnophila sessiliflora*; *Linaria vulgaris*; *Paulownia tomentosa*; *Striga asiatica*; *Strobilanthes hamiltoniana*; *Verbascum thapsus*
Simaroubaceae	*Ailanthus altissima*
Solanaceae	*Cestrum nocturnum, C. parqui*; *Nicotiana glauca*; *Physalis peruviana*; *Solanum mauritianum, S. seaforthianum, S. sisymbriifolium, S. tampicense, S. torvum, S. viarum*
Tamaricaceae	*Tamarix aphylla, T. parviflora, T. ramosissima*
Typhaceae	*Typha latifolia*
Urticaceae	*Boehmeria penduliflora*
Verbenaceae	*Citharexylum spinosum*; *Lantana camara*; *Verbena brasiliensis, V. rigida*; *Vitex rotundifolia*
Vitaceae	*Ampelopsis brevipedunculata*
Zingiberaceae	*Alpinia zerumbet*; *Hedychium coccineum, H. coronarium, H. flavescens, H. gardnerianum*; *Elettaria cardamomum*
Zosteraceae	*Zostera japonica, Z. marina*

incidental plant introductions, whereas horticulture and nursery commerce are the primary conduits for purposeful introductions (Ward et al. 2020; Beaury et al. 2021). Additionally, a study of Natura 2000 habitats indicates that freshwater and grasslands were the most invaded habitats, followed by coastal dunes and forests (Lazzaro et al. 2020). After establishment, invasive species affect the area's varied ecological and socio-economic characteristics leading to unnatural environmental modifications.

1.4 Ecological and Socio-economic Impacts of Invasive Alien Plants

Invasive plant growth endangers key socio-economic resources in addition to endangering ecosystem processes and natural biodiversity (Lazzaro et al. 2020; Rai and Singh 2020). Ecologically sensitive ecosystems, biodiversity hotspots, and protected areas are more susceptible to invaders and their accompanying impacts as these regions are already facing threats due to habitat loss and global climate change (Bhatta et al. 2020; Fenouillas et al. 2021; Rai and Singh 2021; Brock and Daehler 2022). Similarly, on tropical coral islands, where ecosystems are relatively vulnerable with a high number of endemic species, invasive alien species are a significant component contributing to the deterioration of native flora (Cai et al. 2020). Though it is challenging to determine the exact extent of the harm that invasive plants have caused to the invaded habitat; nonetheless, financial losses due to the impairment of ecological and socio-economic services as well as the imposition of management measures can be calculated.

1.4.1 Ecological Impacts

It is evident that invasive plant species impact ecological processes such as soil chemical properties, biogeochemical cycling, water and fire regimes, climatic conditions, biotic/abiotic interactions, soil microbial assembly, vegetational diversity, and advanced trophic levels, both directly and indirectly (Livingstone et al. 2020; Reilly et al. 2020; Hansen et al. 2021; Sampaio et al. 2021; Torres et al. 2021; Yu et al. 2021; Faccenda and Daehler 2022; Litt and Pearson 2022; Maan et al. 2022; Nasto et al. 2022; Singh et al. 2022; Xu et al. 2022; Zhang et al. 2022; Mehal et al. 2023b; Sharma et al. 2023; Fig. 1.3). These also interfere with the mutualistic connections, such as those between plants and their pollinators or mycorrhiza (Parra-Tabla and Arceo-Gómez 2021; Řezáčová et al. 2021). Native liana and tree communities bear negative effects on their structure, network, and topological roles due to invasive tree species in the natural forests (Addo-Fordjour et al. 2022). Increased hybridization, disease transmission, and obstruction of forest regeneration are some other risks accentuated by invasive plants in woodlands (Langmaier and Lapin 2020).

Apart from these notable effects, ecosystem modifications by invasive species can also have auxiliary cascading impacts on plant and soil communities, which magnify the overall impacts of plant invasion (Carboni et al. 2021). Also, most of the invasive species alter one or more components of the inhabited ecosystem, thus creating novel

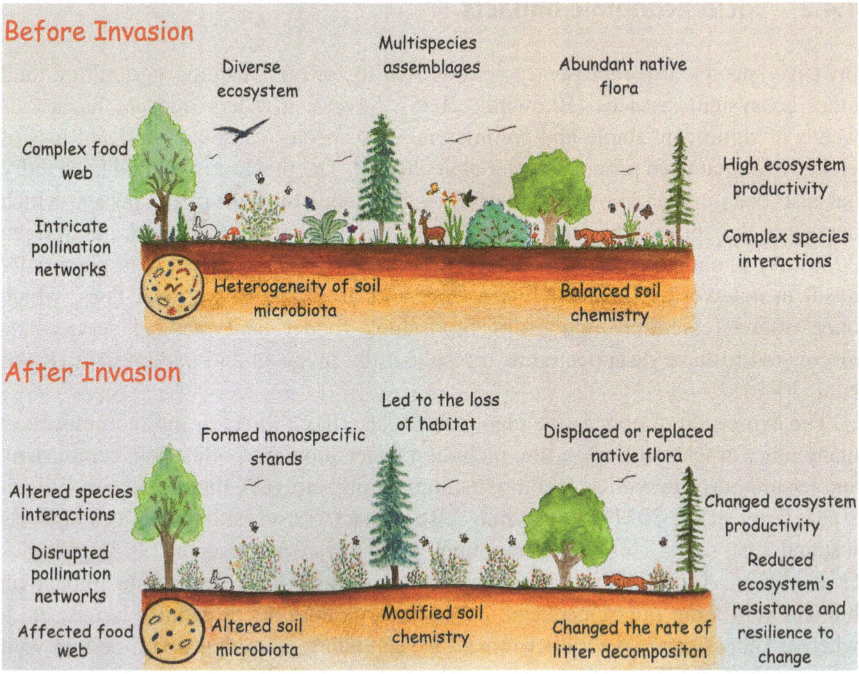

Fig. 1.3 Diagrammatic representation of changes in ecosystem structure and functions after invasion by an alien plant species

niches that continue to exist over longer periods even after the elimination of invading species (known as legacy impact), thereby inhibiting the resurgence of indigenous species and restoration of invaded habitats (Zhang et al. 2021). Moreover, when numerous invaders coexist, the effects on native plant diversity and soil characteristics are comparatively more severe (Vujanović et al. 2022).

There are very few statistical reports that show how much of a hazard invasive plant species are to the local biodiversity; nonetheless, several regional studies offer intriguing insights. Researchers contend that rather than causing species extinctions, non-native plants seem to result in the displacement of native biota, thus resulting in community-level changes. For instance, these decreased the species number from 602 to 410 in case of plants and from 68 to 19 in case of birds in the United States, working in conjunction with natural and anthropogenic disruptions (Gurevitch and Padilla 2004). Likewise, 166 native plant taxa are now categorized as endangered and 113 as vulnerable in New South Wales owing to the ongoing expansion of alien plants (Coutts-Smith and Downey 2006). Both plant and animal invasions can be blamed for nearly one-fourth of the extinct and extinct in the wild endemic flora to a certain extent (Bellard et al. 2016). Although non-native flora interventions do not straightforwardly cause species disappearance, they are clearly liable for changing the extinction trajectory of the species (Downey and Richardson 2016).

1.4.2 Socio-Economic Impacts

Invasive species also represent a serious risk to forestry, fishing, agriculture, and other ecosystem services (Bhowmik 2005). Several invaders that are hazardous weeds of significant staple and commercial crop species cause a significant loss of yield, if not managed properly (Kaur et al. 2022b). The practice of animal husbandry has also been impacted by the invasion of grasslands by alien plant species, which decreases the accessibility of pastures for the animals (O'Connor and van Wilgen 2020). By endangering ecosystem services, invasive plant species may potentially result in noteworthy financial losses (Szabó et al. 2019). Also, local flora, which once offered essential supplies of food, fuel, fodder, and medical services, is discovered to have disappeared as a result of the invasion and habitat drift (Kohli et al. 2006).

The expansion of non-native plant species also has a direct or indirect impact on many other aspects of human life, including water supplies, pollination, ecotourism, and leisure pursuits such as boating, fishing, hiking, etc. (O'Connor and van Wilgen 2020; Ginn et al. 2021). These may also impact conservation practices such as wetland restoration, forest regeneration, etc. (Lázaro-Lobo et al. 2021; Charles et al. 2022). Additionally, certain invasive plants have an immediate impact on human well-being (allergic reactions, dermatic conditions, breathing issues, etc.), whereas others have an indirect impact by spreading pests that infect people with diseases (Rai and Singh 2020; Bernard-Verdier et al. 2022).

Further, managing invasive species requires significant financial resources, which may not even fit into the budgets of nations with weak economies. The over 5000 invasive plants found in the United States cause an annual economic suffering of nearly 35 billion USD (Pimentel et al. 2005). According to research, approximately 38 million USD were exhausted to manage non-native plants in the Cape Floristic's reserved areas, and another 11–175 million USD would be needed in the succeeding times to handle the problem (van Wilgen et al. 2016). Post-removal restoration practices are even more challenging, demanding handsome investments (Adams et al. 2020). However, the enormous data gaps typically found in financial assessments indicate that these projections are considerably understated (Cuthbert et al. 2020). Moreover, it has been confidently projected that these numbers will be dramatically increased in the near future given the unchecked spread of invasive species and upcoming environmental concerns.

1.5 Future Climate Change and Invasive Plants

Invasive plant species are anticipated to be directly impacted by changes in climatic characteristics (temperature, precipitation, atmospheric CO_2 concentrations, etc.), seasonal fluctuations, and any ensuing and extreme weather event (Shrestha and Shrestha 2019; Wang et al. 2022). At the same time, studies have also shown the stimulative impact of invasive plants on volatile emissions, eutrophication, and greenhouse gas emissions (Sage 2020; Bezabih Beyene et al. 2022). Climate change

Table 1.2 Studies predicting the probable impacts of the most significant components of climate change on the plant invaders

Global change	Impact	References
Increased temperature	Positive (enhanced growth, competitive ability and resistance; increased habitat suitability, phenological, and ecophysiological adaptations)	Blumenthal et al. (2016), Liu et al. (2017), Cavieres et al. (2018), Peng et al. (2019), Howell et al. (2020), Nguyen et al. (2020), Duell et al. (2021), Bao et al. (2022), Sun et al. (2022), Adhikari et al. (2023)
	Negative (reduced plant growth, tolerance, plasticity, and defence)	Johnson and Hartley (2018), He and He (2020), Birnbaum et al. (2021)
Increased carbon dioxide concentrations	Positive (increased growth, performance and reproductive potential; improved herbicidal resistance)	Liu et al. (2017), Johnson and Hartley (2018), Bajwa et al. (2019), Cowie et al. (2020)
Increased precipitation	Positive (enhanced growth and competitiveness; niche width expansion)	Blumenthal et al. (2008), Irl et al. (2021), Bao et al. (2022), Li et al. (2022b), Ren et al. (2022), Adhikari et al. (2023)
	Negative (reduced habitat suitability)	Bradley (2009)
Decreased precipitation	Positive (enhanced germination, growth performance, phenotypic plasticity, and resilience to abiotic stress)	Gao et al. (2018), Vetter et al. (2019), Mojzes et al. (2020), Duell et al. (2021), Leal et al. (2022)
	Negative (low seed dormancy)	LaForgia et al. (2018)
Nitrogen deposition	Positive (enhanced growth and competitiveness)	Valliere et al. (2017), Cavieres et al. (2018), Liu et al. (2018), Peng et al. (2019)

usually eases invasions, and invasive species in turn magnify the negative effects of climate change (Sage 2020). For example, an invasive plant *Pueraria montana* (Lour.) Merr. colonizes aggressively with global warming, raised CO_2 levels, and eutrophication (Sage 2020). In turn, the expansion of *P. montana* promotes the emission of volatile organics and CO_2, thereby impacting the microclimatic conditions and promoting climate change (Sage 2020). Although invasive species and climate change both represent a serious risk to the ecological functions, biodiversity, and agronomic systems, an understanding of the interactions between these phenomena and their synergistic effects on ecosystem health and productivity will strongly affect our perception of the potential environmental consequences. Such research aspects demand specific acknowledgements rather than being disregarded or incorporated into conventional invasion science studies.

The futuristic projections and forecasts about the spatiotemporal distribution of non-indigenous species under potential climate change scenarios rely mainly on sophisticated modelling techniques (Table 1.2). The documented spread of invasive plant species in their indigenous and non-indigenous geographical range is used by habitat suitability models to quantify key niche dimensions and foretell new possible invasion locations (Adhikari et al. 2019). The most influential predictors in such studies generally include mean temperature, water deficit, precipitation periodicity, and fire regimes (McMahon et al. 2021). Also, cold deserts and prairies are suspected to be the most vulnerable ecoregions and invasive forb and grass species are expected to demonstrate the maximum expansion (McMahon et al. 2021).

Habitat suitability projections put forward by different studies vary in relation to geographical attributes and target species, with some emphasizing invasive species' range expansion, while others implicating shifts in their habitats. A study by Shrestha and Shrestha (2019) predicted that nearly 75% of the plant invaders in Nepal are going to expand in terms of distribution as well as intensity under future climate change scenarios. Another study examined the probable spread of six plant invaders in North America in response to projected climate change and noticed a shift in their suitable habitats in the coming years (Wang et al. 2022). A study by Fulgêncio-Lima et al. (2021) suggests that the potential distribution and impacts of current invasive plants will not be exacerbated by climate change, but novel invasive species may invade previously uninhabited or lesser-inhabited ecosystems. Certain findings also provide a contradictory opinion that climate change and invasion may individually impact native vegetation; however, climate change will not make invasions worse and it might even lessen the consequences of invasive species (He and He 2020; Kelso et al. 2020; Birnbaum et al. 2021).

In addition to the modelling approach, significant advancements have been accomplished in the latest years in understanding how invasive alien species' establishment, spread, and influence will be altered by climate change and rising carbon dioxide levels by propagating these plants under simulated environments (Ziska 2022; Table 1.2). When the populations of an invasive weed, *Ambrosia artemisiifolia* L. were exposed to two factors, i.e., simulated treatment of temperature warming and herbivory by a biocontrol agent, it was observed that high temperatures diminished the effects of biocontrol agent by producing vigorous and more defensive plants, instigated via genetic changes and transgenerational induction of defenses (Sun et al. 2022). Similarly, the results of a common garden experiment demonstrated that climate warming increased phenological overlapping between native and invasive species, consequently increasing the level of competition for pollination (Giejsztowt et al. 2020). Another multispecies experiment established that while soil fauna may aid in native ecosystems' resistance to alien plant invasions, this benefit may be diminished during periods of drought (Jin et al. 2022).

Even though we now understand more about how different environmental conditions affect the process and distribution of invasive plant species, there are still many information gaps. To make precise predictions about the possible effects of invaders, ecologists must develop frameworks that take species' abundance and not just their presence into account (Funk et al. 2020). Additionally, it is crucial to comprehend how the local plant communities adapt to climate change and whether this results in increased or decreased resistance to invasion. According to Luan et al. (2021), a mechanistic and more accurate prediction regarding the impact of plant invaders on ecosystem functioning can be provided by interpreting changes in macrofauna, along with their relationship with litter traits under changing climatic conditions in comparison to a sole functional trait-based approach. Also, instead of climate extremes, research has mostly concentrated on how species would react to average changes in climatic parameters. Through a number of understudied mechanisms, extreme climatic events (such as floods and droughts) can increase the acclimatization, spread, and effects of invasive species, and such detailing needs more attention from researchers worldwide (Diez et al. 2012). Furthermore, predictions on invasion trends of exotic plants along with simulation of suitable plantations that can block invasion can be of high practical significance for the anticipation and management of plant invasion (Fang et al. 2021). Using a combination of computational modelling and experimental research methods to create predictions about the potential trait adaptations, niche-width expansion, and favourable recruitment sites for invasive plants might serve as the most pragmatic, rational, and reliable approach to deal with the issue (Guerra-Coss et al. 2021).

1.6 Conclusions

Plant invasion is a significant driver of environmental change on a worldwide scale. Invasive species damage invaded locale in a way that triggers a series of ecological changes, which end up modifying most of the habitat components and consequently altering the landscape, productivity, resistance, and resilience of an ecosystem. These threats posed by invasive plants will probably grow as a result of the existing paradigm of global climate change. By altering the seasonal patterns, particularly in the most fragile ecosystems that provide crucial ecosystem services, climate change has the potential to generate conditions and localities in the future that are more conducive for invading species. Therefore, it is imperative for conservation managers to provide a framework for accurately predicting and managing alien plant invasions and/or reducing their cascading impacts on the local biodiversity and ecology. Future studies should concentrate on the mechanisms, complex interactions, and positive/negative feedbacks by which extreme climatic events can hasten the distribution, establishment, and consequences of invasive alien plants and vice-versa.

Acknowledgements AS is thankful to University Grants Commission, New Delhi for financial assistance in form of Senior Research Fellowship. AK is thankful to University Grants Commission, New Delhi for financial assistance in form of DS Kothari Postdoctoral Fellowship.

References

Adams SN, Jennings S, Warnock N (2020) Plant invasion depresses native species richness, but control of invasive species does little to restore it. Plant Ecol Divers 13:257–266. https://doi.org/10.1080/1755087420201817998

Addo-Fordjour P, Ofosu-Bamfo B, Mbroh E et al (2022) Plant invasion drives liana and tree community assemblages and liana-tree network structure in two moist semi-deciduous forests in Ghana. Biol Invasions 25:611. https://doi.org/10.1007/s10530-022-02933-0

Adhikari P, Jeon JY, Kim HW et al (2019) Potential impact of climate change on plant invasion in the Republic of Korea. J Ecol Environ 43:36. https://doi.org/10.1186/s41610-019-0134-3

Adhikari P, Lee YH, Poudel A et al (2023) Predicting the impact of climate change on the habitat distribution of *Parthenium hysterophorus* around the world and in South Korea. Biology 12:84. https://doi.org/10.3390/biology12010084

Bajwa AA, Wang H, Chauhan BS et al (2019) Effect of elevated carbon dioxide concentration on growth, productivity and glyphosate response of parthenium weed (*Parthenium hysterophorus* L.). Pest Manag Sci 75:2934–2941. https://doi.org/10.1002/ps.5403

Bao XY, Wang ZX, He ZS et al (2022) Enhanced precipitation offsets climate warming inhibition on *Solidago canadensis* growth and sustains its high tolerance. Glob Ecol Conserv 34:e02023. https://doi.org/10.1016/j.gecco.2022.e02023

Beaury EM, Finn JT, Corbin JD et al (2020) Biotic resistance to invasion is ubiquitous across ecosystems of the United States. Ecol Lett 23:476–482. https://doi.org/10.1111/ele.13446

Beaury EM, Patrick M, Bradley BA (2021) Invaders for sale: the ongoing spread of invasive species by the plant trade industry. Front Ecol Environ 19:550–556. https://doi.org/10.1002/fee.2392

Bellard C, Cassey P, Blackburn TM (2016) Alien species as a driver of recent extinctions. Biol Lett 12:20150623. https://doi.org/10.1098/rsbl.2015.0623

Bernard-Verdier M, Seitz B, Buchholz S et al (2022) Grassland allergenicity increases with urbanisation and plant invasions. Ambio 51:2261–2277. https://doi.org/10.1007/s13280-022-01741-z

Bezabih Beyene B, Li J, Yuan J et al (2022) Non-native plant invasion can accelerate global climate change by increasing wetland methane and terrestrial nitrous oxide emissions. Glob Change Biol 28:5453–5468. https://doi.org/10.1111/gcb.16290

Bhatta S, Joshi LR, Shrestha BB (2020) Distribution and impact of invasive alien plant species in Bardia National Park, western Nepal. Environ Conserv 47:197–205. https://doi.org/10.1017/S0376892920000223

Bhowmik PC (2005) Characteristics, significance, and human dimension of global invasive weeds. In: Inderjit (ed) Invasive plants: ecological and agricultural aspects. Birkhäuser, Basel, pp 251–268

Birnbaum C, Waryszak P, Farrer EC (2021) Direct and indirect effects of climate change in coastal wetlands: will climate change influence wetlands by affecting plant invasion? Wetlands 41:59. https://doi.org/10.1007/s13157-021-01456-z

Blossey B, Nötzold R (1995) Evolution of increased competitive ability in invasive nonindigenous plants: a hypothesis. J Ecol 83:887–889

Blumenthal D, Chimner RA, Welker JM et al (2008) Increased snow facilitates plant invasion in mixedgrass prairie. New Phytol 179:440–448. https://doi.org/10.1111/j.1469-8137.2008.02475.x

Blumenthal DM, Kray JA, Ortmans W et al (2016) Cheatgrass is favored by warming but not CO_2 enrichment in a semi-arid grassland. Glob Change Biol 22:3026–3038. https://doi.org/10.1111/gcb.13278

Bonnamour A, Gippet JMW, Bertelsmeier C (2021) Insect and plant invasions follow two waves of globalization. Ecol Lett 24:2418–2426. https://doi.org/10.1111/ele.13863

Bradley BA (2009) Regional analysis of the impacts of climate change on cheatgrass invasion shows potential risk and opportunity. Glob Change Biol 15:196–208. https://doi.org/10.1111/j.1365-2486.2008.01709.x

Brock KC, Daehler CC (2022) Island plant invasions. In: Clements DR, Upadhyaya MK, Joshi S, Shrestha A (eds) Global plant invasions. Springer, Cham

Byrne D, Scheben A, Scott JK et al (2022) Genomics reveals the history of a complex plant invasion and improves the management of a biological invasion from the South African–Australian biotic exchange. Ecol Evol 12:e9179. https://doi.org/10.1002/ece3.9179

Cai H, Lu H, Tian Y et al (2020) Effects of invasive plants on the health of forest ecosystems on small tropical coral islands. Ecol Indic 117:106656. https://doi.org/10.1016/j.ecolind.2020.106656

Callaway RM, Ridenour WM (2004) Novel weapons: invasive success and the evolution of increased competitive ability. Front Ecol Environ 2:436–443. https://doi.org/10.1890/1540-9295(2004)002[0436:NWISAT]2.0.CO;2

Callaway RM, Thelen GC, Rodriguez A et al (2004) Soil biota and exotic plant invasion. Nature 427:731–733. https://doi.org/10.1038/nature02322

Carboni M, Livingstone SW, Isaac ME et al (2021) Invasion drives plant diversity loss through competition and ecosystem modification. J Ecol 109:3587–3601. https://doi.org/10.1111/1365-2745.13739

Cavieres LA, Sanhueza AK, Torres-Mellado G et al (2018) Competition between native Antarctic vascular plants and invasive *Poa annua* changes with temperature and soil nitrogen availability. Biol Invasions 20:1597–1610. https://doi.org/10.1007/s10530-017-1650-7

Charles B, Chase M, Pociask G et al (2022) Restored wetlands are greatly influenced by hydrology and non-native plant invasion. Wetl Ecol Manag 31:129. https://doi.org/10.1007/s11273-022-09905-0

Coutts-Smith AJ, Downey PO (2006) The impact of weeds on threatened biodiversity in New South Wales technical series no 11. CRC for Australian Weed Management, Adelaide

Cowie BW, Venter N, Witkowski ET et al (2020) Implications of elevated carbon dioxide on the susceptibility of the globally invasive weed, *Parthenium hysterophorus*, to glyphosate herbicide. Pest Manag Sci 76:2324–2332. https://doi.org/10.1002/ps.5767

Cuthbert RN, Bacher S, Blackburn TM et al (2020) Invasion costs, impacts, and human agency: response to Sagoff 2020. Conserv Biol 34:1579–1582. https://doi.org/10.1111/cobi.13592

Dai ZC, Wan LY, Qi SS et al (2020) Synergy among hypotheses in the invasion process of alien plants: a road map within a timeline. Perspect Plant Ecol Evol Syst 47:125575. https://doi.org/10.1016/j.ppees.2020.125575

Davis MA, Grime JP, Thompson K (2000) Fluctuating resources in plant communities: a general theory of invisibility. J Ecol 88:528–534. https://doi.org/10.1046/j.1365-2745.2000.00473.x

Diez JM, D'Antonio CM, Dukes JS (2012) Will extreme climatic events facilitate biological invasions? Front Ecol Environ 10:249–257. https://doi.org/10.1890/110137

Downey PO, Richardson DM (2016) Alien plant invasions and native plant extinctions: a six-threshold framework. AoB Plants 8:plw047. https://doi.org/10.1093/aobpla/plw047

Duell EB, Londe DW, Hickman KR et al (2021) Superior performance of invasive grasses over native counterparts will remain problematic under warmer and drier conditions. Plant Ecol 222:993–1006. https://doi.org/10.1007/s11258-021-01156-y

Dyderski MK, Gdula AK, Jagodziński AM (2015) "The rich get richer" concept in riparian woody species–a case study of the Warta River Valley (Poznań, Poland). Urban For Urban Green 14:107–114. https://doi.org/10.1016/j.ufug.2014.12.003

Early R, Bradley BA, Dukes JS et al (2016) Global threats from invasive alien species in the twenty-first century and national response capacities. Nat Commun 7:12485. https://doi.org/10.1038/ncomms12485

Elton CS (1958) The ecology of invasions by animals and plants. Springer, Boston

Enders M, Hütt MT, Jeschke JM (2018) Drawing a map of invasion biology based on a network of hypotheses. Ecosphere 9:e02146. https://doi.org/10.1002/ecs2.2146

Ernst AR, Barak RS, Hipp AL et al (2022) The invasion paradox dissolves when using phylogenetic and temporal perspectives. J Ecol 110:443–456. https://doi.org/10.1111/1365-2745.13812

Eschtruth AK, Battles JJ (2009) Acceleration of exotic plant invasion in a forested ecosystem by a generalist herbivore. Conserv Biol 23:388–399. https://doi.org/10.1111/j.1523-1739.2008.01122.x

Faccenda K, Daehler CC (2022) A screening system to predict wildfire risk of invasive plants. Biol Invasions 24:575–589. https://doi.org/10.1007/s10530-021-02661-x

Fang Y, Zhang X, Wei H et al (2021) Predicting the invasive trend of exotic plants in China based on the ensemble model under climate change: a case for three invasive plants of Asteraceae. Sci Total Environ 756:143841. https://doi.org/10.1016/j.scitotenv.2020.143841
Feng YL, Li YP, Wang RF et al (2011) A quicker return energy-use strategy by populations of a subtropical invader in the non-native range: a potential mechanism for the evolution of increased competitive ability. J Ecol 99:1116–1112. https://doi.org/10.1111/j.1365-2745.2011.01843.x
Fenouillas P, Ah-Peng C, Amy E et al (2021) Quantifying invasion degree by alien plants species in Reunion Island. Austral Ecol 46:1025–1037. https://doi.org/10.1111/aec.13048
Fuentes-Lillo E, Lembrechts JJ, Cavieres LA et al (2021) Anthropogenic factors overrule local abiotic variables in determining non-native plant invasions in mountains. Biol Invasions 23: 3671–3686. https://doi.org/10.1007/s10530-021-02602-8
Fulgêncio-Lima LG, Andrade AFA, Vilela B et al (2021) Invasive plants in Brazil: climate change effects and detection of suitable areas within conservation units. Biol Invasions 23:1577–1594. https://doi.org/10.1007/s10530-021-02460-4
Funk JL, Parker IM, Matzek V et al (2020) Keys to enhancing the value of invasion ecology research for management. Biol Invasions 22:2431–2445. https://doi.org/10.1007/s10530-020-02267-9
Gao X, Zhao X, Li H et al (2018) Exotic shrub species (*Caragana korshinskii*) is more resistant to extreme natural drought than native species (*Artemisia gmelinii*) in a semiarid revegetated ecosystem. Agric For Meteorol 263:207–216. https://doi.org/10.1016/j.agrformet.2018.08.029
Giejsztowt J, Classen AT, Deslippe JR (2020) Climate change and invasion may synergistically affect native plant reproduction. Ecology 101:e02913. https://doi.org/10.1002/ecy.2913
Gimeno I, Vila M, Hulme PE (2006) Are islands more susceptible to plant invasion than continents? A test using *Oxalis pes-caprae* L. in the western Mediterranean. J Biogeogr 33:1559–1565. https://doi.org/10.1111/j.1365-2699.2006.01525.x
Ginn BK, Dias EF, Fleischaker T (2021) Trends in submersed aquatic plant communities in a large, inland lake: impacts of an invasion by starry stonewort (*Nitellopsis obtusa*). Lake Reserv Manag 37:199–213. https://doi.org/10.1080/10402381.2020.1859025
Global Invasive Species Database (2023) List of the invasive plant species of the world. http://www.iucngisd.org/gisd/search.php. Accessed 12 Jan 2023
Guerra-Coss FA, Badano EI, Cedillo-Rodríguez IE et al (2021) Modelling and validation of the spatial distribution of suitable habitats for the recruitment of invasive plants on climate change scenarios: an approach from the regeneration niche. Sci Total Environ 777:146007. https://doi.org/10.1016/j.scitotenv.2021.146007
Guo C, Zhao C, Li F et al (2022) Phylogenetic relationships and disturbance explain the resistance of different habitats to plant invasions. Life 12:1785. https://doi.org/10.3390/life12111785
Gurevitch J, Padilla DK (2004) Are invasive species a major cause of extinctions? Trends Ecol Evol 19:470–474. https://doi.org/10.1016/j.tree.2004.07.005
Hansen W, Wollny J, Otte A et al (2021) Invasive legume affects species and functional composition of mountain meadow plant communities. Biol Invasions 23:281–296. https://doi.org/10.1007/s10530-020-02371-w
He ZS, He WM (2020) Asymmetric climate warming does not benefit plant invaders more than natives. Sci Total Environ 742:140624. https://doi.org/10.1016/j.scitotenv.2020.140624
Hobbs RJ, Huenneke LF (1992) Disturbance, diversity, and invasion—implications for conservations. Conserv Biol 6:324–337. https://doi.org/10.1046/j.1523-1739.1992.06030324.x
Holzmueller EJ, Jose S (2011) Invasion success of cogongrass, an alien C4 perennial grass, in the southeastern United States: exploration of the ecological basis. Biol Invasions 13:435–442. https://doi.org/10.1007/s10530-010-9837-1
Howell A, Winkler DE, Phillips ML et al (2020) Experimental warming changes phenology and shortens growing season of the dominant invasive plant *Bromus tectorum* (cheatgrass). Front Plant Sci 11:570001. https://doi.org/10.3389/fpls.2020.570001

Huston MA (2004) Management strategies for plant invasions: manipulating productivity, disturbance, and competition. Divers Distrib 10:167–178. https://doi.org/10.1111/j.1366-9516.2004.00083.x

Ibáñez I, Liu G, Petri L et al (2021) Assessing vulnerability and resistance to plant invasions: a native community perspective. Invasive Plant Sci Manag 14:64–74. https://doi.org/10.1017/inp.2021.15

Irl SD, Schweiger AH, Steinbauer MJ et al (2021) Human impact, climate and dispersal strategies determine plant invasion on islands. J Biogeogr 48:1889–1903. https://doi.org/10.1111/jbi.14119

Jeschke JM (2008) Across islands and continents, mammals are more successful invaders than birds. Divers Distrib 14:913–916. https://doi.org/10.1111/j.1472-4642.2008.00488.x

Jin H, Chang L, van Kleunen M et al (2022) Soil mesofauna may buffer the negative effects of drought on alien plant invasion. J Ecol 110:2332–2342. https://doi.org/10.1111/1365-2745.13950

Johnson SN, Hartley SE (2018) Elevated carbon dioxide and warming impact silicon and phenolic-based defences differently in native and exotic grasses. Glob Change Biol 24:3886–3896. https://doi.org/10.1111/gcb.13971

Kalisz S, Kivlin SN, Bialic-Murphy L (2021) Allelopathy is pervasive in invasive plants. Biol Invasions 23:367–371. https://doi.org/10.1007/s10530-020-02383-6

Kaur A, Batish DR, Kaur S, Singh HP, Kohli RK (2017) Phenological behaviour of *Parthenium hysterophorus* in response to climatic variations according to the extended BBCH scale. Ann Appl Biol 171:316–326. https://doi.org/10.1111/aab.12374

Kaur A, Kaur S, Singh HP et al (2019) Phenotypic variations alter the ecological impact of invasive alien species: lessons from *Parthenium hysterophorus*. J Environ Manage 241:187–197. https://doi.org/10.1016/j.jenvman.2019.03.129

Kaur A, Kaur S, Singh HP et al (2022a) Alterations in phytotoxicity and allelochemistry in response to intraspecific variation in *Parthenium hysterophorus*. Ecol Complex 50:100999. https://doi.org/10.1016/j.ecocom.2022.100999

Kaur A, Batish DR, Kohli RK (2022b) Challenges on account of invasive alien terrestrial plants. In: Kaur S, Batish DR, Singh HP, Kohli RK (eds) Biodiversity in India: status, issues and challenges. Springer, Singapore, pp 495–515

Keane RM, Crawley MJ (2002) Exotic plant invasions and the enemy release hypothesis. Trends Ecol Evol 17:164–170. https://doi.org/10.1016/S0169-5347(02)02499-0

Kelso MA, Wigginton RD, Grosholz ED (2020) Nutrients mitigate the impacts of extreme drought on plant invasions. Ecology 101:e02980. https://doi.org/10.1002/ecy.2980

Kohli RK, Batish DR, Singh HP et al (2006) Status, invasiveness and environmental threats of three tropical American invasive weeds (*Parthenium hysterophorus* L, *Ageratum conyzoides* L, *Lantana camara* L) in India. Biol Invasions 8:1501–1510. https://doi.org/10.1007/s10530-005-5842-1

Kolar CS, Lodge DM (2001) Progress in invasion biology: predicting invaders. Trends Ecol Evol 16:199–204. https://doi.org/10.1016/S0169-5347(01)02101-2

LaForgia ML, Spasojevic MJ, Case EJ et al (2018) Seed banks of native forbs, but not exotic grasses, increase during extreme drought. Ecology 99:896–903. https://doi.org/10.1002/ecy.2160

Lal R, Kaur A, Kaur S et al (2021) Nature of phytotoxic interference of alien weed '*Calyptocarpus vialis*' against some crop plants. Environ Monit Assess 193:334. https://doi.org/10.1007/s10661-021-09092-0

Langmaier M, Lapin K (2020) A systematic review of the impact of invasive alien plants on forest regeneration in European temperate forests. Front Plant Sci 11:524969. https://doi.org/10.3389/fpls.2020.524969

Lázaro-Lobo A, Lucardi RD, Ramirez-Reyes C et al (2021) Region-wide assessment of fine-scale associations between invasive plants and forest regeneration. For Ecol Manag 483:118930. https://doi.org/10.1016/j.foreco.2021.118930

Lazzaro L, Bolpagni R, Buffa G et al (2020) Impact of invasive alien plants on native plant communities and natura 2000 habitats: state of the art, gap analysis and perspectives in Italy. J Environ Manag 274:111140. https://doi.org/10.1016/j.jenvman.2020.111140

Leal RP, Silveira MJ, Petsch DK et al (2022) The success of an invasive Poaceae explained by drought resilience but not by higher competitive ability. Environ Exp Bot 194:104717. https://doi.org/10.1016/j.envexpbot.2021.104717

Li SP, Jia P, Fan SY et al (2022a) Functional traits explain the consistent resistance of biodiversity to plant invasion under nitrogen enrichment. Ecol Lett 25:778–789. https://doi.org/10.1111/ele.13951

Li XQ, Tang SC, Pan YM et al (2022b) Increased precipitation magnifies the effects of N addition on performance of invasive plants in subtropical native communities. J Plant Ecol 15:473–484. https://doi.org/10.1093/jpe/rtab103

Liao H, Pal RW, Niinemets Ü et al (2021) Different functional characteristics can explain different dimensions of plant invasion success. J Ecol 109:1524–1536. https://doi.org/10.1111/1365-2745.13575

Litt AR, Pearson DE (2022) A functional ecology framework for understanding and predicting animal responses to plant invasion. Biol Invasions 24:2693–2705. https://doi.org/10.1007/s10530-022-02813-7

Liu Y, Oduor AM, Zhang Z et al (2017) Do invasive alien plants benefit more from global environmental change than native plants? Glob Change Biol 23:3363–3370. https://doi.org/10.1111/gcb.13579

Liu G, Yang YB, Zhu ZH (2018) Elevated nitrogen allows the weak invasive plant *Galinsoga quadriradiata* to become more vigorous with respect to inter-specific competition. Sci Rep 8:3136. https://doi.org/10.1038/s41598-018-21546-z

Liu XS, Cai JF, Liu L et al (2022) Nitrogen addition frequency and propagule pressure influence *Solidago canadensis* invasion into native plant community. Ecosyst Health Sustain 8:2052763. https://doi.org/10.1080/20964129.2022.2052763

Livingstone SW, Isaac ME, Cadotte MW (2020) Invasive dominance and resident diversity: unpacking the impact of plant invasion on biodiversity and ecosystem function. Ecol Monogr 90:e01425. https://doi.org/10.1002/ecm.1425

Lockwood JL, Hoopes MF, Marchetti MP (2013) Invasion ecology. Wiley-Blackwell, Chichester

Luan J, Liu S, Li S et al (2021) Functional diversity of decomposers modulates litter decomposition affected by plant invasion along a climate gradient. J Ecol 109:1236–1249. https://doi.org/10.1111/1365-2745.13548

Lucero JE, Schaffner U, Asadi G et al (2019) Enemy release from the effects of generalist granivores can facilitate *Bromus tectorum* invasion in the Great Basin desert. Ecol Evol 9:8490–8499. https://doi.org/10.1002/ece3.5314

Maan I, Kaur A, Sharma A et al (2022) Variations in leaf litter decomposition explain invasion success of *Broussonetia papyrifera* over confamilial non-invasive *Morus alba* in urban habitats. Urban For Urban Green 67:127408. https://doi.org/10.1016/j.ufug.2021.127408

MacArthur R (1970) Species packing and competitive equilibrium for many species. Theor Popul Biol 1:1–11. https://doi.org/10.1016/0040-5809(70)90039-0

MacArthur R, Levins R (1967) The limiting similarity convergence and divergence of coexisting species. Am Nat 101:377–385

Mack RN, Simberloff D, Lonsdale WM et al (2000) Biotic invasions: causes, epidemiology, global consequences and control. Ecol Appl 10:689–710. https://doi.org/10.1890/1051-0761(2000)010[0689:BICEGC]2.0.CO;2

McMahon DE, Urza AK, Brown JL et al (2021) Modelling species distributions and environmental suitability highlights risk of plant invasions in western United States. Divers Distrib 27:710–728. https://doi.org/10.1111/ddi.13232

Mehal KK, Kaur A, Singh HP et al (2023a) Investigating the phytotoxic potential of *Verbesina enceliodes*: effect on growth and performance of co-occurring weed species. Protoplasma 260:77–87. https://doi.org/10.1007/s00709-022-01761-2

Mehal KK, Sharma A, Kaur A et al (2023b) Modelling the ecological impact of invasive weed Verbesina encelioides on vegetation composition across dryland ecosystems of Punjab, north-western India. Environ Monit Assess 195:725. https://doi.org/10.1007/s10661-023-11299-2

Mojzes A, Ónodi G, Lhotsky B et al (2020) Experimental drought indirectly enhances the individual performance and the abundance of an invasive annual weed. Oecologia 193:571–581. https://doi.org/10.1007/s00442-020-04711-y

Montesinos D (2022) Fast invasives fastly become faster: invasive plants align largely with the fast side of the plant economics spectrum. J Ecol 110:1010–1014. https://doi.org/10.1111/1365-2745.13616

Nasto MK, McLeod ML, Bullington L (2022) The effect of plant invasion on soil microbial carbon-use efficiency in semi-arid grasslands of the Rocky Mountain West. J Ecol 110:479–493. https://doi.org/10.1111/1365-2745.13815

Nguyen HM, Yadav NS, Barak S et al (2020) Responses of invasive and native populations of the seagrass *Halophila stipulacea* to simulated climate change. Front Mar Sci 6:812. https://doi.org/10.3389/fmars.2019.00812

Ni M, Deane DC, Li S et al (2021) Invasion success and impacts depend on different characteristics in non-native plants. Divers Distrib 27:1194–1207. https://doi.org/10.1111/ddi.13267

O'Connor TG, van Wilgen BW (2020) The impact of invasive alien plants on rangelands in South Africa. In: van Wilgen B, Measey J, Richardson D, Wilson J, Zengeya T (eds) Biological invasions in South Africa invading nature, Springer series in invasion ecology. Springer, Cham, pp 459–487

Omer A, Fristoe T, Yang Q (2022) The role of phylogenetic relatedness on alien plant success depends on the stage of invasion. Nat Plants 8:906–914. https://doi.org/10.1038/s41477-022-01216-9

Park DS, Feng X, Maitner BS et al (2020) Darwin's naturalization conundrum can be explained by spatial scale. Proc Natl Acad Sci 117:10904–10910. https://doi.org/10.1073/pnas.1918100117

Parra-Tabla V, Arceo-Gómez G (2021) Impacts of plant invasions in native plant–pollinator networks. New Phytol 230:2117–2128. https://doi.org/10.1111/nph.17339

Pearson DE, Eren Ö, Ortega YK (2022) Combining biogeographical approaches to advance invasion ecology and methodology. J Ecol 110:2033–2045. https://doi.org/10.1111/1365-2745.13945

Peng Y, Yang JX, Zhou XH et al (2019) An invasive population of *Solidago canadensis* is less sensitive to warming and nitrogen-addition than its native population in an invaded range. Biol Invasions 21:151–162. https://doi.org/10.1007/s10530-018-1812-2

Pimentel D, Zuniga R, Morrison D (2005) Update on the environmental and economic costs associated with alien-invasive species in the United States. Ecol Econ 52:273–288. https://doi.org/10.1016/j.ecolecon.2004.10.002

Price JN, Pärtel M (2013) Can limiting similarity increase invasion resistance? A meta-analysis of experimental studies. Oikos 122:649–656. https://doi.org/10.1111/j.1600-0706.2012.00121.x

Pyšek P, Pergl J, Essl F et al (2017) Naturalized alien flora of the world. Preslia 89:203–274. https://doi.org/10.23855/preslia.2017.203

Pyšek P, Pergl J, Dawson W (2022) European plant invasions. In: Clements DR, Upadhyaya MK, Joshi S, Shrestha A (eds) Global plant invasions. Springer, Cham

Rai PK, Singh JS (2020) Invasive alien plant species: their impact on environment, ecosystem services and human health. Ecol Indic 111:106020. https://doi.org/10.1016/j.ecolind.2019.106020

Rai PK, Singh JS (2021) Plant invasion in protected areas, the Indian Himalayan region, and the north East India: progress and prospects. Proc Indian Natl Sci Acad 87:19–35. https://doi.org/10.1007/s43538-021-00013-w

Rathee S, Ahmad M, Sharma P et al (2021) Biomass allocation and phenotypic plasticity are key elements of successful invasion of *Parthenium hysterophorus* at high elevation. Environ Exp Bot 184:104392. https://doi.org/10.1016/j.envexpbot.2021.104392

Reilly MJ, McCord MG, Brandt SM et al (2020) Repeated, high-severity wildfire catalyzes invasion of non-native plant species in forests of the Klamath Mountains, northern California, USA. Biol Invasions 22:1821–1828. https://doi.org/10.1007/s10530-020-02227-3

Reinegger RD, Oleksy RZ, Gazagne E et al (2022) Foraging strategies of invasive *Macaca fascicularis* may promote plant invasion in Mauritius. Int J Primatol. https://doi.org/10.1007/s10764-022-00324-9

Rejmánek M, Richardson DM (1996) What attributes make some plant species more invasive? Ecology 77:1655–1661

Ren G, Du Y, Yang B et al (2022) Influence of precipitation dynamics on plant invasions: response of alligator weed (*Alternanthera philoxeroides*) and co-occurring native species to varying water availability across plant communities. Biol Invasions 25:519. https://doi.org/10.1007/s10530-022-02931-2

Řezáčová V, Řezáč M, Gryndler M et al (2021) Plant invasion alters community structure and decreases diversity of arbuscular mycorrhizal fungal communities. Appl Soil Ecol 167:104039. https://doi.org/10.1016/j.apsoil.2021.104039

Richardson DM, Pyšek P (2008) Fifty years of invasion ecology–the legacy of Charles Elton. Divers Distrib 14:161–168. https://doi.org/10.1111/j.1472-4642.2007.00464.x

Richardson DM, Pyšek P, Rejmánek M et al (2000) Naturalization and invasion of alien plants: concepts and definitions. Divers Distrib 6:93–107. https://doi.org/10.1046/j.1472-4642.2000.00083.x

Richardson DM, Wit AB, Pergl J et al (2022) Plant invasions in Africa. In: Clements DR, Upadhyaya MK, Joshi S, Shrestha A (eds) Global plant invasions. Springer, Cham

Roiloa SR, Yu FH, Barreiro R (2020) Plant invasions: mechanisms, impacts and management. Flora 267:151603. https://doi.org/10.1016/j.flora.2020.151603

Sage RF (2020) Global change biology: a primer. Glob Change Biol 26:3–30. https://doi.org/10.1111/gcb.14893

Sampaio JAG, Reis CRG, Cunha-Lignon M et al (2021) Plant invasion affects vegetation structure and sediment nitrogen stocks in subtropical mangroves. Mar Environ Res 172:105506. https://doi.org/10.1016/j.marenvres.2021.105506

Sax DF, Stachowicz JJ, Brown JH et al (2007) Ecological and evolutionary insights from species invasions. Trends Ecol Evol 22:465–471. https://doi.org/10.1016/j.tree.2007.06.009

Shackleton RT, Shackleton CM, Kull CA (2019) The role of invasive alien species in shaping local livelihoods and human wellbeing: a review. J Environ Manage 229:145–157. https://doi.org/10.1016/j.jenvman.2018.05.007

Sharma A, Kaur A, Kohli RK et al (2023) Bidens pilosa (Asteraceae) invasion reshapes the pattern of plant communities and edaphic properties across the north-western Himalayan landscape. Ecol Inform 78:102281. https://doi.org/10.1016/j.ecoinf.2023.102281

Shrestha UB, Shrestha BB (2019) Climate change amplifies plant invasion hotspots in Nepal. Divers Distrib 25:1599–1612. https://doi.org/10.1111/ddi.12963

Simberloff D, von Holle B (1999) Positive interactions of nonindigenous species: invasional meltdown? Biol Invasions 1:21–32. https://doi.org/10.1023/A:1010086329619

Singh KK, Bhattarai N, Vukomanovic J (2022) Landscape-scale hydrologic response of plant invasion relative to native vegetation in urban forests. Sci Total Environ 802:149903. https://doi.org/10.1016/j.scitotenv.2021.149903

Stohlgren TJ, Jarnevich C, Chong GW et al (2006) Scale and plant invasions: a theory of biotic acceptance. Preslia 78:405–426

Sun Y, Züst T, Silvestro D et al (2022) Climate warming can reduce biocontrol efficacy and promote plant invasion due to both genetic and transient metabolomic changes. Ecol Lett 25:1387–1400. https://doi.org/10.1111/ele.14000

Szabó S, Peeters ETHM, Várbíró G et al (2019) Phenotypic plasticity as a clue for invasion success of the submerged aquatic plant *Elodea nuttallii*. Plant Biol 21:54–63. https://doi.org/10.1111/plb.12918

Tershy B, Donlan CJ, Keitt BS et al (2002) Island conservation in north-west Mexico: a conservation model integrating research, education and exotic mammal eradication. In: Veitch CR, Clout MN (eds) Turning the tide: the eradication of invasive species. IUCN Species Survival Commission, Gland, pp 293–300

Theoharides KA, Dukes JS (2007) Plant invasion across space and time: factors affecting nonindigenous species success during four stages of invasion. New Phytol 176:256–273. https://doi.org/10.1111/j.1469-8137.2007.02207.x

Torres N, Herrera I, Fajardo L (2021) Meta-analysis of the impact of plant invasions on soil microbial communities. BMC Ecol Evol 21:172. https://doi.org/10.1186/s12862-021-01899-2

Valliere JM, Irvine IC, Santiago L et al (2017) High N, dry: experimental nitrogen deposition exacerbates native shrub loss and nonnative plant invasion during extreme drought. Glob Change Biol 23:4333–4345. https://doi.org/10.1111/gcb.13694

van Kleunen M, Dawson W, Essl F et al (2015) Global exchange and accumulation of non-native plants. Nature 525:100–103. https://doi.org/10.1038/nature14910

van Wilgen BW, Fill JM, Baard J et al (2016) Historical costs and projected future scenarios for the management of invasive alien plants in protected areas in the cape floristic region. Biol Conserv 200:168–177. https://doi.org/10.1016/j.biocon.2016.06.008

Vetter VMS, Walter J, Wilfahrt PA et al (2019) Invasion windows for a global legume invader are revealed after joint examination of abiotic and biotic filters. Plant Biol 21:832–843. https://doi.org/10.1111/plb.12987

Vujanović D, Losapio G, Milić S et al (2022) The impact of multiple species invasion on soil and plant communities increases with invasive species co-occurrence. Front Plant Sci 13:875824. https://doi.org/10.3389/fpls.2022.875824

Wagner V, Večeřa M, Jiménez-Alfaro B et al (2021) Alien plant invasion hotspots and invasion debt in European woodlands. J Veg Sci 32:13014. https://doi.org/10.1111/jvs.13014

Wang A, Melton AE, Soltis DE et al (2022) Potential distributional shifts in North America of allelopathic invasive plant species under climate change models. Plant Divers 44:11–19. https://doi.org/10.1016/j.pld.2021.06.010

Ward SF, Taylor BS, Dixon Hamil KA et al (2020) Effects of terrestrial transport corridors and associated landscape context on invasion by forest plants. Biol Invasions 22:3051–3066. https://doi.org/10.1007/s10530-020-02308-3

Williams DG, Mack RN, Black RA (1995) Eco-physiology of introduced *Pennisetum setaceum* on Hawaii: the role of phenotypic plasticity. Ecology 76:1569–1580. https://doi.org/10.2307/1938158

Wolkovich EM, Cleland EE (2011) The phenology of plant invasions: a community ecology perspective. Front Ecol Environ 9:287–294. https://doi.org/10.1890/100033

Xiao L, Ding J, Zhang J et al (2020) Chemical responses of an invasive plant to herbivory and abiotic environments reveal a novel invasion mechanism. Sci Total Environ 741:140452. https://doi.org/10.1016/j.scitotenv.2020.140452

Xiong W, Tao J, Liu C et al (2019) Invasive aquatic plant (*Alternanthera philoxeroides*) facilitates the invasion of western Mosquitofish (*Gambusia affinis*) in Yangtze River, China. Aquat Ecosyst Health Manag 22:408–416. https://doi.org/10.1080/14634988.2019.1700090

Xu H, Liu Q, Wang S et al (2022) A global meta-analysis of the impacts of exotic plant species invasion on plant diversity and soil properties. Sci Total Environ 810:152286. https://doi.org/10.1016/j.scitotenv.2021.152286

Young SL, Kimball S, Novak SJ (2022) Invasion of plant communities. In: Clements DR, Upadhyaya MK, Joshi S, Shrestha A (eds) Global plant invasions. Springer, Cham

Yu H, Le Roux JJ, Jiang Z et al (2021) Soil nitrogen dynamics and competition during plant invasion: insights from *Mikania micrantha* invasions in China. New Phytol 229:3440–3452. https://doi.org/10.1111/nph.17125

Zenni RD, Herrera I, de Sá DM et al (2022) Plant invasions in South America. In: Clements DR, Upadhyaya MK, Joshi S, Shrestha A (eds) Global plant invasions. Springer, Cham

Zhang Z, Bhowmik PC, Suseela V et al (2021) Effect of soil carbon amendments in reversing the legacy effect of plant invasion. J Appl Ecol 58:181–191. https://doi.org/10.1111/1365-2664.13757

Zhang G, Bai J, Tebbe CC et al (2022) Plant invasion reconstructs soil microbial assembly and functionality in coastal salt marshes. Mol Ecol 31:4478–4494. https://doi.org/10.1111/mec.16600

Zhao YZ, Liu MC, Feng YL et al (2020) Release from below-and aboveground natural enemies contributes to invasion success of a temperate invader. Plant Soil 452:19–28. https://doi.org/10.1007/s11104-020-04520-5

Ziska LH (2022) Plant invasions, rising CO_2, and global climate change. In: Clements DR, Upadhyaya MK, Joshi S, Shrestha A (eds) Global plant invasions. Springer, Cham

Impacts of Plant Invasions on Ecosystem Functionality: A Perspective for Ecosystem Health and Ecosystem Services

2

Adrián Lázaro-Lobo, Álvaro Alonso, Romina D. Fernández, Elena Granda, Alberto Romero-Blanco, Asunción Saldaña-López, and Pilar Castro-Díez

Abstract

Invasive plants affect the capacity of ecosystems to perform key functions, including primary production, nutrient and water cycling, decomposition, energy flow through food webs, or control of disturbance regimes, hydrology, and sedimentation. Invasive plants can also change the composition and structure of the resident community through different mechanisms, including direct competition, allelopathy, habitat alterations, and hybridization. Both changes in ecosystem functionality and community structure affect the capacity of ecosystems to deliver the three categories of services that contribute to human well-being: provisioning (e.g., food, water, wood, medicines, etc.), regulating and maintenance (e.g., climate regulation, erosion control, flood regulation, fire protection, regulation of soil fertility and water quality, etc.), and cultural (e.g., spiritual, intellectual, or symbolic assets). Invasive plants can also increase the negative effects of ecosystems on human well-being (i.e., ecosystem disservices, such as allergies and infrastructure damage). Impacts on ecosystem services may vary in magnitude and direction depending on the type of invader, the invasion scenario, and the spatio-temporal scale. Also, synergies and trade-offs between ecosystem services may arise when invasive species promote many services simultaneously

A. Lázaro-Lobo (✉)
Department of Life Sciences, University of Alcalá, Alcalá de Henares, Madrid, Spain

Department of Biology and Geology, King Juan Carlos University, Móstoles, Spain

Á. Alonso · E. Granda · A. Romero-Blanco · A. Saldaña-López · P. Castro-Díez
Department of Life Sciences, University of Alcalá, Alcalá de Henares, Madrid, Spain
e-mail: alvaro.alonso@uah.es; elena.granda@uah.es; alberto.romerob@uah.es; asuncion.saldana@uah.es; mpilar.castro@uah.es

R. D. Fernández
Institute of Regional Ecology, National University of Tucumán-CONICET, Tucumán, Argentina

S. Tripathi et al. (eds.), *Plant Invasions and Global Climate Change*,
https://doi.org/10.1007/978-981-99-5910-5_2

or favor some services at the expense of impairing others. For example, some invasive plants can act as C sinks, increase timber provision, and contribute to the formation and protection of soil against erosion, while simultaneously increasing fire risk through increased fuel input, declining water provision through high water consumption, or reducing landscape aesthetics. Climate change may create opportunities for some invasive species and alter the severity of their impacts on ecosystem services, through alterations in species distributions, biological interactions, and ecosystem processes. Indeed, the synergistic effect of invasive species and climate change often cause the most detrimental outcomes for ecosystems.

In this chapter, we first compile information regarding the impacts of plant invasions on ecosystem functionality, focusing on key functions that regulate the fluxes of energy and cycles of matter. Then, we examine how those changes affect the delivery of provisioning, regulating, and cultural ecosystem services. Lastly, we analyze the role of climate change in altering the impacts of invasive plants on ecosystem functionality and ecosystem service delivery. We recommend that future studies investigate how climate change affects the impact of invasive plants on multiple ecosystem processes and services, rather than considering them in isolation. This would improve decision-making on invasive species management under climate change.

Keywords

Biodiversity · Climate change · Cultural services · Ecological processes · Provisioning services · Regulating services

2.1 Introduction

Ecosystem functionality is commonly defined as the capacity of ecosystems to provide key functions that control the fluxes of energy and cycles of matter, including primary production, nutrient and water cycling, decomposition, energy flow through food webs, community dynamics, disturbance regimes, hydrology, and sedimentation (Bennett et al. 2009; Freudenberger et al. 2012; Palmeri et al. 2013). Through ecological interactions, ecosystems build structural and functional networks that increase their regulation capacity of ecological processes, which is fundamental for buffering against environmental changes and disturbances (Freudenberger et al. 2012; Norris et al. 2012). Ecosystem functionality is related to the term "ecosystem health." Healthy ecosystems are characterized by having essential functions and attributes that keep them stable and sustainable over time, maintaining their organization and resilience to stress (Costanza 1992; Rapport et al. 1998; Lu et al. 2015). Biotic diversity, resulting from evolutionary processes, contributes to the complexity of functional networks through niche complementarity, which provides ecosystems with the necessary resilience and adaptive capacity to change (Cardinale et al. 2012; Freudenberger et al. 2012). Indeed, previous research found that ecosystem functionality increases with diversity (Hector and Bagchi 2007; Lefcheck et al. 2015; Gamfeldt and Roger 2017).

Ecosystem functionality is the basis of ecosystem service delivery (Freudenberger et al. 2012). Ecosystem services are the benefits that human populations obtain from ecosystem functions (Costanza et al. 1997; Haines-Young and Potschin-Young 2018). Ecosystem services are of fundamental importance to human well-being, health, livelihoods, and survival (Costanza et al. 2014). According to the Common International Classification of Ecosystem Services (CICES, v5.1; Haines-Young and Potschin-Young 2018), there are three broad categories of ecosystem services: "Provisioning services" are tangible resources (material and energetic) obtained from ecosystems, such as food, water, wood, medicines, and fuel. "Regulating and maintenance services" result from the capacity of ecosystems to regulate key ecological processes for human well-being, including climate regulation, erosion control, flood regulation, fire protection, regulation of soil fertility and water quality, etc. "Cultural services" refer to the non-material (e.g., spiritual, intellectual, or symbolic) benefits obtained from ecosystems, such as education, opportunities for leisure and recreation, science, inspiration, cultural heritage, sense of place, etc. (Balvanera et al. 2017; Costanza et al. 2017; Haines-Young and Potschin-Young 2018). Ecosystems can also cause negative effects on human well-being (i.e., ecosystem disservices). For example, damages to infrastructures caused by plant growth, seasonal allergies caused by pollen, or pests reducing crop production (Von Döhren and Haase 2015; Blanco et al. 2019).

Biological invasions are a widespread and significant component of global change (Vitousek et al. 1997). The introduction of novel species into a region can alter ecosystem structure and functionality (Levine et al. 2003; Dukes and Mooney 2004; Vilà et al. 2011), which, in turn, may also affect the capacity of ecosystems to deliver services, causing environmental, economic, and social impacts (Charles and Dukes 2007; Lázaro-Lobo and Ervin 2021; Fig. 2.1). Previous research suggests that most introduced species do not significantly affect ecosystem functionality and service delivery (Thompson 2014). Only a small portion of non-native species undergo substantial demographic increase (often referred to as a demographic explosion) and rapid expansion in the introduced region, having the potential to cause a variety of ecological, social, or economic impacts (hereafter "invasive species"; Richardson et al. 2000; Catford et al. 2009; Blackburn et al. 2011). However, some non-native species may have an impact even not being invasive. The classical example is that of allergenic non-native plants planted in gardens that have a negative impact on human health, even if they do not have the capacity to establish in natural areas.

In this chapter, we first compile information regarding the impacts (positive and negative) of plant invasions on ecosystem functionality, focusing on key functions that regulate the fluxes of energy and cycles of matter. Then, we examine how those changes affect the delivery of provisioning, regulating, and cultural ecosystem services. Lastly, we analyze the role of climate change in altering the impacts of invasive plants on ecosystem functionality and ecosystem service delivery. Throughout the chapter, we review case studies conducted around the world to

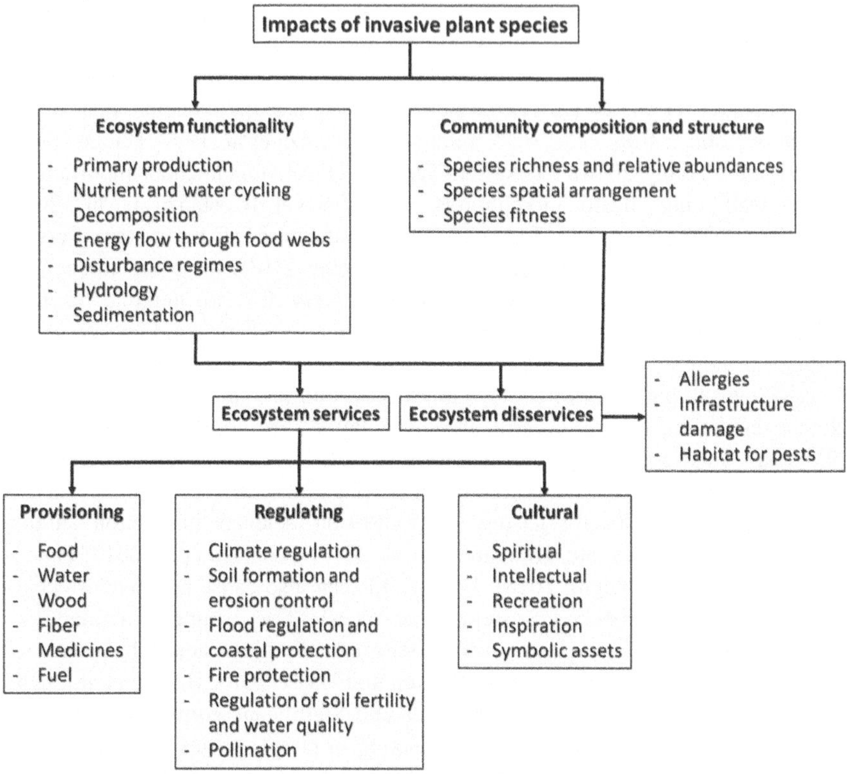

Fig. 2.1 General framework for the impacts of plant invasions on ecosystem functionality, community composition and structure, and ecosystem services and disservices, including some examples

deepen our understanding of how biological invasions and climate change are altering ecosystem functionality and ecosystem service delivery in multiple ecosystems.

2.2 Impacts of Invasive Plants on Community Composition and Structure

Invasive plants alter the community structure of terrestrial, freshwater, and marine ecosystems (Hejda et al. 2009; Gallardo et al. 2016; Schirmel et al. 2016; Abgrall et al. 2019; Anton et al. 2019). Invasive plants usually affect multiple taxonomic assemblages and levels of the ecosystem food web. For example, in South America, invasion by *Pinus* spp. and *Pseudotsuga menziesii* affected the composition and structure of the native plant, bird, and soil arthropod assemblages, displaced endemic native species, and promoted invasion by other alien species (León-Gamboa et al. 2010; de Abreu and Durigan 2011; Zenni and Ziller 2011; Pauchard et al. 2015).

Similarly, the invasion of *Baccharis halimifolia* in western Europe, eastern Australia, and New Zealand caused changes on plant, bird, arthropod assemblages (Lázaro-Lobo et al. 2021a; Table 2.1). Particularly, arthropod diversity loss due to plant invasions may have drastic effects on food web dynamics (van Hengstum et al. 2014). However, most studies have focused on the impacts of plant invasions on a few groups of organisms, as we explain below.

2.2.1 Impacts of Invasive Plants on Plant Assemblages

Invasive plants can change the composition and structure of resident plant assemblages through different mechanisms, including direct competition, allelopathy, habitat alterations, and hybridization. As a result, taxonomic and functional richness, diversity, and evenness can be reduced in invaded areas when compared to uninvaded ones (Hejda et al. 2009; Marchante et al. 2015; Castro-Díez et al. 2016). Many invasive plants outcompete native plant species in resource acquisition due to their rapid growth rates (Richardson and Kluge 2008; Fernández et al. 2020). Invasive plants can also negatively affect native plants via allelopathy. For example, Aguilera et al. (2015) found that aqueous extracts of the tree *Acacia dealbata* can interfere with the establishment of native herbaceous and tree species. Similarly, the invasive herb *Parthenium hysterophorus* releases allelochemical components that decreased both the germination and growth of native plants in India (Dogra and Sood 2012). Invasive plants can also alter plant assemblages through physical changes, which can make the environment less suitable for native plant species. For example, invasive floating-leaved plants such as *Nelumbo lutea* and *Brasenia schreberi* and free-floating plants such as *Eichhornia crassipes* and *Salvinia molesta* can form dense floating mats that shade out submerged vegetation (Lázaro-Lobo and Ervin 2021; Fig. 2.2). Lastly, invasive plants can alter the species composition of the recipient communities by hybridization and introgression with native flora (Rhymer and Simberloff 1996).

2.2.2 Impacts of Invasive Plants on Animal Assemblages

Although less studied than plant assemblages, invasive plants also cause impacts on animal assemblages (Ortega et al. 2014; Schirmel et al. 2016). In a global meta-analysis, Schirmel et al. (2016) found that invasive plants reduced animal abundance, diversity, and fitness, although the impacts were most evident in riparian ecosystems and the most affected taxonomic groups were birds and insects. Invasive plants can decrease macroinvertebrate abundance, richness, biomass, and diversity in aquatic ecosystems by reducing sunlight penetration and dissolved oxygen content, and altering water temperature, pH, and nutrient concentrations (Gerber et al. 2008; Havel et al. 2015; Seeney et al. 2019; Wahl et al. 2021). Another meta-analysis also showed that invasive plants globally reduced native animal abundance, but such effect was not evident for non-native animals (Fletcher et al. 2019). Abgrall

Table 2.1 Examples of impacts of plant invasions on community structure and composition

Invasive plant/s	Affected assemblage/s	Type of impact	Geographical area	Reference
Albizia julibrissin, *Ligustrum* spp., *Lonicera japonica*, *Lygodium japonicum*, *Microstegium vimineum*, *Rosa* spp., and *Triadica sebifera*	Trees	Reduced forest regeneration	Southeastern United States	Lázaro-Lobo et al. (2021b)
Acacia longifolia	Coastal dune plant assemblage	Decreased plant diversity	Portugal	Marchante et al. (2015)
Ligustrum lucidum	Anurans	Reduced anuran diversity	Argentina	Segura et al. (2021)
Ailanthus altissima	Bacteria	Alteration of soil bacterial assemblages	Spain	Medina-Villar et al. (2016)
Bromus tectorum	Arbuscular mycorrhizal fungi	Changes on arbuscular mycorrhizal fungal assemblage	Western United States	Busby et al. (2013)
Parthenium hysterophorus	Plants	Reduced regeneration	India	Dogra and Sood (2012)
Salvinia molesta	Insects	Decreased richness and abundance of aquatic insects	Southeastern United States	Wahl et al. (2021)
Baccharis halimifolia	Plants, birds, and arthropods	Changes on plant, bird, and arthropod assemblages	Western Europe, eastern Australia, and New Zealand	Lázaro-Lobo et al. (2021a)
Carpobrotus spp.	Plants	Reduced species and functional richness	Mediterranean islands	Castro-Díez et al. (2016)
Spartina alterniflora	Bacteria	Decreased bacterial diversity	Eastern China	Gao et al. (2019)
Eichhornia crassipes	Submerged plants, fish and invertebrates	Reduction of submerged plant, fish, and invertebrate abundance	Ethiopia	Dechassa and Abate (2020)

(continued)

Table 2.1 (continued)

Invasive plant/s	Affected assemblage/s	Type of impact	Geographical area	Reference
Alliaria petiolata	Fungi	Fungal assemblage homogenization	Northeastern United States	Anthony et al. (2017)
Impatiens glandulifera	Soil microbes	Changes on soil microbial assemblage	United Kingdom	Pattison et al. (2016)
Juncus acutus	Invertebrates	Changes on invertebrate assemblage	Southeasteam Australia	Harvey et al. (2014)
Prosopis julifora	Plants	Reduced regeneration, diversity, and abundance	Saharan and southern Africa, the Middle East, Pakistan, India, and Hawaii	Hussain et al. (2021)

Fig. 2.2 An invasive population of water hyacinth (*Eichornia crassipes*) in Mississippi (USA). Inset shows one of the authors (Lázaro-Lobo) for scale

et al. (2019) assessed the global impacts of invasive plants on soil invertebrates and found that the abundance of detritivores tended to decline in invaded open habitats, however, they did not find impacts for other functional groups. Contrastingly, McCary et al. (2016) showed that invasive plants alter the trophic structure of invertebrates globally, especially in wetlands and woodlands. By altering the physical environment, invasive plants can affect animal movement, safety, and reproduction (Ortega et al. 2014; Stewart et al. 2021). For example, invasion by the shrub *Ligustrum lucidum* in a semiarid subtropical forest resulted in dark environments with less native anuran richness (Segura et al. 2021). In South Africa, the thorny *Opuntia ficus-indica* hinders livestock access to forage (van Wilgen et al. 2020).

2.2.3 Impacts of Invasive Plants on Microbial Assemblages

Invasive plants may also affect both above- and below-ground microbial assemblages, but these impacts are deeply underestimated (Torres et al. 2021). In some cases, invasive plants can decrease both microbiomes and microbial diversity (bacteria and fungi; Malacrinò et al. 2020). However, they can also increase productivity and abundance of soil microbial assemblages involved in N cycling, as found in invaded grasslands (McLeod et al. 2021). Invasive plants can also alter soil bacterial assemblages (Medina-Villar et al. 2016), increasing bacterial richness and diversity in some instances (Rodríguez-Caballero et al. 2017; Torres et al. 2021). The few studies conducted on the effect of invasive plants on microbial assemblages suggest that such impacts might be context-dependent.

2.3 Impacts of Invasive Plants on Ecosystem Functionality

2.3.1 Impacts on Primary Production

Invasive plants can drastically affect standing biomass and net primary production (C acquisition) by differing from native species in overall size, morphology, phenology, or productivity (Ehrenfeld 2003; Dassonville et al. 2008; Fig. 2.3). Several meta-analyses have shown that plant invasions usually enhance primary production in invaded ecosystems compared with native ecosystems (Ehrenfeld 2003; Liao et al. 2008; Vilà et al. 2011; Castro-Díez et al. 2019). However, biomass production of the resident biota is usually reduced by plant invasions (Pyšek et al. 2012; Helsen et al. 2018). Invasive plants can also alter primary production through their effects on nutrient retention and turnover, site water balance, and disturbance frequency, as we explain below (Walker and Smith 1997; Peltzer et al. 2010).

2.3.2 Impacts on Nutrient Cycling and Organic Matter Decomposition

Plant invasions can modify nutrient cycling in different ways, including changes in soil nutrient content and microbial activity, changes in the timing of nutrient availability, alteration of N fixation rates, and production of litter of differing quality than co-occurring native plants (Ehrenfeld 2003; Zhou and Staver 2019; Lázaro-Lobo and Ervin 2021; Xu et al. 2022, Fig. 2.3). Several studies have shown that

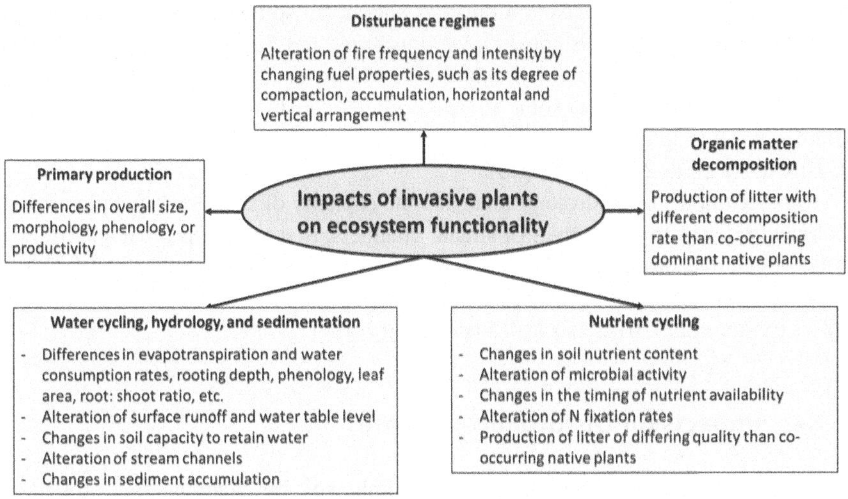

Fig. 2.3 Impacts of plant invasions on ecosystem functionality, including mechanisms driving the alterations

invasive plants accelerate N cycling by having greater litter production with a high N content and by increasing plant N uptake (via greater fine root production and specific root length; Liao et al. 2008; Jo et al. 2017; Incerti et al. 2018). However, particular impacts on N pools and fluxes are highly context-dependent and therefore difficult to predict (Castro-Díez et al. 2014; Castro-Díez and Alonso 2017). Moreover, litter of invasive plants usually decomposes more rapidly than that of native ones (Allison and Vitousek 2004; Liao et al. 2008). But there are also examples showing the opposed trend, as non-native *Eucalyptus* spp. and *Pinus* spp. producing litter that is poor in N or resistant to soil microbial decomposition (due to the high concentration of phenols and other secondary compounds), which slows down nutrient cycling (Zhang et al. 2019; Castro-Díez et al. 2021). Such tree invaders can also acidify the soil, which can greatly reduce microbial activity and nutrient cycling (Soumare et al. 2016). Nutrient cycling can also be reduced when the leaves/shoots of invasive plants, such as *Pinus* spp. and *Bromus tectorum*, have a high lignin content (Evans et al. 2001; Levine et al. 2003). This molecule can sequester protein N and form molecular complexes that are difficult for soil microorganisms to access, reducing its availability to many plants (Castro-Díez et al. 2014).

2.3.3 Impacts on Water Cycling, Hydrology, and Sedimentation

Water cycles may be disrupted when invasive plants significantly differ from natives in key functional traits related to the use and economy of water, such as evapotranspiration and water consumption rates, rooting depth, phenology, leaf area, root: shoot ratio, etc. (Levine et al. 2003; Charles and Dukes 2007, Fig. 2.3). The high transpiration and water consumption rates shown by some invasive woody plants, such as *Acacia* spp., *Pinus* spp., *Eucalyptus* spp., and *Tamarix* spp., can decrease surface runoffs and lower water tables (Le Maitre et al. 1996, 2002; Cronk and Fennessy 2001). However, some invasive plants can increase the soil capacity to retain water through their high supply of organic matter, especially in degraded soils (Castro-Díez et al. 2019; Lal 2020).

Another impact on hydrology relates to the capacity of some invasive plants such as *Arundo donax* to block river or stream channels, due to their profuse growth and rapid vegetative propagation. This may decrease water flow velocity and facilitate the accumulation of fine sediments, altering channel structure (Charles and Dukes 2007; Eviner et al. 2012; Lázaro-Lobo and Ervin 2021).

2.3.4 Impacts on Disturbance Regimes

Invasive plants can alter fire frequency and intensity by changing fuel properties, such as its degree of compaction, accumulation, horizontal and vertical arrangement, etc. (Brooks et al. 2004, Fig. 2.3). The invasion of ecosystems by annual grasses often leads to the accumulation of more persistent fuel beds, which increases fire frequency (Levine et al. 2003). For example, the invasion of shrublands in the western United States by *Bromus tectorum* has increased fire frequency to such an extent that native shrublands cannot recover (Whisenant 1990; D'Antonio and Vitousek 1992; Brooks et al. 2004). Invasive plants can also increase fire intensity when they accumulate more fuel in the environment than native species (Fig. 2.4), due to invasive species having greater biomass and lower litter decomposition rates (Levine et al. 2003; Taylor et al. 2017). Some taxa that are considered as fire-promoters include *Eucalyptus* spp., *Pinus* spp., *Acacia* spp., *Andropogon virginicus*, *Bromus rubens*, *Melaleuca quinquenervia*, or *Melinis minutiflora* (Richardson et al. 2000; Brooks et al. 2004; Fernandes 2009; Silva et al. 2009; Le Maitre et al. 2011). On the contrary, other invasive species such as the trees *Schinus terebinthifolius* and *Triadica sebifera* can reduce fire frequency and severity when forming dense populations. These invasive plants produce low-flammable biomass that requires very high temperatures to burn, interrupting the vertical and/or horizontal fire transmission, or competitively displacing native fire-promoting species (Grace 1998; Brooks et al. 2004; Stevens and Beckage 2010).

2.4 Impacts of Invasive Plants on the Delivery of Ecosystem Services

2.4.1 Impacts of Invasive Plants on Provisioning Services

Human populations demand a wide variety of resources to sustain their livelihoods, such as food, water, medicines, wood, fibers, or fuel. This demand has exponentially increased through time, promoting the introduction of non-native species worldwide

Fig. 2.4 Biomass accumulation by non-native eucalypts (*Eucalyptus globulus*) in the Cíes Islands, Spain

at an increasing frequency and abundance (Shackleton et al. 2007; Kull et al. 2011; Castro-Díez et al. 2019). Well-known examples are those of common crop plants, such as wheat, rice, corn, or potato, which nowadays are grown worldwide. Also, some tree species able to provide wood, wood-derived or other resources (cork, resin, fiber, etc.) have been planted worldwide (Castro-Díez et al. 2019). Obviously, these non-native species have increased the provisioning ecosystem services, promoting human well-being. Yet, the worldwide movement of species also has drawbacks for resource provisioning, sometimes due to the unintended introduction of weeds accompanying traded species, but others by the own traded non-natives.

The introduction of non-native plants inevitably involves the unintentional introduction of invasive weeds, associated with them in their native areas (Davis and Landis 2011). These weeds may have great success because they are likely to be introduced in a suitable climate, similar to that of their region of origin (Fried et al. 2017), and because they are benefited by the conditions created in agricultural areas (e.g., suppression of competition and high levels of fertilization and disturbance). Agricultural areas are also prone to be invaded by invasive plants that have been unintentionally dispersed by motor vehicles along surrounding areas (e.g., roadsides and field margins; Lázaro-Lobo and Ervin 2019). Indeed, a great proportion of invasive plants can be found in croplands (DAISIE 2009). For instance, tropical C_4 grasses showed great success in many agricultural areas around the globe because of their drought tolerance and efficient photosynthetic pathway (Maillet and López-

García 2000). Invasive weeds reduce crop yields due to competition, but also due to allelopathy or parasitism (Fried et al. 2017). For instance, the parasitic plant, *Striga hermonthica* causes annual losses in maize of US $7 billion in Africa (Burgiel and Muir 2010). A full review of examples and impacts of invasive weeds on agriculture can be found in Fried et al. (2017).

Sometimes the own traded species may escape from cultivation and establish in the wild, and a few of these naturalized species may also escape from natural mechanisms controlling for population growth and become invasive. Both naturalized and invasive plants may alter the equilibrium and functioning of the native ecosystem, and thus the ability of the ecosystem to provide resources. One example is that of Northern Hemisphere pines introduced in treeless high-altitude or latitude areas of the Southern Hemisphere for erosion control and/or timber provision (Simberloff et al. 2010). Some of them turned to be successful invaders, declining the supply of some services (Ledgard 2001; Rundel et al. 2014). For instance, the North American *Pinus contorta* was introduced to New Zealand at the end of the nineteenth century and rapidly spread across mountain grasslands, particularly those heavily grazed, declining their capacity to provide food for livestock (Ledgard 2001). Also, the conversion of Andean Páramo grasslands of Ecuador into *Pinus radiata* plantations has drastically decreased the ability of soils to retain water, threatening water provision to highland cities and towns (Farley et al. 2004). Also remarkable is the water yield reduction in catchments invaded by non-native trees (mostly *Pinus* spp. and *Acacia* spp.) in the Western Cape Province of South Africa. These trees were introduced by European colonists to cover their wood demands in a treeless region, but they consumed much more water than the native shrubby vegetation, decreasing water yield and threatening water supply to urban and rural populations (Le Maitre et al. 1996; van Wilgen et al. 1998). Several species of *Prosopis* spp. were introduced globally because of their capacity to provide a wide variety of resources (wood, firewood, fodder, etc.) in arid regions, where they became an important source of income for local rural communities (Chikuni et al. 2004; Choge et al. 2012; Shackleton et al. 2014). However, *Prosopis* spp. became invasive along temporary streams due to their ability to reach deep soil water (down to 50 m), where they pump high quantities of deep water, contributing to the drying of wells. Additionally, *Prosopis* spp. thorns negatively affect livestock health, reducing access to water sources and causing flesh wounds (Shackleton et al. 2014).

2.4.2 Impacts of Invasive Plants on Regulating Services

2.4.2.1 Impacts on Climate Regulation

High productivity and fast growth are key traits that often determine the selection of non-native trees for introduction and their impacts on regulating ecosystem services (Richardson 1998; Castro-Díez et al. 2019). *Pinus* spp., *Eucalyptus* spp., and *Acacia* spp. are some of the trees that have been planted all over the world (Richardson 1998). Fast-growing invasive plants are often considered as important carbon sinks due to their high gross primary production; however, these species often have less durable C stocks than slow-growing native species (Suryaningrum et al. 2022). It has also been shown that areas invaded by exotic plants can promote C loss by having higher respiration rates, leaching, and disturbance frequency and intensity than uninvaded areas (Peltzer et al. 2010). Furthermore, the effect of invasive plants on the soil ability to store C is often neglected, and the few case studies that exist suggest a negative impact (Wu et al. 2020; Zarafshar et al. 2020). Thus, the balance between C inputs and outputs could either increase or decrease net C sequestration in the invaded area. Overall, previous research suggests that plant invasions generally promote climate regulation (via carbon uptake; Liao et al. 2008; Castro-Díez et al. 2019). However, invasive plants not only influence C sequestration over short-term scales (weeks to years) by directly affecting rates of primary production or decomposition but also over long-term scales (decades and centuries) by causing compositional changes in the dominant tree species or by altering the resilience of ecosystems to disturbances and climate change (Bunker et al. 2007; Peltzer et al. 2010).

2.4.2.2 Impacts on Soil Fertility

Invasive N-fixing plants are a clear example of how the introduction of new functional traits, such as the ability to fix N, can significantly alter regulating ecosystem services, especially when the recipient ecosystem lacks native N-fixers. Invasive N-fixers have been introduced repeatedly to improve soil fertility and promote agriculture and forestry (Kozlowski and Pallardy 1997). The increase in nutrients has dramatic impacts on native communities adapted to infertile soils, which are outcompeted by fast-growing invasive species benefiting from the nutrient surplus (González-Muñoz et al. 2012). These changes in community structures may lead to further disruptions in regulating ecosystem services. Other invaders can reduce soil fertility by displacing native N-fixing species or inhibiting the activity of soil biota, which may negatively affect agricultural yields. Evidence of this has been found in the invasive thistle *Carduus nutans* in New Zealand, whose decomposing leaves interfere with the fixation activity of the herb *Trifolium repens* (Wardle et al. 1994; Dukes and Mooney 2004). Invasive plants may also reduce soil fertility by increasing fire frequency, which intensifies long-term N depletion (Dukes and Mooney 2004).

2.4.2.3 Impacts on Soil Formation and Erosion Control

Multiple invasive plants have been deliberately introduced worldwide to promote soil formation, erosion control, and sediment stabilization, due to their high root growth and/or high supply of organic matter to the soil (Castro-Díez et al. 2019). For example, *Acacia* spp. have been widely planted in dunes for sand dune binding (Breton et al. 2008; Marchante et al. 2008), *Cynodon dactylon* has been introduced to protect riverbanks against erosion caused by flooding (Dukes and Mooney 2004; Chen et al. 2015), and *Baccharis halimifolia* has been planted as windbreaks along field perimeters to reduce wind erosion (Lázaro-Lobo et al. 2021a). Invasive plants that increase fire frequency and intensity can also increase erosion and decrease soil formation and water retention when the organic matter that binds soil particles is burned (Le Maitre et al. 2011). Fires can also increase soil erosion by removing vegetation and leaf litter cover (Swanson 1981).

2.4.2.4 Impacts on Fire Protection

Invasive plants that promote fire through increased fuel input into the ecosystem may increase fire frequency, intensity, and severity, which can have dramatic effects on the fire protection (Levine et al. 2003; Castro-Díez et al. 2019). The alteration of this ecosystem service often leads to catastrophic consequences, such as irreversible changes in the structure and functioning of ecosystems and socio-economic impacts (Brooks et al. 2004; Gaertner et al. 2014). For instance, increments in fire incidence, intensity, or rate of spread by invasive species can negatively affect human health and safety, as well as the use of natural resources by people.

2.4.2.5 Impacts on Air Quality

Air quality can also be affected by invasive plants. Some of these effects are mediated by an increase in fire frequency, which causes the emission of carbon monoxide and dioxide, as well as nitrogen oxides (Hickman et al. 2010). Furthermore, some species, such as *Eucalyptus* spp. or the vine *Pueraria montana* can produce high quantities of isoprene, a volatile organic compound that can form ozone and smog when reacting with nitrogen oxides (Wolfertz et al. 2003; Forseth and Innis 2004; Hickman et al. 2010). Increased tropospheric ozone decreases air quality, particularly in areas with low rates of ozone formation (e.g., areas far from urban centers; Hickman et al. 2010; Eviner et al. 2012). Through these emissions, invasive plants can alter atmospheric composition and the ability of ecosystems to regulate climate (Dukes and Mooney 2004).

2.4.2.6 Impacts on Water Quality

Depletion of water resources by some invasive species (e.g., *Eucalyptus* sp. and *Acacia* sp.) leads to a decrease in dilution capacity and, consequently, to an increase in salinity and the concentration of nutrients and pollutants (Chamier et al. 2012). *Impatiens glandulifera*, which is widespread in several river basins of the northern hemisphere, in addition, to promote the erosion of invaded riparian zones, can decrease water quality by eutrophication through the incorporation of dead plant material and organic matter (Greenwood and Kuhn 2014; Coakley and Petti 2021). Invasive aquatic plants, such as *Azolla filiculoides*, *Eichhornia crassipes*, *Pistia stratiotes*, *Alternanthera philoxeroides*, or *Elodea canadensis*, also alter water quality by forming dense mats, which reduce dissolved oxygen in the water column, either by direct consumption, by reducing water flow, or by suppressing submersed aquatic plants that release oxygen into the water (Wang et al. 2016; Zahari 2021). In addition, the large amount of organic matter they produce promotes eutrophication and enhances microbial growth (Wang et al. 2016; Zahari 2021).

2.4.2.7 Impacts on Flood Control and Coastal Protection

Some invasive plants, such as *Tamarix* spp. and *Arundo donax*, may increase sediment accumulation and bank stabilization, which narrow stream channels and make them less effective in the evacuation of water during flood events (Charles and Dukes 2007; Eviner et al. 2012). Moreover, some invasive plants, such as *Casuarina equisetifolia* and *Cocos nucifera* trees have been planted along coasts for protection from cyclones, tsunamis, and tidal water damage (Mattsson et al. 2009; Feagin et al. 2010; Das and Sandhu 2014). However, native vegetation, such as mangroves, generally provides better protection against storm surges (Das and Sandhu 2014).

2.4.2.8 Impacts on Pollination

Invasive plants can cause disruptions to the pollination of native plants and crops (Morales and Traveset 2009; Goodell and Parker 2017; Nel et al. 2017). In general, studies on invasive plant–pollinator mutualisms show negative effects of invasive plants on reproduction in co-flowering native plants, particularly when the invaders are more abundant (Morales and Traveset 2009; Vanbergen et al. 2018; Ojija et al. 2019). However, some invasive plants can positively affect the pollination of native plants via the attraction of native pollinators (e.g., Bezemer et al. 2014; Vanbergen et al. 2018). In agricultural areas, invasive plants are often found on the edges of crop fields and might facilitate or compete with crops for pollinators (Nel et al. 2017). Nel et al. (2017) found that the invasive plant *Lantana camara* had a positive effect on mango flower visitation at low to medium mango flower density, but not at high mango flower densities. Thus, invasive plants may attract insects, or support crop flower visitors, when crop flower density is low, but lure pollinating insects away from crops when crop flower density is high (Nel et al. 2017).

2.4.3 Impacts of Invasive Plants on Cultural Services

Many invasive plants have been introduced for ornamental use and aesthetic purposes due to their abundant or colorful flowers, especially in urban parks and gardens where they are generally valued by people (Guo et al. 2019). However, the establishment and spread of invasive plants in natural environments have a negative impact on nature route users seeking for experience in such environments, thus decreasing recreation services (Vaz et al. 2018; Castro-Díez et al. 2019). Plant invasions are also a challenge for the conservation of cultural heritage sites, which have been protected from urban development and have become important components of the urban green infrastructures in growing megacities (Gopal et al. 2018; Celesti-Grapow and Ricotta 2021).

2.4.4 Impacts of Invasive Plants on Ecosystem Disservices

Invasive plants can increase the negative effects of some ecosystem functions on human well-being. For example, some plants such as *Ambrosia artemisiifolia* L.,

Ailanthus altissima, and *Eucalyptus* spp., originate allergenic issues and asthma (Belmonte and Vilà 2004; Nentwig et al. 2017). Others contain toxins that can be fatal if ingested, such as *Nerium oleander* which produces cardiac glycosides affecting the heart, the gastrointestinal system, and the central nervous system. The sap of *Ailanthus altissima* is also toxic and *Opuntia* spp. causes strong dermatitis (Nentwig et al. 2017). Furthermore, invasive macrophytes such as *Eichhornia crassipes* and *Salvinia molesta* can expand the habitat for vectors (e.g., mosquitos) of human and other animal diseases, exacerbating problems related to human health (Lázaro-Lobo and Ervin 2021). Infrastructures can also be damaged by invasive plants that cover their surfaces, such as *Hedera helix* and *Pueraria montana* (Von Döhren and Haase 2015; Blanco et al. 2019). Invasive trees can also damage infrastructures through mechanical and chemical action. For example, the tree *Ailanthus altissima* was the most damaging and widespread species in the monuments of Rome (Trotta et al. 2020; Celesti-Grapow and Ricotta 2021).

2.5 Interaction Between Climate Change and Plant Invasions on the Delivery of Ecosystem Services

Throughout this chapter, we have shown that invasive species are a major threat to biodiversity and ecosystem service delivery. Climate change might exacerbate the magnitude of these threats by altering species distributions, biological interactions, and ecosystem processes (Dukes and Mooney 1999; Burgiel and Muir 2010; Goldstein and Suding 2014). Indeed, the synergistic effect of invasive species and climate change often cause the most detrimental outcomes for ecosystems (Caldeira et al. 2015; Vilà et al. 2021; López et al. 2022). For instance, in Cape region (South Africa), the invasion of non-native trees (which transpire more than the native vegetation) along with climate aridification acts synergistically, reducing the water supply for the human population. These invasions will also likely increase fire intensity and erosion, as well as diminish water quality (Le Maitre et al. 1996). Similarly, the increase in fire frequency and/or intensity due to climate change may increase the chances of the spread of invasive plants which depend on fire (e.g., *Acacia* spp., *Eucalyptus* spp.; Sage 1996). The spread of such species can make ecosystems even more prone to fire (Brooks et al. 2004) with strong consequences for the ecosystem functionality and, thus, ecosystem service delivery.

Invasive species usually have different traits from native ones, including greater competitive capacity, reproductive success, broader climatic tolerance, etc. (Qian and Ricklefs 2006; van Kleunen et al. 2010). These traits might positively affect invasive species responses to climate change and facilitate range shifts, thus increasing their future distributions and impacts (Hellmann et al. 2008; Bradley et al. 2015). However, reduced impacts of invasive species with climate change have also been shown, with drought being one of the most important limiting factors (e.g., Liu et al. 2017). Extreme events might also decline invasive species performance. For instance, after the Filomena snowstorm in January 2021 in central Spain, many introduced non-native plants not adapted to snow were swept (e.g., *Eucalyptus camaldulensis*, *Leucaena* spp.). Thus, the effect of climate change on the impacts of plant invaders is highly dependent on the site conditions and on the species physiological properties.

Previous research suggests that climate change differently affects the impacts of invasive plants on ecosystem functioning and service provision in different regions around the world. Ecosystems at higher elevations and latitudes will probably be more dramatically affected, given that climate change might decrease the climatic filters that prevent many plant invasions in these areas (Pauchard et al. 2009; Marini et al. 2012).

2.6 Conclusion and Future Perspectives

In this chapter, we have shown that the magnitude and direction of impacts of invasive plants on community structure, ecosystem functionality, and service delivery depend on the type of invader, the invasion scenario, and the spatio-temporal scale. Also, synergies and trade-offs between ecosystem services may arise when invasive species promote many services simultaneously or favor some services at the expense of impairing others. For example, we have shown that the introduction of productive fast-growing plants can act as C sinks, increase timber provision, and contribute to the formation and protection of soil against erosion. However, such invasive plants can also increase fire risk through increased fuel input into the ecosystem, alter water cycles through high water consumption, modify soil properties and microbial communities, and impact aesthetic cultural values.

Climate change may create opportunities for some invasive species and alter the severity of their impacts on ecosystem services, through alterations in species distributions, biological interactions, and ecosystem processes. Future studies should focus on how climate change affects the impact of invasive plants on multiple ecosystem processes and services, rather than considering them in isolation. This would improve decision-making on invasive species management under climate change.

Acknowledgments This research was supported by the Ministry of Science and Innovation of Spain, the Spanish Research Agency, the European Regional Development Fund [EXARBIN (RTI2018-093504-B-I00) and InvaNET (RED2018-102571-T) projects], and by the REMEDINAL projects of the Community of Madrid (TE-CM S2018/EMT-4338 and TE-CM, S2018/EMT-4338).

References

Abgrall C, Forey E, Chauvat M (2019) Soil fauna responses to invasive alien plants are determined by trophic groups and habitat structure: a global meta-analysis. Oikos 128:1390. https://doi.org/10.1111/oik.06493

Aguilera N, Becerra J, Guedes LM, Villasenor-Parada C, Gonzalez L, Hernandez V (2015) Allelopathic effect of the invasive *Acacia dealbata* link (Fabaceae) on two native plant species in south-central Chile. Gayana Bot 72:231–239

Allison SD, Vitousek PM (2004) Rapid nutrient cycling in leaf litter from invasive plants in Hawai'i. Oecologia 141:612–619

Anthony MA, Frey SD, Stinson KA (2017) Fungal community homogenization, shift in dominant trophic guild, and appearance of novel taxa with biotic invasion. Ecosphere 8:e01951

Anton A, Geraldi NR, Lovelock CE, Apostolaki ET, Bennett S, Cebrian J, Krause-Jensen D, Marba N, Martinetto P, Pandolfi JM, Santana-Garcon J, Duarte CM (2019) Global ecological impacts of marine exotic species. Nat Ecol Evol 3:787–800. https://doi.org/10.1038/s41559-019-0851-0

Balvanera P, Quijas S, Karp DS, Ash N, Bennett EM, Boumans R, Brown C, Chan KMA, Chaplin-Kramer R, Halpern BS, Honey-Rosés J, Kim CK, Cramer W, Martínez-Harms MJ, Mooney H, Mwampamba T, Nel J, Polasky S, Reyers B, Roman J, Turner W, Scholes RJ, Tallis H, Thonicke K, Villa F, Walpole M, Walz A (2017) Ecosystem services. In: The GEO handbook on biodiversity observation networks. Springer, Cham

Belmonte J, Vilà M (2004) Atmospheric invasion of non-native pollen in the Mediterranean region. Am J Bot 91:1243–1250. https://doi.org/10.3732/ajb.91.8.1243

Bennett AF, Haslem A, Cheal DC, Clarke MF, Jones RN, Koehn JD, Lake PS, Lumsden LF, Lunt ID, Mackey BG, Nally RM, Menkhorst PW, New TR, Newell GR, O'Hara T, Quinn GP, Radford JQ, Robinson D, Watson JEM, Yen AL (2009) Ecological processes: a key element in strategies for nature conservation. Ecol Manag Restor 10:192–199

Bezemer TM, Harvey JA, Cronin JT (2014) Response of native insect communities to invasive plants. Annu Rev Entomol 59:119–141. https://doi.org/10.1146/annurev-ento-011613-162104

Blackburn TM, Pyšek P, Bacher S, Carlton JT, Duncan RP, Jarošík V, Wilson JRU, Richardson DM (2011) A proposed unified framework for biological invasions. Trends Ecol Evol 26:333–339

Blanco J, Dendoncker N, Barnaud C, Sirami C (2019) Ecosystem disservices matter: towards their systematic integration within ecosystem service research and policy. Ecosyst Serv 36:100913

Bradley BA, Early R, Sorte CJB (2015) Space to invade? Comparative range infilling and potential range of invasive and native plants. Glob Ecol Biogeogr 24:348–359. https://doi.org/10.1111/geb.12275

Breton C, Guerin J, Ducatillion C, Medail F, Kull CA, Berville A (2008) Taming the wild and "wilding" the tame: tree breeding and dispersal in Australia and the Mediterranean. Plant Sci 175:197–205. https://doi.org/10.1016/j.plantsci.2008.05.013

Brooks ML, D'Antonio CM, Richardson DM, Grace JB, Keeley JE, DiTomaso JM, Hobbs RJ, Pellant M, Pyke D (2004) Effects of invasive alien plants on fire regimes. Bioscience 54:677–688. https://doi.org/10.1641/0006-3568(2004)054[0677:EOIAPO]2.0.CO;2

Bunker DE, DeClerk F, Bradford JC, Colwell RK, Perfecto I, Phillips OL, Sankaran M, Naeem S (2007) Species loss and aboveground carbon storage in a tropical forest. Science 310:1029–1031

Burgiel S, Muir A (2010) Invasive species, climate change and ecosystem-based adaptation: addressing multiple drivers of global change. Global Invasive Species Programme

Busby RR, Stromberger ME, Rodriguez G, Gebhart DL, Paschke MW (2013) Arbuscular mycorrhizal fungal community differs between a coexisting native shrub and introduced annual grass. Mycorrhiza 23:129–141

Caldeira MC, Lecomte X, David TS, Pinto JG, Bugalho MN, Werner C (2015) Synergy of extreme drought and shrub invasion reduce ecosystem functioning and resilience in water-limited climates. Sci Rep 5:15110. https://doi.org/10.1038/srep15110

Cardinale BJ, Duffy JE, Gonzalez A, Hooper DU, Perrings C, Venail P, Narwani A, Mace GM, Tilman D, Wardle DA, Kinzig AP, Daily GC, Loreau M, Grace JB, Larigauderie A, Srivastava DS, Naeem S (2012) Biodiversity loss and its impact on humanity. Nature 486:59–67

Castro-Díez P, Alonso A (2017) Alteration of nitrogen cycling as a result of invasion. In: Vilà M, Hulme PE (eds) Impact of biological invasions on ecosystem services. Springer International Publishing, Cham, pp 49–62

Castro-Díez P, Godoy O, Alonso A, Gallardo A, Saldaña A (2014) What explains variation in the impacts of exotic plant invasions on the nitrogen cycle? A meta-analysis. Ecol Lett 17:1–12. https://doi.org/10.1111/ele.12197

Castro-Díez P, Pauchard A, Traveset A, Vilá M (2016) Linking the impacts of plant invasion on community functional structure and ecosystem properties. J Veg Sci 27:1233–1242

Castro-Díez P, Vaz AS, Silva JS, van Loo M, Alonso Á, Aponte C, Bayón Á, Bellingham PJ, Chiuffo MC, DiManno N, Julian K, Kandert S, La Porta N, Marchante H, Maule HG, Mayfield MM, Metcalfe D, Monteverdi MC, Núñez MA, Ostertag R, Parker IM, Peltzer DA, Potgieter L, Raymundo M, Rayome D, Reisman-Berman O, Richardson DM, Roos RE, Saldaña A, Shackleton RT, Torres A, Trudgen M, Urban J, Vicente JR, Vilà M, Ylioja T, Zenni RD, Godoy O (2019) Global effects of non-native tree species on multiple ecosystem services. Biol Rev 94:1477–1501

Castro-Díez P, Alonso Á, Saldaña-López A, Granda E (2021) Effects of widespread non-native trees on regulating ecosystem services. Sci Total Environ 778:146141. https://doi.org/10.1016/j.scitotenv.2021.146141

Catford JA, Jansson R, Nilsson C (2009) Reducing redundancy in invasion ecology by integrating hypotheses into a single theoretical framework. Divers Distrib 15:22–40

Celesti-Grapow L, Ricotta C (2021) Plant invasion as an emerging challenge for the conservation of heritage sites: the spread of ornamental trees on ancient monuments in Rome, Italy. Biol Invasions 23:1191–1206. https://doi.org/10.1007/s10530-020-02429-9

Chamier J, Schachtschneider K, Le Maitre DC, Ashton PJ, van Wilgen BW (2012) Impacts of invasive alien plants on water quality, with particular emphasis on South Africa. Water SA 38: 345–356. https://doi.org/10.4314/wsa.v38i2.19

Charles H, Dukes JS (2007) Impacts of invasive species on ecosystem services. In: Nentwig W (ed) Biological invasions. Springer, Berlin, pp 217–237

Chen F, Zhang J, Zhang M, Wang J (2015) Effect of *Cynodon dactylon* community on the conservation and reinforcement of riparian shallow soil in the Three Gorges Reservoir area. Ecol Process 4:1–8. https://doi.org/10.1186/s13717-014-0029-2

Chikuni MF, Dudley CO, Sambo EY (2004) *Prosopis glandulosa* Torry (Leguminosae-Mimosoidae) at Swang'oma, Lake Chilwa plain: a blessing in disguise. MJST 7:10–16

Choge SK, Clement N, Gitonga M, Okuye J (2012) Status report on commercialization of *Prosopis* tree resources in Kenya. KEFRI, Nairobi

Coakley S, Petti C (2021) Impacts of the invasive *Impatiens glandulifera*: lessons learned from one of Europe's top invasive species. Biology 10:619. https://doi.org/10.3390/biology10070619

Costanza R (1992) Toward an operational definition of ecosystem health. In: Costanza R, Norton BG, Haskell B (eds) Ecosystem health: new goals for environmental management. Island Press, pp 239–256

Costanza R, de Groot R, Farber S, Grasso M, Hannon B, Limburg K, Naeem S, O'Neill RV, Paruelo J, Raskin RG, Sutton P, Van Den Belt M (1997) The value of the world's ecosystem services and natural capital. Nature 387:253–260

Costanza R, De Groot R, Sutton P, Van der Ploeg S, Anderson SJ, Kubiszewski I, Farber S, Turner RK (2014) Changes in the global value of ecosystem services. Glob Environ Change 26:152–158

Costanza R, De Groot R, Braat L, Kubiszewski I, Fioramonti L, Sutton P, Farber S, Grasso M (2017) Twenty years of ecosystem services: how far have we come and how far do we still need to go? Ecosyst Serv 28:1–16

Cronk JK, Fennessy MS (2001) Wetland plants: biology and ecology. CRC Press, Boca Raton, FL

DAISIE (2009) Handbook of alien species in Europe. Springer Publishers

D'Antonio CM, Vitousek PM (1992) Biological invasions by exotic grasses, the grass/fire cycle, and global change. Annu Rev Ecol Evol Syst 23:63–87

Das S, Sandhu H (2014) Role of exotic vegetation in coastal protection: an investigation into the ecosystem services of casuarina in Odisha. Econ Polit Wkly 49:42–50

Dassonville N, Vanderhoeven S, Vanparys V, Hayez M, Gruber W, Meerts P (2008) Impacts of alien invasive plants on soil nutrients are correlated with initial site conditions in NW Europe. Oecologia 157:131–140

Davis AD, Landis DA (2011) Agriculture. In: Simberloff D (ed) Encyclopeia of biological invasions. University of California Press, Berkeley, pp 7–10

de Abreu RC, Durigan G (2011) Changes in the plant community of a Brazilian grassland savannah after 22 years of invasion by *Pinus elliottii* Engelm. Plant Ecol Divers 4:269–278

Dechassa N, Abate B (2020) Current status of water hyacinth (*Eichhornia crassipes*) in Ethiopia: achievements, challenges and prospects: a review. J Environ Earth Sci 10:1–13

Dogra KS, Sood SK (2012) Phytotoxicity of *Parthenium hysterophorus* residues towards growth of three native plant species (*Acacia catechu* willd, *Achyranthes aspera* L. and *Cassia tora* L.) in Himachal Pradesh, India. Int J Plant Physiol Biochem 4:105–109

Dukes JS, Mooney HA (1999) Does global change increase the success of biological invaders? Trends Ecol Evol 14:135–139

Dukes JS, Mooney HA (2004) Disruption of ecosystem processes in western North America by invasive species. Rev Chil de Hist Nat 77:411–437. https://doi.org/10.4067/S0716-078X2004000300003

Ehrenfeld JG (2003) Effects of exotic plant invasions on soil nutrient cycling processes. Ecosystems 6:503–523. https://doi.org/10.1007/s10021-002-0151-3

Evans RD, Rimer R, Sperry L, Belnap J (2001) Exotic plant invasion alters nitrogen dynamics in an arid grassland. Ecol Appl 11:1301–1310. https://doi.org/10.1890/1051-0761(2001)011[1301:EPIAND]2.0.CO;2

Eviner VT, Garbach K, Baty JH, Hoskinson SA (2012) Measuring the effects of invasive plants on ecosystem services: challenges and prospects. Invasive Plant Sci Manag 5:125–136. https://doi.org/10.1614/IPSM-D-11-00095.1

Farley KA, Kelly EF, Hofstede RGM (2004) Soil organic carbon and water retention following conversion of grasslands to pine plantations in the Ecuadoran Andes. Ecosystems 7:729–739

Feagin RA, Mukherjee N, Shanker K, Baird AH, Cinner J, Kerr AM, Koedam N, Sridhar A, Arthur R, Jayatissa LP, Lo Seen D, Menon M, Rodriguez S, Shamsuddoha M, Dahdouh-Guebas F (2010) Shelter from the storm? Use and misuse of coastal vegetation bioshields for managing natural disasters. Conserv Lett 3:1–11

Fernandes PM (2009) Combining forest structure data and fuel modelling to classify fire hazard in Portugal. Ann For Sci 66:1–9. https://doi.org/10.1051/forest/2009013

Fernández RD, Ceballos SJ, Aragón R, Malizia A, Montti L, Whitworth-Hulse JI, Castro Diez P, Grau HR (2020) A global review of *Ligustrum lucidum* (OLEACEAE) invasion. Bot Rev 86: 93–118

Fletcher RA, Brooks RK, Lakoba VT, Sharma G, Heminger AR, Dickinson CC, Barney JN (2019) Invasive plants negatively impact native, but not exotic, animals. Glob Chang Biol 25:3694–3705. https://doi.org/10.1111/gcb.14752

Forseth IN, Innis AF (2004) Kudzu (*Pueraria montana*): history, physiology, and ecology combine to make a major ecosystem threat. CRC Crit Rev Plant Sci 23:401–413. https://doi.org/10.1080/07352680490505150

Freudenberger L, Hobson PR, Schluck M, Ibisch PL (2012) A global map of the functionality of terrestrial ecosystems. Ecol Complex 12:13–22

Fried G, Chauvel B, Reynaud P, Sache I (2017) Decreases in crop production by non-native weeds, pests, and pathogens. In: Vila M, Hulme PE (eds) Impact of biological invasions on ecosystem services. Springer International Publishing AG, Cham, pp 83–101

Gaertner M, Biggs R, Te Beest M, Hui C, Molofsky J, Richardson DM (2014) Invasive plants as drivers of regime shifts: identifying high-priority invaders that alter feedback relationships. Divers Distrib 20:733–744. https://doi.org/10.1111/ddi.12182

Gallardo B, Clavero M, Sanchez MI, Vila M (2016) Global ecological impacts of invasive species in aquatic ecosystems. Glob Chang Biol 22:151–163. https://doi.org/10.1111/gcb.13004

Gamfeldt L, Roger F (2017) Revisiting the biodiversity–ecosystem multifunctionality relationship. Nat Ecol Evol 1:0168

Gao GF, Li PF, Zhong JX, Shen ZJ, Chen J, Li YT, Isabwe A, Zhu XY, Ding QS, Zhang S, Gao CH (2019) *Spartina alterniflora* invasion alters soil bacterial communities and enhances soil N2O emissions by stimulating soil denitrification in mangrove wetland. Sci Total Environ 653:231–240

Gerber E, Krebs C, Murrell C, Moretti M, Rocklin R, Schaffner U (2008) Exotic invasive knotweeds (*Fallopia* spp.) negatively affect native plant and invertebrate assemblages in European riparian habitats. Biol Conserv 141:646–654

Goldstein LJ, Suding KN (2014) Intra-annual rainfall regime shifts competitive interactions between coastal sage scrub and invasive grasses. Ecology 95:425–435. https://doi.org/10.1890/12-0651.1

González-Muñoz N, Costa-Tenorio M, Espigares T (2012) Invasion of alien *Acacia dealbata* on Spanish *Quercus robur* forests: impact on soils and vegetation. For Ecol Manag 269:214–221. https://doi.org/10.1016/j.foreco.2011.12.026

Goodell K, Parker IM (2017) Invasion of a dominant floral resource: effects on the floral community and pollination of native plants. Ecology 98:57–69

Gopal D, von der Lippe M, Kowarik I (2018) Sacred sites as habitats of culturally important plant species in an Indian megacity. Urban For Urban Green 32:113–122. https://doi.org/10.1016/j.ufug.2018.04.003

Grace JB (1998) Can prescribed fire save the endangered coastal prairie ecosystem from Chinese tallow invasion. Endangered Species Update 15:70–76

Greenwood P, Kuhn NJ (2014) Does the invasive plant, *Impatiens glandulifera*, promote soil erosion along the riparian zone? An investigation on a small watercourse in northwest Switzerland. J Soils Sediments 14:637–650. https://doi.org/10.1007/s11368-013-0825-9

Guo WY, van Kleunen M, Pierce S, Dawson W, Essl F, Kreft H, Maurel N, Pergl J, Seebens H, Weigelt P, Pyšek P (2019) Domestic gardens play a dominant role in selecting alien species with adaptive strategies that facilitate naturalization. Glob Ecol Biogeogr 28:628–639

Haines-Young R, Potschin-Young M (2018) Revision of the common international classification for ecosystem services (CICES V5. 1): a policy brief. One Ecosyst 3:e27108

Harvey KJ, Britton DR, Minchinton TE (2014) Detecting impacts of non-native species on associated invertebrate assemblages depends on microhabitat. Aust Ecol 39:511–521

Havel JE, Kovalenko KE, Thomaz SM, Amalfitano S, Kats LB (2015) Aquatic invasive species: challenges for the future. Hydrobiologia 750:147–170

Hector A, Bagchi R (2007) Biodiversity and ecosystem multifunctionality. Nature 448:188–190

Hejda M, Pyšek P, Jarošík V (2009) Impact of invasive plants on the species richness, diversity and composition of invaded communities. J Ecol 97:393–403. https://doi.org/10.1111/j.1365-2745.2009.01480.x

Hellmann JJ, Byers JE, Bierwagen BG, Dukes JS (2008) Five potential consequences of climate change for invasive species. Conserv Biol 22:534–543

Helsen K, Kapás RE, Rakvaag G, Speed JD, Graae BJ (2018) Impacts of an invasive plant on primary production: testing a functional trait-based framework with a greenhouse experiment. J Veg Sci 29:157–166

Hickman JE, Wu S, Mickley LJ, Lerdau MT (2010) Kudzu (*Pueraria montana*) invasion doubles emissions of nitric oxide and increases ozone pollution. PNAS 107:10115–10119. https://doi.org/10.1073/pnas.0912279107

Hussain MI, Shackleton R, El-Keblawy A, González L, Trigo M (2021) Impact of the invasive *Prosopis juliflora* on terrestrial ecosystems. Sustain Agric Rev 52:223–278

Incerti G, Cartenì F, Cesarano G, Sarker TC, Abd El-Gawad AM, D'ascoli R, Bonanomi G, Giannino F (2018) Faster N release, but not C loss, from leaf litter of invasives compared to native species in mediterranean ecosystems. Front Plant Sci 9:00534. https://doi.org/10.3389/fpls.2018.00534

Jo I, Fridley JD, Frank DA (2017) Invasive plants accelerate nitrogen cycling: evidence from experimental woody monocultures. J Ecol 105:1105–1110

Kozlowski TT, Pallardy SG (1997) Cultural practices and vegetative growth. In: Kozlowski TT, Pallardy SG (eds) Growth control in woody plants. Elsevier, pp 352–393

Kull CA, Shackleton CM, Cunningham PJ, Ducatillon C, Dufour-Dror J, Esler KJ, Friday JB, Gouveia AC, Griffin AR, Marchante E, Midgley SJ, Pauchard A, Rangan H, Richardson DM, Rinaudo T, Tassin J, Urgenson LS, von Maltitz GP, Zenni RD, Zylstra MJ (2011) Adoption, use and perception of Australian acacias around the world. Divers Distrib 17:822–836. https://doi.org/10.1111/j.1472-4642.2011.00783.x

Lal R (2020) Soil organic matter and water retention. Agron J 11:3265–3277. https://doi.org/10.1002/agj2.20282

Lázaro-Lobo A, Ervin GN (2019) A global examination on the differential impacts of roadsides on native vs. exotic and weedy plant species. Glob Ecol Conserv 17:e00555

Lázaro-Lobo A, Ervin GN (2021) Wetland invasion: a multi-faceted challenge during a time of rapid global change. Wetlands 41:1–16

Lázaro-Lobo A, Ervin GN, Caño L, Panetta FD (2021a) Biological invasion by Baccharis. In: Fernandes GW, Oki Y, Barbosa M (eds) Baccharis: from evolutionary and ecological aspects to social uses and medicinal applications. Springer Nature, Cham, pp 185–214

Lázaro-Lobo A, Lucardi RD, Ramirez-Reyes C, Ervin GN (2021b) Region-wide assessment of fine-scale associations between invasive plants and forest regeneration. For Ecol Manag 483:118930

Le Maitre DC, van Wilgen BW, Chapman RA, McKelly DH (1996) Invasive plants and water resources in the Western Cape Province, South Africa: modelling the consequences of a lack of management. J Appl Ecol 33:161–172. https://doi.org/10.2307/2405025

Le Maitre DC, van Wilgen BW, Gelderblom CM, Bailey C, Chapman RA, Nel JA (2002) Invasive alien trees and water resources in South Africa: case studies of the costs and benefits of management. For Ecol Manag 160(1–3):143–159

Le Maitre DC, Gaertner M, Marchante E, Ens EJ, Holmes PM, Pauchard A, O'Farrell PJ, Rogers AM, Blanchard R, Blignaut J, Richardson DM (2011) Impacts of invasive Australian acacias: implications for management and restoration. Divers Distrib 17:1015–1029. https://doi.org/10.1111/j.1472-4642.2011.00816.x

Ledgard N (2001) The spread of lodgepole pine (*Pinus contorta*, Dougl.) in New Zealand. For Ecol Manag 141:43–57

Lefcheck JS, Byrnes JE, Isbell F, Gamfeldt L, Griffin JN, Eisenhauer N, Hensel MJS, Hector A, Cardinale BJ, Duffy JE (2015) Biodiversity enhances ecosystem multifunctionality across trophic levels and habitats. Nat Commun 6:1–7

León-Gamboa AL, Ramos C, García MR (2010) Efecto de plantaciones de pino en la artropofauna del suelo de un bosque Altoandino. Rev Biol Trop 58:1031–1048

Levine JM, Vila M, Antonio CMD, Dukes JS, Grigulis K, Lavorel S (2003) Mechanisms underlying the impacts of exotic plant invasions. Proc R Soc B 270:775–781. https://doi.org/10.1098/rspb.2003.2327

Liao CZ, Peng RH, Luo YQ, Zhou XH, Wu XW, Fang CM, Chen JK, Li B (2008) Altered ecosystem carbon and nitrogen cycles by plant invasion: a meta-analysis. New Phytol 177:706–714. https://doi.org/10.1890/1051-0761(2001)011[1301:EPIAND]2.0.CO;2

Liu Y, Oduor AMO, Zhang Z, Manea A, Tooth IM, Leishman MR, Xu X, van Kleunen M (2017) Do invasive alien plants benefit more from global environmental change than native plants? Glob Chang Biol 23:3363–3370. https://doi.org/10.1111/gcb.13579

López BE, Allen JM, Dukes JS, Lenoir J, Vilà M, Blumenthal DM, Beaury EM, Fusco EJ, Laginhas BB, Morelli TL, O'Neill MW, Sorte CJB, Maceda-Veiga A, Whitlock R, Bradley BA (2022) Global environmental changes more frequently offset than intensify detrimental effects of biological invasions. PNAS 119:e2117389119

Lu Y, Wang R, Zhang Y, Su H, Wang P, Jenkins A, Ferrier RC, Bailey M, Squire G (2015) Ecosystem health towards sustainability. Ecosyst Health Sustain 1:1–15

Maillet J, López-García C (2000) What criteria are relevant for predicting the invasive capacity of a new agricultural weed? The case of invasive American species in France. Weed Res 40:11–26

Malacrinò A, Sadowski VA, Martin TK, Cavichiolli de Oliveira N, Brackett IJ, Feller JD, Harris KJ, Heredia OC, Vescio R, Bennett AE (2020) Biological invasions alter environmental microbiomes: a meta-analysis. PLoS One 15:e0240996

Marchante E, Kjøller A, Struwe S, Freitas H (2008) Short-and long-term impacts of *Acacia longifolia* invasion on the belowground processes of a Mediterranean coastal dune ecosystem. Appl Soil Ecol 40:210–217. https://doi.org/10.1016/j.apsoil.2008.04.004

Marchante H, Marchante E, Freitas H, Hoffmann JH (2015) Temporal changes in the impacts on plant communities of an invasive alien tree, *Acacia longifolia*. Plant Ecol 216:1481–1498

Marini L, Battisti A, Bona E, Federici G, Martini F, Pautasso M, Hulme PE (2012) Alien and native plant life-forms respond differently to human and climate pressures. Glob Ecol Biogeogr 21: 534–544. https://doi.org/10.1111/j.1466-8238.2011.00702.x

Mattsson E, Ostwald M, Nissanka SP, Holmer B, Palm M (2009) Recovery and protection of coastal ecosystems after tsunami event and potential for participatory forestry CDM–examples from Sri Lanka. Ocean Coast Manag 52:1–9

McCary MA, Mores R, Farfan MA, Wise DH (2016) Invasive plants have different effects on trophic structure of green and brown food webs in terrestrial ecosystems: a meta-analysis. Ecol Lett 19:328–335

McLeod ML, Bullington L, Cleveland CC, Rousk J, Lekberg Y (2021) Invasive plant-derived dissolved organic matter alters microbial communities and carbon cycling in soils. Soil Biol Biochem 156:108191

Medina-Villar S, Rodríguez-Echeverría S, Lorenzo P, Alonso A, Pérez-Corona E, Castro-Díez P (2016) Impacts of the alien trees *Ailanthus altissima* (Mill.) Swingle and *Robinia pseudoacacia* L. on soil nutrients and microbial communities. Soil Biol Biochem 96:65–73

Morales CL, Traveset A (2009) A meta-analysis of impacts of alien vs. native plants on pollinator visitation and reproductive success of co-flowering native plants. Ecol Lett 12:716–728

Nel L, Pryke JS, Carvalheiro LG, Thébault E, van Veen FF, Seymour CL (2017) Exotic plants growing in crop field margins provide little support to mango crop flower visitors. Agric Ecosyst Environ 250:72–80

Nentwig W, Mebs D, Vilà M (2017) Impacts of non-native animals and plants to human health. In: Vilà M, Hulme PE (eds) Impact of biological invasions on ecosystem services. Springer, Heidelberg, pp 277–293

Norris C, Hobson P, Ibisch PL (2012) Microclimate and vegetation function as indicators of forest thermodynamic efficiency. J Appl Ecol 49:562–570

Ojija F, Arnold SEJ, Treydte AC (2019) Impacts of alien invasive *Parthenium hysterophorus* on flower visitation by insects to co-flowering plants. Arthropod Plant Interact 13:719–734. https://doi.org/10.1007/s11829-019-09701-3

Ortega YK, Benson A, Greene E (2014) Invasive plant erodes local song diversity in a migratory passerine. Ecology 95:458–465

Palmeri L, Barausse A, Jorgensen SE (2013) Ecological processes handbook. CRC Press

Pattison Z, Rumble H, Tanner RA, Jin L, Gange AC (2016) Positive plant–soil feedbacks of the invasive *Impatiens glandulifera* and their effects on above-ground microbial communities. Weed Res 56:198–207

Pauchard A, Kueffer C, Dietz H, Daehler CC, Alexander J, Edwards PJ, Arévalo JR, Cavieres LA, Guisan A, Haider S, Jakobs G, McDougall K, Millar CI, Naylor BJ, Parks CG, Rew LJ, Seipel T (2009) Ain't no mountain high enough: plant invasions reaching new elevations. Front Ecol Environ 7:479–486. https://doi.org/10.1890/080072

Pauchard A, García R, Zalba S, Sarasola M, Zenni R, Ziller S, Nuñez MA (2015) Pine invasions in South America: reducing their ecological impacts through active management. In: Clode C (ed) Biological invasions in changing ecosystems. De Gruyter Open Ltd., Berlin, pp 318–342

Peltzer DA, Allen RB, Lovett GM, Whitehead D, Wardle DA (2010) Effects of biological invasions on forest carbon sequestration. Glob Change Biol 16:732–746

Pyšek P, Jarošik V, Hulme PE, Pergl J, Hejda M, Schaffner U, Vilà M (2012) A global assessment of invasive plant impacts on resident species, communities and ecosystems: the interaction of impact measures, invading species' traits and environment. Glob Change Biol 18:1725–1737

Qian H, Ricklefs RE (2006) The role of exotic species in homogenizing the north American flora. Ecol Lett 9:1293–1298. https://doi.org/10.1111/j.1461-0248.2006.00982.x

Rapport DJ, Costanza R, McMichael AJ (1998) Assessing ecosystem health. Trends Ecol Evol 13: 397–402

Rhymer JM, Simberloff D (1996) Extinction by hybridization and introgression. Annu Rev Ecol Evol Syst 27:83–109

Richardson DM (1998) Forestry trees as invasive aliens. Conserv Biol 12:18–26. https://doi.org/10.1111/j.1523-1739.1998.96392.x

Richardson DM, Pyšek P, Rejmanek M, Barbour MG, Panetta FD, West CJ (2000) Naturalization and invasion of alien plants: concepts and definitions. Divers Distrib 6(2):93–107

Richardson DM, Kluge RL (2008) Seed banks of invasive Australian *Acacia* species in South Africa: role in invasiveness and options for management. Perspect Plant Ecol Evol Syst 10:161–177

Rodríguez-Caballero G, Caravaca F, Alguacil MM, Fernández-López M, Fernández-González AJ, Roldán A (2017) Striking alterations in the soil bacterial community structure and functioning of the biological N cycle induced by *Pennisetum setaceum* invasion in a semiarid environment. Soil Biol Biochem 109:176–187

Rundel PW, Dickie IA, Richardson DM (2014) Tree invasions into treeless areas: mechanisms and ecosystem processes. Biol Invasions 16:663–675

Sage EF (1996) Modification of fire disturbance by elevated CO2. In: Korner C, Bazzaz FA (eds) Carbon dioxide, populations, and communities. Academic Press, San Diego, pp 230–249

Schirmel J, Bundschuh M, Entling MH, Kowarik I, Buchholz S (2016) Impacts of invasive plants on resident animals across ecosystems, taxa, and feeding types: a global assessment. Glob Change Biol 22:594–603

Seeney A, Eastwood S, Pattison Z, Willby NJ, Bull CD (2019) All change at the water's edge: invasion by non-native riparian plants negatively impacts terrestrial invertebrates. Biol Invasions 21:1933–1946

Segura EM, Giorgis MA, Lescano JN (2021) Native anurans threatened by the alien tree *Ligustrum lucidum* in a seasonal subtropical forest. Biol Invasions 23:3859–3869

Shackleton CM, McGarry D, Fourie S, Gambiza J, Shackleton SE, Fabricius C (2007) Assessing the effects of invasive alien species on rural livelihoods: case examples and a framework from South Africa. Hum Ecol 35:113–127

Shackleton RT, Le Maitre DC, Pasiecznik NM, Richardson DM (2014) Prosopis: a global assessment of the biogeography, benefits, impacts and management of one of the world's worst woody invasive plant taxa. AoB Plants 6:18

Silva JS, Moreira F, Vaz P, Catry F, Godinho-Ferreira P (2009) Assessing the relative fire proneness of different forest types in Portugal. Plant Biosyst 143:597–608. https://doi.org/10.1080/11263500903233250

Simberloff D, Nunez MA, Ledgard NJ, Pauchard A, Richardson DM, Sarasola M, van Wilgen BW, Zalba SM, Zenni RD, Bustamante R, Pena E, Ziller SR (2010) Spread and impact of introduced conifers in South America: lessons from other southern hemisphere regions. Austral Ecol 35: 489–504

Soumare A, Sall SN, Sanon A, Cissoko M, Hafidi M, Ndoye I, Duponnois R (2016) Changes in soil pH, polyphenol content and microbial community mediated by *Eucalyptus camaldulensis*. Appl Ecol Environ Res 14:1–19

Stevens JT, Beckage B (2010) Fire effects on demography of the invasive shrub Brazilian pepper (*Schinus terebinthifolius*) in Florida pine savannas. Nat Areas J 30:53–63. https://doi.org/10.3375/043.030.0106

Stewart PS, Hill RA, Stephens PA, Whittingham MJ, Dawson W (2021) Impacts of invasive plants on animal behaviour. Ecol Lett 24(4):891–907

Suryaningrum F, Jarvis RM, Buckley HL, Hall D, Case BS (2022) Large-scale tree planting initiatives as an opportunity to derive carbon and biodiversity co-benefits: a case study from Aotearoa New Zealand. New For 53:589–602
Swanson FJ (1981) Fire and geomorphic processes. In: Mooney HA, Bonnicksen NL, Christensen NL, Lotan JE, Reiners WA (eds) Fire regimes and ecosystem properties. USDA Forest Service, pp 401–420
Taylor KT, Maxwell BD, McWethy DB, Pauchard A, Nuñez MA, Whitlock C (2017) Pinus contorta invasions increase wildfire fuel loads and may create a positive feedback with fire. Ecology 98:678–687
Thompson K (2014) Where do camels belong?: why invasive species aren't all bad. Greystone Books
Torres N, Herrera I, Fajardo L, Bustamante RO (2021) Meta-analysis of the impact of plant invasions on soil microbial communities. BMC Ecol Evol 21:172. https://doi.org/10.1186/s12862-021-01899-2
Trotta G, Savo V, Cicinelli E, Carboni M, Caneva G (2020) Colonization and damages of *Ailanthus altissima* (Mill.) Swingle on archaeological structures: evidence from the Aurelian Walls in Rome (Italy). Int Biodeterior Biodegrad 153:105054. https://doi.org/10.1016/j.ibiod.2020.105054
van Hengstum T, Hooftman DA, Oostermeijer JGB, van Tienderen PH (2014) Impact of plant invasions on local arthropod communities: a meta-analysis. J Ecol 102:4–11
van Kleunen M, Weber E, Fischer M (2010) A meta-analysis of trait differences between invasive and non-invasive plant species. Ecol Lett 13:235–245. https://doi.org/10.1111/j.1461-0248.2009.01418.x
van Wilgen BW, Le Maitre DC, Cowling RM (1998) Ecosystem services, efficiency, sustainability and equity: South Africa's working for water programme. Trends Ecol Evol 13:378–378
Van Wilgen BW, Measey J, Richardson DM, Wilson JR, Zengeya TA (2020) Biological invasions in South Africa. Springer Nature, p 975
Vanbergen AJ, Espíndola A, Aizen MA (2018) Risks to pollinators and pollination from invasive alien species. Nat Ecol Evol 2:16–25. https://doi.org/10.1038/s41559-017-0412-3
Vaz AS, Castro-Díez P, Godoy O, Alonso A, Vilà M, Saldaña A, Marchante H, Bayon A, Silva JS, Vicente JR, Honrado JP (2018) An indicator-based approach to analyse the effects of non-native tree species on multiple cultural ecosystem services. Ecol Indic 85:48–56. https://doi.org/10.1016/j.ecolind.2017.10.009
Vilà M, Espinar JL, Hejda M, Hulme PE, Jarošik V, Maron JL, Pergl J, Schaffner U, Sun Y, Pyšek P (2011) Ecological impacts of invasive alien plants: a meta-analysis of their effects on species, communities and ecosystems. Ecol Lett 14:702–708
Vilà M, Beaury EM, Blumenthal DM, Bradley BA, Early R, Laginhas BB, Trillo A, Dukes JS, Sorte CJB, Ibáñez I (2021) Understanding the combined impacts of weeds and climate change on crops. Environ Res Lett 16:034043
Vitousek PM, D'Antonio CM, Loope LL, Rejmanek M, Westbrooks R (1997) Introduced species: a significant component of human-caused global change. N Z J Ecol 21:1–16
Von Döhren P, Haase D (2015) Ecosystem disservices research: a review of the state of the art with a focus on cities. Ecol Indic 52:490–497
Wahl C, Diaz R, Kaller M (2021) Invasive floating Fern limits aerial colonization and alters community structure of aquatic insects. Wetlands 41:1–11
Walker LR, Smith SD (1997) Impacts of invasive plants on community and ecosystem properties. In: Luken JO, Thieret JW (eds) Assessment and management of plant invasions. Springer, New York, pp 69–86
Wang H, Wang Q, Bowler PA, Xiong W (2016) Invasive aquatic plants in China. Aquat Invasions 11:1–9. https://doi.org/10.3391/ai.2016.11.1.01
Wardle DA, Nicholson KS, Ahmed M, Rahman A (1994) Interference effects of the invasive plant *Carduus nutans* L. against the nitrogen fixation ability of *Trifolium repens* L. Plant Soil 163:287–297. https://doi.org/10.1007/BF00007978

Whisenant SG (1990) Changing fire frequencies on Idaho's Snake river plains: ecological and management. In: Proceedings-symposium on cheatgrass invasion, shrub die-off and other aspects of shrub biology and management, vol 276. Intermountain Research Station, Forest Service, US Department of Agriculture, Las Vegas, p 4

Wolfertz M, Sharkey TD, Boland W, Kühnemann F, Yeh S, Weise SE (2003) Biochemical regulation of isoprene emission. Plant Cell Environ 26:1357–1364. https://doi.org/10.1046/j.0016-8025.2003.01059.x

Wu M, He Z, Fung S, Cao Y, Guan D, Peng Y, Lee SY (2020) Species choice in mangrove reforestation may influence the quantity and quality of long-term carbon sequestration and storage. Sci Total Environ 714:136742

Xu H, Liu Q, Wang S, Yang G, Xue S (2022) A global meta-analysis of the impacts of exotic plant species invasion on plant diversity and soil properties. Sci Total Environ 810:152286

Zahari AFM (2021) The review on impacts of invasive plants on the physico-chemical characteristic of water body quality. RTCEBE 2:293–304

Zarafshar M, Bazot S, Matinizadeh M, Bordbar SK, Rousta MJ, Kooch Y, Enayatid K, Abbasi A, Negahdarsaber M (2020) Do tree plantations or cultivated fields have the same ability to maintain soil quality as natural forests? Appl Soil Ecol 151:103536

Zenni RD, Ziller SR (2011) An overview of invasive plants in Brazil. Braz J Bot 34:431–446

Zhang X, Wang B, Liu Z (2019) Impacts of plant secondary metabolites from conifer litter on the decomposition of *Populus purdomii* litter. J For Res 30:2237–2245. https://doi.org/10.1007/s11676-018-0766-7

Zhou Y, Staver AC (2019) Enhanced activity of soil nutrient-releasing enzymes after plant invasion: a meta-analysis. Ecology 100:e02830

Menace of Plant Invasion: A View from Ecological Lens

3

Abhishek Raj, Manoj Kumar Jhariya, Arnab Banerjee, S. Dinesha, Ramesh Kumar Jha, Krishna Kumar, and Taher Mechergui

Abstract

Plant invasion is the biggest challenge for ecologist that affects biodiversity and environmental health. Forecasting of invasive plant species, its identification, early detection and distribution mapping are necessary for making plan of actions against negative consequences of alien invasive species. An invasive plant affects biodiversity along with ecosystem health and services. However, very few studies are available on plant invasion dynamics and its impacts on ecosystem. Invasive plant invades natural ecosystem including forest and agriculture which affects soil, food and climate security. Human and animals are also affected by dynamic intervention of invasive species. Mostly, invasive species also fix atmospheric carbon (C) through C sequestration process which helps in mitigating C footprint and climate change issues. However, many invasive species change its distribution and other mechanisms under changing climate scenario. Climate change amplifies the population dynamics and diversity of invasive species which is a

A. Raj · S. Dinesha · R. K. Jha · K. Kumar
Pt. Deendayal Upadhyay College of Horticulture and Forestry, Dr. Rajendra Prasad Central Agriculture University, Pusa, Samastipur, Bihar, India
e-mail: ramesh.jha@rpcau.ac.in

M. K. Jhariya (✉)
Department of Farm Forestry, Sant Gahira Guru Vishwavidyalaya, Sarguja, Ambikapur, Chhattisgarh, India

A. Banerjee
Department of Environmental Science, Sant Gahira Guru Vishwavidyalaya, Sarguja, Ambikapur, Chhattisgarh, India

T. Mechergui
Faculty of Sciences of Bizerte (FSB), Jarzouna, Tunisia

Sylvo-Pastoral Resources Laboratory of Tabarka, Tabarka, Tunisia

S. Tripathi et al. (eds.), *Plant Invasions and Global Climate Change*,
https://doi.org/10.1007/978-981-99-5910-5_3

major ecological risk. However, it is needed for accurately prediction of invasive plant distributions and its varying impacts on desire species that can be changed under projected climate change scenarios. This study helps in understanding effective control and preventive measures against spreading of plant invasions. A sound scientific strategy and policy framework are required for invasive plant management which would be helpful in conservation of desired natural resources. A link must exist between local and global policy networks to address plant invasions in changing climate. Therefore, an effective policy framework and plan of actions are employed to control and prevent plant invasions which build ecological stability and environmental sustainability.

Keywords

Carbon dynamics · Ecological risk · Ecosystem services · Plant invasion · Sustainability

Abbreviations

C	Carbon
ETM+	Enhanced Thematic Mapper Plus
IAPS	Invasive alien plant species
IAS	Invasive alien species
IPBES	The Intergovernmental Science-Policy Platform on Biodiversity and Ecosystem Services
MODIS	The moderate resolution imaging spectroradiometer
SDGs	Sustainable development goals

3.1 Introduction

Invasive plant emergence in any ecosystem enforces a threat to biodiversity and environmental health (Seebens et al. 2018). Alteration in land use and climate change induces biological invasions which impede human and environmental health (Ebi et al. 2017). Environmental pollutions, habitat destruction and human-based global climate changes are key threats to biodiversity (Raj et al. 2018). Invasion ecologists debated enormously on deadly plant invasion and its consequences on the environment (Young and Larson 2011). However, some anthropogenic perturbations have accelerated global problems of plant invasions (Young and Larson 2011; Jhariya et al. 2022a). IPBES (2019) has also declared plant invasion as a key driver of biodiversity losses and ecosystem unbalances. Moreover, intensive agricultural practices also accelerate the invasion of plant species besides assurance of food security (Rai et al. 2018; Jhariya et al. 2021a, b). Plant invasion affects flora and faunal diversity along with soil and water resource depletion in any ecosystem

(Gichua et al. 2013). It not only affects our environment but sometimes it also influences humans in both positive and negative ways. Impacts of plant invasion could be evaluated in both socio-ecological and socioeconomic points of view. However, invasions lead to biodiversity losses that modify climatic (temperature, humidity and others) parameters which indirectly affect human health and society (Jones 2019).

Many national and international organizations are involved in controlling and managing plant invasion and its consequences. The Convention on Biological Diversity emphasizes on controlling and management of global plant invasions which negatively affect the ecosystem and human health which is further discussed in Biosafety and Cartagena Protocol (Pysek and Richardson 2010). Plant invasion has been further recognized by Earth Summit in Rio de Janeiro, 1992 under forestry and agroforestry sectors due to its harmful effects on biodiversity, the environment and public health. Therefore, plant invasion and its ecology are regarded as the trans-disciplinary subject which is greatly linked with many topics such as land use changes, global change biology, restoration and conservation biology along with health sciences (Heshmati et al. 2019).

The present chapter explores the plant invasion ecology and its impacts on the environment and natural resources. Mapping, detecting and monitoring of invasive plants through geospatial tools including remote sensing are also described. Plant invasion impacts on human health by affecting food security and ecosystem services are also included which directly or indirectly influence the socio-economic status of the people. However, climate change impacts on invaders and its role in carbon (C) dynamics are also discussed.

3.2 Plant Invasion Ecology: Science and Mechanisms

"Invasion ecology" is not a very old discipline but highly discussed among researchers and ecologists at a constant pace since twentieth century (Richardson 2011). Although, Grinnell (2000) has reported the first paper on species invasions in the year 1919. He has also reported many European plants which were flourishing as alien's species in the region of South America. He has reported the invading plant species and their rapid spread in native ranges are possible due to escape from the parasites and diseases attack on them (Sax et al. 2005). Different terminology and views of invasion ecology are mentioned in the book of Davis (2009).

"Invasion ecology" term emphasized the study of human-induced plant invasion outside the native areas through different mechanisms such as transport, establishment, colonization and spread in any landscape. Plant invasion is considered as an ecological phenomenon that destroys global biodiversity and leads to species extinction in many island regions (Sharma et al. 2005). The characteristic features, advantages and disadvantages of invasive plants are depicted in Fig. 3.1 (Langmaier and Lapin 2020; Rai and Singh 2020).

Invasive plants invade both aquatic and terrestrial ecosystems (Totland et al. 2005). Besides human-mediated, there are many other mechanisms involved in

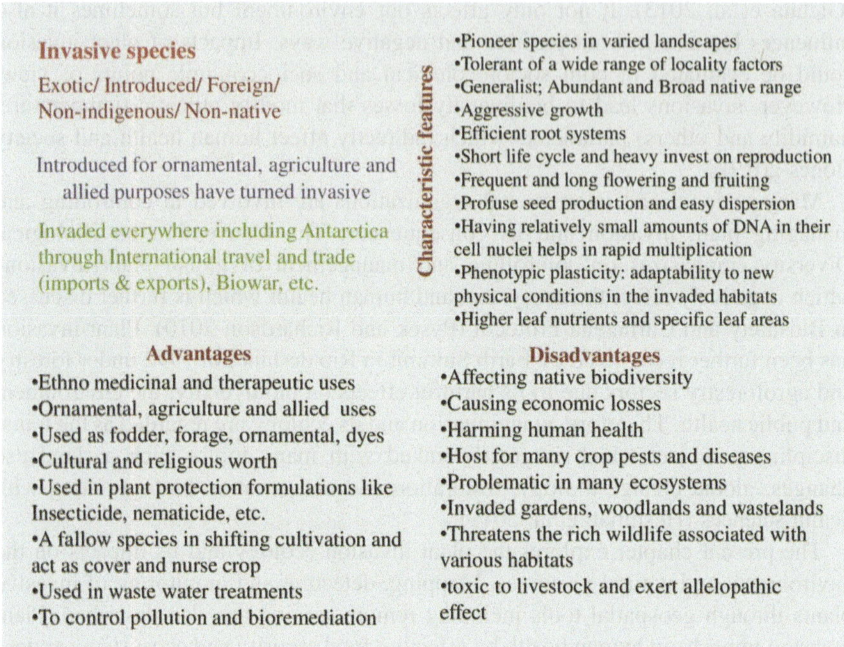

Fig. 3.1 The characteristic features, advantages and disadvantages of invasive plants. (Based on Rai and Singh 2020; Langmaier and Lapin 2020)

invasive plant introductions in any region. For example, ocean currents have been worked for the introduction of *Limnocharis flava* in the Kerala region of India (Abhilash et al. 2008). Similarly, ocean currents also involve in coconut dispersion in many regions of the world and islands (Harries and Clement 2014). Therefore, the horizon of plant invasions has been expanded gradually. However, invasive alien plant species (IAPS) also directly or indirectly affect the ecology of native plant species of invaded areas. These species evolved without any human intrusion and flourishing by natural means (Rai and Singh 2020). Therefore, the concept of plant invasion is considered as a kind of ecological explosion in the current period of ecological sciences. Exotic species invasion in any natural ecosystem can be treated as an ecological perturbation that accelerates many dynamic ecosystems (Khan et al. 2019).

3.3 Invasive Plant Mapping Through Remote Sensing

Plant invasion is continuously increased at a high pace over the centuries, and its impacts have been seen on biodiversity, the economy and human health (Pysek and Richardson 2010). Half of India's geographical area is at risk of being occupied by IAPS (Mainali et al. 2015). Furthermore, less diverse areas with fewer species are more vulnerable to invasion. Therefore, invasive alien species (IAS) of India including Bihar and their advantages and disadvantages are depicted in Table 3.1. However, it is very difficult to detect, map and monitor the established invasive plants in any region. In this context, scientific monitoring and early detection of plant invasion are very important due to their detrimental impact on the environment, economy and human health. The remote sensing method is used for early detection and monitoring of invasive plants which replaced traditional or old field survey methods due to more efficient and cheaper quality (Sladonja and Damijanic 2021). Remote sensing-based detection is a more reliable, faster and less resource-intensive based monitoring system of invasion (Underwood et al. 2003). The use of unmanned aerial vehicles as drones is a highly preferable remote sensing method in environmental biology (Nowak et al. 2018). However, aerial detection of any invasive plants becomes more useful during different growth stages such as flowering, ripening and other stages (Müllerová et al. 2017).

Several authors have emphasized on importance and scope of remote sensing and its different methods for identifying and mapping the spread and presence of invasive species. A multi-temporal coverage, synoptic view, multispectral data and profitability are identified as the significance of remote sensing which helps in plant invasion study (Joshi et al. 2004). Although satellite data and its uses for the study of plant invasions are very limited around the globe (Müllerová et al. 2017). MODIS-based satellite data provide very less spatial resolution as compared to QuickBird, WorldView and Pleiades which are very high-resolution satellites and highly expensive. These expensive satellite data depend on cloud cover during acquisition. Similarly, MODIS sensors delivered high spectral resolution (Nowak et al. 2018). Furthermore, Sentinel 2 and Landsat provide the high spatial image with lower spectral and temporal resolution. A moderate spatial and temporal resolution preferred the use of satellite imagery (Alvarez-Taboada et al. 2017). Landsat 7 ETM-based multispectral satellites were used for tracking many invasive plants. Three HySpex hyperspectral datasets were collected for mapping of three invasive plants such as *Rubus* species, *Solidago* species and *Calamagrostis epigejos*. F1 scores of *Rubus* species, *Solidago* species and *C. epigejos* were reported after mapping which varied from 0.89–0.97, 0.99 and 0.87–0.89, respectively (Sabat-Tomala et al. 2022). Similarly, both multiple end member SMA (MESMA) and Maxent can help in the understanding of understory vegetation and its distribution. Also, these contribute to biodiversity conservation and eco-restoration and ensure uncountable ecosystem services in sustainable ways (Dai et al. 2020a).

3.4 Plant Invasion Dynamics and Its Impacts on Natural Resources

Natural resources play a key role in ecosystem maintenance and ecological stability (Prasad et al. 2021a, b). Plant invasion spreads enormously and has been recognized as a global environmental problem that affects various resources and its dynamics (Jhariya et al. 2022b). Invasive plants destroy the ecology and economy of any

Table 3.1 Invasive alien plant species of India including Bihar and their advantages and disadvantages

Scientific name and family	Common name	Origin	Advantages	Disadvantages	References
Trees					
Acacia farnesiana (L.) wild.; Mimosaceae	Sweet acacia	South America	Used as fodder and forage, ornamental, dyes, and traditional medicine	Affecting native biodiversity	Hernández-García et al. (2019)
Acacia mearnsii De wild.; Mimosaceae	Black wattle	Southeast Australia	The bark is one of the world highest yielding sources of condensed tannin	Native biodiversity loss	Ogawa and Yazaki (2018)
Acer saccharinum L.; Sapindaceae	Silverleaf maple	Eastern and Central America	Wood pulp and medicine used	Causes economic and ecological impacts	Patykowski et al. (2018)
Couroupita guianensis Aubl.; Lecythidaceae	Cannonball tree	North and South America	Used as animal feed, medicine, cultural and religious worth	Affecting native biodiversity	Barooah and Ahmed (2014)
Prosopis juliflora (Sw.) DC.; Caesalpinioideae	Mesquite	Central and south America	Used as forage, firewood need. Sweet pods are edible and nutritious	Noxious invader causes severe economic and ecological impacts in many countries	Barooah and Ahmed (2014)
Ziziphus jujuba Mill.; Rhamnaceae	Jujube	South Eastern Europe	Used in different traditional medicine to treat asthma and laryngitis. Fruits edible.	Affecting native biodiversity	Ji et al. (2017)
Shrub					
Cassia alata L.; Fabaceae	Ringworm bush	South America	Traditionally used to cure typhoid, diabetes, malaria and ringworms	Toxic to animals and humans	Panda et al. (2018)
Hibiscus hispidissimus Griff.; Malvaceae	Mupparacham	South East Asia	Used after cooking	Anti-inflammatory, anthelmintic and used to treat liver	Ajesh et al. (2012)
Ipomoea carnea Jacq.; Convolvulaceae	Bush Morning Glory	Central America	Used in paper making and traditional medicine	Toxic plant causes poisoning of livestock	Panda et al. (2018)

Parthenium hysterophorus L.; Asteraceae	Congress grass	Central and south America	To treat fever, diarrhoea, neurologic disorders, malaria	Affecting livestock, crop and human health	Panda et al. (2018)
Solanum torvum Sw.; Solanaceae	Turkey berry	West Indies	Fruits used to make chutney	Medicinal	Mahapatra et al. (2012)
Herb					
Acanthospermum hispidum DC.; Asteraceae	Bristly starbur/Goat's head	Central and south America	Used in malaria jaundice and snake bite	Reduces yields, host for many crop pests and diseases	Panda et al. (2018)
Ageratina adenophora (Spreng.) King & H.Rob.	Crofton weed	Central America	Used for various health conditions in traditional medicines	Toxic to livestock. Leads to chronic lung disease in horses	Panda et al. (2018)
Bidens pilosa L.; Asteraceae	Cobblers pegs	South America	A source of food and medicine	Invaded gardens, woodlands and waste areas.	Panda et al. (2018)
Chromolaena odorata (L.) R.M.King & H.Rob.; Asteraceae	Siam weed	America	Ethno medicinal, fungicidal, nematicidal use. A fallow species in shifting cultivation	Detrimental impact on young plantations, pastures and native vegetation	Panda et al. (2018)
Cyperus iria L.; Cyperaceae	Rice flat sedge	Tropical America	It is astringent, stimulant, used in menstrual ailments	Most problematic and economically damaging weed	
Solanum americanum Mill.; Solanaceae	Kalabegun	Tropical America	Fruits are used in traditional medicine	Native diversity loss	Saha et al. (2013)

nation by affecting natural resource dynamics. However, many invasive plants invaded both terrestrial and aquatic ecosystems drastically which threatens the natural ecosystem and affects human and environmental health. IAPS also threatens wetlands throughout the world. Plant invasion impacts on wetlands and its comprehensive national inventory have been developed by 40% of Ramsar Parties (Ramsar Convention 2018; IPBES 2019). Therefore, invasive plants affect many natural resources such as forests, agriculture, human, animals and soils which is comprehensively elaborated.

3.4.1 Forests

Invasive plants threaten forest ecosystems through various mechanisms such as species competition, hybridizations and disease transmission (Langmaier and Lapin 2020). Plant invasion in any region and its positive socioeconomic effects lead to the spread of few IAS in forest ecosystems (Castro-Díez et al. 2019). However, various policies, legislations and identified risk assessments help in regulating IAS spread in forest ecosystems (Pötzelsberger et al. 2020). Several forests related activities may affect the invasion of alien plant species in vegetation. Clearing or cutting of forests regulates light conditions and affects other resource availability that is suitable for IAS. Moreover, the movement of constructed materials and contaminated soil during forest road construction promotes the spreading of invasive plant seeds. Climate change-mediated flood, storm and anthropogenic forest fires also accelerate the invasion of alien plants (Lake and Leishman 2004; Jhariya 2017; Jhariya and Singh 2021). The spreading of alien plant species also reduces species diversity, richness and understory composition in forest ecosystems (Navarro et al. 2018). Also, the introduction of invasive plants in natural forests also affects the successful regeneration of tree species. Similarly, the existence of *Impatiens parviflora* which is a shade-tolerant invasive plant species has been observed under temperate broadleaf forests of European countries (Lapin et al. 2019).

3.4.2 Agriculture

Practicing modern farming promises food security but also increases the spreading of invasive plants (Rai et al. 2018). These invasive plants affect overall agricultural health and productivity. They reduce crop diversity, productivity, soil health and the entire plant ecosystem. These IAS disturb the environment and related services such as water regulation, food security, soil fertility, human health and overall agricultural sustainability which is prime towards sustainable development goals (SDGs) (Pysek and Richardson 2010; Ebi et al. 2017). Interestingly, these invasive plants are the source of food as well as have detrimental effects that have been reported in terms of poor crop productivity (Shackleton et al. 2019). Cheat grass (*Bromus tectorum*) as an invasive plant can promote fungal pathogen's outbreaks which strongly affects the health of the native plants (Beckstead et al. 2010). Similarly, some pathogenic alien species such as *Cryphonectria parasitica* removed *Castanea dentata* which is earlier dominant native plant species (Andersen et al. 2004). Further, *Parthenium hysterophorus* is IAS which is used for demonstration for the spreading of phytoplasmas a vegetable pathogen. A similar genetic lineage of *Parthenium* has been observed for phytoplasmas infecting vegetables (Cai et al. 2016).

3.4.3 Human

Invasive plants relocate some important native plants and affect ecosystems and human health drastically (Xie et al. 2020). Invasive plants can displace native plants, degrade ecosystems, and negatively impact human health (Xie et al. 2020). Very few studies are available on plant invasion impacts on human health and the ecosystem. Invasive plants affect humans in several ways such as it causing infectious diseases, exposing humans to wounds, injuries and even death, and negatively affecting human livelihood. Moreover, some invasive plants affect human health by producing a toxin that harms drastically (Mazza et al. 2014). These alien plant species disturb human life by affecting social, economic and ecological perspectives. Thus, an invasive plant reduces biodiversity and alters climatic variables (temperature and humidity) which directly or indirectly affect human health (Jones 2019). Although, many invasive plants are introduced for ornamental and decorative purposes but they affect human and ecosystem health drastically (Rai 2015; Keshri et al. 2016). Most of the invasive species contributed to environmental contamination that affects human health (Jones and McDermott 2018). Some invasive species destroy native plants which act as a valuable source for human health. For example, the invasion of emerald ash borer which is well-known invasive plant pathogen causes losses of ash trees (*Fraxinus excelsior*) in the United States. These trees are considered as a good sink of air pollutants that protect humans from the dangerous effects of air pollution (Jones and McDermott 2018).

3.4.4 Animals

Invasive plants also affect animals health and its productivity. *Parthenium hysterophorus* is a common IAS that causes many diseases after ingestion. Many invasive plants cause dermatitis and skin disease or even death due to excessive salivation (Thiel et al. 2018). It also causes animal death to a significant level due to excessive consumption (Mawal and Patil 2019). These invasive weeds cause numerous deaths of animals in the region of Kangra district of Himachal Pradesh, India. Excessive use of herbicides for depriving *Lantana* plant causes carcinogenic effects on humans and animals (González et al. 2017). However, using biological control method for suppressing invasive plants becomes safer and eco-friendlier than chemical methods. An invasive plant adversely affects animal's diversity and population. *Scirpus mariqueter* and *Phragmites australis* are two native macrophytes found in the wetland region of China which is replaced by the invasive plant "*Spartina alterniflora*". This caused a decrease in the avian population due to feeding restrictions and its unavailability (Gan et al. 2009).

3.4.5 Soil

Invasive plants also affect soil health and quality. These plants directly or indirectly influence soil physico-chemical properties which determine nutrient status and productivity. Invasive species also affect C storage and sequestration potential of the soil ecosystem (Martin et al. 2017). IAS also affects biotic and abiotic components of soil attributes in both spatial and temporal manner (Gibbons et al. 2017). Many studies have reported a significant change in soil attributes while the invasion of alien plants over native regions. For example, a higher soil nitrate level was reported under invasive plants like *Ageratina adenophora* which is closely associated with soil microbial diversity (Kong et al. 2017). Also, soil bacterial and fungal populations were increased under *Impatiens glandulifera* (Gaggini et al. 2018). Similarly, nitrogen-fixing invasive species like *Prosopis pallida* regulate water resources to a significant level which alters the soil environment (Dudley et al. 2014). In grassland ecosystems, some IAS such as *B. tectorum* (cheat grass), *Centaurea stoebe* (spotted knapweed) and *Euphorbia esula* (leafy spurge) have profound effects on soil quality (Gibbons et al. 2017). Also, invasive plants of the Mediterranean ecosystem such as *Acacia dealbata* strongly affect the soil chemistry and microbial diversity that reduces the diversity of native plants (Lazzaro et al. 2014). However, invasive plants alter soil attributes but *Pueraria montana* (Kudzu) plant has the capacity to minimize the chances of soil erosion at certain extent (Forseth and Innis 2004).

3.5 Plant Invasion and Food Security

The biological invasion has significant impacts on social, economic and ecological parameters which are considered as the greatest driver for global environmental change (Millennium Ecosystem Assessment 2005). These invasive plants affect global food security by disturbing agriculture productivity to a certain extent (Fleming et al. 2017). The economic losses due to poor plant productivity after the invasion of alien plants are a great concern of today (Seebens et al. 2017). Plant invasion causes the loss of billions of dollars due to poor agricultural productivity in many countries (Sinden et al. 2004). The USA has invested one-fourth of the gross national product of agriculture in the management of invasive plants (Simberloff 1996). IAS causes higher cost investment for their management which causes more economic losses due to poor agricultural productivity. Invasive plant species is a big challenge to achieving SDGs-2 which focuses on ensuring food security by reducing hunger and improvement in nutrition with sustainable agricultural production. The significant impacts of invasive plants on agriculture have been studied under climate suitability maps that access the agricultural vulnerability due to plant invasions (Kariyawasam et al. 2021). Therefore, plant invasion is a major curse of agricultural productivity by affecting biodiversity and ecosystem services. Poor food production and less nutritive fruits, and food are the results of exotic plant invaders in the native agricultural system (Cook et al. 2011). Allelopathy, parasitism, and competition for natural resources (water, light, nutrients) are different mechanisms behind agricultural losses due to the invasion of exotic plants (Bajwa et al. 2019).

3.6 Plant Invasion Impacts on the Environment and Ecosystem Services

Biotic invaders drastically reduce the variety of ecosystem services that affect environmental health (Bartz and Kowarik 2019). Plant invasion affects overall native flora and faunal diversity which are key sources of uncountable ecosystem services. A remarkable change was observed in soil attributes including microbial diversity due to invasive plants. These diversified soils deliver environmental services which maintain ecosystem health and ecological sustainability. Native flora and fauna reduction due to exotic plant invaders are a prime concern today (Pysek et al. 2012). The IAS has adverse impacts on native plant diversity that disturbs environmental functioning and ecosystem services along with the promotion of global climate change (Heshmati et al. 2019). Many invasive plants have well-known impacts on regulatory, aesthetic, cultural and recreational-based ecosystem services (Pejchar and Mooney 2009). Invasive plants adversely affect tourism and recreational services due to impeding water navigation (Eiswerth et al. 2005). Thus, alien plants invade native areas where they affect flora and faunal diversity which reduces ecosystem services and affects overall environmental health and sustainability (Mechergui et al. 2021). However, invasive plants also provide ecosystem services more or less, and are not much significant as native species. Different studies have been compiled on invasive plants and its ecosystem services for ensuring global sustainability. Potgieter et al. (2017) have also recorded 10 invasive plants after mining of 335 papers from 27 countries comprising 58 urban cities of the world which delivered varying percentages of ecosystem services. The highest percentage of ecosystem services was delivered by provisioning and cultural services, whereas least percentage was recorded under supporting and regulating services.

3.7 Plant Invasion and Socio-Economic Losses

Plant invasion drastically affects overall floral health and productivity which causes certain economic losses. Plant invasion-mediated economic losses (€ billion/year) in the world are depicted in Fig. 3.2 (European Commission 2013; Haubrock et al. 2021). The US has invested 600 million US dollar for minimizing the losses caused by invasive plants to agriculture and the environment (Andersen et al. 2004). In

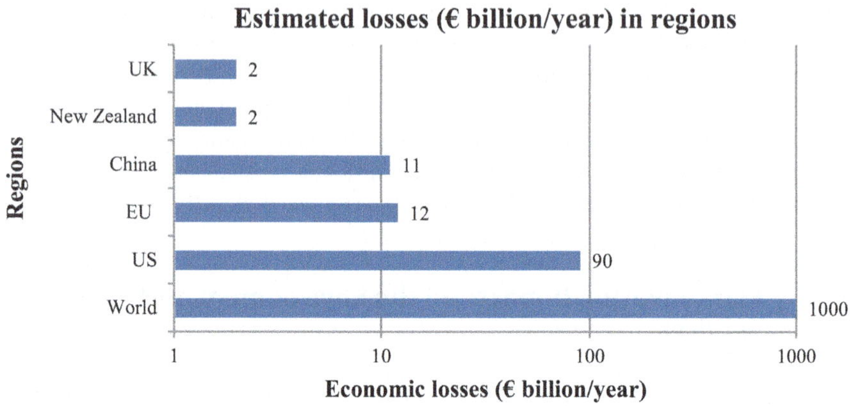

Fig. 3.2 Plant invasion-mediated economic losses (€ billion/year) in the world (European Commission 2013; Haubrock et al. 2021)

China, a total of 283 invasive flora and fauna hampered agriculture, grassland, forest and wetland productivity caused economic losses of 14.45 billion US dollars which is highly linked with human well-being (Xu et al. 2006). Similarly, invasive plants resulted in 1.0 billion US dollar of economic losses due to agricultural crop damage in African regions (Sileshi et al. 2019). Approximately 1.85 billion US dollar of economic losses was reported due to disease-spreading alien invasions to human health in Southeast Asia (Nghiem et al. 2013). Similarly, *Opuntia stricta* invasions drastically affect the economy and environment in the region of Africa. This invader also affects people livelihoods by poor fodder production and livestock health (Shackleton et al. 2017).

3.8 Plant Invasion Dynamics Under Changing Climate

Climate change is the biggest challenge which is further induced by drastically spreading of invaders. Changing climate and extreme weather directly influence the diversity and spread of alien plant species in native regions. Environmental changes have been reported due to climate change which directly influences the spread and distribution of plant invasion (Demertzis and Iliadis 2018). However, the topic of plant invasions and its interaction with global climate change has been gaining wider recognition from the last two to three decades. Climate change has both positive and negative impacts on IAS in any agroecosystem (Dai et al. 2020b). Extreme weather and climatic variability (temperature, rainfall and humidity) affect alien species diversity and its pattern of shifting or invaders into their native place (IPCC 2007). Uncertain rainfall and temperature variation have long-lasting impacts on the survival and growth of plants including IAS (Thuiller et al. 2008). Heavy rainfall also promotes the growth and development of IAS in arid savannah zone of

South Africa (Richardson et al. 2000). Similarly, global warming affects the germination rate and seed longevity of invasive plants (Bernareggi et al. 2015).

3.9 Invasive Plant and Carbon Dynamics

Many identified invasive plants are better sources of bio-energy, bio-polymers and are used as animal feeding materials which promote the concept of the green economy. Invasive plant such as *Spartina alterniflora* has great potential of C sequestration and works as bio-agent for the phytoremediation of heavy metals (Prabakaran et al. 2019). Similarly, invasive grasses have replaced sagebrush ecosystems which drastically reduce C sequestration potential in the Great and Amazon Basins of the US (Pejchar and Mooney 2009). However, invasive tree species such as *Prosopis glandulosa* also tend to increase C sequestration potential (32% increment) by replacing grasses species in the region of the US (Hughes et al. 2006). Therefore, grassland invasions with woody alien species can promote climate change mitigation. Although microbial priming and litter chemistry of invasive plants also affect C sequestration potential in invaded ecosystems (Tamura and Tharayil 2014). Furthermore, McKenzie et al. (2014) have reported some anthropogenic disturbance and mediated climate change which is strongly linked with invasive species that exist in marine meadows (McKenzie et al. 2014). Therefore, climate change has profound effects on IAS, its diversity, regeneration, seed germination, longevity and its global distribution.

Climate change induces distribution and establishment of many alien plant species which becomes invasive in due course of time. Changing climate and global warming minimizes the habitat resilience to biological invasions throughout the world. Rising CO_2 induces global warming that alters corridors for movement of plant invasion at global scale. For example, climate change induces dramatic niche shift for *Centaurea maculosa* (spotted knapweed) which became aggressive plant invaders in the region of Western North America (Broennimann et al. 2007; Walther et al. 2009). However, it is need for accurately prediction of invasive plant distributions and its varying impacts on desire species that can be changed under projected climate change scenarios. This study helps in understanding an effective control and preventive measures against spreading of plant invasions (Finch et al. 2021).

3.10 Strategic Plan and Policy for Plant Invasion Management

Invasive alien species (IAS) adversely affects native plant diversity and its growth in agroecosystem. Therefore, strategic plans must be framed for managing plant invasion and its negative consequences on native biodiversity and related ecosystem services. Also, a plan and policy must be developed for identifying alien-introduced species which may be problematic for native plants. However, remote sensing is also effective for mapping, detecting and monitoring of IAS. These technologies must be included in future plans and policies due to its higher recommendation for invasive plant management. Moreover, enlisting of both positive and negative consequences of an invasive plant on environmental health, socioeconomic and ecosystem services are generated. Accordingly, a judicious policy is framed for the management of IAS in order to check biodiversity loss, and ecosystem degradation and ensure human health (Rai and Singh 2020).

3.11 Scientific Research and Future Recommendation

Plant invasion is an ecological menace and greater environmental challenges. Invasion of alien plants into native regions induces global change which is highly complex to study (IPCC 2007; Vilà et al. 2007). Climate change and its variability also affect plant invasions in positive or negative ways. Invasion may be enhanced or hindered under different abiotic conditions. For example, higher CO_2 favours invasion, whereas the sudden change in temperature and humidity hinders invasions in any region (Bradley et al. 2010). Scientific research and policy for long-term experiments are needed for understanding the impacts of certain climatic changes on invasion ecology at different stages (Catford et al. 2020). Using trans-disciplinary research on IAS is an integrated approach that helps in the sustainable management of invasive plants.

3.12 Conclusion

Plant invasion is considered a burning topic of today which is rigorously discussed by scientists, academicians, ecologists, policymakers and stakeholders at various national and international platforms. Invasive plants threaten to biodiversity, natural resources (forests, agriculture, soil, human, animals) and environmental health. Therefore, forecasting of invasive plant species, its identification, early detection and distribution mapping can be possible through geospatial tools including remote sensing. These are necessary for making a plan of action against the negative consequences of IAS. The impact of plant invasion on human health by depriving food quality and other ecosystem services also influences socioeconomic status or livelihood security. However, changing climate and adverse weather also influence invasive plant distribution and diversity. Besides the negative consequences, IAS has good potential for C sequestration and maintaining C dynamics in the ecosystem. Thus, a strategic plan for scientific research and a judicious policy must be framed for the management of invasive plants which promise environmental sustainability and ecological stability.

References

Abhilash PC, Singh N, Sylas VP, Kumar BA, Mathew JC, Satheesh R, Thomas AP (2008) Eco-distribution mapping of invasive weed *Limnocharis flava* (L.) Buchenau using geographical information system: implications for containment and integrated weed management for ecosystem conservation. Taiwania 53(1):30–41

Ajesh TP, Naseef SA, Kumuthakalavalli R (2012) Ethnobotanical documentation of wild edible fruits used by Muthuvan tribes of Idukki, Kerala-India. Int J Pharma Bio Sci 3(3):479–487

Alvarez-Taboada F, Paredes C, Julian Pelaz J (2017) Mapping of the invasive species Hakea sericea using unmanned aerial vehicle (UAV) and WorldView-2 imagery and an object-oriented. Remote Sens 9(9):913–929

Andersen MC, Adams H, Hope B, Powell M (2004) Risk assessment for invasive species. Risk Anal 24(4):787–793

Bajwa AA, Farooq M, Nawaz A, Yadav L, Chauhan BS, Adkins S (2019) Impact of invasive plant species on the livelihoods of farming households: evidence from *Parthenium hysterophorus* invasion in rural Punjab, Pakistan. Biol Invasions 21(11):3285–3304

Barooah C, Ahmed I (2014) Plant diversity of Assam—a checklist of angiosperms and gymnosperms. Assam Science Technology and Environment Council, Guwahati, 599 pp

Bartz R, Kowarik I (2019) Assessing the environmental impacts of invasive alien plants: a review of assessment approaches. NeoBiota 43:69–99

Beckstead J, Meyer SE, Connolly BM, Huck MB, Street LE (2010) Cheatgrass facilitates spillover of a seed bank pathogen onto native grass species. J Ecol 98(1):168–177

Bernareggi G, Carbognani M, Petraglia A, Mondoni A (2015) Climate warming could increase seed longevity of alpine snowbed plants. Alp Bot 125:69–78

Bradley BA, Blumenthal DM, Wilcove DS, Ziska LH (2010) Predicting plant invasions in an era of global change. Trends Ecol Evol 25:310–318

Broennimann O, Treier UA, Müller-Schärer H, Thuiller W, Peterson AT, Guisan A (2007) Evidence of climatic shifts during biological invasion. Ecol Lett 10:701–709

Cai H, Wang LC, Yang ZX, Wan QL, Wei W, Davis RE, Zhao Y (2016) Evidence for the role of an invasive weed in widespread occurrence of phytoplasma diseases in diverse vegetable crops: implications from lineage-specific molecular markers. Crop Prot 89:193–201

Castro-Díez P, Vaz AS, Silva JS, Van Loo M, Alonso Á, Aponte C, Bayón Á, Bellingham PJ, Chiuffo MC, DiManno N, Julian K (2019) Global effects of non-native tree species on multiple ecosystem services. Biol Rev 94(4):1477–1501

Catford JA, Dwyer JM, Palma E, Cowles JM, Tilman D (2020) Community diversity outweighs effect of warming on plant colonization. Glob Chang Biol 26:3079–3090

Cook DC, Fraser RW, Paini DR, Warden AC, Lonsdale WM, De Barro PJ (2011) Biosecurity and yield improvement technologies are strategic complements in the fight against food insecurity. PLoS One 6(10):e26084

Dai J, Roberts DA, Stow DA, An L, Hall SJ, Yabiku ST, Kyriakidis PC (2020a) Mapping understory invasive plant species with field and remotely sensed data in Chitwan, Nepal. Remote Sens Environ 250:112037

Dai ZC, Wan LY, Qi SS, Rutherford S, Ren GQ, Wan JS, Du DL (2020b) Synergy among hypotheses in the invasion process of alien plants: a road map within a timeline. Perspect Plant Ecol Evol Syst 47:125575

Davis AM (2009) Invasion biology. Oxford University Press, p 244

Demertzis K, Iliadis L (2018) The impact of climate change on biodiversity: the ecological consequences of invasive species in Greece. In: Leal Filho W, Manolas E, Azul A, Azeiteiro U, McGhie H (eds) Handbook of climate change communication, Climate change management, vol 1. Springer, Cham

Dudley BD, Hughes R, Ostertag R (2014) Groundwater availability mediates the ecosystem effects of an invasion of *Prosopis pallida*. Ecol Appl 24(8):1954–1971

Ebi KL, Frumkin H, Hess J (2017) Protecting and promoting population health in the context of climate and other global environmental changes. Anthropocene 19:1–12

Eiswerth ME, Darden TD, Johnson WS, Agapoff J, Harris TR (2005) Input–output modeling, outdoor recreation, and the economic impacts of weeds. Weed Sci 53(1):130–137

European Commission (2013) Flash Eurobarometer 379—attitudes towards biodiversity. https://ec.europa.eu/environment/eurobarometers_en.htm

Finch DM, Butler JL, Runyon JB, Fettig CJ, Kilkenny FF, Jose S, Frankel SJ, Cushman SA, Cobb RC, Dukes JS, Hicke JA (2021) Effects of climate change on invasive species, pp 57–83. https://doi.org/10.1007/978-3-030-45367-1_4

Fleming PJ, Ballard G, Reid NC, Tracey JP (2017) Invasive species and their impacts on agri-ecosystems: issues and solutions for restoring ecosystem processes. Rangeland J 39(6):523–535

Forseth I, Innis A (2004) Kudzu (*Pueraria montana*): history, physiology, and ecology combine to make a major ecosystem threat. Crit Rev Plant Sci 23:401–413

Gaggini L, Rusterholz H, Baur B (2018) The invasive plant *Impatiens glandulifera* affects soil fungal diversity and the bacterial community in forests. Appl Soil Ecol 124:335–343

Gan X, Cai Y, Choi C, Ma Z, Chen J, Li B (2009) Potential impacts of invasive Spartina alterniflora on spring bird communities at Chongming Dongtan, a Chinese wetland of international importance. Estuar Coast Shelf Sci 83(2):211–218

Gibbons SM, Lekberg Y, Mummey DL, Sangwan N, Ramsey PW, Gilbert JA (2017) Invasive plants rapidly reshape soil properties in a grassland ecosystem. MSystems 2(2):e00178-16

Gichua M, Njorage G, Shitanda D, Ward D (2013) Invasive species in East Africa: current status for informed policy decisions and management. J Agric Sci Technol 15(1):45–55

González E, Sher AA, Anderson RM, Bay RF, Bean DW, Bissonnete GJ et al (2017) Secondary invasions of noxious weeds associated with control of invasive Tamarix are frequent, idiosyncratic and persistent. Biol Conserv 213:106–114

Grinnell J (2000) The English sparrow has arrived in Death Valley: an experiment in nature. Am Nat 53:468

Harries HC, Clement CR (2014) Long-distance dispersal of the coconut palm by migration within the coral atoll ecosystem. Ann Bot 113(4):565–570

Haubrock PJ, Turbelin AJ, Cuthbert RN, Novoa A, Taylor NG, Angulo E, Ballesteros-Mejia L, Bodey TW, Capinha C, Diagne C, Essl F (2021) Economic costs of invasive alien species across Europe. NeoBiota 67:153–190

Hernández-García E, García A, Garza-González E, Avalos-Alanís FG, Rivas-Galindo VM, Rodríguez-Rodríguez J, del Rayo Camacho-Corona M (2019) Chemical composition of Acacia farnesiana (L) wild fruits and its activity against Mycobacterium tuberculosis and dysentery bacteria. J Ethnopharmacol 230:74–80

Heshmati I, Khorasani N, Shams-Esfandabad B, Riazi B (2019) Forthcoming risk of Prosopis juliflora global invasion triggered by climate change: implications for environmental monitoring and risk assessment. Environ Monit Assess 191(2):1–12

Hughes RF, Archer SR, Asner GP, Wessman CA, McMurtry CHAD, Nelson JIM, Ansley RJ (2006) Changes in aboveground primary production and carbon and nitrogen pools accompanying woody plant encroachment in a temperate savanna. Glob Chang Biol 12(9): 1733–1747

IPBES (2019) In: Díaz S, Settele J, Brondizio ES, Ngo HT, Guèze M, Agard J, Arneth A, Balvanera P, Brauman KA, Butchart SHM et al (eds) Summary for policymakers of the global assessment report on biodiversity and ecosystem services of the Intergovernmental Science-Policy Platform on Biodiversity and Ecosystem Services. IPBES Secretariat, Bonn. Accessed 27 Sept 2019

IPCC (2007) Climate change 2007: the physical science basis. Contribution of working group I to the fourth assessment report of the Intergovernmental Panel on Climate Change. Cambridge University Press, Cambridge

Jhariya MK (2017) Vegetation ecology and carbon sequestration potential of shrubs in tropics of Chhattisgarh, India. Environ Monit Assess 189(10):518. https://doi.org/10.1007/s10661-017-6246-2

Jhariya MK, Singh L (2021) Herbaceous diversity and biomass under different fire regimes in a seasonally dry forest ecosystem. Environ Dev Sustain 23(5):6800–6818. https://doi.org/10.1007/s10668-020-00892-x

Jhariya MK, Meena RS, Banerjee A (2021a) Ecological intensification of natural resources for sustainable agriculture. Springer Nature, Singapore, p 655. eISBN: 978-981-334-203-3, Hardcover ISBN: 978-981-334-206-6. https://doi.org/10.1007/978-981-33-4203-3

Jhariya MK, Banerjee A, Meena RS, Kumar S, Raj A (2021b) Sustainable intensification for agroecosystem services and management. Springer Nature, Singapore. eISBN: 978-981-16-3207-5, Hardcover ISBN: 978-981-16-3206-8. https://doi.org/10.1007/978-981-16-3207-5

Jhariya MK, Banerjee A, Raj A, Meena RS, Khan N, Kumar S, Bargali SS (2022a) Species invasion and ecological risk. In: Natural resources conservation and advances for sustainability. Elsevier, Rio de Janeiro. ISBN: 9780128229767, pp 503–531. https://doi.org/10.1016/B978-0-12-822976-7.00014-4

Jhariya MK, Meena RS, Banerjee A, Meena SN (2022b) Natural resources conservation and advances for sustainability. Elsevier, Academic Press. ISBN: 9780128229767. https://doi.org/10.1016/C2019-0-03763-6

Ji X, Peng Q, Yuan Y, Shen J, Xie X, Wang M (2017) Isolation, structures and bioactivities of the polysaccharides from jujube fruit (*Ziziphus jujuba* Mill.): a review. Food Chem 227:349–357

Jones BA (2019) Tree shade, temperature, and human health: evidence from invasive species-induced deforestation. Ecol Econ 156:12–23

Jones BA, McDermott SM (2018) Health impacts of invasive species through an altered natural environment: assessing air pollution sinks as a causal pathway. Environ Resour Econ 71(1): 23–43

Joshi C, De Leeuw J, Van Duren IC (2004) Remote sensing and GIS applications for mapping and spatial modeling of invasive species. Int J Geo Inform 35:669–677

Kariyawasam CS, Kumar L, Ratnayake SS (2021) Potential risks of invasive alien plant species on agriculture under climate change scenarios in Sri Lanka. Curr Res Environ Sustain 3:100051

Keshri H, Yadav DK, Jhariya MK (2016) Invasive alien species in urban ecosystem of Sarguja. J Plant Dev Sci 8(2):65–69

Khan N, Jhariya MK, Yadav DK (2019) Invasion and impact of non-native species on indigenous flora in an urban ecosystem of India: a case study. Int J Ecol Dev 34(4):103–118

Kong Y, Kong J, Wang D, Huang H, Geng K, Wang Y, Xia Y (2017) Effect of Ageratina adenophora invasion on the composition and diversity of soil microbiome. J Gen Appl Microbiol 63(2):114–121

Lake JC, Leishman MR (2004) Invasion success of exotic plants in natural ecosystems: the role of disturbance, plant attributes and freedom from herbivores. Biol Conserv 117(2):215–226

Langmaier M, Lapin K (2020) A systematic review of the impact of invasive alien plants on forest regeneration in European temperate forests. Front Plant Sci 11:524969

Lapin K, Oettel J, Steiner H, Langmaier M, Sustic D, Starlinger F, Kindermann G, Frank G (2019) Invasive alien plant species in unmanaged forest reserves, Austria. NeoBiota 48:71

Lazzaro L, Giuliani C, Fabiani A, Agnelli AE, Pastorelli R, Lagomarsino A, Benesperi R, Calamassi R, Foggi B (2014) Soil and plant changing after invasion: the case of Acacia dealbata in a Mediterranean ecosystem. Sci Total Environ 497:491–498

Mahapatra AK, Mishra S, Basak UC, Panda PC (2012) Nutrient analysis of some selected wild edible fruits of deciduous forests of India: an explorative study towards non conventional bio-nutrition. Adv J Food Sci Technol 4(1):15–21

Mainali KP, Warren DL, Dhileepan K, McConnachie A, Strathie L, Hassan G, Karki D, Shrestha BB, Parmesan C (2015) Projecting future expansion of invasive species: comparing and improving methodologies for species distribution modeling. Glob Chang Biol 21(12): 4464–4480

Martin PA, Newton AC, Bullock JM (2017) Impacts of invasive plants on carbon pools depend on both species' traits and local climate. Ecology 98(4):1026–1035

Mawal S, Patil S (2019) Assessment of allelopathic potential of Lantana species on some selected agricultural crops. Assessment 4(1):15

Mazza G, Tricarico E, Genovesi P, Gherardi F (2014) Biological invaders are threats to human health: an overview. Ethol Ecol Evol 26(2–3):112–129

McKenzie LJ, Yoshida RL, Unsworth RKF (2014) Disturbance influences the invasion of a seagrass into an existing meadow. Mar Pollut Bull 86:186–196

Mechergui T, Pardos M, Jhariya MK, Banerjee A (2021) Mulching and weed management towards sustainability. In: Ecological intensification of natural resources for sustainable agriculture. Springer, Singapore. eBook ISBN: 978-981-334-203-3. Hardcover ISBN: 978-981-334-202-6, pp 255–287. https://doi.org/10.1007/978-981-33-4203-3_8

Millennium Ecosystem Assessment (2005) Ecosystems and human well-being, vol 5. Island Press, Washington, DC, pp 563–563

Müllerová J, Brůna J, Bartaloš T, Dvořák P, Vítková M, Pyšek P (2017) Timing is important: unmanned aircraft vs. satellite imagery in plant invasion monitoring. Front Plant Sci 8:887

Navarro L, Morin H, Bergeron Y, Girona MM (2018) Changes in spatiotemporal patterns of 20th century spruce budworm outbreaks in eastern Canadian boreal forests. Front Plant Sci 9:1905

Nghiem LT, Soliman T, Yeo DC, Tan HT, Evans TA, Mumford JD, Keller RP, Baker RH, Corlett RT, Carrasco LR (2013) Economic and environmental impacts of harmful non-indigenous species in Southeast Asia. PLoS One 8(8):e71255

Nowak MM, Dziób K, Bogawsk P (2018) Unmanned aerial vehicles (UAVs) in environmental biology: a review. Eur J Ecol 4(2):56–74

Ogawa S, Yazaki Y (2018) Tannins from Acacia mearnsii De Wild. Bark: Tannin determination and biological activities. Molecules 23(4):837

Panda T, Mishra N, Pradhan BK, Mohanty RB (2018) Expansive alien flora of Odisha, India. J Agric Environ Int Dev (JAEID) 112(1):43–64

Patykowski J, Kołodziejek J, Wala M (2018) Biochemical and growth responses of silver maple (*Acer saccharinum* L.) to sodium chloride and calcium chloride. PeerJ 6:e5958

Pejchar L, Mooney HA (2009) Invasive species, ecosystem services and human wellbeing. Trends Ecol Evol 24(9):497–504

Potgieter LJ, Gaertner M, Kueffer C, Larson BMH, Livingstone S, O'Farrell P, Richardson DM (2017) Alien plants as mediators of ecosystem services and disservices in urban systems: a global review. Biol Invasions 19:3571. https://doi.org/10.1007/s10530-017-1589-8

Pötzelsberger E, Lapin K, Brundu G, Adriaens T, Andonovski V, Andrašev S, Bastien JC, Brus R, Čurović M, Čurović Ž, Cvjetković B (2020) Mapping the patchy legislative landscape of non-native tree species in Europe. Forestry 93(4):567–586

Prabakaran K, Li J, Anandkumar A, Leng Z, Zou CB, Du D (2019) Managing environmental contamination through phytoremediation by invasive plants: a review. Ecol Eng 138:28–37

Prasad R, Jhariya MK, Banerjee A (2021a) Advances in sustainable development and management of environmental and natural resources: economic outlook and opinions, vol I. CRC Press, Apple Academic Press Inc., Palm Bay, FL. Hardcover ISBN: 9781774910344, pp 1–437

Prasad R, Jhariya MK, Banerjee A (2021b) Advances in sustainable development and management of environmental and natural resources: economic outlook and opinions, vol II. CRC Press, Apple Academic Press Inc., Palm Bay, FL. Hardcover ISBN: 9781774910368, pp 1–428

Pysek P, Richardson DM (2010) Invasive species, environmental change and management, and health. Annu Rev Environ Resour 35:25–55

Pysek P, Jarošík V, Hulme PE, Pergl J, Hejda M, Schaffner U, Vilà M (2012) A global assessment of invasive plant impacts on resident species, communities and ecosystems: the interaction of impact measures, invading species' traits and environment. Glob Chang Biol 18(5):1725–1737

Rai PK (2015) Paradigm of plant invasion: multifaceted review on sustainable management. Environ Monit Assess 187:759

Rai PK, Singh JS (2020) Invasive alien plant species: their impact on environment, ecosystem services and human health. Ecol Indic 111:106020

Rai PK, Kumar V, Tsang YF, Naddem Ok Y, KimJH TYF (2018) Nanoparticle-plant interaction: implications in energy, the environment, and agriculture. Environ Int 119:1–19

Raj A, Jhariya MK, Harne SS (2018) Threats to biodiversity and conservation strategies. In: Sood KK, Mahajan V (eds) Forests, climate change and biodiversity. Kalyani Publisher, Chennai, pp 304–320, 381

Ramsar Convention (2018) Global wetland outlook. Ramsar Secretariat, Gland. Ramsar Convention on Wetlands ('Ramsar'). 1971. https://www.ramsar.org/. Accessed 27 Sept 2019

Richardson DM (2011) Fifty years of invasion ecology: the legacy of Charles Elton. Wiley-Blackwell

Richardson DM, Bond WJ, Dean WRJ, Higgins SI, Midgley GF, Milton SJ, Powrie LW, Rutherford MC, Samways MJ, Schulze RE (2000) Invasive alien species and global change: a south African perspective. In: Invasive species in a changing world. Island Press, Washington, DC, pp 303–349

Sabat-Tomala A, Raczko E, Zagajewski B (2022) Mapping invasive plant species with hyperspectral data based on iterative accuracy assessment techniques. Remote Sens 14(1):64. https://doi.org/10.3390/rs14010064

Saha A, Ray A, Das P (2013) Fungal colonization of the phylloplane of Psidium Guineense Sw. growing in Suryamaninagar, Tripura, northeast India. Int J Basic Appl Chem Sci 3(1): 62–67

Sax DF, Stachowicz JJ, Gaines SD (2005) Species invasions: insights into ecology, evolution, and biogeography. Sinauer, Sunderland, MA, p 509

Seebens H, Blackburn TM, Dyer EE, Genovesi P, Hulme PE, Jeschke JM, Pagad S, Pyšek P, Winter M, Arianoutsou M, Bacher S (2017) No saturation in the accumulation of alien species worldwide. Nat Commun 8(1):1–9

Seebens H, Blackburn TM, Dyer EE, Genovesi P, Hulme PE, Jeschke JM, Pagad S, Pyšek P, van Kleunen M, Winter M, Ansong M (2018) Global rise in emerging alien species results from increased accessibility of new source pools. Proc Natl Acad Sci 115(10):E2264–E2273

Shackleton RT, Witt ABR, Piroris FM, van Wilgen BW (2017) Distribution and socio-ecological impacts of the invasive alien cactus Opuntia stricta in eastern Africa. Biol Invasions 19:2427–2441

Shackleton RT, Shackleton CM, Kull CA (2019) The role of invasive alien species in shaping local livelihoods and human well-being: a review. J Environ Manag 229:145–157. https://doi.org/10.1016/j.jenvman.2018.05.007

Sharma GP, Raghubanshi AS, Singh JS (2005) Lantana invasion: an overview. Weed Biol Manag 5:157–165

Sileshi GW, Gebeyehu S, Mafongoya P (2019) The threat of alien invasive insect and mite species to food security in Africa and the need for a continent-wide response. Food Security 11:763–775

Simberloff D (1996) Impacts of introduced species in the United States. Consequences 2(2):13–22

Sinden J, Jones R, Hester S, Odom D, Kalisch C, James R, Cacho O (2004) The economic impact of weeds in Australia, Technical Series 8. CRC for Australian Weed Management, Glen Osmond, SA

Sladonja B, Damijanic D (2021) Remote sensing in invasive species detection and monitoring. Int J Environ Sci Nat Res 29(1):556255

Tamura M, Tharayil N (2014) Plant litter chemistry and microbial priming regulate the accrual, composition and stability of soil carbon in invaded ecosystems. New Phytol 203:110–124

Thiel CH, de David FA, Galon L, Deuner S, Forte CT, Perin GF, Concenço G (2018) Physiology of weeds in intraspecific competition. J Agric Sci 10:334

Thuiller W, Richardson DM, Midgley GF (2008) Will climate change promote alien plant invasions? In: Biological invasions. Springer, Berlin, pp 197–211

Totland O, Nyeko P, Bjerknes A, Hegland SJ, Nielsen A (2005) Does forest gap size affects population size, plant size, reproductive success and pollinator visitation in *Lantana camara*, a tropical invasive shrub? For Ecol Manag 215:329–338

Underwood E, Ustin S, DiPetro D (2003) Mapping nonnative plants using hyperspectral imagery. Remote Sens Environ 86(2):150–161

Vilà M, Corbin JD, Dukes JS, Pino J, Smith SD (2007) Linking plant invasions to global environmental change. In: Terrestrial ecosystems in a changing world. Springer, Berlin, pp 93–102

Walther GR, Roques A, Hulme PE, Sykes MT, Pyšek P, Kühn I, Zobel M, Bacher S, Botta-Dukát Z, Bugmann H, Czúcz B (2009) Alien species in a warmer world: risks and opportunities. Trends Ecol Evol 24(12):686–693

Xie W, Zhong C, Li X, Guo Z, Shi S (2020) Hybridization with natives augments the threats of introduced species in Sonneratia mangroves. Aquat Bot 160:103166

Xu H, Ding H, Li M, Qiang S, Guo J, Han Z, Huang Z, Sun H, He S, Wu H, Wan F (2006) The distribution and economic losses of alien species invasion to China. Biol Invasions 8(7): 1495–1500

Young AM, Larson BMH (2011) Clarifying debates in invasion biology: a survey of invasion biologists. Environ Res 111:893–898

Role of Extreme Climate Events in Amplification of Plant Invasion

4

Sundari Devi Laishram and Rashmi Shakya

Abstract

Anthropogenic-induced changes in climate and its resultant extreme climatic events, such as changes in the annual cycles of precipitation and fire with accelerating global mean temperature cause rapid alteration in the vegetation. One of the main issues in response to such changes in climatic events and ecosystem communities is the invasion of exotic species. Extreme polarization of annual precipitation and amplification of the hydrological cycle causes more flood events and longer intervals between rainfalls and droughts. With increasing temperatures, the threatening intensity of fire cycles destroys much fire-intolerant vegetation and soil constituents. Moreover, the overall habitat ranges are also moving toward the north in latitude and upward in elevation. These habitat shifts may threaten critical habitats or may stress certain innate species, eventually creating a favorable condition for many invasive species. Most of the invasive species are so well adapted to many diverse and extreme climatic conditions that they can out-compete aggressively their native challengers leading to the destruction of the existing environment and biodiversity. The degree of effects due to extreme climatic conditions may differ between invasive and native species, especially for endemic species. The frequency of extreme climatic conditions is accelerating owing to global warming. Some invasive species may die out, but some may continue to establish causing problems to many vulnerable and sensitive ecosystems. The invasion of exotic species could affect many aspects of ecological, economic, and sociological consequences. Unfortunately, the efforts to control them are expensive and time-consuming and the level of success varies. Therefore, immediate action is necessary to prevent the introduction of

S. D. Laishram (✉) · R. Shakya
Department of Botany, Miranda House, University of Delhi, Delhi, India
e-mail: laishram.sundari@mirandahouse.ac.in; rashmi.shakya@mirandahouse.ac.in

S. Tripathi et al. (eds.), *Plant Invasions and Global Climate Change*,
https://doi.org/10.1007/978-981-99-5910-5_4

such invasive species through early detection and implementation of management approaches. In this chapter, we are discussing the different climatic events which are accelerating due to global warming, the ecological responses, and their roles in the amplification of exotic species.

Keywords

Biodiversity · Climate change · Ecological function · Ecosystem · Extreme events · Invasion · Native species

Abbreviations

°C	Degree Celsius
CO_2	Carbon dioxide
IPCC	The Intergovernmental Panel on Climate Change

4.1 Introduction

Global climate change is a real concern now manifesting with the increase in average global atmospheric and ocean temperatures accelerating the melting of ice, thermal expansion of sea water, and reduction of global biodiversity. One of the major challenges of climate change is the rise in the magnitude of extreme events in different parts of the globe with massive impacts on human communities as well as on natural ecosystems. Extreme climatic events are very rare actions, but they occur with tremendously high intensity and carry several ecological impacts. Such events can cause short-term and strong influences on the ecosystem's functioning (Sanz-Lázaro 2016). The past decade has witnessed an increase in the number of cyclones, floods, droughts, landslides, and many destructive natural events. The Indian subcontinent has encountered intensified frequency of cyclones per year (Mondal et al. 2022; Patri et al. 2022). In 2021, India was hit by six cyclonic storms, namely Tauktae (May 2021), Amphan (May 2021), Yaas (May 2021), Gulaab (September 2021), Shaheen (September 2021), Jawaad (December 2021) causing huge loss of properties and swiping away livelihoods of many people during the global pandemic of Coronavirus (https://www.mha.gov.in/). Moreover, in 2021, Uttarakhand was also hit by two natural disasters, a glacier burst in the Chamoli district in February and a flood in October (Siddique et al. 2022). There are several reports of such natural disasters happening across the globe. The IPCC report 2022 has summarized the overall impacts of climate change into following key points: climate impacts are already more widespread and severe than expected; we are already locked into even worse impacts from climate change; risks will escalate quickly with higher temperatures, often causing irreversible impacts; inequity, conflict, and development challenges heighten vulnerability to climate risks;

adaptation is very crucial and more impacts will reach to vulnerable communities, and some effects are too severe to cope up with.

Despite many efforts to control the emission of greenhouse gases, the global temperatures are expected to increase more in the near future causing many unavoidable negative impacts on several aspects of ecological, economic, and sociological consequences. The increase in temperature is mainly manifested in changes in the minimum temperature, i.e., the increase in winter temperature is more than in summer (Wei et al. 2023). From the ecological point of view, one of the major issues associated with global climate change and the indulgence of extreme climatic conditions is the amplification of invasive species destroying the functioning of many natural ecosystems. During extreme conditions, organisms encounter stress and engage in strategies to cope with the compulsion of rapid changes in the ecosystem structure and function. Invasive species, also termed as alien species, non-native, exotic or introduced species, could significantly disturb or modify the habitat where they colonized (Rafferty 2021).

Predictions and evaluation of the impacts of climate change on ascended invasive species or potentially invasive species are very essential in the current scenario for adapting effective approaches for prevention, control as well as restoration of the natural ecological systems. Climate-related variables play a very important role in determining the distribution, adaptation, reproduction, survival, and many other ecological functions of both native and exotic species (Finch et al. 2021). With average warming atmospheric temperatures, many invasive species have enhanced the selection of many superior traits, and also, as an invader, they might have a higher ability toward positive response to many extreme climatic events (Jarnevich et al. 2014). Climate change affects both native and exotic species; however, the capacity of reproduction with the changing climate determines the continuity of the species. Selection for the superior trait for reproductive success helps the expansion of the species. Moreover, due to extreme climate events and anthropogenic activities, there is an increase in habitat fragmentation or empty niches, providing an opportunity for the primary invaders to establish (Jarnevich et al. 2014; Finch et al. 2021). The successful establishment of invasive species with such extreme climatic conditions includes the aspects of adaptive traits, genetic richness, diversity, and physiological plasticity of the species so that they can grow, survive, and continue their lineages. Climate change impacts invasive species in different ways which can be described as: (1) direct impact on the individual level of the species, (2) indirect impact that could be on resource availability and biological interaction, and (3) others including human activities altering their habitats (Finch et al. 2021). In this chapter, the ecological responses toward climate change and their impacts on the amplification of invasive species have been discussed in detail.

4.2 Extreme Climate Events

The average temperature of the global surface is predicted to be increased by 1.8–3.6 °C by the end of 2100 driven by mainly CO_2 emission from both natural and anthropogenic sources (IPCC 2007), and there will be a big gap between the minimum and maximum temperatures in annual seasonal cycles (Donat et al. 2013). Around the globe, reduction of cold days and nights as well as the global frozen regions is likely to be continued (Ummenhofer and Meehl Gerald 2017). This phenomenon will directly impact the annual seasonal cycles and ecological processes. Due to an increase in the temperature, there will be a drastic change in many climatic events. The Cambridge University Press releases report of IPCC 2012 which highlighted the shifting of temperature distribution curves and their correlation with the occurrence of extreme climatic events (Fig. 4.1). Temperature changes directly influence the global and regional hydrological cycles. The soil water content will be highly reduced increasing the magnitude of drought and affecting soil microflora (Le Houerou 1996).

Water plays an important role in a broad exchange of energy and mass between the atmosphere, hydrosphere, and lithosphere; therefore, it's cycle could be easily influenced by climate (Kundzewicz 2008). Due to an increase in the global climate temperatures, evaporation from many natural water reservoirs has enhanced which has created serious concerns about the regional water budget, especially in arid and semiarid regions. The three major processes of global hydrological balance, evaporation, condensation, and precipitation are highly disrupted. Precipitation is an important variable that lies at the interface of the climatical and hydrological systems (Kundzewicz 2008). Changes in the frequency as well as the magnitude of the precipitation pattern have enhanced prolonged waterlogging and frequent sudden droughts. The overall increase in the precipitation events in most of the global regions by the end of the twentieth century was stated in the IPCC report of 2012. Such changes in hydrological balance in any ecosystem severely affect the existence of biodiversity, especially, species with narrow tolerance ranges (Corli et al. 2021). The atmospheric water vapor along with other greenhouse gases elevated the greenhouse effect ultimately contributing to global warming. With warming, the hydrological cycle is also speeding up by enhancing evapotranspiration and precipitation. Under intense precipitation, runoff water carries the topsoil depleting the quality of the soil. Sometimes in extreme conditions, heavy mudslides and landslides cause much damage in the constancy of the slopes as well as fatalities due to mud or debris flows, rocks fall, etc. causing endless environmental and socioeconomic destructions (Parkash 2023).

Worldwide in the terrestrial ecosystem about 30% land is covered by forests (FAO and UNEP 2020). Forests play an important role in maintaining the global ecosystem and biodiversity. It has been estimated that about 60% of vascular plants are grown in tropical forests, and 68% of mammals, 75% of birds, and 80% of amphibians find their shelter in these forests (Vie et al. 2009). Furthermore, humans are deeply dependent on forest resources from the very beginning of human settlement. Forest shares the total global carbon reservoir of 45% and are under intense

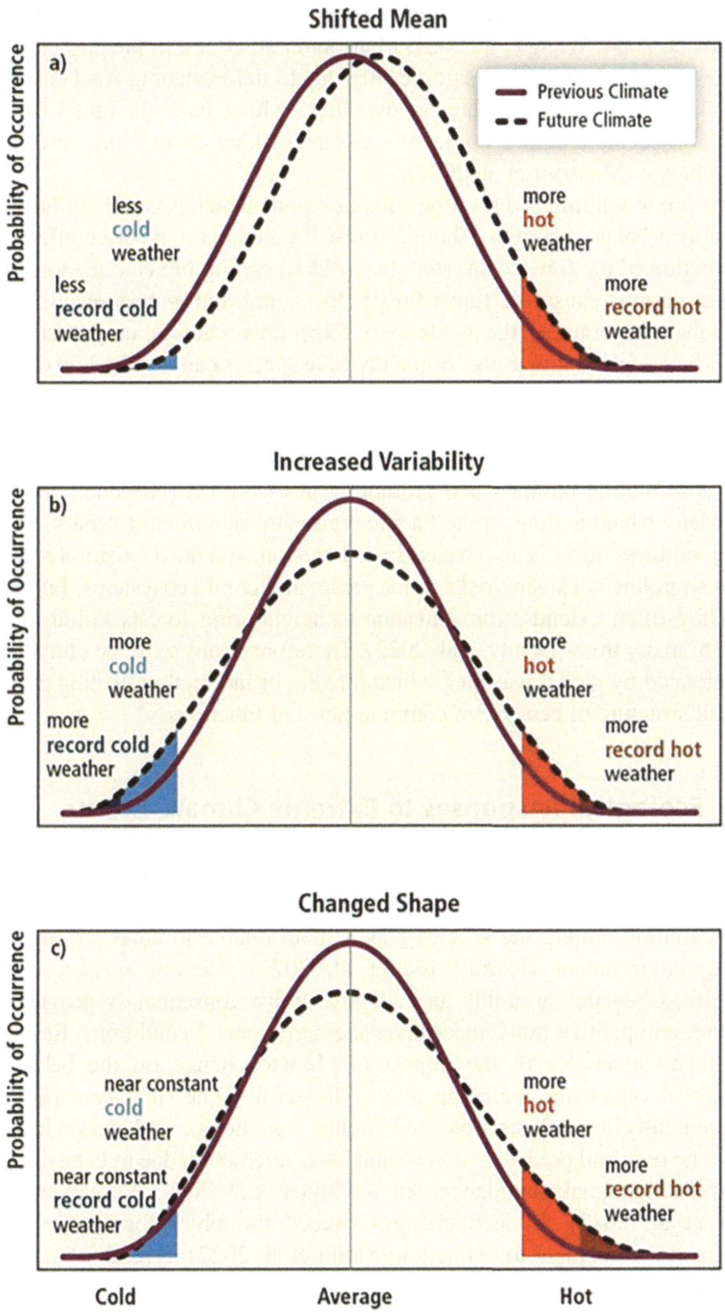

Fig. 4.1 The shifting of temperature distribution curves between present and future climate and their correlation with the occurrence of extreme climate events. (**a**) Shifting of mean of the entire distribution toward a warmer climate; (**b**) increase in the temperature variability without shifting the mean; and (**c**) altering the skewness of the distribution toward the warmer distribution. (Source: IPCC 2012)

pressure due to anthropogenic activities, mainly urbanization and agriculture (Field and Raupach 2004; Wang et al. 2021; Mansoor et al. 2022). In the past decades, the total forest area has decreased significantly due to deforestation, road construction, industrial establishment, and human dwellings which have disrupted the natural ecological systems creating immense pressure making them more susceptible to climate change (Mansoor et al. 2022).

Forest fire or wildfire is one of the climatic events which is greatly influencing the natural forest balance system. Though forest fire has many positive effects on the reconstruction of the forest ecosystem but wild-spreading fire can be more harmful. There are various causes of forest fires both natural and anthropogenic, but with rising global temperatures, the incidence of forest fires is accelerating threatening the green surface of the entire globe. Some invasive species can also fuel up the wildfire by increasing the frequency as well as the severity of the fire (Bubb and Williams 2022; Mansoor et al. 2022). Expert-based studies on 49 species to screen the potential wildfire risk in the Hawaiian Islands showed 21 species at very high fire risk by contributing biomass and promoting the wildfires (Faccenda and Daehler 2022). Many island nations of the Pacific region are rich in biodiversity, but due to frequent wildfires, there is an increase in soil erosion, and the deposition of sediment near the shoreline is causing risks to the steam and coral ecosystems. Furthermore, the wildfire often extends from Savanna to neighboring forests killing the dense overtop of many trees (Dendy et al. 2022). Therefore, many extreme climatic events are accelerated by global warming which directly or indirectly affecting the composition and structure of ecosystem communities and functions.

4.3 Ecological Responses to Extreme Climate Events

Living organisms maintain a balance among themselves to coexist in a common habitat. Their interaction might be neutral, negative, or positive (Cordero et al. 2023). Variation among the species makes them unable to adapt or adjust to the changing environment (Ibarra-Isassi et al. 2022). Certain species have well characterized genetic variability and adaptive traits, consequently providing them with better competitive traits under diverse environmental conditions. Research has shown many examples of the impact of climate change on the behavior and adaptivity of organisms including altered flowering time (mainly earlier), range shifts (generally toward the pole and higher elevations), and desynchronization between the prey and predator, insects, and hosts interaction due to behavior shifting from hibernation, peak abundances, etc. (Walther et al. 2002; Wookey et al. 2009; Nielsen et al. 2012). If such changes exceed the physiological limits, many organisms are vulnerable to extinction (Smith et al. 2022). Overall climate change affects the natural ecological system in several ways. Some species might be severely affected while others remain unharmed, instead, some might become invasive while others might shift their habitats to different geographical ranges. The frequency and intensity of extreme climate events are driving changes in species diversity from individual level to the community level in terms of species richness, dominance, composition, and density (Harris et al. 2020). These changes might be reversible or irreversible. Some of the invasive species from different parts of the globe, and their respective impact on the different types of ecosystems are summarized in Table 4.1.

Changes in environmental conditions could induce many physiological and reproductive processes in plants and other organisms. As mentioned before, the impact could be direct or indirect. For instance, some species directly respond to the increased CO_2 rather than changing temperature and precipitation (Dukes et al. 2011). But some other species like insects are not directly influenced by increased CO_2, instead, they are affected indirectly through plants' alterations with CO_2 levels. Besides, insects can be directly influenced by increasing temperature by changing their behavior, phenological activities, host interaction, growth and dispersal, etc. (Finch et al. 2021). Similarly, in the case of pathogens, changing climate in terms of temperature, precipitation, and humidity can alter their host interaction, sporulation, and other physiological cycles directly disrupting their pathogenicity (Jeger 2021). The impacts of extreme climate events on the structures and functions of ecosystems and the resulting consequences of invasion have been summarized in Fig. 4.2.

4.4 Amplification of Invasive Species with Changing Climate Variables

Predicting the potential geographical distribution of invasive species within their tolerance range requires statistical data or models that could explain their tolerance range with changing climate variables (Broennimann and Guisan 2008). In climatic models, the native range of distribution includes the non-climatic biotic and abiotic factors, such as competition, predation, edaphic factors, etc. (Pearman et al. 2007; Finch et al. 2021). However, the actual prediction of the invasive species in the future concerning climate change is not easy as multiple factors influence their native and non-native distribution (Mainali et al. 2015). Generally, climate change may influence positively some species favoring their establishment and expansion into new habitats but it could also alter the native distribution, abundance, and interaction with existing species (Hellmann et al. 2008; Walther et al. 2009; Poland et al. 2021; Bortolini-Rosales and Reyes-Aldana 2023) and make them susceptible to the newly colonizing invaders. If the ability of the new invaders to compete with native species is affected by climate change, the ecological and economic footprints of the primary invaders are not sufficient to establish themselves as invasive species (Bradley et al. 2010; Bellard et al. 2013). In contrast, the change in the environmental factors can convert or induce the inhabitant non-native species into invaders and could also facilitate an increase in the frequency, abundance, and density of secondary invaders by reducing the competitive aptitude of primary invaders (Richardson et al. 2000; Pearson et al. 2016). The pool of native or non-native species may give rise to secondary invaders (Kuebbing et al. 2013).

Table 4.1 Some invasive species of different global regions and their respective impacts on different ecosystems

Sl. no.	Invasive species	Native	Region/s	Environmental impact/s	Reference/s
1	*Acacia longifolia*	South-eastern Australia and Tasmania	Portugal	Speedy growth and dissemination causing loss of biodiversity in the invasive areas	Goncalves et al. (2021)
2	*Ageratina adenophora*	Mexico Central America	Central Himalaya	Degrading the understory diversity of Chir pine forests	Kumar and Garkoti (2021)
3	*Bromus inermis*	Eurasia	North America	Disrupting grassland communities, especially herbivores	Rosenkranz and McGonigle (2022), Pei et al. (2023)
4	*Chromolaena odorata*	Tropical America	Asia, Austria, West Africa	Produces secondary metabolites and destruct the structure of soil microbial communities	Kato-Noguchi and Kato (2023)
5	*Dolichandra unguis-cati*	America	Himalayan region	Eroding genetic resources of plants	Rawat (2022)
6	*Eichhornia crassipes*	South America	More than 50 countries	Damaging freshwater biodiversity and ecological structure	May et al. (2020)
7	*Eucalyptus* sp.	Australia	Brazil	Biological invasion; host of many insect pests	Mota et al. (2022)
8	*Lantana camara*	South and Central America	India	Reduction of livestock forage, natural resources, and obstruction in the movement of animals	Tiwari et al. (2022)
9	*Lythrum salicaria*	Eurasia	United States	Destruction of wetland biodiversity	Chaudhuri and Mishra (2023)
10	*Melaleuca quinquenervia*	Australia	Florida	Chemical exudates from the leaf litter inhibit the growth of other species around	Lu et al. (2022)
11	*Mikania micrantha*	America	China	Change the structure of soil	Zhao et al. (2023)

(continued)

Table 4.1 (continued)

Sl. no.	Invasive species	Native	Region/s	Environmental impact/s	Reference/s
				microbial communities	
12	*Mimosa pigra* L.	Tropical America	Africa, Asia, Australia	Reduction in native resources, pastoral grazing, ecotourism, etc.	Thi et al. (2022)
13	*Parthenium hysterophorus*	Tropical America	India	Destruction of forest and agricultural areas	Ahmad et al. (2019)
14	*Phragmites australis*	Europe	United States	Reduction of flora and fauna biodiversity of wetlands	Bonello and Judd (2019) Pyšek et al. (2020)
15	*Prosopis juliflora*	Mexico, South America, Caribbean	India, USA	Reduce productivity and diversity of arid and semiarid regions	Kaur et al. (2012)
16	*Poa pratensis*	Europe, North Asia	North America	Disrupting grassland communities, especially herbivores	Pei et al. (2023)
17	*Reynoutria japonica*	Eastern Asia	North America	Huge genetic diversity makes the population very aggressive	VanWallendael et al. (2021)
18	*Saccharum spontaneum*	Indian Subcontinent	Republic of Panama	High ploidy and aggressive expansion causing reduction in biodiversity in the invaded regions	Saltonstall et al. (2021)
19	*Salvinia molesta*	Brazil	Africa	Destruction of aquatic communities	Coetzee and Hill (2020)
20	*Senna spectabilis*	Tropical America	India	High dissemination; destructing the tree forest of Western Ghats	Anoop et al. (2021)

4.4.1 Drifting of Invasive Species

The potential invaders are capable of escaping from many environmental filters or selection hurdles during the process. However, they have to overcome the

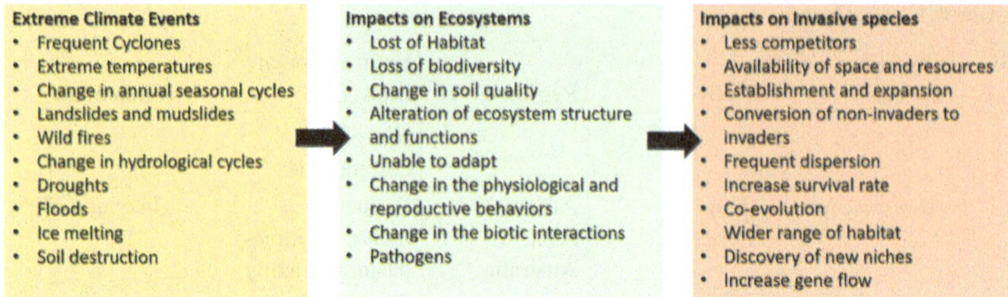

Fig. 4.2 Impacts of extreme climate events on ecosystem and invasion

geographical barriers during the initial stage which is mostly facilitated by human activities and other environmental factors. Many potential invaders are transported through cargo ships during goods and commodities transportation. Therefore, in countries like USA, there are proper regulations for cargo ships against government-listed species (Lehan et al. 2013). With global warming, the melting of ice reduced the period of ice cover on the sea surface making it faster for transportation through ships. This increases the survival rate of propagules enhancing the coincidental establishment into new geographical regions (Pyke et al. 2008). Depletion of ice packs in the sea also increases the frequency of migration as well as the movement of many different marine species (migratory birds, marine mammals, etc.). This could also enhance the expansion and establishment of many species in a wide range of habitats (McKeon et al. 2016; Viana et al. 2016). The short-term, as well as long-term dispersal of many potential invasive species, is also boosted by the frequently occurring high-intensity extreme weather events like cyclones, hurricanes, floods, storms, etc. through seeds, pollens, vectors, insects, larvae, and any other propagules that could further be colonized (Schneider et al. 2005; Walther et al. 2009). Kerala is one Indian state facing frequent landslides and floods in recent years. It is fighting alien species brought through water bodies, such as *Lantana camara*, *Mimosa diplotricha*, *Mikania micrantha,* and *Chromolaena odorata* (Baboo 2020). According to the data released by the National Biodiversity Authority, Ministry of Environment, Forests and Climate change, Government of India (2018), the cumulative number of invasive alien species in different ecosystems has ben reported to be 173 species with 54 species in terrestrial plants, 56 species in the aquatic ecosystem, 47 in agriculture ecosystem, and 14 in an island ecosystem.

Another important factor in the introduction of invasive species is the importing of ornamental and cultivating plants for many human purposes. Although many such purposefully introduced plants do not become invasive, some of them could become primary invaders. Such risk of anthropogenically introducing plants can be controlled by making laws and regulation policies. One of the best examples of becoming invasive from an ornamental plant is the current situation of *Lantana camara* in India, which was introduced in the 1800s as an ornamental plant and now spreading over the entire Indian ecosystem especially taking up 44% of the Indian forest area (Mungi et al. 2020). Another example is *Eichhornia crassipes* (water hyacinth), native to ecosystem of Amazonia (Brazil) that has spread as an invasive species throughout the globe in the past few decades. In India, it was a gift of the British to India during the nineteenth century, and since then, it is aggressively affecting many water bodies in India. The futility of this species can be imagined once 7000 Indian Army personnel had been deployed to clean water hyacinths from Ulsooru lake in Bangalore (Gopal 2018). Generally, plants that are selected for introduction as ornamental and cultivational drives have a broad tolerance range as they could be established and adapted easily. Moreover, with climate change, there is also a high demand for such broad range of plants that could survive in extreme conditions (Bradley et al. 2010, 2012). With extreme polarization of annual precipitation and amplification of extreme hydrological cycles, we have to switch our economical crops that can overcome extreme environmental conditions like floods, droughts, and high salinity to maintain our demand-supply balance. Consequently, in these current circumstances of global climate change, it is required to introduce new varieties of plants for fulfilling productivity from time to time. And if native species have failed to compete and migrate, they will lose the battle of competition to the newly invading species. Therefore, we could carefully design the control measures on such compulsively introduced plants to control future invasions. Predicting and analyzing the potential invaders requires lots of experiments and distribution models. In the past few decades, publication rate of research articles on invasive species has suddenly accelerated. However, most of the case studies were performed in developed countries like USA (Smith et al. 2012), but developing countries like India still have to go miles in this endeavor.

4.4.2 Alteration in Species Distribution During the Invasion

Invasive species have a higher ability to establish and reproduce in different environmental conditions (Broennimann and Guisan 2008). Most of the invasive species have broad tolerance ranges, high growth rates, and dispersal rates with a shorter generation time, that collectively enable them to challenge many extreme conditions of biotic and abiotic pressures (Finch et al. 2021). Invasion of species certainly reduces the local as well as regional biodiversity which is further facilitated by human activities. During invasion, there are lots of changes in the ecological functions, increase in interspecific competition, diversification of trophic niche and co-existence of native and non-native species leading to co-evolutionary changes (Wang et al. 2021). Success in the invasion disturbs the community function, composition, and structure either directly or indirectly, which facilitates further invasion of other species (Leinaas et al. 2015). Decline in the richness of species during invasion may also be due to either displacement of native species or reduction of the population with space and resource limitations.

In Antarctic polar deserts, there are great expanses of dry and saline soils, but due to the melting of ice and glaciers during peak summer, there is formation of transient wetlands and streams increasing the soil water availability. The flow of water changes the structure and function of communities as well as the soil properties, leaching of the salts reduces the osmotic stress of the soil making the soil available to many species and facilitating their colonization and establishment (Nielsen et al. 2012). Avian re-assembly is also directly influenced by extreme events like storms with the shifting of habitat preferences, foods, and flocking behaviors ultimately changing the species richness and abundance of their communities. Such impacts are highly vulnerable to those at higher trophic levels; however, it becomes beneficial for habitat generalist species like bulbuls (Zhang et al. 2016). A similar observation was also seen in the frog community in Costa Rica after events like El Niño Southern Oscillation. However, such changes in the community composition and structure could be recovered in some species like frogs, if, there is no additional stress of disease or loss of habitat (Ryan et al. 2015; Harris et al. 2020). Similarly, on the rocky shores of marine ecosystems, frequent storms favor the restructuring of the biological communities by generating bare patches that can be easily colonized by many invasive species (Paine and Levin 1981; Sanz-Lázaro et al. 2022). However, such colonization favours the abundance of certain early colonizers as compared to the species frequency as well as to the late colonizers (Sanz-Lázaro et al. 2022).

When the impact of invasion interwinds with human activities, there are irreversible changes in the structure and function of the ecological communities. In many parts of grassland in North America, the native plant communities have been replaced by exotic grasses such as *Poa pratensis* and *Bromus inermis* causing historical and landscape disturbances. Changes in the litter depth and type, area covered by grass, bare patches, etc. outspread their impact not only on the native plant diversity but also on the species dependent on the existing communities (Pei et al. 2023). Another case study in Kanha Tiger Reserve, a tropical seasonally dry forest in the Central part of India showed the compositional alteration in the native plant communities in terms of richness, abundance, and soil quality due to invasion by *Lantana camara* and *Pogostemon benghalensis* (Rastogi et al. 2023). In India, lantana has occupied more than 3 lakh km^2 of forest and is predicted to be expanded further with global climate change (Mungi et al. 2020; Rastogi et al. 2023). In addition, some invasive species evolve an efficient defense mechanism against their natural competitors by producing secondary metabolites changing the environmental conditions of their rhizosphere. *Chromolaena odorata* produces toxic substances from the roots inhibiting germination and growth of other plants and increasing the mortality of many soil microorganisms (Kato-Noguchi and Kato 2023).

4.5 Evolutionary Changes During Invasion

Invasion by non-native species acts as a strong selection pressure for the native species through competition (Leger and Espeland 2010). The native species that are successful in the competition continue to survive and enter into the lanes of co-evolution with the invaders. Such existence of co-evolution helps us to understand how ecological communities maintain their resistance and resilience against changes in the environment. With the altering environment both native and non-native species need to face selections pressures, but the non-native species which are already promoted in the competition, and already escaped from many innate competitors are likely to be superior in expansion and establishment as they are also evolving through the interaction, restriction, and competition (Blossey and Nötzold 1995; Finch et al. 2021). The flexible conduct of invaders in terms of resources and co-existence might also be helping in their invasion. During the invasion, many evolutionary pathways of the native species can be altered in terms of predation, displacement, hybridization, genetic drifting, introgression, and eventually extinction (Mooney and Cleland 2001). On the other hand, during the invasion, species experience a harsh bottleneck losing their genetic diversity drastically, however in some rare cases, successful invaders increase diversity at the introduced niches. This might be due to the introduction of invaders from different ecological sources (Genton et al. 2005; Dlugosch and Parker 2008). Reduction of genetic diversity was found during *Mycoplasma* epidemic, attacking the populations of *Carpodacus mexicanus* (Wang et al. 2003; Hawley et al. 2006). Therefore, in some species, introduction from multiple sources and increased gene flow might be helping in the evolution of traits that facilitate the colonization in non-native habitats or new geographical areas (Dlugosch and Parker 2008).

Not all the introduced plants become invaders, some species fail to establish and expand into new environments while others play at the forefront. Polyploidy contributes to higher survival and superior traits to the species during establishment and expansion into new environments, therefore, polyploidy might be contributing to the invasiveness of many grass species (Saltonstall et al. 2021). Change in the ploidy level during invasion also increases the genetic diversity of the invading population. Such conditions of increased ploidy level during invasion can be observed in central Panama which is dominated by *Saccharum spontaneum* which might accidentally escape from cultivated species of sugarcane (Saltonstall et al. 2021). Genome size and ploidy level directly linked with the aggressive invasion are not collective but might be indirectly influencing the intraspecific competition (Pyšek et al. 2020). In the invasive populations of *Spartina alterniflora* along the coastal areas of China, hybridization and intermixing of genetic compositions among the isolated populations generated new superior traits and facilitated their invasion (Xia et al. 2020). Environmental selection of genotypes helping in the growth and reproduction of species under a broad range of climate conditions allows them to be invaders into the diverse environment and the adaptation of phenotypic variation among the population in response to climate change is also determined by various evolutionary and ecological history of the species (Monty et al. 2013; Finch et al. 2021).

4.6 Adaptive Changes of Invasive Species

Many potential invasive species emerge as populations with rapid adaptive dimensions. They keep changing their behavior with time when they arrive in a new environment. From the evolutionary point of view, natural selection and genetic drift are the two important evolutionary processes that happen at an ecologically fitting time scale (Carroll et al. 2007) driving new changes for subsequent generations (Kilkenny and Galloway 2013). These evolutionary changes might have undergone up to 20 generations or less for the adaptation to a new environmental condition (Prentis et al. 2008). When environmental fluctuations happen frequently, many directional evolutionary shifts come across each other maintaining relative stability in the overall characteristics of the population or species (Grant and Grant 2002). Nonetheless, if the environmental fluctuation is happening on a longer time scale, there is a prolonged directional evolutionary change and an increase in the variability within the population or species (Carroll et al. 2007). According to the Darwinian concept of natural selection, evolution is a very slow and regular process over time, therefore, there is a huge time constraint on the invasive species during the gradual process of adaption and establishment in the new ranges of environment. Additionally, there is a huge reduction in the genetic diversity and population due to founder effect limiting the local adaptation of the species. But such limitations are also overcome by many invasive species that show self-incompatibility (Oduor et al. 2016). However, the underlying changes for such evolutionary processes at the genetic and epigenetic levels have not been fully characterized so far. Therefore, it is necessary to understand the sources of variations among the population that could contribute to the adaptive evolution during invasion along with the changing climatic events. In the current scenario, the two main drivers of global environmental changes are climate change and the rising number of invasive species which will undoubtedly impact the eco-evolutionary process in the long run (Finch et al. 2021).

4.7 Management of Invasive Species

Due to significant socioeconomic and environmental impacts, many countries are concerned and adapting the regulating practices for the introduction of non-native species either commercially or personally to control future possible invasions. Many often make "black lists" and "white lists" of many species for introduction into new regions or countries (Genovesi et al. 2015). Black list species are those which have been already identified as destructive while white list species are safe and can be introduced. The development of such lists usually requires detailed analyses and studies regarding their potential invasiveness or risk assessment with much biological and economic evidence (Roy et al. 2017). The European Union (EU) implemented Regulation 1143/2014 on invasive species on 1st January 2015 by listing many species on the "List of Invasive Alien Species of Union concern" and this has been revised periodically to classify species with strong restrictions during possession, exporting/importing, and selling (FAO 2014). In 2021, the list comprised 36 plant species and 30 animal species, including many aquatic invasive species which are extremely unsafe for river water bodies (https://ec.europa.eu/environment/nature/invasivealien). Similarly, in the USA, there are proper regulations and screening of cargo ships against many listed species by the government (Lehan et al. 2013). In India, the Plant Quarantine Order 2003 under Indian Council of Agricultural Research (ICAR), regulate import into the country addressing the threat of invasive species. The Indian government has adopted biological control as a priority and implemented many national policies supporting research institutes like ICAR and NBAIR (National Bureau of Agricultural Insect Resources).

The most effective management measure is the prevention of invasions beforehand as the risk assessment and eradication practices are very costly and time-consuming. Assessment of possible risk and potential invasion helps in making policies and regulatory guidelines. However, once the establishment of invasive species has started, it is very difficult to eradicate. Prevention requires proper legislative regulation or prohibition of unsafe species from importation, border controls, limited transportation, etc. Initially, such preventive measures seem expensive and challenging, but these will avoid the cost for future management for eradication and ecological destruction. Along with above-mentioned preventive actions, there should be a regular examination of potentially invasive species, cleaning of susceptible sites, and arrangement of mandatory educational campaigns (Rothlisberger et al. 2010). The direct or indirect cost of the management of invasive species is very high, so it's time to favor more practical approaches for the management and prevention in light of the challenges of climate change.

4.8 Challenges and Opportunities

Despite the awareness about the huge impacts of anthropogenic climate change on the natural and human systems, the adaptation of standard and systematic scientific methods for the evaluation and detection is a great challenge as it entails many interdisciplinary designs, concepts, and methods. One of the most challenging shortcomings is the limitation of knowledge on the understanding of the underlying mechanism of ecological changes as well as the lack of long-term ecological observations (Stone et al. 2013). Therefore, future research efforts are necessary to understand the driving forces of ecological and evolutionary changes. Extensive investigations on extreme environmental events and climate change have been carried out in the past decades, but recently the focus has been shifted to the impact of such events on the socioecological systems (Stone et al. 2013; Abbas et al. 2022). The major challenge is that the driving factors of all these are so massive and the data is very complex which requires a broad interdisciplinary approach to draw a valid conclusion from the collected information. Furthermore, the control and regulation at the national or centralized level require timely implementation of policies and action plans which will further require more realistic ecological, economic, and sociological parameter assumptions and models to set the socioecological thresholds for the risk assessment.

4.9 Conclusions and Future Recommendations

Global climate change is the real threat causing many extreme climate events. Such climate events are favoring many fast-growing species enhancing the invasion, and inhibiting many slow-growing native species. The species skilled to survive under a wide range of environmental conditions are more likely to become invasive. Management of invasive species is essential to control and help the native species while competing with them. Prior identification of the possible invading species before introducing them outside their natural habitat is very compulsive. To mitigate the challenges of controlling invasive species, a collaborative works of ecologists, policymakers, economists, and many scientists is required to predict and set models as the natural ecosystem becomes more vulnerable to such non-native species together with global climate change.

References

Abbas A, Ekowati D, Suhariadi F, Fenitra RM (2022) Health implications, leaders societies, and climate change: a global review. In: Chatterjee U, Akanwa AO, Kumar S, Singh SK, Dutta Roy A (eds) Ecological footprints of climate change, Springer climate. Springer, Cham. https://doi.org/10.1007/978-3-031-15501-7_26

Ahmad R, Khuroo AA, Hamid M et al (2019) Predicting invasion potential and niche dynamics of *Parthenium hysterophorus* (Congress grass) in India under projected climate change. Biodivers Conserv 28:2319–2344. https://doi.org/10.1007/s10531-019-01775-y

Anoop NR, Sen S, Vinayan PA, Ganesh T (2021) Native mammals disperse the highly invasive *Senna spectabilis* in the Western Ghats, India. Biotropica 54:1310–1314. https://doi.org/10.1111/btp.12996

Baboo SK (2020) Native insect pollinators visiting flowers of selected invasive plant species in three districts of Kerala. Ph.D. thesis. Department of Zoology, University of Calicut. http://scholar.uoc.ac.in/bitstream/handle/20.500.12818/164/1274.pdf?sequence=1

Bellard C, Thuiller W, Leroy B, Genovesi P, Bakkenes M, Courchamp F (2013) Will climate change promote future invasions? Glob Chang Biol 19:3740–3748. https://doi.org/10.1111/gcb.12344

Blossey B, Nötzold R (1995) Evolution of increased competitive ability in invasive nonindigenous plants: a hypothesis. J Ecol 83:887–889. https://doi.org/10.2307/2261425

Bonello JE, Judd KE (2019) Plant community recovery after herbicide management to remove *Phragmites australis* in Great Lakes coastal wetlands. Restor Ecol 28:215–221. https://doi.org/10.1111/rec.13062

Bortolini-Rosales JL, Reyes-Aldana HE (2023) Invasive alien species of invertebrates and fishes introduced into Mexican freshwater habitats. In: Jones RW, Ornelas-García CP, Pineda-López R, Álvarez F (eds) Mexican fauna in the anthropocene. Springer, Cham. https://doi.org/10.1007/978-3-031-17277-9_21

Bradley BA, Blumenthal DM, Wilcove DS, Ziska LH (2010) Predicting plant invasions in an era of global change. Trends Ecol Evol 25:310–318. https://doi.org/10.1016/j.tree.2009.12.003

Bradley BA, Bluementhal DM, Early R, Grosholz ED, Lawler JJ et al (2012) Global change, global trade, and the next wave of plant invasions. Front Ecol Environ 10:20–28. https://doi.org/10.1890/110145

Broennimann O, Guisan A (2008) Predicting current and future biological invasions: both native and invaded ranges matter. Biology Lett 4(5):585–589

Bubb IE, Williams ZB (2022) Spatial and temporal patterns of fire on Saipan, CNMI. Pac Sci 76:1–15. https://doi.org/10.2984/76.1.1
Carroll SP, Hendry AP, Reznick DN, Fox CW (2007) Evolution on ecological time-scales. Funct Ecol 21(3):387–393. https://doi.org/10.1111/j.1365-2435.2007.01289.x
Chaudhuri G, Mishra NB (2023) Detection of aquatic invasive plants in wetlands of the upper Mississippi river from UAV imagery using transfer learning. Remote Sens 15(3):734. https://doi.org/10.3390/rs15030734
Coetzee JA, Hill MP (2020) *Salvinia molesta* D. Mitch. (Salviniaceae): impact and control. CABI Rev. https://doi.org/10.1079/PAVSNNR202015033
Cordero S, Gálvez F, Fontúrbel FE (2023) Ecological impacts of exotic species on native seed dispersal systems: a systematic review. Plants 12(2):261. https://doi.org/10.3390/plants12020261
Corli A, Walter J, Sheppard CS (2021) Invasion success of *Bunias orientalis* (Warty Cabbage) in grasslands: a mesocosm experiment on the role of hydrological stress and disturbance. Front Ecol Evol 9:625587. https://doi.org/10.3389/fevo.2021.625587
Dendy J, Mesubed D, Colin PL, Giardina CP, Cordell S et al (2022) Dynamics of anthropogenic wildfire on Babeldaob Island (Palau) as revealed by fire history. Fire 5(2):45. https://doi.org/10.3390/fire5020045
Dlugosch KM, Parker IM (2008) Founding events in species invasions: genetic variation, adaptive evolution, and the role of multiple introductions. Mol Ecol 17(1):431–449. https://doi.org/10.1111/j.1365-294X.2007.03538.x
Donat MG, Alexander LV, Yang H, Durre I, Vose R et al (2013) Updated analyses of temperature and precipitation extreme indices since the beginning of the twentieth century: the HadEX2 dataset. J Geophys Res Atmos 118(5):2098–2118. https://doi.org/10.1002/jgrd.50150
Dukes JS, Chiariello NR, Loarie SR, Field CB (2011) Strong response of an invasive plant species (*Centaurea solstitialis* L.) to global environmental changes. Ecol Appl 21:18. https://doi.org/10.1890/11-0111.1
Faccenda K, Daehler CC (2022) A screening system to predict wildfire risk of invasive plants. Biol Invasions 24:575–589. https://doi.org/10.1007/s10530-021-02661-x
FAO (2014) Regulation (EU) No. 1143/2014 of the European Parliament and of the Council on the prevention and management of the introduction and spread of invasive alien species. https://www.ecolex.org/
FAO and UNEP (2020) The state of the world's forests 2020: forests, biodiversity and people. FAO and UNEP, Rome. https://www.fao.org/3/ca8642en/ca8642en.pdf
Field CB, Raupach MR (2004) The global carbon cycle: integrating humans, climate, and the natural world. Island Press, Washington. https://islandpress.org/books/global-carbon-cycle
Finch DM, Butler JK, Runyon JB, Fettig CJ, Kilkenny FF et al (2021) Effects of climate change on invasive species. In: Poland TM, Patel-Weynand T, Finch DM, Miniat CF, Hayes DC, Lopez VM (eds) Invasive species in forests and rangelands of the United States. Springer, Cham. https://doi.org/10.1007/978-3-030-45367-1_4
Genovesi P, Carboneras C, Vilà M, Walton P (2015) EU adopts innovative legislation on invasive species: a step towards a global response to biological invasions? Biol Invasions 17:1307–1311. https://doi.org/10.1007/s10530-014-0817-8
Genton BJ, Shykoff JA, Giraud T (2005) High genetic diversity in French invasive populations of common ragweed, *Ambrosia artemisiifolia*, as a result of multiple sources of introduction. Mol Ecol 14:4275–4285. https://doi.org/10.1111/j.1365-294X.2005.02750.x
Goncalves C, Santana P, Brandao T, Guedes M (2021) Automatic detection of *Acacia longifolia* invasive species based on UAV-acquired aerial imagery. Inf Process Agric 9:276–287. https://doi.org/10.1016/j.inpa.2021.04.007
Gopal S (2018) Researchers innovate to make money out of water hyacinth. Mongabay. 23 Jan 2018. https://india.mongabay.com
Grant PR, Grant BR (2002) Unpredictable evolution in a 30-year study of Darwin's finches. Science 296:707–711. https://doi.org/10.1126/science.1070315

Harris RMB, Loeffler F, Rumm A, Fischer C, Horchler P et al (2020) Biological responses to extreme weather events are detectable but difficult to formally attribute to anthropogenic climate change. Sci Rep 10:14067. https://doi.org/10.1038/s41598-020-70901-6
Hawley DM, Hanley D, Dhondt AA, Lovette IJ (2006) Molecular evidence for a founder effect in invasive house finch (*Carpodacus mexicanus*) populations experiencing an emergent disease epidemic. Mol Ecol 15:263–275. https://doi.org/10.1111/j.1365-294X.2005.02767.x
Hellmann JJ, Byers JE, Bierwagen BG, Dukes JS (2008) Five potential consequences of climate change for invasive species. Conserv Biol 22:534–543. https://doi.org/10.1111/j.1523-1739.2008.00951.x
Ibarra-Isassi J, Handa IT, Lessard JP (2022) Community-wide trait adaptation, but not plasticity, explains ant community structure in extreme environments. Funct Ecol 37:139–149. https://doi.org/10.1111/1365-2435.14185
IPCC (2007) Climate change 2007: synthesis report. http://www.ipcc.ch/publications_and_data/ar4/syr/en/contents.html
IPCC (2012) Summary for policymakers. In: Field CB et al (eds) Managing the risks of extreme events and disasters to advance climate change adaptation, A special report of working groups I and II of the Inter-governmental Panel on Climate Change. Cambridge University Press, Cambridge, pp 3–21
IPCC (2022) Climate change 2022: impacts, adaptation and vulnerability. https://www.ipcc.ch/report/ar6/wg2/
Jarnevich CS, Esaias WE, Ma PLA, Morisette JT, Nickeson JE et al (2014) Regional distribution models with lack of proximate predictors: Africanized honeybees expanding north. Divers Distrib 20(2):193–201. https://doi.org/10.1111/ddi.12143
Jeger MJ (2021) The impact of climate change on disease in wild plant populations and communities. Plant Pathol 71:111–130. https://doi.org/10.1111/ppa.13434
Kato-Noguchi H, Kato M (2023) Evolution of the secondary metabolites in invasive plant species *Chromolaena odorata* for the defense and allelopathic function. Plants 12(3):521. https://doi.org/10.3390/plants12030521
Kaur R, Gonzales WL, Llambi LD, Soriana PJ et al (2012) Community impacts of Prosopis juliflora invasion: biogeographic and congeneric comparisons. PLoS One 7:e44966. https://doi.org/10.1371/journal.pone.0044966
Kilkenny FF, Galloway LF (2013) Adaptive divergence at the margin of an invaded range. Evolution 67:722–731. https://doi.org/10.1111/j.1558-5646.2012.01829.x
Kuebbing SE, Nuñez MA, Simberloff D (2013) Current mismatch between research and conservation efforts: the need to study cooccurring invasive plant species. Environ Sci Biol Conserv 160: 121–129. https://doi.org/10.1016/J.BIOCON.2013.01.009
Kumar M, Garkoti SC (2021) Allelopathy effects of invasive alien *Ageratina adenophora* on native shrub species of chir pine forest in the central Himalaya, India. J For Res 27:53–62. https://doi.org/10.1080/13416979.2021.2002505
Kundzewicz ZW (2008) Climate change impacts on the hydrological cycle. Ecohydrol Hydrobiol 8(2–4):195–203. https://doi.org/10.2478/v10104-009-0015-y
Le Houerou HN (1996) Climate change, drought and desertification. J Arid Environ 34:133–185. https://doi.org/10.1006/jare.1996.0099
Leger EA, Espeland EK (2010) Coevolution between native and invasive plant competitors: implications for invasive species management. Evol Appl 3:169–178. https://doi.org/10.1111/j.1752-4571.2009.00105.x
Lehan NE, Murphy JR, Thorburn LP, Bradley BA (2013) Accidental introductions are important source of invasive plants in the continental United States. Am J Bot 100:1287–1293. https://doi.org/10.3732/ajb.1300061
Leinaas HP, Bengtsson J, Janion-Scheepers C, Chown SL (2015) Indirect effects of habitat disturbance on invasion: nutritious litter from a grazing resistant plant favors alien over native Collembola. Ecol Evol 5(16):3462–3471. https://doi.org/10.1002/ece3.1483

Lu Y, DeAngelis DL, Xia J, Jiang J (2022) Modeling the impact of invasive species litter on conditions affecting its spread and potential regime shift. Ecol Model 468:109962. https://doi.org/10.1016/j.ecolmodel.2022.109962

Mainali KP, Warren DL, Dhileepan K, McConnachie A, Strathie L et al (2015) Projecting future expansion of invasive species: comparing and improving methodologies for species distribution modeling. Glob Chang Biol 21:4464–4480. https://doi.org/10.1111/gcb.13038

Mansoor S, Farooq I, Kachroo MM, Mahmoud AED, Fawzy M et al (2022) Elevation in wildfire frequencies with respect to the climate change. J Environ Manag 301:113769. https://doi.org/10.1016/j.jenvman.2021.113769

May L, Dobel AJ, Ongore C (2020) Controlling water hyacinth (*Eichhornia crassipes* (Mart.) Solms): a proposed framework for preventative management. Inland Waters 12:163–172. https://doi.org/10.1080/20442041.2021.1965444

McKeon CS, Weber MX, Alter SE, Seavy NE, Crandall ED et al (2016) Melting barriers to faunal exchange across ocean barriers. Glob Chang Biol 22:465–473. https://doi.org/10.1111/gcb.13116

Mondal M, Biswas A, Haldar S, Mandal S, Bhattacharya S, Paul S (2022) Spatio-temporal behaviours of tropical cyclones over the Bay of Bengal Basin in last five decades. Trop Cyclone Res Rev 11(1):1–15. https://doi.org/10.1016/j.tcrr.2021.11.004

Monty A, Bizoux JP, Escarré J, Mahy G (2013) Rapid plant invasion in distinct climates involves different sources of phenotypic variation. PLoS One 8(1):e55627. https://doi.org/10.1371/journal.pone.0055627

Mooney HA, Cleland EE (2001) The evolutionary impact of invasive species. PNAS 98(10): 5446–5451. https://doi.org/10.1073/pnas.091093398

Mota JDS, Barbosa LR, Marchioro CA (2022) Suitable areas for invasive insect pests in Brazil and the potential impacts for eucalyptus forestry. Pest Manag Sci 78:2596–2606. https://doi.org/10.1002/ps.6891

Mungi NA, Qureshi Q, Jhala YV (2020) Expanding niche and degrading forests: Key to the successful global invasion of *Lantana camara* (sensu lato). Glob Ecol Conserv 23(1):e01080. https://doi.org/10.1016/j.gecco.2020.e01080

Nielsen UN, Wall DH, Adams BJ, Virginia RA, Ball BA et al (2012) The ecology of pulse events: insights from an extreme climatic event in a polar desert ecosystem. Ecosphere 3(2):1–15. https://doi.org/10.1890/ES11-00325.1

Oduor AMO, Leimu R, Kleunen MV (2016) Invasive plant species are locally adapted just as frequently and at least as strongly as native plant species. J Ecol 104(4):957–968. https://doi.org/10.1111/1365-2745.12578

Paine RT, Levin SA (1981) Intertidal landscapes: disturbance and the dynamics of pattern. Ecol Monogr 51:145–178. https://doi.org/10.2307/2937261

Parkash S (2023) Lessons learned from landslides of socio-economic and environmental significance in India. In: Progress in landslide research and technology, vol 1(2). Springer, Cham. https://doi.org/10.1007/978-3-031-18471-0_23

Patri P, Sharma P, Patra SK (2022) A multidimensional model for cyclone vulnerability assessment of urban slum dwellers in India: a case study of Bhubaneswar city. Int J Disaster Risk Reduct 83: 103439. https://doi.org/10.1016/j.ijdrr.2022.103439

Pearman PB, Guisan A, Broennimann O, Randin CF (2007) Niche dynamics in space and time. Trends Ecol Evol 23:149–158. https://doi.org/10.1016/j.tree.2007.11.005

Pearson DE, Ortega YK, Runyon JB, Butler JL (2016) Secondary invasion: the bane of weed management. Biol Conserv 197:8–17. https://doi.org/10.1016/j.biocon.2016.02.029

Pei CK, Hovick TJ, Limb RF, Harmon JP, Geaumont BA (2023) Invasive grass and litter accumulation constrain bee and plant diversity in altered grasslands. Glob Ecol Conserv 4:e02352. https://doi.org/10.1016/j.gecco.2022.e02352

Poland TM, Patel-Weynand T, Finch DM, Miniat CF, Hayes DC et al (2021) Invasive species in forests and rangelands of the United States. Springer International Publishing. https://doi.org/10.1007/978-3-030-45367-1

Prentis PJ, Wilson JR, Dormontt EE, Richardson DM, Lowe AJ (2008) Adaptive evolution in invasive species. Trends Plant Sci 13(6):288–294. https://doi.org/10.1016/j.tplants.2008.03.004
Pyke CR, Thomas R, Porter RD, Hellmann JJ, Dukes JS et al (2008) Current practices and future opportunities for policy on climate change and invasive species. Conserv Biol 22:585–592. https://doi.org/10.1111/j.1523-1739.2008.00956.x
Pyšek P, Čuda J, Šmilauer P, Skálová H, Chumová Z et al (2020) Competition among native and invasive *Phragmites australis* populations: An experimental test of the effects of invasion status, genome size, and ploidy level. Ecol Evol 10(3):1106–1118. https://doi.org/10.1002/ece3.5907
Rafferty ZP (2021) Invasive species, Encyclopedia Britannica. https://www.britannica.com/science/invasive-species
Rastogi R, Qureshi Q, Shrivastava A, Jhala YV (2023) Multiple invasions exert combined magnified effects on native plants, soil nutrients and alters the plant-herbivore interaction in dry tropical forest. For Ecol Manag 531:120781. https://doi.org/10.1016/j.foreco.2023.120781
Rawat S (2022) Negative impact of new invasive species *Dolichandra unguis-cati* on the flora in Balh valley of Mandi district, India. Ecol Environ Conserv:S197–S203. https://doi.org/10.53550/EEC.2022.v28i04s.029
Richardson DM, Pysek P, Rejmanek M et al (2000) Naturalization and invasion of invasive plants: concepts and definitions. Divers Distrib 6:93–107. https://doi.org/10.1046/j.1472-4642.2000.00083.x
Rosenkranz VE, McGonigle TP (2022) Reductions in foliar Hemiptera in portions of a fescue grassland invaded by Smooth Brome (*Bromus inermis*). Can Field-Naturalist 134(2):156–160. https://doi.org/10.22621/cfn.v134i2.2423
Rothlisberger JD, Chadderton WL, McNulty J, Lodge DM (2010) Aquatic invasive species transport via trailered boats: what is being moved, who is moving it, and what can be done. Fisheries 35(3):121–132. https://doi.org/10.1577/1548-8446-35.3.121
Roy HE, Rabitsch W, Scalera R, Stewart A, Gallardo B et al (2017) Developing a framework of minimum standards for the risk assessment of alien species. J Appl Ecol 55(2):526–538. https://doi.org/10.1111/1365-2664.13025
Ryan MJ, Scott NJ, Cook JA, Willink B, Chaves G et al (2015) Too wet for frogs: changes in a tropical leaf litter community coincide with La Nina. Ecosphere 6:5. https://doi.org/10.1890/ES14-00352.1
Saltonstall K, Bonnett GD, Aitken KS (2021) A perfect storm: ploidy and preadaptation facilitate *Saccharum spontaneum* escape and invasion in the Republic of Panama. Biol Invasions 23:1101–1115. https://doi.org/10.1007/s10530-020-02421-3
Sanz-Lázaro C (2016) Climate extremes can drive biological assemblages to early successional stages compared to several mild disturbances. Sci Rep 6:30607. https://doi.org/10.1038/srep30607
Sanz-Lázaro C, Casado-Coy N, Navarro-Ortín A, Terradas-Fernández M (2022) Anthropogenic pressures enhance the deleterious effects of extreme storms on rocky shore communities. Sci Environ 817:152917. https://doi.org/10.1016/j.scitotenv.2022.152917
Schneider RW, Hollier CA, Whitam HK, Palm ME et al (2005) First report of soybean rust caused by *Phakopsora pachyrhizi* in the continental United States. Plant Dis 89:774. https://doi.org/10.1094/PD-89-0774A
Siddique T, Haris PH, Pradhan SP (2022) Unraveling the geological and meteorological interplay during the 2021 Chamoli disaster, India. Nat Hazard Res 2:75–83. https://doi.org/10.1016/j.nhres.2022.04.003
Smith AL, Hewitt N, Klenk N et al (2012) Effects of climate change on the distribution of invasive alien species in Canada: a knowledge synthesis of range change projections in a warming world. Environ Rev 20:1–16. https://doi.org/10.1139/a11-020
Smith CI, Sweet LC, McKain MR, Heyduk K, Cameron B (2022) Dust storms ahead: Climate change, green energy development and endangered species in the Mojave Desert. Biol Conserv 277:109819. https://doi.org/10.1016/j.biocon.2022.109819

Stone D, Auffhammer M, Carey M et al (2013) The challenge to detect and attribute effects of climate change on human and natural systems. Clim Chang 121:381–395. https://doi.org/10.1007/s10584-013-0873-6

Thi NPA, Khang DT, Khoa NT, Le VM, Do HDK (2022) The complete chloroplast genome of *Mimosa pigra* L. (Fabaceae), a notorious invasive plant. Mitochondrial DNA B Resour 7:1063–1065. https://doi.org/10.1080/23802359.2022.2086076

Tiwari S, Mishra SN, Kumar D et al (2022) Modelling the potential risk zone of *Lantana camara* invasion and response to climate change in eastern India. Ecol Process 11:10. https://doi.org/10.1186/s13717-021-00354-w

Ummenhofer CC, Meehl Gerald A (2017) Extreme weather and climate events with ecological relevance: a review. Philos Trans R Soc Lond B Biol Sci 372(1723):20160135. https://doi.org/10.1098/rstb.2016.0135

VanWallendael A, Alvarez M, Franks SJ (2021) Patterns of population genomic diversity in the invasive Japanese knotweed species complex. Am J Bot 108:857–868. https://doi.org/10.1002/ajb2.1653

Viana DS, Gangoso L, Bouten W, Figuerola J (2016) Overseas seed dispersal by migratory birds. Proc R Soc B 283:2015–2046. https://doi.org/10.1098/rspb.2015.2406

Vie JC, Hilton-taylor C, Stuart SN (2009) Wildlife in a changing world: an analysis of the 2008 IUCN Red List of Threatened Species. IUCN

Walther GR, Post E, Convey P, Menzel A, Parmesan C et al (2002) Ecological responses to recent climate change. Nature 216:389–395. https://doi.org/10.1038/416389a

Walther GR, Roques A, Hulme PE et al (2009) Alien species in a warmer world: risks and opportunities. Trends Ecol Evol 24:686–693. https://doi.org/10.1016/j.tree.2009.06.008

Wang Z, Baker AJ, Hill GE, Edwards SV (2003) Reconciling actual and inferred population histories in the house finch (*Carpodacus mexicanus*) by AFLP analysis. Evolution 57:2852–2864. https://doi.org/10.1111/j.0014-3820.2003.tb01526.x

Wang Y, Tan W, Li B, Wen L, Lei G (2021) Habitat alteration facilitates the dominance of invasive species through disrupting niche partitioning in floodplain wetlands. Div Dist 27:1861–1871. https://doi.org/10.1111/ddi.13376

Wei D, Yang L, Bao Z, Lu Y, Yang H (2023) Variations in outdoor thermal comfort in an urban park in the hot-summer and cold-winter region of China. Sustain Cities Soc 77:103535. https://doi.org/10.1016/j.scs.2021.103535

Wookey PA, Aerts R, Bardgett RD, Baptists F, Brathen KA et al (2009) Ecosystem feedbacks and cascade processes: understanding their role in the responses of Arctic and alpine ecosystems to environmental change. Glob Chan Biol 15:1153–1172. https://doi.org/10.1111/j.1365-2486.2008.01801.x

Xia L, Geng Q, An S (2020) Rapid genetic divergence of an invasive species, *Spartina alterniflora*, in China. Front Genet 11(284):1–19. https://doi.org/10.3389/fgene.2020.00284

Zhang Q, Hong Y, Zou F, Zhang M, Lee TM, Song X, Rao J (2016) Avian responses to an extreme ice storm are determined by a combination of functional traits, behavioural adaptations and habitat modifications. Sci Rep 6:22344. https://doi.org/10.1038/srep22344

Zhao P, Liu B, Zhao H et al (2023) Significant changes in soil microbial community structure and metabolic function after *Mikania micrantha* invasion. Sci Rep 13:114. https://doi.org/10.1038/s41598-023-27851-6

Plant Invasion as Gleaned from Parasara's Vrkshayurveda

5

D. A. Patil

Abstract

The zenith of Parasara's revelations to the world community emerged through his masterpiece work 'Vrksayurveda' which reflected the development of botanical science on more rational, scientific and sound footings. This treatise reflects the profound and in-depth knowledge of the author about the flora of ancient India. It was possibly compiled between the first century BC and first century AD. It has sought the attention of biodiversity experts in the past. However, its contents have not yet been evaluated from the point of plant invasion in erstwhile India. The present author divulged as many as 34 exotic plant species belonging to 32 genera 21 families of angiosperms from this basic treatise. These taxa are limelighted in view of plant invasion and discussed from the perspective of economy and socio-religious changes. The ancient Sanskrit treatises such as Vrksayurveda seek new insight into modern scientific thinking to better our understanding of biodiversity and its implications thereof.

Keywords

Exotic species · Parasara · Vrksayurveda

5.1 Introduction

The history of Vrksayurveda and his author Parasara must be known before going into its details. The manuscript is originally written in handwritten Sanskrit (Devanagari script). It is written in the Sutra style with prose and verses elaborating the

D. A. Patil (✉)
Post-Graduate Department of Botany, S.S.V.P. Sanstha's L.K. Dr. P.R. Ghogrey Science College, Dhule, Maharashtra, India

S. Tripathi et al. (eds.), *Plant Invasions and Global Climate Change*,
https://doi.org/10.1007/978-981-99-5910-5_5

text. It is divided into total six Kanda (parts). However, the last part is missing today. Whatever part of it is available, we Indians must thank first J.N. Sircar for discovery of the manuscript just before 1928 and second to N.N. Sircar and Roma Sarkar for its English translation and preservation. The manuscript is exclusively devoted to plant science, and the plants are referred to both by their local names and their chaste names. This is suggestive clearly that the author executed systematic study and observations over a long past, besides his profound and varied knowledge of ancient Indian flora. Projecting functional attributes of plants is the most scientific exposition. Thus, Vrksayurveda by Parasara is a full-fledged treatise on botany.

Parasara wrote Vrksayurveda (the science of plant life) for the purpose of teaching 'Botany' essentially needed for medicinal studies of plants. He is referred in the Charak (C. second century BC to first century AD) and Susruta Samhita (C. before second century BC) and other Ayurvedic texts of the late period. This fact reminds us that the author's personage Parasara is not a mythical figure but a historical personage. According to Majumdar (1951), Vrksayurveda was compiled by him possibly between the first century BC and first century AD.

A literary antique Parasara's Vrksayurveda attracted attention, in recent time, of Sircar (1950), Majumdar (1951) and Bose et al. (1971). Its English translation is made available by Sircar and Sarkar (1996). While going through it, one can trace development of plant science in the ancient erstwhile India. It appeared worth to evaluate this manuscript from the point of plant invasion in those days on Indian subcontinent. The results of the authors' studies are communicated in this chapter.

5.2 Methodology

As stated earlier, the basic manuscript was found in a somewhat mutilated state. Its English translation (Vrksayurveda of Parasara: Indian Medical Science Series No. 38) by Sircar and Sarkar (1996) is the prime source to study the manuscript. It also includes Sanskrit verses mentioning Sanskrit plant names and Sanskrit terms to be used in describing plants. I carefully examined this treatise emphasising particularly exotic plant species. These selected species are verified for their exotic status using recent botanical literature as mentioned against each species in Table 5.1. These literary sources also indicated their native country of the region. The data accrued is discussed in the light of present information.

The identity and nomenclature were updated consulting national and regional floras such as (1) The Flora of British India Vol. I–VII (Hooker 1872–1897), (2) The Flora of Presidency of Bombay, Vol. I–III (Cooke 1958), (3) Flora of Marathwada, Vol. I–II (Naik 1998), (4) Flora of Maharashtra: Monocotyledons (Sharma et al. 1996), (5) Flora of Maharashtra: Dicotyledons Vol. I (Singh et al. 2000), and (6) Flora of Maharashtra: Dicotyledons Vol. II (Singh et al. 2001), etc.

Table 5.1 Exotic plants gleaned from Parasar's Vrksayurveda (P.V.)

Sr. no.	Plant name and family	Sanskrit name	As per P.V. verse class/ chapter	Habit	Status cultivated (C) wild (W)	Nativity
1	2	3	4	5	6	7
1	[a]*Acacia farnesiana* (L.) Willd. Mimosaceae	Ari	Vanaspati Kanda Somi Vargadhyaya Verse 26	Tree	C	Tropical South America Mitra and Mukherjee (2012) Australia Chandra Sekar (2012)
2	[a]*Albizia lebbeck* (L.) Benth. Mimosaceae	Sirisa	Vanaspatyakanda Sami Vargadhyaya Verse 22	Tree	C	Pantropical Africa and Tropical Asia Bhandari (1978)
3	*Allium cepa* L. Liliaceae	Palandu	Vanaspatyakanda Trnavargodhyaya	Herb	C	West Asia Patil (2003), Gaikwad and Garad (2015) Persia Bailey (1928) Western Temperate Asia De Candole (1886)
4	*Allium sativa* L. Liliaceae	Rason	Vanaspatyakanda Trnavargadhyaya	Herb	C	Europe Patil (2003), Gaikwad and Garad (2015), Yadav and Sardesai (2002), Bailey (1949)
5	*Alocasia macrorrhiza* (L.) G. Don Aracee	Hastikarna	Bijotpatti Kanda Vrksangasutriyadhyaya Verse 13	Shrub	C	Tropical Asia Gaikwad and Garad (2015)
6	[a]*Borassus flabellifer* L. Arecaceae	Tala	Vanaspati Kanda Trnavargodhyaya	Tree	C	Tropical Africa Chandra Sekar (2012)
7	[a]*Calatropis gigantea* (L.) R.Br. Asclepiadaceae	Arka	Bijotpatti Kanda Phalangasutriyadhya Verse 29	Shrub	W	Tropical Africa Sudhakar (2008), Patil (2017), Chandra Sekar (2012)

(continued)

Table 5.1 (continued)

Sr. no.	Plant name and family	Sanskrit name	As per P.V. verse class/ chapter	Habit	Status cultivated (C) wild (W)	Nativity
1	2	3	4	5	6	7
8	[b]*Caesalpinia pulcherrima* (L.) Swartz. (Syn. *Poinciana pulcherrima* L.) Caesalpiniaceae	e.g.	Bijotpatti Kanda Vrksangasutriyadhyaya Verse 82	Shrub	C	South America Singh and Karthikeyan (2000) Tropical America Purseglove (1968) West Indies Gupta and Marlange (1961)
9	[a]*Cassia fistula* L. Caesalpiniacee	Aragvadha	Vanaspatya Kanda Sami Varghyaya Verse 17	Tree	W/C	North America Debnath and Debnath (2017)
10	*Citrus reticulata* Blanco Rutaceae	Nagaranga	Vanaspatya Kanda Vanaspatya Vicaranadhyaya Verse 15	Tree	C	Philippines Singh et al. (2000) Asia (Excl. India) Stewart (1972)
11	*Coriandrum sativum* L. Apiaceae	Karaika	Bijotpatti Kanda Phalangasutriyadhyaya Verse 25	Herb	C	South Europe Gaikwad and Garad (2015), Yadav and Sardesai (2002), Bailey (1949) Mediterranean Region Shetty and Singh (1987)
12	*Daucus carota* L. Apiaceae	Grnjanaka	Bijotpatti Kanda Astangasutriyadhyaya Verse 4	Herb	C	Europe Gaikwad and Garad (2015), Patil (2003), Yadav and Sardesai (2002) Europe & Temperate Asia De Candole (1886)
13	[a]*Eclipta prostrata* L. [Syn. *E. alba* (L.) Hassk.] Asteraceae	Kesaraja	Bijotpatti Kanda Phalangasutriyadhyaya Verse 28	Herb	W	South and Tropical America Patil (2017), Sudhakar (2008), Chandra Sekar (2012), Patil (1990)

14	[b]*Foeniculum vulgare* Mill. Apiaceae	e.g.	Bijtpatti Kanda Vrksangasutriadhyaya Verse 60	Herb	C	South Europe Shetty and Singh (1987), Gaikwad and Garad (2015), Coats (1956) Mediterranean Region Purseglove (1968)
15	*Glycyrrhiza glabra* Papilionaceae	Yasthimadhu	Virudha Valli Kanda Valli Tranagandhalika Versa	Climber	C	Mediterranean Region Uzundzhalieva et al. (2014)
16	[b]*Hibiscus rosa-sinensis* L. Malvaceae	e.g.	Bijotpatti Kanda Verse 52, 82	Shrub	C	China Patil (1995, 2003) Shetty and Singh (1987), Paul and Krishnamurthi (1967)
17	*Ipomoea aquatica* Forsk. Convolvulaceae	Kalambi	Virudha Valli Kanda Valli Trnagadhalika Varga	Climber	W	China Debnath and Debnath (2017)
18	*Lagenaria siceraria* (Mol.) Standl. Cucurbitaceae	Tumbi	Bijotpatti Kanda Phalangasutriyadhyaya Verse 6	Climber	C	Africa Singh and Nigam (2017)
19	*Luffa cylindrica* (L.) Roem. Cucurbitaceae	Jalini	Bijotpatti Kanda Phalangasutriyadhyaya Verse 14	Climber	C	Egypt Cameron (1891)
20	*Paspalum scrobiculatum* L. Poaceae	Kokodrava	Vanaspati Kanda Trnavorgadhyaya	Herb	W/C	Tropical Africa Singh and Nigam (2017)
21	[b]*Passiflora lunata* J.E. Smith Passifloraceae	e.g.	Bijotpatti Kanda Vrksangasutriyadhyaya Verse 56	Climber	C	Ecuadorian & Colombian Andes Perez et al. (2007)
22	*Pennisetum americanum* (L.) K. Schum Poaceae	Bijjira	Vanaspati Kanda Trnavargadhyaya	Herb	C	Central Tropical America Naik (1998)

(continued)

Table 5.1 (continued)

Sr. no.	Plant name and family	Sanskrit name	As per P.V. verse class/ chapter	Habit	Status cultivated (C) wild (W)	Nativity
1	2	3	4	5	6	7
23	*Phoenix dactylifera* Arecaceae	Kharjur	Vanaspati Kanda Trnavangadhyaya	Tree	C	Arabian Gulf Vyawahare et al. (2008)
24	*Piper betle* L. Piperaceae	Tambuli	Virudha Valli Kanda Valli Trnagandhalika Varga	Climber	C	Bali and East Indies Graf (1980)
25	*Polianthes tuberosa* L. Agavaceae	Naktagandha	Vanaspati Kanda Trnavergadhyaya	Herb	C	Mexico Patil (2003), Gaikwad and Garad (2015), Sharma et al. (1996), Bailey (1949)
26	*Psidium guajava* L. Myrtaceae	Parevata	Vanaspatya Kanda Vanaspatya Vicaranadhyaya Verse 20	Tree	C	Tropical America Singh et al. (2001), Patil (2003), Yadav and Sardesai (2002) Mexico Shetty and Singh (1987)
27	*Punica granatum* L. Punicaceae	Dadima	Bijatpatti Kanda Phalangasutryyadhyaya Verse 6	Tree	C	South Asia Gaikwad and Garad (2015) Afghanistan, Baluchistan and Persia Patil (2003), De Candolle (1959), Shetty and Singh (1987)
28	*Raphanus sativus* L. Brassicaceae	Mulaka	Bijatpatti Kanda Astangasutriyadhyaya Verse 4	Herb	C	Western Asia Purseglove (1968) Europe and Temperate Asia Singh et al. (1991), Patil (1995)
29	*Rubia cordifolia* L. Rubiaceae	Majistha, Netraparnini, Raktanala	Virudha Valli Kanda Valli Tranagandhalika Verga	Climber	W	Asia (Excl. India) and Africa Kaul (1986)

30	[a]*Saccharum spontaneum* L. Poaceae	Kas	Vanaspati Kanda Trnavargadhyaya	Herb	W	Tropical West Asia Sudhakar (2008), Patil (2017), Chandra Sekar (2012)
31	*Sesbania grandiflora* (L.) Poir. Papilionaceae	Agstya, munidrumah	Vanaspaty Kanda Sami Vergadhyaya Verse 11–12	Tree	C	Indonesia Patil (1995), Shetty and Singh (1987)
32	*Sesbania sesban* (L.) Merr. (Syn. *S. aegyptiaca* Poir.) Papilionaceae	Jayanti	Vanaspatya Kanda Sami Vargadhyaya Verse 12	Tree	C	Tropical Africa Martin et al. (1987) South Africa Rajagopal and Panigrahi (1965)
33	[b]*Solanum melongena* L. Solanaceae	e.g.	Bijoptatti Kanda Vrksangasutriyadhyaya Verse 78	Shrub	C	East Indies Singh et al. (2001) America Gaikwad and Garad (2015)
34	*Spondias pinnata* (L. *f.*) Kurz. Anacardiaceae	Amrataka	Vanaspatya Kanda Vanaspatya Vicaranadhyaya Verse 6	Tree	C	Tropical Asia Martin et al. (1987)
35	*Tamarindus indica* Linn. Caesalpiniaceae	Tintiri	Vanaspatya Kanda Sami Vargadhya Verse 30	Tree	W/C	Tropical America Shetty and Singh (1987), Patil (1990)
36	[b]*Trianthema portulacastrum* L. Aizoaceae	e.g.	Bijotpatti Kanda Verse 55	Herb	W	Tropical America Qureshi et al. (2014)
37	*Triticum aestivum* L. Poaceae	Godhuma	(1) Vanaspati Kanda Trnavargadhyaya (2) Bijotpati Kanda Bhumivargastriyadhya Verse 7–8	Herb	C	Fertile crescent and Middle East Simmons (1987)
38	[a]*Urena lobata* L. Malvaceae	e.g.	Bijotpatti Kanda Astangastriyadhyaya Verse 17	Shrub	W	Tropical Africa Sudhakar (2008), Patil (2017), Chandra Sekar (2012), Rajagopal and Panigrahi (1965)

(continued)

Table 5.1 (continued)

Sr. no.	Plant name and family	Sanskrit name	As per P.V. verse class/ chapter	Habit	Status cultivated (C) wild (W)	Nativity
1	2	3	4	5	6	7
39	*Vernonia anthelmintica* (L.) Willd. [Syn. *Centratherum anthelmintica* (L.) O. Ktze.] Asteraceae	Somraji	Bijoptupatti Kanda Phalangasutriyadhya Verse 27	Herb	W	Malay Archipelago Mitra and Mukherjee (2012)
40	*Vitis vinifera* L. Vitaceae	Draksa	Virudha Valli Kanda Valli Tranagadhalika	Climber	C	West Asia Gaikwad and Garad (2015) Asia (Excl. India) and Europe Stewart (1972) South-East Europe to West Indies Singh et al. (2000)
41	*Ziziphus jujuba* Mill. Rhamnaceae	Kola, Badara	Vanaspatya Kanda Vicaranadhyaya Verse 32	Tree	C	Subtropics and warm temperate zone Martin et al. (1987)

[a] Invasive species
[b] e.g.: Not mentioned in verses but included by Sircar and Sarkar (1996) while elaborating Sanskrit terminology

5.3 Status of Aliens Plants

The present attempt is to study exotic plant species introduced or naturalised intentionally or negligently on Indian subcontinent in ancient period. Such information is commonly available in ancient Sanskrit scripts compiled by sedges, medicine-men or botanists. Perasara worked in different compartments of knowledge. During his compilation of Vrksayurveda, he particularly emphasised plant science with respect to morphology, taxonomy and functioning of plant species particularly integral part of the then Indian biodiversity. The translated treatise by Sircar and Sarkar (1996) was critically examined for exotic taxa included in it. The present author analysed them first into two groups: (1) first group of exotic taxa originally included in Parasara's basic treatise, and (2) another group of exotic taxa not mentioned in Sanskrit verses composed by Parasara but mentioned by Sircar and Sarkar (1996) to elaborate and explain terminology used by Parasara. Parasara's basic treatise contained total 34 exotic species belonging to 32 genera and 21 families of angiosperms. These taxa can be categorised based on habit of plant species as trees (13), herbs (12), shrubs (3) and climbers (6). These can be also analysed based on their utilities as: plantation for shade (2), ornamental (6), multipurpose uses (1), spices (4), fruits (8), vegetable (5), food grains (3) and chewing and religious (1). It also appears pertinent to focus on the nativity of these exotic taxa. These species particularly included by Parasara are obviously hailed from ancient era in India. They belong to various countries, continents or regions of both New and Old Worlds. They belong to Africa (7), Asia (Excl. India) (9), America (9), Europe (3), China and East Indies and Bali (2 each). Other countries or regions such as Malay Archipelago, Mediterranean Region, Egypt, Philippines, Ecuadorian and Columbian Andes, Fertile Crescent and Middle East, Indonesia, Arabian Gulf, Australia, etc. represented by a single species each. Other seven exotic taxa (*) mentioned as, for example, in Table 5.1 cannot straightway considered for their ancient introduction on Indian territory. They need special study for their status on Indian landmass.

5.4 Role of Aliens Plants

Ancient man ceased his nomadic life and settled down to grow plant species needed for his sustenance. Since the dawn of history, agriculture has been the mainstay and backbone of India. Indians largely followed the ancient methods of raising crops. However, the changing times and tastes of people fetched new plant species or crops to cultivate during the past several centuries. The above resume clearly suggests that the exotic but useful plants have been brought under cultivation for aforesaid purposes. Thus, the exotic species are first stabilised on Indian soil and environment. At the same, they also influenced the economy of our country. Even there are social changes. For example, betel leaf (*Piper betle*) is exotic but appropriated for various religious rituals and ceremonies. The spices, e.g., species of *Allium*, *Foeniculum*, *Coriandrum* changed the taste of the Indians. New kind of fruits, e.g., Guava

(*Psidium guajava*), Pomegranate (*Punica granatum*), Date palm (*Phoenix dactylifera*), etc. supplemented our food required and even they became economic sources. Exotic species also added sources of food grains, e.g., Wheat (*Triticum aestivum*), Kodrava or Kodra (*Paspalum scrobiculatum*), Bajara (*Pennisetum americana*) of the present account. The exotic species have long supported every compartment of human life in Indian territory. Thus, most of the cultigens have their own centre of origin and directions of migration. They followed the land route or even the oceanic path. When and how these were introduced, one cannot conjecture but the ancient literary sources such as Vrksayurveda by Parasara serve the purpose and satisfy our inquiry.

5.5 Aliens and Climate Change

Out of 34 exotic species, 8 species pertaining to 8 genera and 7 angiospermic families are found invasive in nature (cf. Sudhakar 2008; Chandra Sekar 2012; Singh and Das 2015; Sheikh and Dixit 2017). They are either wild or brought under cultivation by ancient Indians. Exotic plant invasion emerged as a global problem causing adverse impact on the ecosystems, economy and human health (Inderjit et al. 2008; Rastogi et al. 2015). Invasive species possess characteristic features like 'pioneer species' in varied landscapes, tolerant of a wide range of soil and weather conditions (Sudhakar 2008). They exhibit different modes of vegetative reproduction along with sexual reproduction. Moreover, allelopathy plays a role in invasion acceleration (Yadav et al. 2016). The present era is although witnessed by climate change, such invasive species have no problem, they compete with native flora and also change according to the environment. A better planning is, therefore, needed for early detection to control and reporting of infestation of spread of new invasive species in a region. Such a measure appears workable to check bio-invasion. Change is inevitable in nature; it is so in the milieu of climate. Climate change has an incredible potential to influence plant wealth of a region with multifarious effects. Plants always remain under constant threat of various abiotic and/or biotic stresses, being affected within their ambit of habitats by single or combined stress. In such scenario, ancient literary works offer beginning and evidence of bio-invasion, and hence, need their in-depth study on modern scientific lines.

5.6 Conclusions

As a result, these exotic commodities not only become a daily necessity to satisfy our desire but also made entries in our social and religious life. They attained sanctity for use in rituals and worships connected with death, birth, marriage, etc. A case of Kodrava or Kodra (*Paspalum scrobiculatum*) is interesting. It is an exotic crop species in India. It was introduced in the ancient past. But now, it is marginalised in such a way that to only in tribal regions of India. Some such crop species are affected by the modern wave of 'green evolution' and changing forces of economy. Biodiversity changes in past are reflected in our literary sources. These should be re-examined for the welfare mankind and management of biodiversity.

Acknowledgements I am thankful to the authorities of S.S.V.P. Sanstha, Dhule (Maharashtra) for library facilities.

References

Bailey LH (1928) The standard cyclopedia of horticulture, vol I, England edn. Macmillan, New York

Bailey LH (1949) Manual of cultivated plants, Rev edn. Macmillan, New York

Bhandari MM (1978) Flora of the Indian desert. Scientific Publishers, Jodhpur

Bose DM, Sen SN, Subbarayappa BV (1971) A concise history of science in India. The Indian National Science Academy, New Delhi

Cameron J (1891) Catalogue of plants in the botanical garden, Bangalore and its vicinity, 2nd edn. Mysore Government Central Press, Bangalore

Chandra Sekar K (2012) Invasive alien plants of Indian Himalayan region: diversity and implication. Am J Plant Sci 3:177–184

Coats AM (1956) Flowers and their histories. Hulton, London

Cooke T (1958) The flora of Presidency of Bombay, vol 1–3. Botanical Survey of India, Calcutta, India

De Candolle A (1959) Origin of cultivated plants (Rev. 2nd Ed.). Hafner Publishing Co., London, UK

De Candole AP (1886) Origin of cultivated plants (translated from 2nd ed. in French, 1959). Hafner, New York

Debnath A, Debnath B (2017) Diversity, invasion status and uses of alien plant species in North Eastern hilly states of Tripura: a confluence of Indo-Barman hotspot. Am J Plant Sci 8:212–235

Gaikwad SP, Garad KU (2015) Flora of Solapur district. Laxmi Book Publications, Solapur, Maharashtra

Graf AB (1980) Exotica: pictorial cyclopedia of exotic plants from tropical and near-tropic regions, 10th edn. Roders Company Inc., Rutherford, NJ

Gupta RK, Marlange M (1961) Le Jardin Botanique de Pondicherry. Trav. Sect Scient et Tech. Tom III: Fascicule 1:1–133

Hooker JD (1872–1897) Flora of British India, vols I–VII. Reeves & Co., London

Inderjit, Seastedt TR, Callaway RM, Kaur J (2008) Allelopathy and plant invasions: traditional, congeneric and biogeographical approaches. Biol Invasion 10:875–890

Kaul MK (1986) Weed flora of Kashmir valley. Scientific Publishers, Jodhpur

Majumdar GP (1951) The history of botany and allied sciences in ancient India. AIHS, UNESCO, Paris

Martin FW, Campbell CW, Ruberte RM (1987) Perennial edible plants of the tropics: an inventory. U.S. Department of Agriculture, Agriculture handbook no. 642. 222 p., illus

Mitra S, Mukherjee SK (2012) Invasive alien species of West Bengal used in human health care. In: Proceedings of plant science research in human welfare. Department of Botany, Dhangar College (in collaboration with Bot. Surv. India), pp 38–58

Naik VN (1998) Flora of Marathwada, vol I–II. Amrut Prakashan, Aurangabad, Maharashtra

Patil DA (1990) Exotic elements in the flora of Dhule district (Maharashtra). J Econ Tax Bot 14(3): 721–724

Patil DA (1995) Exotic elements in the flora of Dhule district (Maharashtra)—II. Biojournal 7(1–2): 1–8

Patil DA (2003) Flora of Dhule and Nandurbar districts (Maharashtra). Bishen Singh Mahendra Pal Singh, Dehradun
Patil DA (2017) Invasive alien species in Khandesh region (Maharashtra, India): diversity, implications and measures. Sch Acad J Biosci 5(12):867–876
Paul BP, Krishnamurthi S (1967) Flowering shrubs. ICAR, New Delhi
Perez JC, Eeckenbrugge GC, Restrepo M, Jarvis A, Salazar M, Cactano C (2007) Diversity of Colombian Passifloraceae: biogeography and updated list for conservation. Biota Colombiana 8(1):1–45
Purseglove JW (1968) Tropical crops-dicotyledons. 2 vols. Longmans, London
Qureshi H, Muhammad A, Yasmin B (2014) Invasive flora of Pakistan: a critical analysis. Int J Biosci 4(1):407–424
Rajagopal T, Panigrahi G (1965) 'Aliens' naturalised in the flora of Allahabad. Proc Natl Acad Sci India B 35(4):411–422
Rastogi J, Rawat DS, Chandra S (2015) Diversity of invasive alien species in Pantnagar flora. Trop Plant Res 2(3):282–287
Sharma BD, Karthikeyan S, Sharma BD (1996) Flora of Maharashtra state: monocotyledons. Bot. Surv. India, Calcutta
Sheikh DK, Dixit AK (2017) Occurrence of invasive plant in three phytogeographical region of Bilaspur district of Chaattisgarh. Ann Plant Sci 6(12):1872–1878
Shetty BV, Singh V (1987) Flora of Rajasthan, vol I. Bot. Surv. India, Calcutta
Simmons SR (1987) Growth, development and physiology. In: Heyne EG (ed) Wheat and wheat improvement, 2nd edn. ASA, CSSA, INF and SSS of America Inc., Madison, WI, pp 77–104
Singh THB, Das AK (2015) Study of alien invasive flora of Valley District of Manipur and their control. Int J Innov Res Sci Technol 1(12):616–626
Singh NP, Karthikeyan S (2000) Flora of Maharashtra, vol 1. Botanical survey of India, Culcutta, pp 193–195
Singh AK, Nigam SN (2017) Ancient alien crop introductions integral to Indian agriculture. An overview. Proc Indian Natl Sci Acad 83(3):549–568
Singh V, Parmar PJ, Pandey RP (1991) Flora of Rajasthan, vol II. Bot. Surv. India, Calcutta
Singh NP, Lakshminarsimhan P, Karthikeyan S (2000) Flora of Maharashtra state: dicotyledons, vol I. Bot. Surv. India, Calcutta
Singh NP, Lakshminarasimhan P, Karthikeyan S, Prasanna PV (2001) Flora of Maharashtra state: dicotyledones, vol II. Bot. Surv. India, Calcutta
Sircar NN (1950) An introduction to the Vrksayurveda of Parasara, vol XVII. Yeark-Book, R.A.S. B., p 176
Sircar NN, Sarkar R (eds) (1996) Vṛkṣāyurveda of Parāśara: A Treatise on Plant Science: Sanskrit Text, English Translation, and Notes with Comparative References to Modern Botany, vol No. 38. Sri Satguru Publications
Stewart RR (1972) An annotated catalogue of the vascular plants of West Pakistan and Kashmir. Fakhri Press, Karachi
Sudhakar RC (2008) Catalogue of invasive alien flora of India. Forestry and Ecology Division, National Remote Sensing Agency, Balanagar, Hyderabad
Uzundzhalieva K, Ruseva R, Kachakova S (2014) Ex situ and in vitro conservation of Glycyrrhiza glabra L. crop wild relative from Fabaceae. Ecol Balkan 5:8–13
Vyawahare N, Pujari R, Kshirsager A, Ingawale D, Patil M, Kagathara V (2008) Phoenix dactylifera: an update of its indigenous uses. Phytochem Pharmacol 7:1
Yadav SR, Sardesai MM (2002) Flora of Kolhapur District. Shivaji University, Kolhapur, Maharashtra
Yadav VV, Singh NB, Singh H, Singh A, Hussain I (2016) Allelopathic invasion of alien plant species in India and their management strategies: a review. Trop Plant Res 3(1):87–101

Part II

Plant Invasion in Different Ecosystems: Case Studies

Water, Wind, and Fire: Extreme Climate Events Enhance the Spread of Invasive Plants in Sensitive North American Ecosystems

6

Jennifer Grenz and David R. Clements

Abstract

Most previous research on climate change and invasive plants has focused on the influence of either climate warming or increased CO_2 levels on invasive plants but extreme climate events may often have more immediate consequences. How increases in frequency and severity of extreme climate events will affect vegetation trajectories, particularly whether they will enhance invasion success over native plant species, will depend on the severity of the event, ecosystem health prior to the event, the pre-event level of invasion, and the adaptability of both the invaders and native species. We review responses of invasive plants to three extreme climate event categories: wildfires, floods, and storms in North America. Fires which formerly served to promote diverse ecosystems in the North American Northwest have changed in severity such that they now threaten to promote invasive species better adapted to establish after fires, as seen in research on several different ecosystems, such as Garry oak (*Quercus garryana*) ecosystems, perennial grasslands, and interior forest ecosystems. The last several years have seen some of the worst fire seasons on record in western North America, as exemplified by the 45,000 ha McKay Creek fire in British Columbia in 2021. Likewise, flooding frequency and severity are increasing, as seen in the 100-year event seen in coastal British Columbia in November 2021. Flooding events facilitate the spread of invasive species adapted to riparian areas such as

J. Grenz (✉)
Department of Forest Resources Management, University of British Columbia, Vancouver, BC, Canada
e-mail: jennifer.grenz@ubc.ca

D. R. Clements
Department of Biology, Trinity Western University, Langley, BC, Canada
e-mail: clements@twu.ca

S. Tripathi et al. (eds.), *Plant Invasions and Global Climate Change*,
https://doi.org/10.1007/978-981-99-5910-5_6

knotweeds (*Reynoutria* species) and many other riparian invasives. Finally, more frequent, and severe storms may also provide new or stronger pathways for invasion. The abrupt increase in extreme events in North America calls for more proactive strategies to protect sensitive ecosystems from invasive species in the wake of extreme events, and our research shows that context specific measures are required to restore them.

Keywords

Climate change · Ecological restoration · Floods · Invasive plants · Wildfire

6.1 Introduction

Research on the impacts of climate change on invasive plants has largely focused on whether the evolutionary potential of invasive species exceeds that of native species they co-exist with, and the potential and predictive patterns for invasive plant range expansion under climate change conditions such as warming and drought (Gianoli and Molina-Montenegro 2021). While understanding the evolutionary rates and performance responses of invasive plant species to factors such as increased temperature, increased CO_2, and decreased precipitation are important, the focus of literature to date could be characterized as being on the slow-moving impacts of a changing climate. This disproportionate focus neglects the more immediate potential impacts of catastrophic climate events on invasive plant populations. As climate models project an increase in both the intensity and frequency of extreme climate events such as heat waves, heavy precipitation, intense storms like hurricanes, flooding, debris flows, and wildfires (AghaKouchak et al. 2020), we must be prepared to address both the increased vulnerability to and rapid alteration of landscapes to plant invasion after such events. Research based on traditional risk assessment frameworks tends to consider climate change hazards in isolation, underestimating potential impacts as they neglect to consider the compounding and cascading effects (AghaKouchak et al. 2020). The rapidity and severity with which extreme climate events alter landscapes, whose plant communities were already on climate-altered trajectories, make it difficult to anticipate resulting vegetation trajectories and dissemination of invasive plants in their aftermath (AghaKouchak et al. 2020).

Extreme climate events are increasingly placing invasive plant researchers and practitioners in the challenging position of having to answer questions, often posed by government agencies managing the restoration after major climate events, on how to prevent and manage plant invasion across highly variable landscapes of North America in the wake of these events. These questions are difficult to answer as major climate events alter the context of invasion biology to one of dynamic rapid change, challenging both our knowledge and experience gained from "normal" conditions. There is considerable pressure to answer these questions quickly to guide immediate restoration activities to prevent the potential for rapid colonization by invasive species after such significant and widespread disturbance (Grenz and Clements, personal communications, 2020, 2021, 2022; Pilliod et al. 2021). Meanwhile, there are examples where reactionary approaches such as seeding have caused other challenges such as the introduction of new invasive species (Ott et al. 2019).

Facing these questions ourselves within the context of recent, significant climate events in the province of British Columbia, Canada, in this chapter, we present our review of responses of invasive plants to the three major extreme weather categories: fires, floods, and storms, in the North American context. We also present our own research to provide tangible insights and illuminate important considerations into the complexities of invasive plant prevention and management to inform restoration approaches after each of these extreme climate events. These include examining the current context of increasing frequency and intensity of wildfires, floods, and storms in North America and their impacts to invasive species. We provide insights regarding site-specific variation of vegetation trajectories to these climate events and discuss site-specific factors that influence vegetation responses. We conclude by suggesting monitoring, mitigation and proactive management measures needed in the face of increasing incidence and magnitude of floods and storms, and post-climate event considerations important to informing further studies and restoration planning.

6.2 Climate Change and Wildfires

6.2.1 Increasing Frequency and Intensity of Wildfires in North America in the Twenty-First Century

The widespread suppression of wildfire in the twentieth century resulted in significant decrease in fire occurrence in North American forests not seen in previous millennia (Hagmann et al. 2021). These significantly altered fire regimes tell a story important to understanding the colonial history that has led to this current juncture in climate history. One where mega-wildfires, paired with increasingly long seasons of drought, have become annually anticipated climate events. These mega-wildfires rapidly alter landscapes and are unlike the historic fire regimes that shaped those same landscapes over millennia (Dickson-Hoyle and John 2021). We are only just beginning to understand the resulting impacts of these types of fires to inform ecological restoration in their aftermath. Invasive plants are well suited for rapid dispersal into such altered landscapes (Brooks et al. 2004). This is concerning for land managers as mega-wildfires may provide an opportunity for new and existing "species to colonize or expand their cover in sites they could not previously dominate" (Brooks et al. 2004).

The role of fire in ecosystem health was unrecognized by European settlers to North America who brought with them the perception that fire was dangerous and harmful and should be extinguished regardless of context. This perception gave rise to 100 plus years of celebrated human ingenuity in wildfire management and control in North America. As wildfire management improved, surrounding landscapes and ecosystem functions changed. Fire excluded forested landscapes transitioned from heterogeneous plant community compositions toward highly altered homogenous vegetation spatial patterns (Hagmann et al. 2021). Without wildfire, tree density increased, changes in speciation occurred toward a greater proportion of those that are fire-intolerant, and fuel loads accumulated at the surface, ladder and canopy levels (Hagmann et al. 2021).

The resulting conditions of a legacy of wildfire suppression in the context of climate change is that fire severity has changed. Fire severity, measured as the percentage mortality of tree biomass after a fire event, is categorized as low (<20%), moderate (20–70%), and high (>70%) (Hagmann et al. 2021). Current forest conditions are fostering large, high-severity wildfires not often observed in historical ranges (Hagmann et al. 2021) where more frequent low and moderate intensity fires were once abundant and highly influential on plant communities. Between 1985 and 2017, there was an eight-fold increase in annual area burned at high severity in western US forests (Parks and Abatzoglou 2020). In British Columbia, Canada, where our own research on wildfire and invasive plants is located, there are similar trends. In the summer of 2017, 1.2 million ha of lands in British Columbia were burned in 'mega-fires' such as the 'Elephant Hill Fire' which burned close to 200,000 ha. In 2021, when a historic heat wave broke records across the Pacific Northwest of North America in June of that year, a climate event referred to as the "heat dome", led to an average of 40 new wildfires starting every day in British Columbia during the first 2 weeks of July. By the end of 2021, 869,279 ha of British Columbia's forests and grasslands were burned, including one of our research areas, the 'McKay Creek Fire' (Fig. 6.1).

Fig. 6.1 Highest severity burn area of the McKay Creek Wildfire, 1 year after wildfire in fall of 2022

6.2.2 Site-Specific Variation

6.2.2.1 McKay Creek Fire in British Columbia

In July of 2022, the McKay Creek Fire, located approximately 11 km north of Lillooet British Columbia, burned over 45,000 ha across four different bio-geoclimatic zones. This area provides important habitat to blue listed species, wildlife, birds, and fish. It is important to the agricultural and recreational sectors and is within the traditional territory of the St'át'imc Nation, an Indigenous Nation, who rely upon it for food, social, and ceremonial purposes. Much of the area had been surveyed prior to the wildfire for invasive plant species and it contained species such as burdock (*Arctium* spp.), diffuse knapweed (*Centaurea diffusa*), spotted knapweed (*Centaurea stoebe*), blueweed (*Echium vulgare*), Canada thistle (*Cirsium arvense*), ox-eye daisy (*Leucanthemum vulgare*), orange hawkweed (*Hieracium aurantiacum*), hoary alyssum (*Berteroa incana*), hound's tongue (*Cynoglossum officinale*), and scentless chamomile (*Matricaria maritima*).

The diverse landscapes of the McKay Creek Wildfire and the diversity of invasive plant species within it have provided a unique opportunity to help determine if post-fire invasions are a widespread phenomenon or are habitat-specific and what the sources of variation in native and invasive plant responses to fire are, both of which have not been rigorously tested (Alba et al. 2015). This is important to help inform and prioritize restoration efforts in the region particularly since proposed government approaches to habitat recovery could be characterized as generalized. For example, proposing seeding across vast tracks of land utilizing the same seed mixes regardless of location. A meta-analysis conducted by Alba et al. (2015) confirms what we are observing across the fire area, that how plant communities respond to fire is quite variable, and that our categorizations of land (bio-geoclimatic zones) and burn severities (low, medium, high), are not predictive of a consistent response across those landscapes. We have observed that the variability of plant community responses is particularly pronounced in the highest burn severity areas, where proximity to watercourses and burned trees appear to influence germination and dissemination of both invasive and native species. We concur with the conclusion of a meta-analysis by Alba et al. (2015) that local assessments are important to determining mechanistic drivers and management policy. While it would be much easier to have broadly applied plant community restoration approaches based on existing categorizations, local knowledges, pre- and post-wildfire, are critical to informing stewardship approaches. While this may seem resource intensive, it is possible that this approach would be a more efficient and effective use of resources as from our own observations, thus far within burned sites (post-fire years 1 and 2), areas of high density of desirable plant species, such as those along riparian zones and animal trails, could provide localized seed sources and a nature-based solution to dissemination of desirable species. Relatedly, these plant pathways could also be a source of propagule pressure of invasive species, and thus, monitoring efforts could be prioritized in these areas. Indigenous and local knowledges have been critical to informing these observations as we prioritized research sites. This also aligns with Alba et al. (2015) as they identified future research needs which included adopting "community level metrics that best capture the impact of fire on suites of co-occurring native and exotic species" along with the importance of studying a wide range of habitat types in areas of high risk to mega fires, particularly those that have existing populations of invasive species and those vulnerable to invasion. Understanding the relationships between plant community responses, the characteristics of both the landscapes and fire itself, and the conditions that made burned areas vulnerable to mega-fire are critically important. Very few publications addressing these relationships to inform post-fire ecological restoration can be found.

6.2.2.2 Mt. Maxwell Fire in British Columbia

In June of 2009, a wildfire was ignited on the west side of Mount Maxwell on Saltspring Island, British Columbia, which burned 17 ha (Gulf Islands Driftwood Community Newspaper 2009). Saltspring Island is an 18,534-ha island belonging to a chain of islands, referred to as the Southern Gulf Islands, in the Salish Sea between the mainland of British Columbia and Vancouver Island. These islands range in size and are characterized by their unique climate, compared to the surrounding areas, as warm summer Mediterranean. Climate change has altered seasonal weather patterns such that the number and severity of droughts on these already dry landscapes have increased (Transition Saltspring 2020). With much of the island being classified for risk of wildfire as "extreme" (Transition Saltspring 2020), it is no surprise that the combination of steep terrain, dry conditions, and high winds allowed the Mount Maxwell fire to spread rapidly. The fire was aggressively contained and controlled with aerial and ground supports and left behind a unique opportunity to study the effect of fire and herbivory on plant communities in Garry Oak ecosystems as 6 ha of Garry Oak savanna were burned. Garry Oak ecosystems are one of Canada's most threatened habitats with only 1–5% remaining in near natural condition (Lea 2006). These ecosystems have high species richness, 454 native taxa (MacDougall and Turkington 2005), and provide specialized habitats for 61 plant species considered at risk within Garry Oak ecosystems (GOEs) and 12 plant taxa considered at risk globally (Fuchs 2001). Prior to European settlement, Coast Salish First Nations frequently burned oak woodlands (Agee 1996; Pellatt and Gedalof 2014) to maintain an open vegetation structure that favoured their primary vegetable food, camas (*Camassia quamash* and *C. leichtlinii*) (Fuchs 2001; Lea 2011) which left the oaks with their thick bark and ability to crown sprout unscathed (Agee 1996). Substantial changes to ecosystem composition, structure, and function occurred after European settlement brought 150 years of fire suppression. Fire suppression has been described as a serious threat to GOEs as the absence of fire allows conifers to establish in the understory converting oak stands to coniferous forests (Fuchs 2001; Hoffman et al. 2019). With the suppression of fire in the Pacific northwest region, little recruitment of Garry oak trees has occurred since about 1940, whereas Douglas-fire recruitment in Garry oak ecosystems has been vigorous since about 1900, such that widespread losses of historic Garry oak habitat have occurred and are projected to continue without active intervention (Pellatt and Gedalof 2014).

The reintroduction of fire as a restoration strategy for GOEs in British Columbia is being considered and actively studied. It is difficult to study the reclamation of traditional stewardship practices using fire as most Garry Oak savannas exists within close proximity to, and even within, residential neighbourhoods. Further, these systems have changed drastically since its historical use (over 150 years ago), and thus, should not be considered sufficient guidance (Agee 1996; Pellatt and Gedalof 2014). While the reintroduction of fire may offer an opportunity to meet restoration objectives, reintroducing fire to fire-suppressed systems can have dramatic impacts on structure and function, involving many considerations of a particular habitat, including the presence of invasive plants (Howe 1995; Gedalof et al. 2005; Polster 2011).

Complex considerations must be made such as fire frequency, intensity, and current species composition (MacDougall 2005) as fire could acerbate other factors that threaten GOEs such as invasion by non-native, invasive species (Dennehy et al. 2011). The entire plant structure could have a new quasi-equilibrium where alien plants dominate due to changes in resource availability (Keeley et al. 2003). Knowledge of the effects of fire on non-native species in GOEs is quite limited as documented in a review of more than 100 non-native species of importance in these systems (Dennehy et al. 2011). Non-native species are prevalent in GOEs in British Columbia with up to 82% of herbaceous cover being non-native species with exotic grasses dominating most sites. A number of management attempts have been utilized for the control of non-native plants in GOEs and efforts as having been resource intensive with limited success.

Our research did not provide evidence that fire alone had any significant effect on native and non-native plant species. What it did show was a slight increase in the mean percent cover of non-native plants in the burned areas over the 3-year study period. This information should caution land managers that may use controlled burning to stop canopy closure. Any disturbance introduced to an area could cause other issues such as the introduction of invasive, non-native species. While not all non-native species are invasive, species such as Scotch Broom (*Cytisus scoparius*) that has invaded GOEs in British Columbia can alter vegetation structure and increase risk of high-intensity fire, while capable of recovering from fire via regrowth from roots or germination from the seedbank (Dennehy et al. 2011). Scotch Broom proliferation can be promoted by certain fire regimes, so land managers need to exercise caution that they are not opening or promoting niches for invasive species while attempting to meet other restoration objectives.

This research illuminated the complexities of studying vegetation trajectories after wildfire, including confounding variables such as herbivory, making it clear that longer term study is needed to understand the dynamics of its impact in the modern, fire-suppressed context of these ecosystems. Non-native plant species may represent the new equilibrium in these systems, so focus should be shifted to those species that are truly invasive and their specific responses and possible contributions to the metrics of the estimated burn severity of the invaded Garry Oak system.

6.2.2.3 Role of Site-Specific Factors

Both of our research projects have made clear the necessity of regionally specific data to understand the relationships between wildfires and invasive plants (Grenz and Clements, unpublished). Research conducted in other ecosystems with differing burn severities and differing plant species may not be useful to land managers to directly inform plant responses in their own contexts. Our preliminary research on the McKay Creek Fire and our results from the Mount Maxwell Fire have changed our own expectations of the impact of our research. Contributions to the body of knowledge regarding the relationship between invasive plants and wildfire beyond the stakeholders within these specific fire zones are not to improve predictability elsewhere, but to identify and examine important relationships that may otherwise be overlooked in post-wildfire recovery that could be applicable in other contexts to inform restoration.

6.2.3 Wildfire Hotspots in North America

Over the North American continent, there are a diversity of landscapes/ecosystem types which vary greatly in terms of both the level of wildfire occurrence risk and their sensitivity to wildfire impacts, including potential impacts of invasive plants. The variety of these landscapes and approaches to fire management underline our call for diverse approaches to both understanding and managing fire effects. Table 6.1 summarizes the diversity of these landscapes and specific invasive plant species being studied within the current academic literature.

6.2.4 Impacts of High Intensity Wildfire on Invasive Plants

The size and intensity of wildfires have been increasing globally (Faccenda and Daehler 2022), and the resiliency of forests to high-severity fire may be diminishing (Chileen et al. 2020). Given this, the abilities of invasive plants to have direct and indirect effects on native plants, and change ecosystems through the alteration of soil stability, promotion of erosion, and colonizing open substrates (Brookes et al. 2004), may only be compounded. Diminished resiliency is attributed to lower: post-fire seed availability; nutrient availability; and establishment and survival rates of conifer seedlings (Chileen et al. 2020). A context could provide ideal conditions which promote invasion and contribute to establishment by invasive plant species as they take advantage of both reduced competition and the nutrient rich environments that often result from fire (Alba et al. 2015). The significant and rapid alteration of landscapes caused by the intensity of these wildfires cannot be understated. Transformation from dense, closed-canopy forests, to wide-open, exposed mineral soils is not common. Vast tracks of land are left vulnerable to rapid colonization of invasive plant species.

While factors associated with post-wildfire alteration of growing conditions such as changes in available light, significant releases of soil nutrients, altered hydrology,

Table 6.1 Wildfire issues with invasive species in various North American regions

Region	Ecosystems affected	Invasive species of particular concern	References
Pacific northwest	Rangeland	*Centaurea diffusa* *Centaurea stoebe*	Pokorny et al. (2010)
Western	Semi-arid shrub steppe	*Taeniatherum caput-medusae* *Ventenata dubia* *Bromus tectorum*	Applestein and Germino (2021)
Southwestern	Mixed conifer-hardwood forest	*Senecio sylvaticus* *Cirsium vulgare* *Lactuca serriola* *Aira caryophyllea* *Vulpia myuros* *Bromus tectorum*	Reilly et al. (2020)
Eastern	Deciduous forest	*Microstegium vimineum* *Elaeagnus umbellata* *Schedonorus phoenix*	Emery et al. (2011)
Midwest	North-central Appalachia	*Paulownia tomentosa*	Williams and Wang (2021)
Great Basin	Cold desert	*Bromus tectorum*	Brunson and Tanaka (2011)
Mojave Desert	Hot desert	*Bromus rubens*	Horn et al. (2015)
Hawai'i	Random forest	Addressed 49 species of concern	Faccenda and Daehler (2022)
	All landscape types	*Megathyrsus maximus* *Cenchrus setaceus* *Melinis minutiflora* *Cenchrus ciliaris*	Trauernicht et al. (2015)

changes to soil microbial ecology, altered seedbank dynamics, and newly available niches are more commonly expected. Our on-going research on the McKay Creek Wildfire is showing that the activities associated with wildfire management, subsequent ecological restoration, and other land-use activities/pressures may present the greatest risks to invasability. While this research is in early stages, observations made 1-year post-wildfire include invasive plant species on access roads moving into burned areas (particularly moderate and severely burned areas), changes in wildlife grazing/browsing patterns moving invasive species along their trails (particularly from riparian areas), vectoring along fire guards, and invasive plants in areas where there is increased recreational pressures due to the new opportunities for mushroom picking. These observations are consistent with the hypothesis (Reilly et al. 2020) that low elevation landscapes with legacies of past management such as clearcuts and extensive road networks may be particularly susceptible to invasion and along fire breaks (Reilly et al. 2020) as we observed already at year 1, their facilitation of the spread of invasive plants, particularly in areas with historic presence of those species.

The relationship between wildfire and invasive plants is not necessarily straight forward. There are considerable complexities including pre-wildfire conditions such as the historic presence of invasive plant species, ecosystem health as well as the fire behaviour over the entire landscape (e.g., a single wildfire will not be solely classified as "high-severity" as these fires are made up of a mosaic of burn severities). Further, the relationship between invasive plants and wildfire could be characterized as a "cause and effect dilemma" as we know that in some contexts, specific invasive plant species, such as cheatgrass (*Bromus tectorum*) have been responsible for increasing fuel loads, and thus, intensity of wildfires while climate-caused changes to wildfire regimes may alter post-fire recovery vegetation trajectories in favour of invasive plant species over desirable native plant species (Pilliod et al. 2021).

Given that native and exotic species fundamentally differ in their response to wildfire (Alba et al. 2015), the multiple effects of plant invasions after wildfire can complicate the task of restoring ecosystems. Post-wildfire changes to the ecosystem have resulted in altered vegetation trajectories (Chileen et al. 2020). This has been seen in old-growth Douglas fir (*Pseudotsuga menziesii*), lodgepole pine (*Pinus contorta*), and Engelmann spruce (*Picea engelmannii*) sites in the Flathead National Forest in Montana where a study observed a high number of non-native plant species post-fire (Rew and Johnson 2010). Similarly, a study by Wolfson et al. (2005 in Rew and Johnson 2010) showed that burn severity in Arizona ponderosa pine communities had an impact on the germination of diffuse knapweed (*C. diffusa*) following fire with mean seed germination by month significantly higher over time in the severely burned than unburned or moderately burned sites. Vegetation trajectory is also a concern in addressing resiliency to future wildfires as invasive plants are often an important component of biomass that promotes wildfires and so proactive management to reduce future wildfire risk is an important consideration in dry climates (Faccenda and Daehler 2022). Sagebrush steppe, which once occupied approximately 45 million ha across the western United States, is such an example. Fire-invasion feedback loops, a phenomenon that describes invasive plant invasion that increases wildfire size and frequency then increasing further invasion, are partially responsible for the loss of nearly half of sagebrush steppe lands (Chambers et al. 2014). These lands, which were predominately perennial native bunchgrasses that are considered more resistant to fire as well as invasive plant invasion (Davies et al. 2015), have been transformed into monocultures of exotic annual grasses, which include species *Bromus tectorum* (cheatgrass), *Taeniatherum caput-medusae* (medusahead), and *Ventenata dubia* (ventenata). A transformation has led to soil erosion, alteration of hydrology, changes in nutrient cycling (Bansal et al. 2014), all resulting in the perpetuation of an environment highly susceptible to wildfire.

6.2.5 Post-Fire Considerations

6.2.5.1 Succession After Wildfire

Community succession following disturbance such as high-intensity wildfire is an important consideration that is difficult to predict. Three common mechanisms that can occur following disturbance (Applestein et al. 2018) are (1) early successional species facilitate later successional species; or (2) both early and late successional species co-exist without facilitating or inhibiting each other; or (3) early successional species inhibit establishment or growth of later successional species (Applestein et al. 2018). While consideration of these mechanisms may not be directly helpful in predicting post-wildfire invasion outcomes, as the results of high-intensity wildfire may most accurately be characterized as cascading effects, they do help illuminate the complex web of relational considerations (Grenz 2022) needed to anticipate possible outcomes informing both monitoring and restoration planning. While the focus is often on the invaders themselves in invasive species research, the interactions among invaders in these disturbed systems may deserve more attention as early invaders may facilitate or inhibit establishment of other species (Applestein et al. 2018). The phenomenon "invasional meltdown" has been used to describe how initial invaders can lower landscape resistance such that it becomes vulnerable to secondary invaders via "invader-invader facilitation" (O'Loughlin and Green 2017). Further, vegetation trajectories, whether plants are native or invasive, should be carefully observed as both can have long-term effects on succession (O'Loughlin and Green 2017).

6.2.5.2 Wildfire Restoration Responses

Within our own research area, there is considerable variability in vegetation responses even within the same habitat type at the same burn severity. Observations that support the findings of Applestein and Germino (2022) revealed key differences in invasion patterns and successional mechanisms among exotic annual grasses. They express similar concerns to us that agencies responsible for post-wildfire restoration activities address specific plant communities and/or habitat types uniformly as a whole despite observation that there can be considerable differences in how each invader relates to its environment. This emphasizes the importance of wildfire-specific data on vegetation trajectories and detailed on-the-ground knowledges. Seeding is a typically expressed first restorative action, but these findings (Applestein and Germino 2022), along with our preliminary research results, emphasize the importance in choosing seed mixes with greater specificity to specific areas (rather than one seed mix over what may be considered the same habitat type). Consideration should be given that seed mixes be formulated to include species that best compete with a specific invader of concern and be diversified to compete with different invaders across landscapes to provide the best overall resistance to invasion. This also illuminates the importance of quality baseline invasive plant presence/non-presence data, as well as other land uses (forestry, agriculture), prior to wildfire as invasive plant prevention and management activities (control, increasing landscape resistance) could be prioritized within the wildfire area based on knowledge of historic priority invasive plant populations. This information is also important to preventing human facilitated invasive plant dispersal, where post-wildfire restoration activities such as seeding, replanting, road repairs, fire break maintenance, etc. could spread historic populations as they reappear on the landscape (Reilly et al. 2020).

6.3 Flooding and Storms

Expected trends in extreme weather include extreme precipitation, and recorded extreme precipitation events have increased steadily on a global basis, as well as in the Northern Hemisphere (the context for this chapter), as evaluated over 1964 to 2013 (AghaKouchak et al. 2020). This increase in the intensity of precipitation would naturally be expected to lead to increasing flooding as well, and although flood damage has been increasing over the past 60 years, this is attributed to changes in infrastructure rather than a measurable increase in flood magnitude or frequency at a global level (AghaKouchak et al. 2020). Still, extreme flooding events in particular regions continue apace as a result of extreme precipitation events, and result in unprecedented and large-scale impacts as seen in Pakistan in 2022 when monsoon rains flooded more than a third of the country and affected more than 33 million people (Sarkar 2022).

Storms have been observed to becoming more intense and frequent in many world regions, including areas like North America, such as the increase in both the frequency and magnitudes of hurricanes forming in the North Atlantic Basin (Knutson et al. 2021). For invasive plant species, the ramifications of extreme flooding are generally quite similar to those for extreme flooding, and indeed many extreme storms are accompanied by extensive flooding.

As the frequency and extent of flooding increase with climate change, as expected, a concomitant increase in the spread of most invasive plant species associated with aquatic or riparian areas is generally predicted (Easterling et al. 2000; Diez et al. 2012). Likewise, increasing frequency and severity of storms would be expected to promote greater spread of invasive plants generally (Bhattarai and Cronin 2014; Murphy and Metcalfe 2016). These predicted dynamics could be predicted readily through logical deduction, especially considering the tendency of invasive plants to be highly adapted to disturbance, and yet there is a paucity of available research on the impacts of flooding and storms on the invasion dynamics of invasive plants.

Even without the extremes of flooding predicted by climate change, many bodies of water are prone to periodic flooding, and most aquatic and riparian plant species tend to be adapted to thrive and/or exhibit increased reproduction when seasonal or periodic flooding occurs (Poff 2002; Aronson et al. 2017). Invasive plants tend to be better at utilizing riparian areas subject to frequent flooding as a means of proliferating because of a generally higher tolerance to disturbance. Kercher and Zedler (2004) compared a number of native and invasive wetland angiosperms and found that some plants fared worse in flooding than others because they were adapted to drier, more inland areas. They also found that grasses and graminoids nearly always withstood flooding better than forbs, and that in particular, tall plants which tend to be invasive, such as reed canary grass (*Phalaris arundinacea*) and broadleaf cattail (*Typha latifolia*), tended to take advantage of flooding to become dominant in an area even if native flood-tolerant plants were present (Kercher and Zedler 2004). Under climate change, another invasive plant with a tall stature, phragmites (*Phragmites australis*), is well equipped to thrive in areas where water levels fluctuate widely as periodic dropping of water levels promote seed germination on exposed benthic areas (Tougas-Tellier et al. 2015). It is clear that invasive plants differ greatly in how their dissemination and persistence strategies and ecophysiology are adapted to flooding and other extreme events, and a more comprehensive understanding of the threat of increased spread within the context of such climate events requires more research.

6.3.1 Site-Specific Variation

6.3.1.1 The 100-Year Flood in British Columbia in 2021

In November 2021, the south coast region of British Columbia experienced extreme rainfall due to a high magnitude atmospheric river originating over the Pacific Ocean causing the costliest natural disaster in British Columbia according to the Insurance Bureau of Canada (Gillett et al. 2022). Gillett et al. (2022) calculated that the November 2021 flooding was at least 60% more likely due to human-induced climate change. The damage included extensive flooding in the Fraser Valley including landslides, extensive highway damage (e.g., the trans-Canada highway was closed for several days), flooding of farmland, death of livestock, and many other costs not accounted for in insurance claims (Gillett et al. 2022). Rivers in the area exceeded normal flow rates, and for the Chilliwack River, at the height of the flooding on November 15th, a discharge of 10 times the mean rate was observed, resulting in extensive flooding and erosion of the stream banks (Ham and Church 2000; Government of Canada 2021). The potential spread of invasive plant species due to the flooding represents a cost not accounted for in the insurance claims resulting from this natural disaster.

Invasive knotweed (*Reynoutria* spp.) is one of the costliest invasive plant taxa in British Columbia and other parts of North America and Europe both in terms of its impacts and management (Clements et al. 2016; Gillies et al. 2016; Drazan et al. 2021). It thrives in riparian areas along fast-flowing rivers subject to frequent flooding (Duquette et al. 2016; Colleran et al. 2020) such as the Chilliwack River. Our research is evaluating the impact of the November 2021 flood on knotweed distribution along the Chilliwack River. In 2022, we recorded 1690 infestations of knotweed species along the mainstem of the river, when there had only been 341 surveyed in 2019 (Clements, unpublished data). The majority of the infestations surveyed in 2022 in the riparian zone were very small in both size and stature, with only 24% over one meter in height at the expected time of full plant maturity (approximately half their expected height), indicating that flooding led to rapid

Fig. 6.2 Knotweed (*Reynoutria* spp.) growing in woody debris spread in the Chilliwack River, BC, Canada during the November 2021 100-year flood event

colonization by knotweed rhizome pieces being carried downstream. Many of the new infestations were among debris spread by the flooding or on new land (especially islands) created during the intense rainfall event (Fig. 6.2).

The suitability of riverbanks generally for colonization following flooding is expected to vary with certain areas creating more suitable growing conditions. Microsite characteristics vary considerably on the Chilliwack River and we are currently evaluating what makes certain sites more vulnerable to invasion. Knotweed is known to alter soil pH and nutrients such as nitrogen (Aguilera et al. 2010) and phosphorus (Jones 2022), indicating that it may prefer sites with similar soil conditions. Regeneration rates tend to be reduced in nutrient poor soils (Navratil et al. 2021; Martin 2019), while it grows quickly in areas with high levels of nitrogen present (Parepa et al. 2019). Therefore, temporary pulses of nitrogen on riverbanks from flood runoff could stimulate knotweed population expansion (Parepa et al. 2019) as could the open areas of high disturbance created by flooding conditions given the species light requirements for growth (Martin 2019). Secondary spread caused by flood event recovery activities, such as movement of soil and debris or vegetation management, may facilitate the secondary spread (i.e., distribution from original foci) of knotweed along roadsides and railways (Barney 2006). Studies by Parepa et al. (2013, 2019), which examined the effects of soil biota inoculates and the timing of nitrogen pulses on knotweed growth, revealed highly variable responses, underlining the importance of assessing species response in the contexts of microsite conditions.

6.3.1.2 Tropical Storm Irene in New England

The state of Vermont experienced widespread flooding when Tropical Storm Irene hit in August 2011, with rainfall levels exceeding 100-year levels or even 500-year levels in some areas (Anderson et al. 2017). Many streams experienced record peak flows which facilitated extensive spread of Japanese knotweed (*Reynoutria japonica*) rhizome, and stem fragments (Colleran and Goodall 2014). Colleran and Goodall (2014) analysed the regenerating fragments in several rivers in the season following Tropical Storm Irene. Different streams exhibited different levels of regeneration and knotweed population growth. They found in addition to site-specific factors, root mass and underground growth were important determinants of success early in the regeneration process (Colleran and Goodall 2014).

Like most invasive plants, knotweed species tend to proliferate in the wake of disturbance. Colleran and Goodall (2015) found that not only were the rhizome and stem fragments spread by the Tropical Storm Irene itself, but further spread and regeneration resulted from efforts to clean up after the storm, through dredging and removal of debris. The bare ground exposed by removal of existing vegetation provided an ideal substrate for the regenerating knotweed fragments.

6.3.1.3 Role of Site-Specific Factors

As we have seen in reviewing these two instances of flooding, one on the west coast of North America and one on the east coast, site-specific factors are important in determining how much flooding will stimulate population growth. Similarly, Predick and Turner (2008) found that landscape configuration, including the degree of habitat fragmentation and amount of edge habitat, shaped the response of invasive shrubs in the floodplain forests bordering the Wisconsin River. They found that in areas of the forest experiencing altered flood regimes and more edge habitat due to human influences (e.g., roads) were more invaded by shrubs (e.g., *Lonicera* spp. and *Rhamnus* spp.) than large unfragmented forest expanses. Site-specific factors must be carefully considered in understanding and managing the impacts of increased flooding under climate change on the spread and establishment of invasive plants.

6.3.2 North American Regions Vulnerable to Plant Invasion Linked to Flooding

North American regions most likely vulnerable to invasion after flooding at an intuitive level would be wetter areas that frequently experience high precipitation levels. However, when normally dry areas experience flash flooding, the magnitude of occasional large-scale flooding can be quite large. Thus, it is difficult to rank North American regions by vulnerability to plant invasion due to extreme flooding linked to climate change. Notwithstanding, it is still useful to examine particular ways different regions are vulnerable to invasion after flooding events. Just as site-specific factors are important in assessing response to flooding, region-specific factors may be of value in understanding the potential impacts of climate change.

The southwest corner of North America is known for hot, dry weather, with shortage of water for agriculture and other purposes a major concern and predicted to be increasingly serious (Archer and Predick 2008). Archer and Predick (2008) comment on the predicted paradox, however, that along with longer dry periods, will come more intense rainfall events in future climate change scenarios for the region, producing both more droughts and more floods. Non-native plants are predicted to increase in a dry climate punctuated by flooding, as the diverse native annuals in arid areas giving way to non-native grasses like cheatgrass (*B. tectorum*) and non-native shrubs such as saltcedar (*Tamarix ramosissima*) (Archer and Predick 2008; Polacik and Maricle 2013). Polacik and Maricle (2013) found that saltcedar could become acclimated to flooding with photosynthesis rates returning to normal after 3 weeks of flooded conditions. More research is needed to understand the potential of desert flooding in these areas of the U.S. southwest for dispersing the propagules of invasive plants.

The picture is quite different in the southeast corner of the U.S. While rainfall is generally more abundant, climate-caused alterations of riverine flows- riverine storm flows are expected to increase while base flows decrease, may exacerbate impacts of riparian invasive plants (Flanagan et al. 2015). By comparing riparian vegetation on dammed and undammed rivers, Flanagan et al. (2015) concluded that in this future climate scenario, invasive plants will become more competitive than native plants by taking advantage of the combination of human-altered watershed conditions along with increased disturbance due to flooding. Potential for increasing future storm activity in this region has been dramatically highlighted by the more intense and frequent hurricanes due to warmer ocean temperatures (Ting et al. 2019; Knutson et al. 2021). Invasive *Phragmites* (*P. australis*) stands have been on the increase in wetlands on the Atlantic and Gulf Coasts which take the brunt of this hurricane activity (Bhattarai and Cronin 2014). Bhattarai and Cronin (2014), using historical aerial photography, found that 81% of the variation in infestation size could be explained by the frequency of hurricane-strength winds. They listed several possible biological mechanisms for this success amidst extreme weather including: *Phragmites* may recover and regrow more rapidly than native species after storms; the effects of extreme rainfall on salinity levels and resulting increase in site invasibility; and/or recently increased anthropogenic nutrient loading in wetlands exacerbated by hurricane activity.

The northwest corner of North America includes a whole array of ecosystems created by mountainous topography, and thus it is difficult to generalize about climate trends and flooding frequency. As discussed above with reference to the catastrophic November 2021 floods in British Columbia, the coastal northwest zones are predicted to experience more extreme flooding events in the future due to climate change (Gillett et al. 2022). Interior habitats in the northwest region of the continent tend to be more arid, and yet as seen in southwest, extreme flooding has occurred and is predicted to increase in frequency, potentially impacting the distribution and establishment of invasive plants, although droughts and fires, as discussed previously, are seen to be much more significant (Chambers and Pellant 2008).

Finally, the northeast corner of North America may also be subject to increased flooding due to climate change, as discussed previously in terms of the impact of Tropical Storm Irene on knotweed populations. Utilizing ensemble models to project future habitat suitability of the International Union for the Conservation of Nature's list of "100 of the world's worst invasive species", Bellard et al. (2013) predicted future invasive species hotspots globally. They projected northeastern North America as one such hotspot, which under all modelling scenarios showed an increased number of invasive species. Ghanbari et al. (2021) emphasized the potent hazard from the worsening combination of coastal and riverine flooding along the eastern U.S. coast, but there are few studies examining the effect of this issue on invasive plant proliferation.

6.3.3 Effects of Extreme Weather on the Spread of Invasive Plants

6.3.3.1 Invasive Plants Benefiting from Increasingly Extreme Weather and Flooding

There are a number of invasive plant species in North America whose adaptations to extreme weather and flooding enable them to benefit from such events (Table 6.2). The seven species we have profiled are highly successful, each having invaded much of the continent already. Still, it is evident that most have not occupied all of the habitats they could possibly occupy, and if flooding aids in their dispersal (e.g., *Reynoutria* spp., *Impatiens glandulifera*) or establishment (e.g., *P. australis*), increased flooding will promote their expansion unless effective prevention and management efforts are deployed.

6.3.3.2 Invasive Plants May Increase Flood Risk

Many of the plant species listed in Table 6.2 not only tend to benefit from flooding, but when abundant, they may make flooding worse through increasing erosions. Colleran et al. (2020) refer to knotweeds as "catalysts" for riparian erosion. The large rhizomes they possess may tend to displace other roots of other plants that serve to hold the soil, increasing the risk riverbanks will fail under flooding conditions. With potential increasing frequency and magnitude of floods, a feedback loop could enable knotweed populations to grow continually if unchecked, leading to further riparian degradation. Invasive plants may also jeopardize hydrological processes, increasing flood risk, through the presence of dense growth in waterways, as has been shown for *Arundo donax* in California waterways (Spencer et al. 2013).

Table 6.2 Some well-studied invasive plants in North America adapted to flooding. Distribution information from plants.usda.gov

Species	North American distribution	Adaptation(s) to flooding	Reference
Arundo donax (Giant reed)	Southern U.S. states	Rhizome fragments can be spread via flooding to establish new plants	Boland (2008)
Impatiens glandulifera (Himalayan balsam)	NW and NE U.S., SW and SE Canada	Seeds adapted for water dispersal, effective colonizer of disturbed riverbanks	Čuda et al. (2017)
Lepidium latifolium (Perennial pepperweed)	Western and central U.S., western Canada and Quebec	Maintains relatively high photosynthesis and accumulation of soluble sugar in roots when flooded	Chen et al. (2005)
Phalaris arundinacea (Reed canary grass)	All North America except the Arctic and some southern states	Grows back much faster after flooding than native species	Kercher et al. (2007)
Phragmites australis (Phragmites)	All North America except the Arctic	Fluctuating water levels due to flooding promote seed germination	Tougas-Tellier et al. (2015)
Reynoutria spp. (Knotweed)	U.S. except southern states, southern Canada except prairies	Rhizomes and seeds adapted to floating downstream to colonize new areas	Duquette et al. (2016)
Tamarix ramosissima (Salt cedar)	Southern U.S. and the Dakotas	Capable of acclimatizing to flood conditions despite reduced photosynthesis	Polacik and Maricle (2013)

6.3.4 Post-Flooding Considerations: Best Practices for Restorative Mitigation

In the wake of climate disasters such as flooding, there are many competing priorities and land managers may neglect to monitor invasive plant populations in the process. If greatly expanded populations of invasive plants stem from floods, there may be a narrow temporal window when a rapid response could represent an opportunity to control invasive plant growth in early stages, when efficacy of management strategies are often greatest. Colleran and Goodall (2015) demonstrated that for *Reynoutria japonica*, the plants colonizing from rhizome or stem fragments 1 year or less after flooding, were much easier to remove than large mature patches. Knotweed infestations left to continuously grow over a number of years can generally only be controlled successfully through application of systemic herbicides, which are not permitted within riparian zones in some jurisdictions (Clements et al. 2016).

In mitigating the spread of invasive species, it is important to view control of invasive plants in a holistic, ecosystem context (Zavaleta et al. 2001) including both animal-plant and human-plant relationships. De Jager et al. (2013) demonstrated that although floodplain forests may often be resilient following flooding because of the regenerative ability of native tree seedlings, along the upper Mississippi River where deer were overly abundant, their extensive browsing allowed more invasive plants like *P. arundinacea* to colonize after flooding. Boland (2008) observed that although some recruitment of new *A. donax* plants was occurring in a river in California through rhizomes being fragmented during heavy flooding, far more recruitment occurred through disruption of the plant root systems through bulldozers in the flood recovery process.

In order to create climate resilient systems, it is important to assist ecosystems to endure flooding through ecological restoration practices that utilize native vegetation to minimize ecosystem disturbance. For example the native tree, *Populus fremontii*, has been shown to provide stability to riparian zones threatened by *T. ramosissima* (Sher et al. 2000). In the case of the arid southwestern United States, where the impacts of differing climate events may be compounded- both flooding and fire- restoration strategies must be carefully considered and the impacts of such events not subject to confirmation bias- that they are inherently negative. Ellis (2001) monitored recovery of the vegetation in riparian areas subject to increased fire levels at study sites on the Rio Grande in New Mexico. She concluded that the best prescription for the increasingly severe fires was to remove excess fuel either by re-establishing flooding or manual removal of fuel, especially the invasive *T. ramosissima*. The point that flooding is not always bad is important, and in fact Death et al. (2015) claim that the more extreme floods expected with climate change may in some cases have positive impacts on ecosystems through "overwhelming current anthropogenic constraints and infrastructure to increase habitat complexity and floodplain area". What is clear is that it is critical to understand each ecosystem well enough to know how to create a diverse, functioning, and resilient system. Similarly, regarding confirmation bias, we must approach the presence of invasive species with an epistemic openness as it actually may be better not to eliminate the invasive plants present if they are playing a useful role in the overall system functioning and/or the management techniques themselves may do more harm than good. Both areas of which are deserving of more dedicated research.

6.4 Conclusions and Future Recommendations

While we are often looking within the context of ecological restoration for well-researched, uniform approaches to recovery from disturbances of different types, such as wildfire, it is increasingly clear that context specific approaches rooted in local knowledges are needed. While research specific to vegetation trajectories, both invasive and native plant species, is important to informing severe climate event recovery, we must also focus on the complex relationships influencing these trajectories. Comprehensive baseline data, plant surveys of areas vulnerable to climate events, appear to be critical to informing prioritization of prevention and management activities post-event. While cookie-cutter responses would be easier to apply across multiple landscapes and species, particularly because of the increase in the number and intensity of climate events globally, these will not necessarily lead to effective restoration outcomes. Nor will they consider the diverse needs and values, including socio-cultural considerations, of each of these places. For example, both wildfires and the landscapes they burn today, are not the same as those of the past or given climate projections, those in future. The best we can hope for is for researchers to work together to identify, examine, and illuminate important relationships that influence plant trajectories after each type of severe climate event that are applicable regardless of context, and can be informed by local and Indigenous knowledges. Acknowledging and better understanding these relationships are also critical to provide a preventative focus for land stewardship activities as we work to reduce the risk of impacts of major climate events in the future and improve ecological function through the promotion of desirable plant species- resulting in lands resilient to plant invasion.

References

Agee JK (1996) Achieving conservation biology objectives with fire in the Pacific Northwest. Weed Technol 10:417–421

AghaKouchak A, Chiang F, Huning LS, Love CA, Mallakpour I, Mazdiyasni O, Moftakhari H, Papalexiou SM, Ragno E, Sadegh M (2020) Climate extremes and compound hazards in a warming world. Annu Rev Earth Planet Sci 48:519–548. https://doi.org/10.1146/annurev-earth-071719-055228

Aguilera AG, Alpert P, Dukes JS, Harrington R (2010) Impacts of the invasive plant *Fallopia japonica* (Houtt.) on plant communities and ecosystem processes. Biol Invasions 12:1243–1252. https://doi.org/10.1007/s10530-009-9543-z

Alba C, Skálová H, McGregor KF, D'Antonio C, Pyšek P (2015) Native and exotic plant species respond differently to wildfire and prescribed fire as revealed by meta-analysis. J Veg Sci 26(1): 102–113

Anderson I, Rizzo DM, Huston DR, Dewoolkar MM (2017) Analysis of bridge and stream conditions of over 300 Vermont bridges damaged in Tropical Storm Irene. Struct Infrastruct Eng 13(11):1437–1450. https://doi.org/10.1080/15732479.2017.1285329

Applestein C, Germino MJ (2021) Detecting shrub recovery in sagebrush steppe: comparing Landsat-derived maps with field data on historical wildfires. Fire Ecol 17(1). https://doi.org/10.1186/s42408-021-00091-7

Applestein C, Germino MJ (2022) Patterns of post-fire invasion of semiarid shrub-steppe reveals a diversity of invasion niches within an exotic annual grass community. Biol Invasions 24(3): 741–759

Applestein C, Germino MJ, Fisk MR (2018) Vegetative community response to landscape-scale post-fire herbicide (Imazapic) application. Invasive Plant Sci Manag 11(3):127–135

Archer SR, Predick KI (2008) Climate change and ecosystems of the southwestern United States. Rangelands 30(3):23–28. https://doi.org/10.2111/1551501X(2008)30[23:CCAEOT]2.0.CO;2

Aronson MF, Patel MV, O'Neill KM, Ehrenfeld JG (2017) Urban riparian systems function as corridors for both native and invasive plant species. Biol Invasions 19:3645–3657. https://doi.org/10.1007/s10530-017-1583-1

Bansal S, Sheley RL, Blank B, Vasquez EA (2014) Plant litter effects on soil nutrient availability and vegetation dynamics: changes that occur when annual grasses invade shrub-steppe communities. Plant Ecol 215(3):367–378. https://doi.org/10.1007/s11258-014-0307-1

Barney JN (2006) North American history of two invasive plant species: phytogeographic distribution, dispersal vectors, and multiple introductions. Biol Invasions 8:703–717. https://doi.org/10.1007/s10530-005-3174-9

Bellard C, Thuiller W, Leroy B, Genovesi P, Bakkenes M, Courchamp F (2013) Will climate change promote future invasions? Glob Chang Biol 19(12):3740–3748. https://doi.org/10.1111/gcb.12344

Bhattarai GP, Cronin JT (2014) Hurricane activity and the large-scale pattern of spread of an invasive plant species. PLoS One 9(5):e98478. https://doi.org/10.1371/journal.pone.0098478

Boland JM (2008) The roles of floods and bulldozers in the break-up and dispersal of *Arundo donax* (giant reed). Madrono 55(3):216–222. https://doi.org/10.3120/0024-9637-55.3.216

Brookes ML, D'Antonio CM, Richardson DM, Grace JB, Keeley JE, DiTomaso JM, Hobbs RJ, Pellant M, Pyke D (2004) Effects of invasive alien plants on fire regimes. Bioscience 54(7):677–688. https://doi.org/10.1641/0006-3568(2004)054[0677:EOIAPO]2.0.CO;2

Brooks ML, D'Antonio CM, Richardson DM, Grace JB, al et. (2004 Jul) Effects of invasive alien plants on fire regimes. Bioscience 54(7):677–688

Brunson MW, Tanaka J (2011) Economic and social impacts of wildfires and invasive plants in American deserts: lessons from the Great Basin. Rangeland Ecol Manag 64(5):463–470

Chambers JC, Pellant M (2008) Climate change impacts on northwestern and intermountain United States rangelands. Rangelands 30(3):29–33. https://doi.org/10.2111/1551-501X(2008)30[29:CCIONA]2.0.CO;2

Chambers JC, Bradley BA, Brown CS, D'Antonio C, Germino MJ, Grace JB, Hardegree SP, Miller RF, Pyke DA (2014) Resilience to stress and disturbance, and resistance to Bromus tectorum L. invasion in cold desert shrublands of western North America. Ecosystems 17:360–375

Chen H, Qualls RG, Blank RR (2005) Effect of soil flooding on photosynthesis, carbohydrate partitioning and nutrient uptake in the invasive exotic *Lepidium latifolium*. Aquat Bot 82(4):250–268. https://doi.org/10.1016/j.aquabot.2005.02.013

Chileen BV, McLauchlan KK, Higuera PE, Parish M, Shuman BN (2020) Vegetation response to wildfire and climate forcing in a Rocky Mountain lodgepole pine forest over the past 2500 years. The Holocene 30(11):1493–1503

Clements DR, Larsen T, Grenz J (2016) Knotweed management strategies in North America with the advent of widespread hybrid bohemian knotweed, regional differences, and the potential for biocontrol via the psyllid *Aphalara itadori* Shinji. Invasive Plant Sci Manag 9(1):60–70. https://doi.org/10.1614/IPSM-D-15-00047.1

Colleran BP, Goodall KE (2014) In situ growth and rapid response management of flood-dispersed Japanese knotweed (*Fallopia japonica*). Invasive Plant Sci Manag 7(1):84–92. https://doi.org/10.1614/IPSM-D-13-00027.1

Colleran BP, Goodall KE (2015) Extending the timeframe for rapid response and best management practices of flood-dispersed Japanese knotweed (*Fallopia japonica*). Invasive Plant Sci Manag 8(2):250–253. https://doi.org/10.1614/IPSM-D-14-00046.1

Colleran B, Lacy SN, Retamal MR (2020) Invasive Japanese knotweed (*Reynoutria japonica* Houtt.) and related knotweeds as catalysts for streambank erosion. River Res Appl 36(9):1962–1969. https://doi.org/10.1002/rra.3725

Čuda J, Rumlerová Z, Brůna J, Skálová H, Pyšek P (2017) Floods affect the abundance of invasive Impatiens glandulifera and its spread from river corridors. Divers Distrib 23(4):342–354. https://doi.org/10.1111/ddi.12524

Davies KW, Boyd CS, Johnson DD, Nafus AM, Madsen MD (2015) Success of seeding native compared with introduced perennial vegetation for revegetating medusahead invaded sagebrush rangeland. Rangel Ecol Manag 68(3):224–230

De Jager NR, Cogger BJ, Thomsen MA (2013) Interactive effects of flooding and deer (*Odocoileus virginianus*) browsing on floodplain forest recruitment. For Ecol Manag 303:11–19. https://doi.org/10.1016/j.foreco.2013.02.028

Death RG, Fuller IC, Macklin MG (2015) Resetting the river template: the potential for climate-related extreme floods to transform river geomorphology and ecology. Freshw Biol 60(12): 2477–2496. https://doi.org/10.1111/fwb.12639

Dennehy C, Alverson ER, Anderson HE, Clements DR, Gilbert R, Kaye TN (2011) Management strategies for invasive plants in Pacific northwest prairies, savannas, and oak woodlands. Northwest Sci 85(2):329–351. https://doi.org/10.3955/046.085.0219

Dickson-Hoyle S, John C (2021) Elephant Hill: Secwе́pemc leadership and lessons learned from the collective story of wildfire recovery. Secwepemcul'ecw Restoration and Stewardship Society

Diez JM, D'Antonio CM, Dukes JS, Grosholz ED, Olden JD, Sorte CJ, Blumenthal DM, Bradley BA, Early R, Ibáñez I, Jones SJ (2012) Will extreme climatic events facilitate biological invasions? Front Ecol Environ 10(5):249–257. https://doi.org/10.1890/110137

Drazan D, Smith AG, Anderson NO, Becker R, Clark M (2021) History of knotweed (*Fallopia* spp.) invasiveness. Weed Sci 69(6):617–623. https://doi.org/10.1017/wsc.2021.62

Duquette MC, Compérot A, Hayes LF, Pagola C, Belzile F, Dubé J, Lavoie C (2016) From the source to the outlet: understanding the distribution of invasive knotweeds along a North American river. River Res Appl 32(5):958–966. https://doi.org/10.1002/rra.2914

Easterling DR, Meehl GA, Parmesan C, Changnon SA, Karl TR, Mearns LO (2000) Climate extremes: observations, modeling, and impacts. Science 289(5487):2068–2074. https://doi.org/10.1126/science.289.5487.2068

Ellis LM (2001) Short-term response of woody plants to fire in a Rio Grande riparian forest, Central New Mexico, USA. Biol Conserv 97(2):159–170. https://doi.org/10.1016/S0006-3207(00)00108-7

Emery SM, Uwimbabazi J, Flory SL (2011) Fire intensity effects on seed germination of native and invasive eastern deciduous forest understory plants. For Ecol Manag 261(8):1401–1408

Faccenda K, Daehler CC (2022) A screening system to predict wildfire risk of invasive plants. Biol Invasions 24(2):575–589

Flanagan NE, Richardson CJ, Ho M (2015) Connecting differential responses of native and invasive riparian plants to climate change and environmental alteration. Ecol Appl 25(3): 753–767. https://doi.org/10.1890/14-0767.1

Fuchs MA (2001) Towards a recovery strategy for Garry oak and associated ecosystems in Canada: ecological assessment and literature review. Technical Report GBEI/EC-00-030. Environment Canada, Canadian Wildlife Service, Pacific and Yukon Region

Gedalof Z, Peterson DL, Mantua NJ (2005) Atmospheric, climatic, and ecological controls on extreme wildfire years in the northern United States. Ecol Appl 15:154–174

Ghanbari M, Arabi M, Kao SC, Obeysekera J, Sweet W (2021) Climate change and changes in compound coastal-riverine flooding hazard along the US coasts. Earth Future 9(5): e2021EF002055. https://doi.org/10.1029/2021EF002055

Gianoli E, Molina-Montenegro MA (2021) Evolution of physiological performance in invasive plants under climate change. Evolution 75(12):3181–3190

Gillett NP, Cannon AJ, Malinina E, Schnorbus M, Anslow F, Sun Q, Kirchmeier-Young M, Zwiers F, Seiler C, Zhang X, Flato G, Wan H, Li G, Castellan A (2022) Human influence on the 2021 British Columbia floods. Weather Clim Extremes 36:100441. https://doi.org/10.1016/j.wace.2022.100441

Gillies S, Clements DR, Grenz J (2016) Knotweed (*Fallopia* spp.) invasion of North America utilizes hybridization, epigenetics, seed dispersal (unexpectedly), and an arsenal of physiological tactics. Invasive Plant Sci Manag 9(1):71–80. https://doi.org/10.1614/IPSM-D-15-00039.1

Government of Canada (2021) Real-time hydrometric data graph for Chilliwack river at Vedder crossing (08MH001) [BC]. Official Website of the Government of Canada. Accessed 16 Dec 2021. https://wateroffice.ec.gc.ca/report/real_time_e.html?stn=08MH001&mode=Graph&startDate=2021-11-10&endDate=2021-prm1Min=&prm2=47&y2Max=&y2Min=

Grenz J (2022) The responsibility of relational consideration: invasive and non-native species in the context of indigenous food security and food sovereignty in a changing climate. In: Jacobs L (ed) Indigenous voices: critical reflections on traditional ecological knowledge. Oregon State University Press. Manuscript with Publisher

Gulf Islands Driftwood Community Newspaper: Salt Spring Archives. 2009. https://saltspringarchives.com/driftwood/2009/v49n24June17-2009.pdf

Hagmann RK, Hessburg PF, Prichard SJ, Povak NA, Brown PM, Fulé PZ et al (2021) Evidence for widespread changes in the structure, composition, and fire regimes of western north American forests. Ecol Appl 31(8):e02431

Ham D, Church M (2000) Bed-material transport estimated from channel morphodynamics: Chilliwack River, British Columbia. Earth Surf Process Landf 25:1123–1142. https://doi.org/10.1002/1096-9837(200009)25:10<1123::AID-ESP122>3.0.CO;2-9

Hoffman KM, Wickham SB, McInnes WS, Starzomski BM (2019) Fire exclusion destroys habitats for at-risk species in a British Columbia protected area. Fire 2(3):48, 1–13. https://doi.org/10.3390/fire2030048

Horn KJ, Nettles R, Clair SB (2015) Germination response to temperature and moisture to predict distributions of the invasive grass red brome and wildfire. Biol Invasions 17(6):1849–1857

Howe HE (1995) Succession and fire season in experimental prairie plantings. Ecology 76:1917–1925

Jones V (2022) Understanding plant-soil-management interactions of *Reynoutria* spp (knotweeds) to inform ecological restoration. M.Sc. thesis, University of British Columbia

Keeley JE, Keeley DJ, Fotheringham CJ (2003) Fire and grazing impacts on plant diversity and alien plant invasions in the southern Sierra Nevada. Ecol Appl 13:1355–1374

Kercher SM, Zedler JB (2004) Flood tolerance in wetland angiosperms: a comparison of invasive and noninvasive species. Aquat Bot 80(2):89–102. https://doi.org/10.1016/j.aquabot.2004.08.003

Kercher SM, Herr-Turoff A, Zedler JB (2007) Understanding invasion as a process: the case of *Phalaris arundinacea* in wet prairies. Biol Invasions 9:657–665. https://doi.org/10.1007/s10530-006-9066-9

Knutson TR, Chung MV, Vecchi G, Sun J, Hsieh TL, Smith AJ (2021) Climate change is probably increasing the intensity of tropical cyclones. In: Critical issues in climate change science, Science brief. https://doi.org/10.5281/zenodo

Lea T (2006) Historical Garry oak ecosystems of Vancouver Island, British Columbia, pre-European contact to the present. Davidsonia 17:34–50

Lea T (2011) Garry oak and associated ecosystems: description and distribution. In: Clements DR, Hooper TD, Martell K, Masson C, Garry Oak Ecosystems Recovery Team (eds) Restoring British Columbia's Garry oak ecosystems: principles and practices, 520 pp. https://goert.ca/wp/wp-content/uploads/GOERT-Restorating-BC-GOE-2011.pdf

MacDougall AS (2005) Responses of diversity and invasibility to burning in a Northern Oak Savanna. Ecology 86:3354–3363

MacDougall AS, Turkington R (2005) Are invasive species the drivers or passengers of change in degraded ecosystems? Ecology 86(1):42–55. https://doi.org/10.1890/04-0669

Martin FM (2019) The study of the spatial dynamics of Asian knotweeds (*Reynoutria* spp.) across scales and its contribution for management improvement. Doctoral dissertation, Université Grenoble Alpes

Murphy HT, Metcalfe DJ (2016) The perfect storm: weed invasion and intense storms in tropical forests. Austral Ecol 41(8):864–874

Navratil O, Brekenfeld N, Puijalon S, Sabastia M, Boyer M, Pella H, Lejot J, Piola F (2021) Distribution of Asian knotweeds on the Rhône River basin, France: a multi-scale model of invasibility that combines biophysical and anthropogenic factors. Sci Total Environ 1(763):142995. https://doi.org/10.1016/j.scitotenv.2020.142995

O'Loughlin LS, Green PT (2017) Secondary invasion: when invasion success is contingent on other invaders altering the properties of recipient ecosystems. Ecol Evol 7(19):7628–7637

Ott JE, Kilkenny FF, Summers DD, Thompson TW (2019) Long-term vegetation recovery and invasive annual suppression in native and introduced postfire seeding treatments. Rangel Ecol Manag 72(4):640–653. https://doi.org/10.1016/j.rama.2019.02.001

Parepa M, Schaffner U, Bossdorf O (2013) Help from under ground: soil biota facilitate knotweed invasion. Ecosphere 4(2):1–1. https://doi.org/10.1890/ES13-00011.1

Parepa M, Kahmen A, Werner RA, Fischer M, Bossdorf O (2019) Invasive knotweed has greater nitrogen-use efficiency than native plants: evidence from a 15 N pulse-chasing experiment. Oecologia 191:389–396. https://doi.org/10.1007/s00442-019-04490-1

Parks SA, Abatzoglou JT (2020) Warmer and drier fire seasons contribute to increases in area burned at high severity in western US forests from 1985 to 2017. Geophys Res Lett 47(22): e2020GL089858

Pellatt MG, Gedalof ZE (2014) Environmental change in Garry oak (*Quercus garryana*) ecosystems: the evolution of an eco-cultural landscape. Biodivers Conserv 23:2053–2067. https://doi.org/10.1007/s10531-014-0703-9

Pilliod DS, Jeffries MA, Welty JL, Arkle RS (2021) Protecting restoration investments from the cheatgrass-fire cycle in sagebrush steppe. Conserv Sci Pract 3(10):10.1111/csp2.508

Poff NL (2002) Ecological response to and management of increased flooding caused by climate change. Philos Trans A Math Phys Eng Sci 360(1796):1497–1510. https://doi.org/10.1098/rsta.2002.1012

Pokorny ML, Mangold JM, Hafer J, Denny MK (2010) Managing spotted knapweed (Centaurea stoebe)–infested rangeland after wildfire. Invasive Plant Sci Manag 3(2):182–189

Polacik KA, Maricle BR (2013) Effects of flooding on photosynthesis and root respiration in saltcedar (*Tamarix ramosissima*), an invasive riparian shrub. Environ Exp Bot 89:19–27. https://doi.org/10.1016/j.envexpbot.2012.12.005

Polster D (2011) Restoration strategies. In: Clements DR, Hooper TD, Martell K, Masson C, Garry Oak Ecosystems Recovery Team (eds) Restoring British Columbia's Garry oak ecosystems: principles and practices, 520 pp. https://goert.ca/wp/wp-content/uploads/GOERT-Restoring-BC-GOE-2011.pdf

Predick KI, Turner MG (2008) Landscape configuration and flood frequency influence invasive shrubs in floodplain forests of the Wisconsin River (USA). J Ecol 96(1):91–102. https://doi.org/10.1111/j.1365-2745.2007.01329.x

Reilly MJ, McCord MG, Brandt SM, Linowksi KP, Butz RJ, Jules ES (2020) Repeated, high-severity wildfire catalyzes invasion of non-native plant species in forests of the Klamath Mountains, northern California, USA. Biol Invasions 22(6):1821–1828

Rew LJ, Johnson MP (2010) Reviewing the role of wildfire on the occurrence and spread of invasive plant species in wildland areas of the intermountain western United States. IPSM 3(4): 347–364

Sarkar S (2022) Pakistan floods pose serious health challenges. BMJ 378:o2141. https://doi.org/10.1136/bmj.o2141

Sher AA, Marshall DL, Gilbert SA (2000) Competition between native *Populus deltoides* and invasive *Tamarix ramosissima* and the implications for reestablishing flooding disturbance. Conserv Biol 14(6):1744–1754. https://doi.org/10.1111/j.1523-1739.2000.99306.x

Spencer DF, Colby L, Norris GR (2013) An evaluation of flooding risks associated with giant reed (*Arundo donax*). J Freshw Ecol 28(3):397–409. https://doi.org/10.1080/02705060.2013.769467

Ting M, Kossin JP, Camargo SJ, Li C (2019) Past and future hurricane intensity change along the US East Coast. Sci Rep 9(1):1–8. https://doi.org/10.1038/s41598-019-44252-w

Tougas-Tellier MA, Morin J, Hatin D, Lavoie C (2015) Freshwater wetlands: fertile grounds for the invasive *Phragmites australis* in a climate change context. Ecol Evol 5(16):3421–3435. https://doi.org/10.1002/ece3.1576

Transition Saltspring (2020) Climate action plan 2020 to 2030 executive summary. https://transitionsaltspring.com/wp-content/uploads/2021/02/1.-CAP-2.0-Executive-Summary.pdf

Trauernicht C, Pickett E, Giardina CP, Litton CM, Cordell S, Beavers A (2015) The contemporary scale and context of wildfire in Hawai'i. Pacific Sci 69(4):427–444

Williams R, Wang H (2021) Effects of wildfire and the presence of the invasive Paulownia tomentosa on the regeneration of native tree species in north-central Appalachia. Fire 4(3):60

Wolfson BAS, Kolb TE, Sieg CH, Clancy KM (2005) Effects of post-fire conditions on germination and seedling success of diffuse knapweed in northern Arizona. For Ecol Manag 216(1–3) 342–358. https://doi.org/10.1016/j.foreco.2005.05.047

Zavaleta ES, Hobbs RJ, Mooney HA (2001) Viewing invasive species removal in a whole-ecosystem context. Trends Ecol Evol 16(8):454–459. https://doi.org/10.1016/S0169-5347(01)02194-2

Understanding Eco-Geographical Relationship in Invaded Ranges by *Acacia longifolia* (Andrews) Willd.: An Intercontinental Case Study on *Acacia* Invasions

7

Jorge Luis P. Oliveira-Costa, Rui Ferreira de Figueiredo, Vânia Regina Pivello, Lia Montti, and César Fagúndez-Pachón

Abstract

Biological invasion is a process of ecosystem degradation caused by the proliferation of exotic species. The success of this process depends on the biological characteristics of the species (invasiveness) and the abiotic characteristics of

Manuscript resulting from the Doctoral thesis entitled "Biological invasion and environmental susceptibility in temperate regions: determinants and impacts of spread and invasion by *Acacia longifolia* (Andr.) Willd." carried out at the Faculty of Letters and Humanities of the University of Coimbra (Portugal).

J. L. P. Oliveira-Costa (✉)
Center of Studies in Geography and Spatial Planning (CEGOT), University of Coimbra, Coimbra, Portugal

R. F. de Figueiredo
Department of Geography and Tourism, CEGOT, Faculty of Letters and Humanities, University of Coimbra, Coimbra, Portugal
e-mail: ruiff@ci.uc.pt

V. R. Pivello
Department of Ecology, Institute of Biosciences, University of São Paulo (USP), São Paulo, Brazil
e-mail: vrpivel@ib.usp.br

L. Montti
Instituto de Investigaciones Marinas y Costeras (IIMyC), FCEyN, Universidad Nacional de Mar del Plata-CONICET, Mar del Plata, Argentina

Instituto de Geología de Costas y del Cuaternario (IGCyC), FCEyN, Universidad Nacional de Mar del Plata-CIC, Mar del Plata, Argentina

C. Fagúndez-Pachón
Departamento Interdisciplinario de Sistema Costero y Marino, Centro Universitario Regional del Este, Universidad de La Republica Oriental del Uruguay, Rocha, Uruguay
e-mail: cfagundez@cure.edu.uy

S. Tripathi et al. (eds.), *Plant Invasions and Global Climate Change*,
https://doi.org/10.1007/978-981-99-5910-5_7

ecosystems (invasibility). Concern about this process has been growing in recent times but, in most cases, decisions on its management have been taken without sufficient knowledge of the impacts involved. Invasive species have a wide phenotypic diversity which, associated with the ecological-geographic conditions of the territories, seems to increase their ability to invade both natural and humanized environments.

This communication aimed to (1) define and map the global and local cores (Portugal, Brazil, Argentina, and Uruguay) invaded by *Acacia longifolia*; (2) identify its impacts on coastal ecosystems; (3) understand the relationships between the patterns and processes responsible for the proliferation and invasion of this species; (4) and, finally, this information pretend to be a support to the future evaluation of the distribution patterns and ecological processes of the species to develop national and transnational invasive alien species (IAS) management. *Acacia longifolia* is a coastal shrub native to Australia and shows significant invasive potential. It was introduced first in Portugal at the end of the nineteenth century, about a century before it was brought to South America (Brazil, Argentina, Uruguay). We did a global analysis of the invasive species dynamics (regarding its invasibility/invasiveness), focusing on areas invaded by the species with different natural and socio-ecological characteristics. Considering that total eradication in the short-medium term is technically unfeasible, more pragmatic management solutions are required whose adaptation to local realities greatly benefits from these global analyses.

Keywords

Acacia · Biological invasions · Environmental susceptibility · Temperate regions

7.1 Research Approach and Statement of Problem

In our contemporary human society, globalization processes and the development of transport on a global scale have created conditions for the much easier spatial spread of species (Vitousek 1990; D'Antonio et al. 1996; Theoharides and Dukes 2007; Simberloff and Rejmànek 2011). Many of these "alien" species are harmless, but carry profoundly negative consequences for ecosystems and biodiversity (Czech and Krausman 1997; Mooney and Hobbs 2000; Vilà et al. 2021). In the European Union alone, the losses associated with invasion processes by alien species are estimated to amount to 12 billion euros per year (Pimentel et al. 2001), with the prospect that these processes will continue to increase in the immediate future (Vilà et al. 2011; Hulme et al. 2013; van Kleunen et al. 2015; Seebens et al. 2017, 2021a). The study of biological invasion is of great importance in environmental studies. Scientifics has been investigating the phenomenon of biological invasions for over 100 years, since a first study in 1882 on the invasion by *Mangifera indica* in Jamaica (Espínola and Ferreira 2007), however only after the emergence of modern biogeography (twentieth century) this theoretical model became systematized, having among its main precursors Albert De Candolle, Charles Darwin, and Charles Elton (Vavilov 1992; Davis 2006; Jeschke 2014).

A pioneering milestone in the reflection about the concerns of biological invasions dates back to 1964 with the symposium of the *International Union of Biological Sciences* in the United States, which incorporated, in the core of the discussions about invasive species, genetic and evolutionary aspects, which resulted in the work "The Genetics of Colonizing Species", by Hebert Baker and Leyard Stebbins (Baker and Stebbins 1965; Davis 2006; Barrett et al. 2016). The conceptual language adopted in the work by Baker and Stebbins (1965) is different from that used by Charles Elton in *The Ecology of Invasions by Animals and Plants* (Elton 1958), one of the first works on biological invasions, which has strong influence today because it contains concerns in the field of biological conservation. Another landmark of the discussions on invasions was the emergence of *Biological Conservation* as a scientific research area, in the 1970s, which allowed a paradigm shift and, thus, the creation of a scientific division committee focused on the impacts of invasions, with SCOPE (*Scientific Committee on Problems of the Environment*) (Cadotte et al. 2006; Davis 2006; Richardson 2011; Jeschke 2014).

Biological invasion is mainly understood as the process of ecosystem degradation resulting from the establishment of exotic populations, after the transfer of individuals, by human action, to areas that generally keep similarities (climatic, edaphic, altitudinal, among other aspects) with the place of origin (Elton 1958; Williamson 1996; Blackburn et al. 2011). In general, studies on the invasion of alien species seek to integrate data on the environment and the species in question, so as to translate the complexity of invasiveness and invasibility (Rejmànek et al. 2005; Rejmànek 2011, 1999). Invasiveness, or invasive capacity, refers to the biological traits that enhance species establishment, while invasibility, or susceptibility to invasion, refers to the attributes that characterize the vulnerability of individuals and receptive ecosystems (Pyšek et al. 1995; Falk-Petersen et al. 2006; Drenovsky and James 2010; Bacon et al. 2014).

The development process of the invasion phenomenon congregates different stages (Williamson 1996; Pyšek 2001; Colautti and MacIsaac 2004; Richardson and Pyšek 2006; Theoharides and Dukes 2007), beginning with the translocation of potentially invasive species by humans to other ecosystems where these species are not found (introduction). The process follows with an initial slow phase of the establishment of individuals, where the exotic species successfully produces viable individuals according to the probability of their survival in the new environment (establishment). This stage is followed by a phase of exponential population growth, which can result in the success of the biological invasion, making the invasive alien species dominant (expansion). Some of this species also can cause changes in the functioning of ecosystems from which they may not recover naturally to the original situation (Blackburn et al. 2011). Combining the main three stages of the invasion process, four general categories for identifying invasive alien species have been proposed (Richardson and Pyšek 2006). The species is said to be "casual" when it is unable to reproduce and propagate in the new environment and is said to be

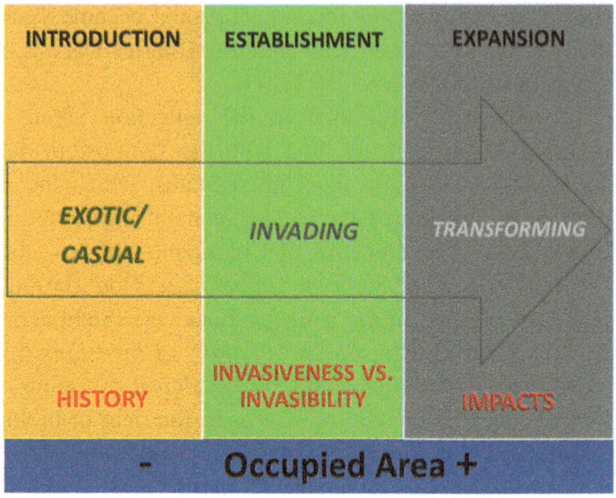

Fig. 7.1 The theoretical model of biological invasions highlighting the degrees/steps of invasion process (introduction—establishment—spreading), and the status of exotic species into the invasion process (exotic/casual—invasive—transforming)

"naturalized" or "established" when it is able to form autonomous populations in the new environment. The invasive species, on the other hand, rapidly expands away from the epicentre of introduction, with stable populations of numerous individuals; during this process some species are capable of profoundly altering the environment, causing damage to the abiotic environment, the disappearance of native species, alteration of nutrient cycles, or disease transmission (Fig. 7.1) (Pyšek et al. 1995; Espinosa-García et al. 2004).

Among the main properties of plant invasive species associated with their impacts on non-native areas, the following are commonly related: large seed production with high viability and longevity; the rapid growth of the root system; adaptation to fire; the ability to interfere with the growth of neighbouring plants; high production of propagules; morphological/physiological similarity with native species; the ability to fix nitrogen by symbiont bacteria (Cronk and Fuller 1995; Caley et al. 2008). As for general predictors of environments more prone to invasions, attributes such as low latitude, water availability, mesic climate, and human intervention (generating disturbance in the environment and facilitating the establishment of invasive alien species) are highlighted (Levine et al. 2003; Guo 2006; Gallien et al. 2010; Rejmànek and Richardson 2013; Crall et al. 2013). Therefore, island areas are considered more prone to biological invasions than continental areas; temperate zones are more prone than tropical zones; wetlands are more than arid zones; and disturbed areas are more prone to invasions than intact areas (Richardson 2004; Enders et al. 2020).

Among the groups of invasive alien plants most widespread around the world, the genus *Acacia* (Tourn.) Mill. (Miller 1754; Bentham 1875; Pedley 1978) stands out. The native range of this group is Australia, with 1012 species (González-Orozco et al. 2013; Richardson et al. 2011). Various species of *Acacia*, originating from Australia, were introduced into numerous regions between the eighteenth and nineteenth centuries and are now found in the catalogues of invasive species of

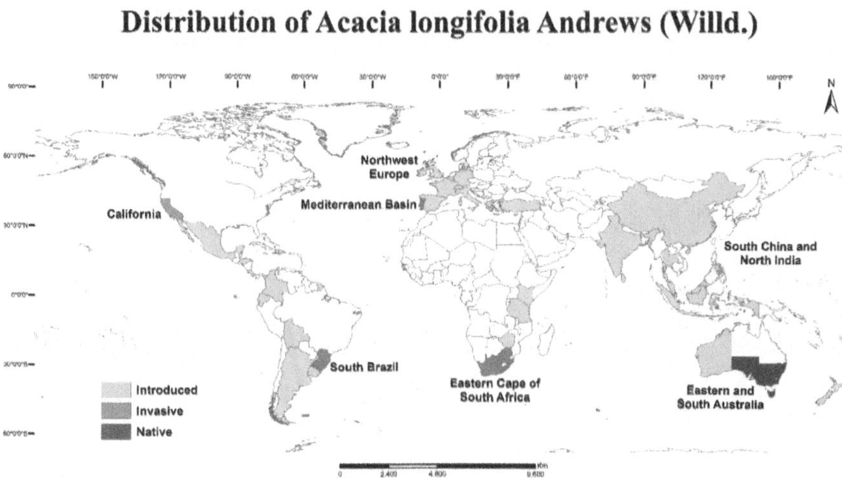

Fig. 7.2 Global distribution of *Acacia longifolia* (Andrews) Willd.

several countries, given their proven invasive character, affecting, above all, areas of high ecological value. Adapted to the different environmental conditions of Australia, Acacias have shown ample capacity to penetrate different regions of the world (Le Maitre et al. 2011). Among its attributes, the leaf endowed with numerous evergreen phyllodes stands out. These phyllodes help the plants in functions such as nutrient fixation, making the species more efficient in the acquisition and concentration of resources (Robinson and Harris 2000; Bairstow et al. 2010). In many species, the seeds have reduced size and mass, with a capacity for dormancy, which amplifies the potential for fertility/reproduction (Orians and Milewski 2007; Burrows et al. 2018). In the roots, they have a symbiotic association with bacteria of the genus *Rhizobium*, which assists in the fixation of atmospheric nitrogen facilitating the establishment and growth of the species (Burdon et al. 1999; Rodríguez-Echeverría 2010; Rodríguez-Echeverría et al. 2011). Other functional attributes provide greater survival to Acacias, such as the trunk of many species that allows them to resist fire, and the allocation of little energy to the production of floral nectar, which can enhance pollen reproduction (Maslin et al. 2003; Orians and Milewski 2007; Richardson et al. 2011). All these characteristics help explain the competitive vigour and high resilience and colonization capacity of Australian Acacias in different regions of the world.

Acacia longifolia (Andrews) Willd. (Fabaceae)—popularly known as golden wattle—seems to be representative of these theoretical assumptions, because it has a wide distribution in various types of environments and in several regions of the world (Fig. 7.2). It is a shrub of up to 8 m, whose introduction into Europe and America since the late eighteenth century seems to be due to its use as a fixer of coastal dunes (Nunes et al. 2019). From Australia, *A. longifolia* was transported to countries in Europe (late eighteenth century) and America (nineteenth century) and began to be cultivated as an ornamental plant and in plantations to fix coastal dunes

Fig. 7.3 Australian *Acacia longifolia* (Andrews) Willd. and *A. sophorae* (Labill.) Court. (*Source*: Photos by Jorge Luis P. Oliveira-Costa (South Brazil, July 2015, and West Portugal, July 2021))

and for purposes of combating soil erosion. Today it is considered one of the most problematic invaders for the conservation of environments in most areas of its distribution outside Australia. In most regions, the species potentially occur in coastal systems, throughout the oceans coastlines, on coastal dunes, in degraded areas and the midst of native vegetation. There are some taxonomic varieties of this species recorded in global databases, found in various parts of the world, two of them being the most common, called *A. longifolia* var. *longifolia* and *A. longifolia* var. *sophorae* (Fig. 7.3) (Pieterse 1987; Hooper and Maslin 1978). These two subspecies appear to be the most abundant worldwide, especially in Australia (Emeny 2009; Manea et al. 2019).

There are isolated efforts to control the species *A. longifolia*. In South Africa and Portugal, physical methods (manual and mechanical) and chemical methods (herbicides, e.g., glyphosate) have been successfully tested (Pieterse and Cairns 1986; Pieterse 1987; Marchante et al. 2008), and there has been the introduction of an agent for biological control of the invasive populations (the African wasp *Trichilogaster acaciaelongifoliae*), which has apparently been successful. In most regions invaded by this species, there is still no large-scale programme for the recovery of areas invaded by *A. longifolia*, or protection of areas not yet colonized by the species (recently physical and chemical methods were tested in the control of the species' populations in South Brazil) (Oliveira-Costa et al. 2020, 2021a, b).

Among the various characteristics of its invasiveness, *A. longifolia* is quite aggressive competitively and able to colonize native environments, with rapid proliferation in coastal ecosystems, and can even cover extensive surfaces, with the formation of dense mats, on coastal dunes, limiting the arrival of light, preventing the gaseous exchange between air and soil, increasing nitrogen levels (Marchante et al. 2011; Welgama et al. 2019). Other characteristics of the species are the high capacity for reproduction and dispersal, in addition to rapid growth, producing a large number of small seeds with high viability, which gives it a high invasion capacity (Correia et al. 2015). In the southern region of Brazil and central Portugal, *Acacia longifolia* seems to fix more nitrogen in the aerial part (leaves) and in the soil than species native to the coastal zones of these countries (Marchante 2001, 2007; Oliveira-Costa et al. 2020, 2021a, b).

All these characteristics give *A. longifolia* the ability to alter the composition and microbiology of the soil and prevent the development of the "restinga" vegetation, causing a decrease in local diversity, in environments highlighted by poverty in nutrient availability, which would increase its possibility of resistance and colonization (Carvalho et al. 2010; Welgama et al. 2019). Moreover, it is a species with seminal reproduction, generating a viable seed bank for up to 10 years (Marchante 2007; Correia et al. 2015).

The invasion capacity of *A. longifolia* also seems to be strongly determined by the fire regime (Pieterse and Cairns 1986; Marchante et al. 2008). In coastal and surrounding areas disturbed by fire may favour *A. longifolia* over native species, since the latter, unlike the invasive species, do not have high resistance to fire (Pieterse and Cairns 1986; Pieterse 1987). However, in coastal environments with little disturbance, the probability of invasion by *A. longifolia* seems to be reduced, where native species seem to take competitive advantage (Carvalho et al. 2010).

The present work, conceived within the scope of the project entitled "Biological Invasion and Environmental Susceptibility in Temperate Regions: Determinants and Impacts of Spread and Invasion by *Acacia longifolia* (Andr.) Willd.", was developed in a partnership among researchers from Portugal, Brazil, Argentina, and Uruguay. In these countries, the field of study of the biological invasion by *A. longifolia* initially needed to develop the work of cartographic and bibliographic revision, in view of the limitations encountered linked to bibliographic and cartographic aspects, on the distribution and strong impacts caused by this tree. Moreover, the literature of the ecology of the invasion of this species, besides being fragmented, is generalized, covering mainly its native range in the Australian subcontinent.

Thus, this study contemplates a proposal for the definition and mapping of global and local cores (Portugal, Brazil, Argentina, and Uruguay) invaded by *A. longifolia*, as a support to the future evaluation of the distribution patterns and ecological processes of the species, and the verification of its impact on coastal ecosystems. This work involved the survey of studies already carried out (that somehow refer to the invasion by *A. longifolia*), aiming not only the detection and location of *A. longifolia*, but also the knowledge of the relationships between the patterns and processes responsible for the proliferation and invasion of this species.

The purposes and justifications of the present study on the invasions by *A. longifolia* express relationships among several components, implying the possibility of numerous tests, aiming at obtaining real evidence within the complexity of invasion relationships. In formulating the ideas for the contextualization of the problematization of a work on biological invasion, some difficulties need to be highlighted, since the result of an interpretation in terms of the invasiveness and invasibility processes will depend on the amplitude of environmental changes occurred during a given period of time, and the effects of human activities during and after the occupation of the area by the invasive exotic species.

Portugal, Brazil, Argentina, and Uruguay have one of the highest concentrations of Australian *Acacia* plantations in the European and American continents (Attias et al. 2013; ISSG 2021; I3N BRASIL 2021). Invasive *Acacia* species are found on coastal sedimentary rim soils with abundant coastal dunes (*Acacia cyanophylla*, *A. longifolia*, *A. pycnantha*, *A. podalyriifolia*, *A. retinodes*), on marine calcareous soils (*A. cyclops*, *A. farnesiana*, *A. karroo*), on soils derived from alkaline granites or carbonic shales with marine facies (*A. dealbata*, *A. mearnsii*), besides the areas with sedimentary soils of the sublittoral plains and soils of the Atlantic slopes (*A. melanoxylon*) (Oliveira-Costa et al. 2020, 2021a, b).

Some *Acacia* species are considered global models of invasion science (Richardson et al. 2011), most notably *Acacia longifolia* (Andrews) Willd. The oldest introduction seems to have occurred in Portugal, with records dating back up to 200 years (Fernandes 2012). Despite being the object of several studies (Le Maitre et al. 2011), the species is little known in Brazil, Argentina, and Uruguay, where concern about the invasion by Australian Acacias is recent (Attias et al. 2013; Oliveira-Costa et al. 2020, 2021a, b).

In this work, the Australian *A. longifolia* is chosen as the focal species, a coastal legume of wide geographical distribution with a high capacity to colonize different environments when evaluating its potential distribution both in its area of origin and in invaded areas (Le Maitre et al. 2011), and we have as areas of analysis some sites invaded by it in the Australian Extra-Territories (Portugal, Brazil, Argentina, and Uruguay). *A. longifolia* is considered as one of the most aggressive invasive species in Portugal, by Decree-Law No. 565/99. In Brazil, it is considered invasive in Paraná (IAP Resolution No. 192/05), Santa Catarina (COSEMA Resolution No. 08/12), and in the state of Rio Grande do Sul (SEMA Resolution No. 79/13). *A. longifolia* is also considered invasive in Argentina and Uruguay (ISSG 2021; I3N BRASIL 2021). Data obtained in more than 15 years of research with *Acacia longifolia* in Portugal reveal signs of the presence of the species for more than 200 years in the country

(Nunes et al. 2019). Research with *Acacia longifolia* in Portugal was initially focused on studying the establishment and behaviour of the species. Systematic studies continue to be conducted for the development of management strategies, in an attempt to monitor and control populations in Portugal (Marchante et al. 2018). In the case of Brazil, Argentina, and Uruguay, the introduction of the species seems to have been more recent than in Portugal (approximately 100 years' difference) (Attias et al. 2013; ISSG 2021; I3N BRASIL 2021); however, there are still few specific works on *Acacia longifolia* invasion in these countries (Oliveira-Costa et al. 2020, 2021a, b).

The choice of research area for the coastal zones of Portugal, Brazil, Argentina, and Uruguay is due to the significant ecological importance of these areas, with invasion processes by *A. longifolia* still little known, both from the perspective of species and environments. The importance of this work, with a broad approach to the scope of the invasion of *A. longifolia*, can be justified in the light of Guo's (2006) words: "understanding the behaviour of the same exotic species in different types of invaded environments can bring crucial information about its invasion capacity, which will help to predict which conditions could favour the establishment of this and other exotic species". Also, by bringing together four of its invasive ranges in the world (Portugal, Brazil, Argentina, and Uruguay), the present study of *A. longifolia* seeks to understand the factors that facilitate invasion by the species and the environment, establishing a network of joint efforts, which constitutes a unique opportunity to assess the characteristics and behaviour of this species in different invasive situations.

Human activities interfere with the processes of establishment of *A. longifolia* in the coastal zone, and consequently with the protection of the coastal environmental system. In coastal systems (beaches, dune fields, river plains, fluvial-marine plains, fluvial-lacustrine plains, coastal tablelands), terrestrial and marine environments are integrated systems and their parts are interrelated, where the change in one component influences other parts and the dysfunction of the system as a whole (Nordstrom 2010; Carter 2013). The cumulative impact due to the pressures exerted by the presence of invasive species can produce several negative outcomes, causing changes in hydrological potential and support structure and modification of surface runoff, for example.

Integrated research among four invaded areas help to the extent that data to be compared from the detection phase until the establishment of management actions, through the characterization of species, communities, and invaded habitats. With the interdisciplinary vision effected during the course of this research, with previous experiences in various regions invaded by *A. longifolia*, this research project may broaden the development of new research methods, directed towards the planning and management of species and territories. Previous research aimed to obtain data on the species *A. longifolia* and the invaded environments, including the search for closer relations between researchers and the research institutions involved, potentially creating a network of studies on invasions by *Acacia* in temperate environments, as well as subsidizing public policies for the management and control of *Acacia longifolia*.

7.2 The Importance of *Acacia* Mill. for Biological Invasion Studies

This work is directed at an important group of invasive exotic plant species—the genus *Acacia* Mill. (Miller 1754)—considered of medium and high invasive potential, which, in general, do not require intervention by human to acquire their naturalization and establishment, and have resources for adaptability to new environments, having to coexist with situations of disturbance/degradation in the vast majority of times, due to the ecologically negative behaviour they promote to the conditions of the territories (Fig. 7.3).

The genera *Acacia* Mill., *Eucalyptus* L'Hér. and *Pinus* L. represent the three tree genera in the world with the highest rates of plantation out of their native distribution, in addition to standing out in the representation of global lists of invasive alien species, as well as in investments in research on their specific ecological processes (Richardson et al. 2011). The origin of scientific taxonomic description of the genus *Acacia* is from 1754, established according to Philip Miller's taxonomic classification (Miller 1754; Pedley 1978).

For centuries, Acacias have been planted outside their natural regions, where varied species of the genus assume different behaviour when non-native, even under environmental conditions similar to the native range (Richardson et al. 2011; Burrows et al. 2018). Today, several landscapes around the world are dominated by *Acacia* plantations. Some are among the largest dispersers of all invasive plants, others are only exotic species, and there are those that are not known to be invasive (Fig. 7.4). Human perception of Australian Acacias differs between regions of the world, which implies different forms of management according to legislation, environmental, cultural and socio-political factors (Fig. 7.4) (Le Maitre et al. 2011).

Fig. 7.4 Aspects and habits of some Australia Acacia Species (AAS) invasive around the world. (*Source*: Photos by Jorge Luis P. Oliveira-Costa (Central Portugal, May 2014))

Richardson et al. (2011) explain that the long history of transfer of Australian Acacias around the world implies the creation of a global experimental model considering the opportunities of the determinants of their use: (1) the pathways/ vectors by which species are introduced have been correlated with ecosystem characteristics, the value of the systems (energy, climatology), and how it has been changing in recent years under different circumstances; (2) why species have shown different degrees of invasiveness in new environments; (3) why certain ecosystems are more susceptible to invasion by *Acacia*; (4) the function of Acacias in the receiving ecosystem and their ability to alter ecosystem services; and (5) the factors influencing the evolution of environmental response to *Acacia* invasions in different regions of the world. The multiple dimensions of this model have provided important theoretical and methodological contributions to invasion science, especially as a support to the field of biogeographic conservation, with principles, theories, and analysis of the problems linked to biodiversity conservation (Turnbull 1987; Richardson et al. 2011). Biological invasion, for this group of species, is considered, under certain criteria, to be a case of "ecological imperialism". Richardson et al. (2011) discussed the concept of ecological imperialism, which can be considered as such in function

- of the intercontinental movement that occurred in the colonial era between the New World and the Old World, with Australian groups (*Acacia* Mill., *Pinus* L., *Eucalyptus* L'Hér) representing a special case of "ecological imperialism" (Richardson et al. 2011);
- of the export/translocation of Australian Acacias to other parts of the world following the arrival of European settlers in Australia in 1788 (in Europe, for example, many species of Australian Acacias experienced development and naturalization in the first half of the nineteenth century following the arrival of settlers);
- of after spreading around the world, the adaptability of these species to the abiotic conditions of the territories has increased, being, in the case of Acacias, classified in 4 groups: Acacias from cold climates—*Acacia melanoxylon*—Acacia tropical—Acacia from arid zones;
- of the biological constitution of the species, favouring those that have more efficient characteristics, such as (1) symbiotic bacterial association in the root system with rhizobia, present in numerous legumes, which assists in nitrogen fixation by the species; (2) resistance to fire; (3) use of animals in seed dispersal; (4) particular pollination of Acacias, called the pollination syndrome, with the allocation of little energy to floral nectar, which maximizes pollen reproduction and seed bank after pollination (Bond et al. 2001).

The native distribution area of the genus *Acacia* is Australia with 1012 species, 71 species of which are considered naturalized and another 23 species already identified as invasive, in a geographical region classified as the *Australian Acacia Domain* (Richardson et al. 2011) (Table 7.1). Native species also occur in the Americas (185 species), Africa (144 species), and Asia (89 species), with an

Table 7.1 General aspects about *Acacia* around the world and Australia

Acacia Mill.	General aspects
Native species in Australia	1012 species
Time clipping of the translocation of species	For the last 250 years
Species transported out of Australia	386 species
Species classified as naturalized	71 species
Species categorized as invasive	23 species
Australian native species outside the country	17–20 of the 1022 *phyllodineae* species
Extra-Australian native species	10 species unique to the Indo-Pacific
Native species in America, Africa and Asia	185 species; 144 species; 89 species
Total approx. (species)	**1022 species**

emphasis on 10 *Acacia* species unique to the Indo-Pacific region, territories categorized as the *Extra-Australian Acacia Domain* (Richardson et al. 2011) (Table 7.1). In Europe, there are no native species (Table 7.1). *Acacia* species have been translocated from Australia over the past 250 years, with a total of approximately 366 species transported out of Australia (Table 7.1). In Australia, *Acacia* species occupy the country from east to west, with an important evolutionary feature given the climatic and soil conditions, which are similar to regions such as the Mediterranean, with common climatic conditions (there are 50 common species between these regions). The most characteristic feature of the *Acacias* is the leaf and its evergreen phyllodes (scleromorphic phyllodes with xeromorphic mechanisms), of different sizes and veins.

There are several international initiatives involving all phases of biological invasion processes by Australian Acacias, with the goal of implementing a global model for managing invasion by Acacias in countries invaded by species of this group. The determination of a global model to promote management programs for environments invaded by Australian Acacias seeks to improve the quality and productivity of environments prone to invasions by these species. Among the main authors of these initiatives, in the Extra-Australian Domain, the Centre of Excellence for Invasion Biology, CIB, are highlighted, as well as the participation of other South African institutions such as the University of Stellenbosch, University of Kwazulu-Natal, University of Cape Town, and the University of Pretoria. In the Australian Domain, the highlight is the Flora of Australia Online Project, which is coordinated by the Government of Western Australia and the Western Australian Herbarium (Department of Biodiversity, Conservation and Attractions), as well as the World Wide Wattle Project (Australian database of the genus Acacia). Richardson et al. (2011) list 12 reasons why *Acacia* can be considered a global model for invasion science:

1. A high number (1012 species), being the third most widespread outside its natural region, with 23 confirmed "invasive" species, and many others naturalized (González-Orozco et al. 2013).

2. Established taxonomy and phylogeny allowing invasive potential to be associated with phylogenetic significance, where numerous invasive species come from the major clades of the genus (Turnbull 1987; Miller et al. 2011; Richardson et al. 2011).
3. Present in most biogeographic regions of Australia, a dynamic that benefits adaptation in diverse parts of the world (Le Maitre et al. 2011; González-Orozco et al. 2013).
4. Knowledge of the original distribution range of native species, with local data available, facilitating macroecological and biogeographic analyses, and species distribution modelling (Pohlman et al. 2005; González-Orozco et al. 2013).
5. High levels of intraspecific divergence and variation (genetic diversity, for example) exhibited by populations in new regions of introduction (Le Roux et al. 2011).
6. Variation in land use motivating natural species selections for various factors (growth level, robustness, environmental tolerance), with invasion success (Bui et al. 2014).
7. Use for numerous purposes in areas outside the native range, benefiting the understanding of the need for the introduction of the species, and their assimilation into the crop exchange and other elements of regional systems (Le Maitre et al. 2011).
8. Vast existing documentation of the introduction of Acacias (Richardson et al. 2011).
9. Massively planted for commercial reasons in various parts of the world and are now a component of ecosystems across much of the globe (Le Maitre et al. 2011).
10. Vast existing literature on several native Australian species as exotic species, facilitating intraspecific comparison across many regions (Monks and Coates 2002).
11. The introduction of Acacias allows many opportunities by exploring the interplay between the ecological relationships that contribute to invasion success, through their influences on native biota, and the integrity effects of commercially important species.
12. The long management history of some countries, compared to other countries, creates an ideal situation for the widespread construction of best practices (Richardson et al. 2011).

The focal species of this study—*Acacia longifolia* (Andrews) Willd.—also known as golden wattle (United States), acacia-de-espigas (Portugal) and acacia marítima (Brazil), is a shrub or small tree of up to 8 m, belonging to the Fabaceae family and of Australian origin. Among the various characteristics of its invasion, the rapid proliferation of this species in coastal ecosystems stands out, which can reach extensive areas, with the formation of dense mats, on the coastal dunes, limiting the arrival of light, preventing the gaseous exchange between air and soil, increasing nitrogen levels (Carvalho et al. 2010; Marchante et al. 2008, 2011). Other characteristics of the invasion by *A. longifolia* are its high capacity for reproduction and dispersal, in addition to its rapid growth and its seminal reproduction, generating a viable seed bank for up to 10 years (Emeny 2009; Rascher et al. 2011; Vicente et al. 2018). Within the scope of control techniques, physical methods (manual and mechanical), chemical methods (herbicides, such as glyphosate), and biological control (African wasp *Trichilogaster acaciaelongifoliae*), have been successfully tested in countries such as South Africa, Portugal, and Australia. In the next topic of this chapter, some aspects of the invasion of *Acacia longifolia* on a global and regional scale (Portugal, Brazil, Argentina, and Uruguay) will be highlighted.

7.3 The Statement of Nature Conservation in Invaded Ranges by *Acacia longifolia*

The establishment of ecological-geographical parameters that evaluate the environmental susceptibility and the performance of invasive species in the study areas (and that guide decision-makers in relation to the best economic and ecological solutions to the problem of invasions) is one of the focuses of this work, which may contribute to the improvement of the environmental conditions of the invaded spaces, and to greater nature conservation, considering the degree of invasion severity with which the areas under analysis coexist. Data from ISPM (International Standards for Phytosanitary Measures), Plant Resources Protection—From Harmful Pests (Pest Risk Analysis), and International Trade and Invasive Alien Species, International Plant Protection Convention IPPC (with support from FAO's Plant Protection Service, the World Trade Organization and the Council of Europe COE), are three of the most important international documents and publications that establish the guidelines for the areas invaded by invasive alien species in the world, based on a zoning that groups the global flora as a whole into regions, according to the conservation requirements for each region (Fig. 7.5).

This zoning, called Regional Plant Protection Organizations, divides the world into eight (8) regions (Fig. 7.4): EPPO (European Union); COSAVE (Mercosur); NAPPO (NAFTA); OIRSA (Caribbean); CAN (South America); IAPSC (Sub-Saharan Africa); NEPPO (North Africa and Middle East); APPPC (East and Oceania) (Fig. 7.5). For each region, guidelines are presented for the most invasive species (Red List; Alert List; Observation List), recommendations for pests (Pest Recommended Regulation), the habitats and land uses, life and growth forms of the species. In the case of Portugal, its territory is included in the EPPO (European and Mediterranean Plant Protection Organization), and it is one of the 50 member countries of the Regional Plant Protection Organization for Europe (RPPO). Brazil, Argentina, and Uruguay are part of the COSAVE REGION (Fig. 7.5).

Regarding studies and regional initiatives in Portugal, Brazil, Argentina, and Uruguay, considering the theme of invasions by one of the main plant species of invasive alien plants in these countries—*Acacia longifolia*—the Southern Region of Brazil, the Central Region of Portugal, and the coastal zones of Uruguay and Argentina, are highlighted as the main nuclei of invasions of this group, with an expressive degree of invasibility and invasiveness (Marchante et al. 2008, 2011;

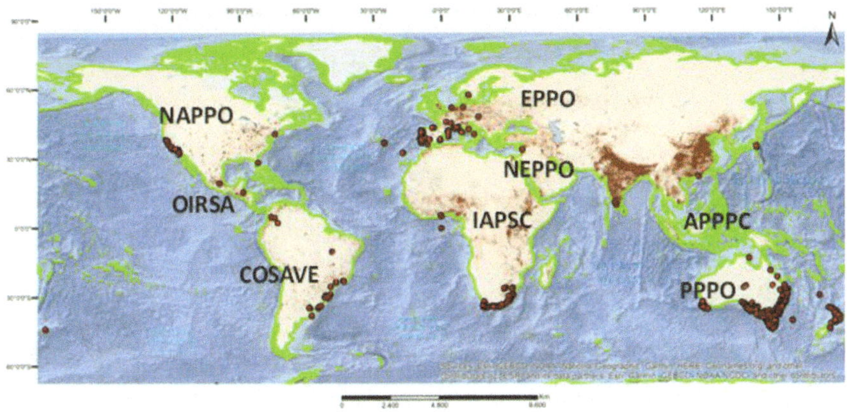

Fig. 7.5 Distribution of *Acacia longifolia* according to the world map of the Regional Plant Protection Organizations. (*Source*: The authors (based on the International Plant Protection Convention, IPPC International—https://www.ippc.int/en/))

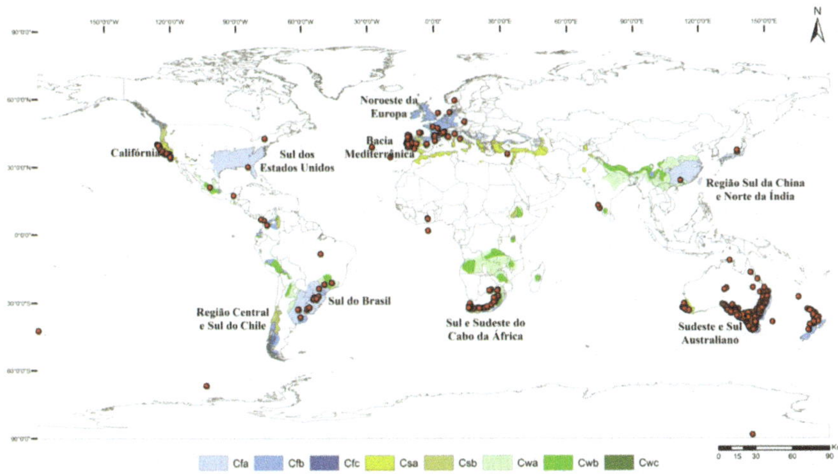

Fig. 7.6 Map of the temperate regions and world distribution of *Acacia longifolia*

Oliveira-Costa and Pivello 2017; Oliveira-Costa et al. 2019, 2020). The climatic regions that contemplate these areas are categorized, according to Köppen and Geiger (1954) and Köppen (1900), as mesothermic and humid (Csa/Cfb), with the indefinite dry season and winter rains, corresponding to the temperate Mediterranean and temperate subtropical climate, being located in the mid-latitudes climatic ranges (Fig. 7.6). These bands cover the whole of Portuguese territory, part of central Brazil and all of its southern portion, and the whole of the territory of Argentina and Uruguay. The medium latitudes are distinguished by a temperate climate, with four well-defined seasons, under the influence of warm air masses coming from the

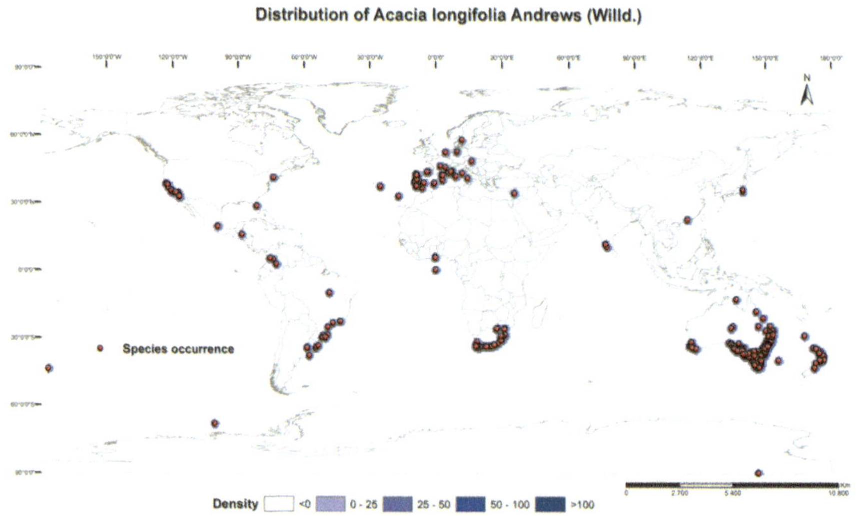

Fig. 7.7 Map of global distribution of *Acacia longifolia* and species occurrence rate

tropical regions, as well as cold air currents coming from the poles (Köppen and Geiger 1954). Average annual temperatures are below 18 °C, and precipitation is above 30 mm rainfall (Köppen 1900). As references of the biological invasions by *Acacia* in the temperate regions, the Central Region of Portugal, and the states of Santa Catarina and Rio Grande do Sul both in the South Region of Brazil, and the coastal zones of Uruguay and Argentina, stand out (Fig. 7.7).

7.3.1 Portugal

Figures from the last century point to a significant increase in the introduction and accumulation of alien species in Europe (Seebens et al. 2021b). Regarding the invasion situation in Portugal, the species considered potentially invasive are estimated at 670 exotic species (18% of the total species of the Portuguese territory, being 15% considered invasive). About the invasion situation in Portugal is highlighting: (1) the main biological groups (where terrestrial plants are highlighted); (2) the causes of transport and introduction of exotic species in Portugal (where the ornamental cause is highlighted); and (3) the types of vectors and means of transport (where the accidental means is highlighted as the main means of translocation of exotic species to Portugal) (EUROPE-ALIENS 2021).

In accordance with European directives on the management and control of biological invasions, Annex I of the Portuguese legislation of 21 December 1999 (Portuguese Decree-Law 565/99) recognized the problem of invasions in Portugal through a decree-law regulating the introduction of species, with the creation of a list of introduced exotic species with invasive behaviour. The calculation of biological invasions in Portugal is based on the number of species considered naturalized in Portugal, distinguishing them with regard to the environment in which they thrive, their use, their place of origin, dates and reasons for introduction, taxonomic diversity, and habitats where they occur. Also, the environmental policy on biological invasions in force in Portugal addresses issues such as the prohibition on the introduction of new species, and the detection, breeding, cultivation, and marketing of species considered invasive and/or of ecological risk (Marchante et al. 2014).

The management of exotic species in Portugal has been happening under three main fronts of intervention: prevention, early detection and rapid response, and control (Marchante et al. 2014). Prevention contemplates activities such as the creation of regulatory legislation, pest exclusion, investments in environmental education, and awareness actions. Early detection contemplates monitoring activities, aiming at rapid response and eradication of invasive species shortly after introduction (where costs may increase with time). Control occurs when eradication is not possible, focusing on reducing the impacts, mainly through the choice of appropriate methodology, and recovery of the invaded areas. In Portugal, mechanical, manual, and chemical controls and biological control have been tested and applied (Marchante et al. 2014, 2018).

Invasive species are found in all provinces of mainland Portugal and also in the archipelagos of Madeira and Azores (Marchante et al. 2014). In the case of some groups, such as terrestrial plants, there is a preference for the cool environments of valleys, mountainous inland areas, and margins of waterways and communication routes, invading after fires (Lourenço 2009; Marques 2010; Marchante et al. 2014; Oliveira-Costa 2014; Oliveira-Costa and Sousa 2015). There are records of the introduction of invasive species in Portugal of up to 200 years (Fernandes 2012; Nunes et al. 2019). Thus, considering that the Portuguese environments had significantly different conditions from the current ones, it is assumed that exotic species certainly had to adapt to the needs arising from the ecological changes that have occurred. However, the changes that have occurred in Portuguese environments in recent decades seem to benefit the establishment of invasive alien species, especially because of the intense anthropic activity that has occurred, which has influenced the reduction of the native component in Portugal, making room for invasive alien species. Among the main practices developed in Portugal within this context, are the fires, the intervention for correction of torrential streams, and reforestation. Another anthropic activity in Portugal that contributes significantly to biological invasions is the fuel management strips deployed on the sides of Portuguese roads.

In the case of Portugal, as is well documented (Marchante 2001, 2007; Almeida and Freitas 2006, 2012; Marchante et al. 2008, 2011, 2014, 2018; Carvalho et al. 2010), the country falls within the context of regions invaded by Australian species of the genus *Acacia*. The first mapping in this scope corresponds to the *Acacia* and *Eucalyptus* Distribution Chart, on a scale of 1/50,000, from 1978, of the Agrarian Reconnaissance and Planning Service (SROA) (Fig. 7.8). This map characterizes the most widespread species of the genus *Acacia* in the country, considering the species *Acacia melanoxylon*, *Acacia dealbata*, and *Acacia longifolia* as the most

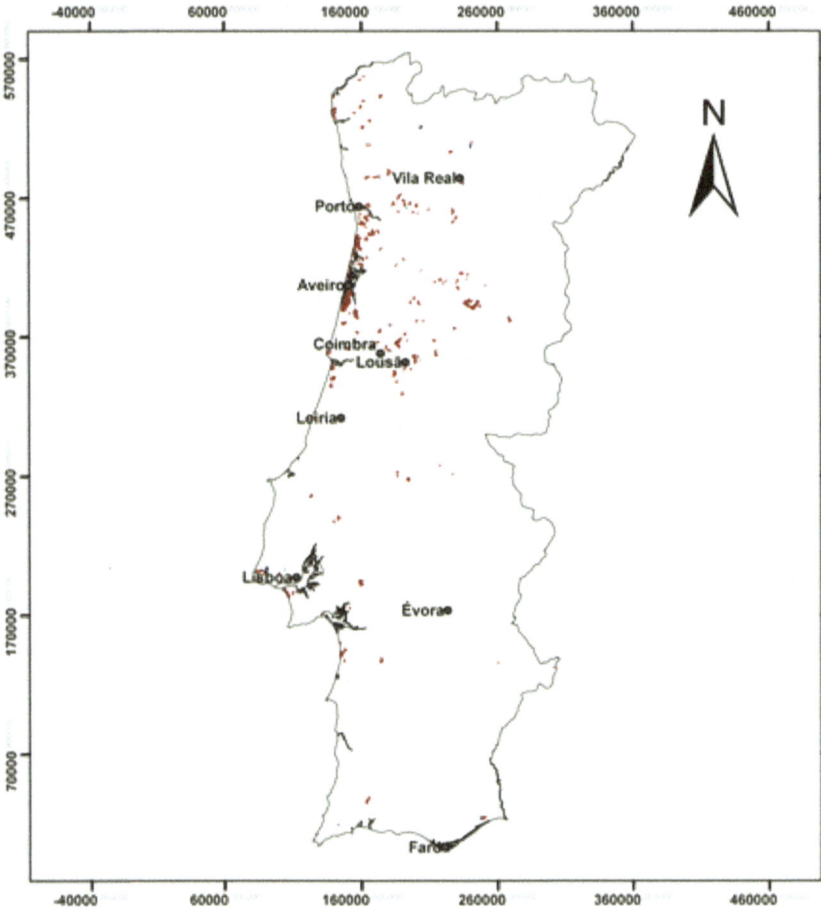

Fig. 7.8 Map of distribution of *Acacia* in Portugal. (*Source*: The authors (based on the Agrarian Reconnaissance and Planning Service, SROA Portugal 1978))

representative. Regarding the occurrence of these species in Portugal, *Acacia* species are identified in the districts of Faro, Beja, Setúbal, Lisbon, Coimbra, Santarém, Leiria, Guarda, Aveiro, Viseu, Porto, Vila Real, Braga, Viana do Castelo. In the 1978 chart, there is no sign of Acacias for the districts of Évora, Portalegre, Castelo Branco, and Bragança.

As references of the invasion by *Acacia longifolia* in Portugal, the District of Leiria, in the Central Region of Portugal, stands out. Leiria, the capital of the District of Leiria, is located at 39°46′00″ N latitude and 53°00′00″ W longitude, within the Central Region of Portugal, sub-region Leiria (Fig. 7.9). The climate in this region is characterized by having Mediterranean influence. During the first semester, in Leiria, the climate is cold and humid, with average temperatures between 15 and 7 °C, with minimum temperatures reaching 0 °C. High precipitations can occur in this period,

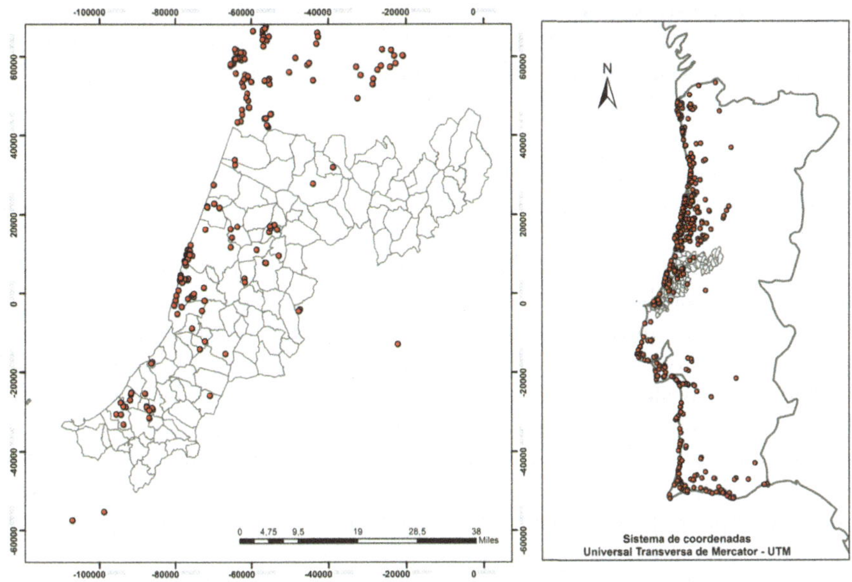

Fig. 7.9 Map of the Leiria District and range expansion of *Acacia longifolia*

from April to June. In the second semester, high temperatures are registered, where the climate is hot and humid, with average temperatures between 25 and 12 °C, with autumn rains.

Taking the municipality of Leiria as an example in the political-administrative context of the Central Region of Portugal, Leiria currently has approximately 127,000 inhabitants, according to INE Portugal. The population of this area has increased by almost 100% in 100 years, which is reflected in the changes in the use and occupation of the land. Leiria has a population density of 225 inhabitants/km^2, with more than 90% of the population living in urban areas—31.9 km^2, respectively. Leiria, in central Portugal, has a human development index (HDI) index of 0.9, which, like the other Portuguese districts, is a high HDI (above 0.8), and per capita gross domestic production (GDP) of 103.18 euros, a value considered to be above the national average in Portugal. In the Central Region of Portugal, characterized in its southern portion by sedimentary formations, sandy soils, with coasts, rocky foothills, and cliffs, and in the northern portion with more compact soils (where the coastal systems are significantly more invaded by *Acacia*), the species *Acacia longifolia* is present in all its districts (Leiria, Coimbra, Aveiro, Viseu, Guarda, Castelo Branco, Santarém, Lisbon).

7.3.2 South America

Southern Latin American countries, like many other developing economies worldwide, harbour highly diverse natural habitats (Mittermeier et al. 2011). Since conquer time but particularly after 1800, the number and distribution of invasive alien species have increased in this region (IPBES 2019), resulting in concurrent impacts upon their amazing native biodiversity, natural ecosystems, and local economies (Early et al. 2016; Seebens et al. 2017). According to data from the I3N Inter-American Biodiversity Information Network, environmental policy in the field of biological invasions in South America mainly addresses issues such as prohibition in the introduction, breeding, cultivation, and marketing of species considered invasive and/or of ecological risk. Despite this wide socio-ecological problem, low availability of context-specific data and scarce knowledge about invasive species' ecology, as like as limited funds for addressing their impacts is still common in this area (Nuñez and Pauchard 2010; IPBES 2019). With the objective to framing the areas of analysis and the objects of this work, we present a brief description about the stage of knowledge of biological invasions, particularly of *A. longifolia*, in Brazil, Argentina, and Uruguay (Figs. 7.10 and 7.11).

In the context of the situation of biological invasion in the Brazilian territory, it can be seen that the Brazilian government and governmental actions, the relevant legislation and the private sector have taken little account of the serious damage resulting from biological invasions in the country, a process that may become irreversible with costs to the available (and not always renewable) natural resources. According to the work done by Zenni and Ziller (2011), in light of initiatives on biological invasions in Brazil in recent years, invasive species in Brazil are estimated at 459 species.

As in Portugal, the calculation of biological invasions in Brazil is based on the number of species considered "naturalized exotic species", distinguishing them relatively as to the ecosystems in which they thrive, their use, area of origin, dates and reasons for introduction, taxonomic diversity, and habitats where they occur. According to the data from the Horus Institute for Environmental Development and Conservation (in cooperation with the I3N Inter-American Biodiversity Information Network) (I3N BRASIL 2021), environmental policy in the field of biological invasions in Brazil mainly addresses issues such as prohibition in the introduction, breeding, cultivation, and marketing of species considered invasive and/or of ecological risk. Since the 1990s, Brazilian municipalities have been responsible for implementing local programs aimed at combating invasion by invasive alien species, being the municipalities responsible for the regulation of most of the rules aimed at combating this problem.

At the national level in Brazil, the Ministry of Environment (MMA) and the Ministry of Agriculture, Livestock and Supply (MAPA) are currently the managers of policies on biological invasions in the country, either through partnerships with municipalities or through their own deliberations (I3N BRASIL 2021). The Presidency of the Republic, since 2002, redirected investments in the scope of nature conservation in Brazil, through the regulation of the National Biodiversity Policy

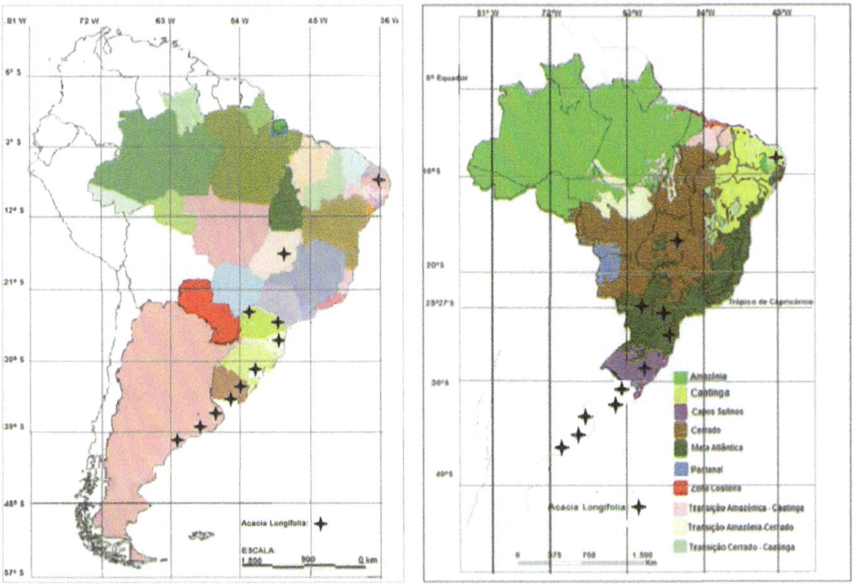

Figs. 7.10 and 7.11 Map of occurrence of *Acacia longifolia* in South America

and the National System of Conservation Units (SNUC). From this, measures were taken to increase the resources allocated to biological invasions, deliberated by bodies such as the *Comissão Nacional da Biodiversidade* (Brazilian National Biodiversity Commission—CONABIO), the *Instituto Brasileiro do Meio Ambiente e dos Recursos Naturais Renováveis* (Brazilian Institute of Environment and Renewable Natural Resources—IBAMA), the *Instituto Chico Mendes de Conservação da Biodiversidade* (Chico Mendes Institute for Biodiversity Conservation—ICMBio), the *Agência Nacional de Vigilância Sanitária* (National Health Surveillance Agency—ANVISA), and the *Concelho Nacional do Meio Ambiente* (National Environment Council—CONAMA) (I3N BRASIL 2021).

Covered by government or private initiatives, biological invasions and invasive alien species nowadays are part of everyday environmental problems in Brazil, given the space they occupy and the impacts they have caused. Invasive alien species are found in all Brazilian states, and also in its archipelagos, where some species have records of introduction dating back to the eighteenth century (I3N BRASIL 2021). The environmental quality of invaded territories may have repercussions on the nuclei where the epicentres of biological invasions are located, as well as on the region as a whole. Therefore, work on biological invasions in Brazil has focused essentially on fostering actions that lead to the balanced relationship of exotic species with the environment, aiming at the sustainability of the development process of invasions, and encouraging sustainable strategies for the management of territories. The management of exotic species in Brazil, as occurs in Europe, is developed through prevention, early detection and control of species, contemplating

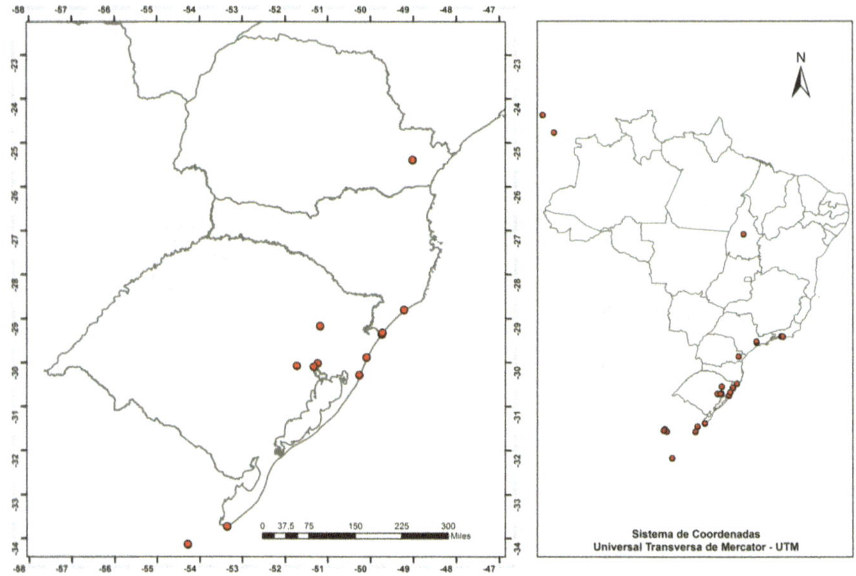

Fig. 7.12 Map of the South Brazil and regional distribution of *Acacia longifolia*

actions and activities from the creation of regulatory legislation, investments in awareness-raising actions, monitoring and eradication of invasive species shortly after introduction, to the control of populations in invaded areas. In Brazil, mechanical, manual and chemical, and biological controls are tested and applied (I3N BRASIL 2021).

Regarding the regional context of Brazil, considering the theme of invasions by *Acacia*, it is known that there are several Brazilian states invaded by Acacias, however, little is known about the invasiveness of the species, besides data on the ecosystems susceptible to its dominance. Despite being the object of some studies in Brazil, many *Acacia* species are little known, because the concern about the invasion by Australian Acacias in Brazil is still recent (Attias et al. 2013). International databases on biological invasions, such as the Invasive Species Specialist Group (ISSG), the Centre for Agriculture and Bioscience International (CABI), and the I3N Inter-American Biodiversity Information Network (I3N-BRASIL), indicate the species *A. mangium*, *A. mearnsii,* and *A. longifolia* as the most widespread Australian Acacias in the Brazilian territory, with expressive potential for invasiveness.

As references to the invasion by *A. longifolia* in this region, the states of Santa Catarina and Rio Grande do Sul, both in the South Region of Brazil, stand out. Porto Alegre, the capital of the State of Rio Grande do Sul, is at 30°01′00″ South latitude and 51°13′00″ West longitude, within the Mesoregion of Grande Porto Alegre, Porto Alegre Microregion (Fig. 7.12). Florianópolis, the capital of the State of Santa Catarina, is located at 27°35′49″ South latitude and 48°32′56″ West longitude, inserted in the Mesoregion of Greater Florianópolis, Microregion of Florianópolis

Fig. 7.13 Mosaic of pictures showing some habits of *Acacia longifolia* across South Brazil. (*Source*: Photos by Jorge Luis P. Oliveira-Costa (Santa Catarina-South Brazil, July 2015))

(Fig. 7.12). The climate in these regions is characterized by having oceanic influence. In Porto Alegre and Florianópolis, in the first semester, the climate is hot and humid, with average temperatures between 25 and 17 °C. The high precipitation is concentrated in this period, in the months of January to March. In the second semester, the lowest temperatures are recorded in these regions, with average temperatures between 17 and 12 °C.

Porto Alegre and Florianópolis, as two examples in the context of the political-administrative situation of the South Region of Brazil, have currently approximately 1,500,000 inhabitants and 508,000 inhabitants, respectively, according to data from the Brazilian Institute of Geography and Statistics (IBGE). The population of these areas has increased by almost 100% in 100 years, which is reflected in the changes in the use and occupation of the land. Porto Alegre has a population density of 3000 inhabitants/km^2, and Florianópolis of 764 inhabitants/km^2, with more than 90% of the population living in urban areas—565 km^2 and 30,000 km^2, respectively. According to the UNDP and IBGE census, Porto Alegre and Florianópolis, in the South of Brazil, have an HDI index of 0.8, classified as "very high" on the global scale of comparability, with a per capita GDP of 68.1 reais.

In Brazil, *A. longifolia* is present in all states of the Southern Region: *Paraná* (Curitiba), *Santa Catarina* (Florianópolis/Ponta das Aranhas/Parque Estadual do Rio Vermelho/Parque Municipal das Dunas da Lagoa da Conceição; Itapema/Restinga; Laguna; Ararnaguá), and in *Rio Grande do Sul* (Pelotas/Estrada para Praia do Laranjal/Lotamento das Acácias/Vila Assunção II; Santa Vitória do Palmar/Hermenegildo/Barra do Chuí; Chuí; Rio Grande/Praia do Cassino; Mostardas/Parque Nacional da Lagoa do Peixe; Torres/Restinga; Tramandaí/Horto Florestal do Litoral Norte/SEMA) (Figs. 7.12 and 7.13).

Fig. 7.14 Mosaic of pictures showing some habits of *Acacia longifolia* across Argentina. (*Source*: Photos by Lia Montti (Mar de Los Pampas-Buenos Aires Province, 2023))

In Argentina, more than 654 invasive exotic species are present, of which 319 are known to cause negative ecological impacts (Zalba et al. 2020) and important economic cost (Duboscq-Carra et al. 2021). Nowadays, Argentina presents a flowering scientific community working on different socio-ecological aspects of biological invasions. Moreover, this problem start to become addressed of the Argentina government, through the National Strategy of Invasive Species (Estrategia Nacional sobre Especies Exóticas Invasoras) that list invasive species to prevent arriving and establishment of them, and promote their management and public policies to minimize their impact (MAyDS and FAO 2019). However, is still necessary improve interactions among scientific and official organisms to communicate the importance of increasing invasive species risk. Regarding regional initiatives on biological invasions in Argentina, considering the theme of invasions by *Acacia* and *A. longifolia*—the coastal zones of Uruguay (and its neighbouring regions) are highlighted as the main nuclei of invasions of this group of specific species in this country (I3N South America). Like in Portugal and Brazil, in Argentina and Uruguay this species was mainly introduced, and is still promoting, to fix costal dunes but also for decorative purposes, in an attempt to improve the attractiveness of seaside resorts (Figs. 7.14 and 7.15) (Zalba and Villamil 2002). Although its wood does not have important applications to forestry, their value as a fuel should not be discounted. In Argentina, *A. longifolia* has already been cited as an invader for the coast of the Pampa Austral (Fig. 7.14) (Lecanda and Cuevas 2013; Stellatelli et al. 2013; Alberio and Comparatore 2014), however, this species is not considered yet as invasive risk for the Argentine National Pest Surveillance and Monitoring System (https://www.sinavimo.gob.ar/plaga/acacia-longifolia). As well

Fig. 7.15 Mosaic of pictures showing some habits of *Acacia longifolia* across Uruguay. (*Source*: Photos by César Fagundez (La Paloma-Uruguay, 2023))

as the situation in Argentina, in Uruguay *A. longifolia* has already been cited as an invader since the north to the south of the "Uruguayan coast zone" (Fig. 7.15).

As in Portugal and Brazil, the calculation of biological invasions in these countries is based on the number of species considered "exotic", or "naturalized exotic species", distinguishing them relatively as to the ecosystems in which they thrive, their use, area of origin, dates and reasons for introduction, taxonomic diversity, and habitats where they occur.

In this work we showed how the *A. longifolia* invasion niche seems to cover a significant part of the coastal areas of Argentina and Uruguay (Figs. 7.14 and 7.15). Theses coastal zones of South America are categorized, according to Köppen (1900), as mesothermic and humid (Csa/Cfb), with indefinite dry season, corresponding to the temperate subtropical climate. Particularly, it is distinguished by a temperate climate, with four well-defined seasons, under the influence, mainly, of cold air currents coming from the poles. Average annual temperatures are below 18 °C, and precipitation is above 30 mm rainfall (Köppen 1900). This species can be an ecosystem engineer with potentially serious ecological and economic impacts in this fragile ecosystem of Argentina and Uruguay. Given the potential magnitude of its impacts and the lack of detailed studies, future research considering the perceptions that different stakeholders may have, such as like, the ecological and economic impacts to have a better picture of *A. longifolia* invasion in Argentina and Uruguay. This knowledge also may be useful to alert the public and policy-makers about the magnitude of the invasion problem in those countries. Furthermore, we encourage the development of collaboratively projects between decision-makers and scientists from different countries where *A. longifolia* is a problem to contribute to mitigate their negative impacts but maintain their benefices. As references of the biological invasions by *A. longifolia* in these regions, the "Mar de Los Pampas" (Fig. 7.14), located in the Buenos Aires Province region (Argentina), and the "La

Paloma" (Fig. 7.15), as also the Montevideo, Rocha, Colonia, Maldonado and Canelones, all areas situated in the coastal region of Uruguay, stand out.

7.4 Final Considerations and Future Perspectives

This is the general picture of the study areas invaded by *Acacia longifolia,* in the Central Portugal, the Southern Region of Brazil, and the coastal zones of Argentina and Uruguay, where, in the future of this project, more detailed data will present the natural dynamics and environmental impacts in these territories. In Portugal, Brazil, Argentina, and Uruguay, due to the already mentioned physical and socio-economic characteristics, the problem of invasions cannot be disregarded in the local space planning and nature conservation project.

In recent decades, there has been a permanent search for a greater occupation of the coastal space in the study regions (invaded by *A. longifolia)*, with modifications, above all, in their dune systems and adjacencies, due to the broader transformations that have affected these regions, such as the processes of urbanization and demand for services. However, what we see in these regions is a total lack of awareness and adaptation of land use and occupation to ecological-geographical conditions (with rare exceptions), resulting in territories that are geographically and ecologically more susceptible to biological invasions. In most cases, invasive plant control and monitoring activities aimed at the areas under analysis have sought only to solve the problem of existing infestation and population proliferation, without considering issues related to the quality of the environments involved, the local socio-environmental conditions, and the coastal system as a whole, which is made up of flows and interactions that take place at the regional scale. For this reason, as the main future perspective of this research, this project on the *Acacia* invasion in Europe and South America (Portugal, Brazil, Uruguay, and Argentina) will evaluate the environmental nuclei occupied and disturbed by *A. longifolia*, from nuclei with a low degree of occupation by the species (very common in some portions in the study regions, with the presence of few individuals), to nuclei under a high degree of its invasion.

Acknowledgements This research received support from the Centre of Studies in Geography and Spatial Planning (CEGOT), funded by national funds through the Foundation for Science and Technology (FCT) under the reference UIDP/GEO/04084/2020_UC.

References

Alberio C, Comparatore V (2014) Patterns of woody plant invasion in an Argentinean coastal grassland. Acta Oecol 54:65–71

Almeida JD, Freitas H (2006) Exotic naturalized flora of continental Portugal—a reassessment. Botanica Complutensis 30:117–130

Almeida JD, Freitas H (2012) Exotic flora of continental Portugal—a new assessment. Bocconea 24:231–237

Attias N et al (2013) Acácias Australianas no Brasil: histórico, formas de uso e potencial de invasão. Biodiversidade Brasileira 3(2):74–96

Bacon SJ, Aebi A, Calanca P, Bacher S (2014) Quarantine arthropod invasions in Europe: the role of climate, hosts and propagule pressure. Divers Distrib 20(1):84–94

Bairstow KA, Clarke KL, Mcgeoch MA, Andrew NR (2010) Leaf miner and plant galler species richness on Acacia: relative importance of plant traits and climate. Oecologia 163(2):437–448

Baker R, Stebbins L (1965) The genetics of colonizing species. Academic Press, New York

Barrett S et al (2016) Invasion genetics: the Baker and Stebbins Legacy. Wiley-Blackwell

Bentham G (1875) Revision of the suborder mimosae. Trans Linn Soc Lond 30:335–664

Blackburn TM, Pyšek P, Bacher S, Carlton JT, Duncan RP, Jarošík V et al (2011) A proposed unified framework for biological invasions. Trends Ecol Evol 26(7):333–339

Bond WJ, Smythe KA, Balfour DA (2001) Acacia species turnover in space and time in an African savanna. J Biogeogr 28(1):117–128

Bui EN, González-Orozco CE, Miller JT (2014) Acacia, climate, and geochemistry in Australia. Plant Soil 381(1):161–175

Burdon JJ, Gibson AH, Searle SD, Woods MJ, Brockwell J (1999) Variation in the effectiveness of symbiotic associations between native rhizobia and temperate Australian Acacia: within-species interactions. J Appl Ecol 36(3):398–408

Burrows GE, Alden R, Robinson WA (2018) The lens in focus–lens structure in seeds of 51 Australian Acacia species and its implications for imbibition and germination. Aust J Bot 66(5): 398–413

Cadotte MW, McMahon SM, Fukami T (2006) Conceptual ecology and invasion biology: reciprocal approaches to nature. Springer

Caley P, Groves RH, Barker R (2008) Estimating the invasion success of introduced plants. Divers Distrib 14(2):196–203

Carter RWG (2013) Coastal environments: an introduction to the physical, ecological, and cultural systems of coastlines. Elsevier

Carvalho LM et al (2010) Disturbance influences the outcome of plant—soil biota interactions in the invasive Acacia longifolia and in native species. Oikos 119:1172–1180

Colautti RI, Macisaac HJ (2004) A neutral terminology to define invasive species. Divers Distrib 10:135–141

Correia M, Castro S, Rodríguez-Echeverría S (2015) Reproductive success of Acacia longifolia (Fabaceae, Mimosoideae) in native and invasive populations. Aust J Bot 63(5):387–391

Crall AW et al (2013) Using habitat suitability models to target invasive plant species surveys. Ecol Appl 23:60–72

Cronk Q, Fuller JL (1995) Plant invaders: the threat to natural ecosystems. Springer, 241 p

Czech B, Krausman P (1997) Distribution and causation of species endangerment in the United States. Science 277(5329):1116–1117

D'Antonio C et al (1996) Biological invasions as global environmental change. Am Sci 84(5)

Davis M (2006) Invasion biology 1958–2005: the pursuit of science and conservation. In: Conceptual ecology and invasions biology: reciprocal approaches to nature. Springer, London

Drenovsky R, James J (2010) Designing invasion-resistant plant communities: the role of plant functional traits. Rangelands 32:32

Duboscq-Carra VG, Fernandez RD, Haubrock PJ, Dimarco RD, Angulo E, Ballesteros-Mejia L, Diagne C, Courchamp F, Nuñez MA (2021) Economic impact of invasive alien species in Argentina: a first national synthesis. In: Zenni RD, McDermott S, García-Berthou E, Essl F (eds) The economic costs of biological invasions around the world, NeoBiota, vol 67, pp 329–348

Early R, Bradley BA, Dukes JS, Lawler JJ, Olden JD, Blumenthal DM et al (2016) Global threats from invasive alien species in the twenty-first century and national response capacities. Nat Commun 7(1):12485

Elton C (1958) The ecology of invasions by animals and plants. Methuen, London

Emeny J (2009). Acacia longifolia spread in the Glenelg plain bioregion: pattern and process. Doctoral dissertation, Deakin University

Enders M, Havemann F, Ruland F, Bernard Verdier M, Catford JA, Gómez Aparicio L et al (2020) A conceptual map of invasion biology: integrating hypotheses into a consensus network. Glob Ecol Biogeogr 29(6):978–991
Espínola LA, Ferreira H (2007) Espécies invasoras: conceitos, modelos e atributos. Interciencia 32(9):580–585
Espinosa-García FJ, Villaseñor JL, Vibrans H (2004) The rich generally get richer, but there are exceptions: correlations between species richness of native plant species and alien weeds in Mexico. Divers Distrib 10(5–6):399–407
EUROPE-ALIENS (2021) DAISIE—inventory of alien invasive species in Europe. DAISIE International. Available at: http://www.europe-aliens.org/. Accessed on 19 Sept 2021
Falk-Petersen J et al (2006) On the numerous concepts in invasion biology. Biol Invasions 8(6): 1409–1424
Fernandes MM (2012) Acácias e geografia histórica: rotas de um percurso global. Cadernos do curso de doutoramento em Geografia Faculdade de Letras da Universidade do Porto, 23–40 p
Gallien L et al (2010) Predicting potential distributions of invasive species: where to go from here? Divers Distrib 16:331–342
González-Orozco CE, Laffan SW, Knerr N, Miller JT (2013) A biogeographical regionalization of Australian Acacia species. J Biogeogr 40(11):2156–2166
Guo QF (2006) Intercontinental biotic invasions: what can we learn from native populations and habitats? Biol Invasions 8:1451–1459
Hooper S, Maslin B (1978) Phytogeography of Acacia in Western Australia. Aust J Bot 26:63–78
Hulme PE, Pyšek P, Jarošík V, Pergl J, Schaffner U, Vilà M (2013) Bias and error in understanding plant invasion impacts. Trends Ecol Evol 28(4):212–218
I3N BRASIL (2021) Acacia. In: Red de Información sobre Especies Invasoras. IABIN International. Available at: http://www.institutohorus.org. Accessed on 19 Sept 2021
IPBES (2019) In: Brondizio ES, Settele J, Díaz S, Ngo HT (eds) Global assessment report on biodiversity and ecosystem services of the intergovernmental science-policy platform on biodiversity and ecosystem services. IPBES Secretariat, Bonn
ISSG (2021) Acacia longifolia. In: Global invasive species database. ISSG International. Available at: http://www.issg.org/database/welcome. Accessed on 19 Sept 2021
Jeschke JM (2014) General hypotheses in invasion ecology. Divers Distrib 20(11):1229–1234
Köppen W (1900) Versuch einer Klassifikation der Klimate, vorzugsweise nachihren Beziehungen zur Pflanzenwelt. Geogr Z 6:593–611, 657–679
Köppen W, Geiger R (1954) Klima der Erde. Justus Perthes, Darmstadt, DE
Le Maitre D et al (2011) Impacts of invasive Australian acacias: implications for management and restoration. Divers Distrib 17(5):1015–1029
Le Roux JJ, Brown GK, Byrne M, Ndlovu J, Richardson DM, Thompson GD, Wilson JR (2011) Phylogeographic consequences of different introduction histories of invasive Australian Acacia species and Paraserianthes lophantha (Fabaceae) in South Africa. Divers Distrib 17(5):861–871
Lecanda J, Cuevas YA (2013) Aspectos biológicos de Acacia longifolia, una especie exótica potencialmente invasora de pastizales psamófilos. Revista de la Facultad de Agronomía 22 (Suppl. 2):99–104
Levine J et al (2003) Mechanisms underlying the impacts of exotic plant invasions. Proc Biol Sci 270(1517):775–781
Lourenço D (2009) Avaliação de áreas invadidas por espécies de Acácia na paisagem protegida da arriba fóssil da Costa de Caparica. Tese de Mestrado, Universidade de Lisboa
Manea A, Tabassum S, Carthey AJ, Cameron DN, Leishman MR (2019) Evidence for a shift in defence driving the invasion success of Acacia longifolia in Australia. Biol Invasions 21(6): 2211–2220
Marchante H (2001) Invasão dos ecossistemas dunares portugueses por Acacia: uma ameaça para a biodiversidade nativa. Tese de Doutoramento, Universidade de Coimbra
Marchante E (2007) Invasion of Portuguese coastal dunes by Acacia longifolia: impacts on soil ecology. Tese de Doutoramento, Universidade de Coimbra

Marchante E et al (2008) Short and long-term impacts of Acacia longifolia invasion on the belowground processes of a Mediterranean coastal dune ecosystem. Appl Soil Ecol 40:210–217

Marchante H et al (2011) The potential role of seed banks in the recovery of dune ecosystems after removal of invasive plant species. Appl Vegetat Sci 14:107–119

Marchante H et al (2014) Guia prático para a identificação de plantas invasoras em Portugal. Imprensa da Universidade de Coimbra, 208 p

Marchante E, Freitas H, Hoffmann J, Marchante H (2018) Trying to introduce a successful biological control agent to control an invasive plant in Portugal: the case study of the Australian bud-galling wasp (Trichilogaster acaciaelongifoliae) and Sydney golden wattle (Acacia longifolia). IOBC/WPRS Bull 137:40–42

Marques DN (2010) Influência espacial da invasora Acacia longifolia (Andrews) Willd. num ecossistema dunar português. Tese de Mestrado, Universidade de Lisboa, Lisboa

Maslin B et al (2003) Overview of the generic status of Acacia (Leguminosae: Mimosoideae). Aust Syst Bot 16:1–18

MAyDS and FAO (2019) Ministerio de Ambiente y Desarrollo Sostenible y Organización de las Naciones Unidas para la Agricultura y la Alimentación. Propuesta Preliminar de Estrategia Nacional sobre Especies Exóticas Invasoras. Informe de Progreso de Proyecto (IPP) noviembre 2019. Proyecto Fortalecimiento de la Gobernanza para la Protección de la Biodiversidad mediante la Formulación e Implementación de la Estrategia Nacional sobre Especies Exóticas Invasoras (ENEEI) (GCP/ARG/023/GFF)

Miller P (1754) The gardeners dictionary, 4th edn. Self-Published, London

Miller JT, Murphy DJ, Brown GK, Richardson DM, González-Orozco CE (2011) The evolution and phylogenetic placement of invasive Australian Acacia species. Divers Distrib 17(5): 848–860

Mittermeier RA, Turner WR, Larsen FW, Brooks TM, Gascon C (2011) In: Zachos FE (ed) Global biodiversity conservation: the critical role of hotspots, Biodiversity hotspots. Springer, 546 pp

Monks L, Coates D (2002) The translocation of two critically endangered Acacia species. Conserv Sci Western Australia 4(3)

Mooney HA, Hobbs RJ (2000) Invasive species in a changing world. Island Press, Washington DC

Nei M (1973) Analysis of gene diversity in subdivided populations. Proc Natl Acad Sci U S A 70: 3321–3323

Nordstrom K (2010) Recuperação de Praias e Dunas. Oficina de Textos, São Paulo

Nunes LJ, Meireles CI, Pinto Gomes CJ, Almeida Ribeiro NM (2019) Historical development of the Portuguese forest: the introduction of invasive species. Forests 10(11):974

Nuñez MA, Pauchard A (2010) Biological invasions in developing and developed countries: does one model fit all? Biol Invasions 12:707–714

Oliveira-Costa JLP (2014) Os Caminhos da Invasão do Género Acacia Mill. na Bacia do Rio Arouce. Tese de Mestrado—Faculdade de Letras da Universidade de Coimbra

Oliveira-Costa JLP, Pivello VR (2017). Acacia ecology on Brazilian dune systems—first data on the ecology of Australian *Acacia longifolia* (Fabaceae) across coastal dunes of Florianopolis, Santa Catarina, South Brazil. In: 14th international conference on ecology and management of alien plant invasions, EMAPI 14th, Lisbon—Portugal. Book of abstracts of EMAPI 14th

Oliveira-Costa JLP, Sousa RJ (2015) Organização geossistêmica e invasões por Acacia sp. (Fabaceae: Mimosoideae) na Bacia Hidrográfica do Rio Arouce. Uma visão das invasões biológicas à escala do Geossistema, vol 8. GOT—Revista de Geografia e Ordenamento do Território

Oliveira-Costa JLP, Figueiredo RF, Pivello VR (2019) Determinants and impacts of Acacia longifolia Andrews (Willd.) spread: a comparative study between Portugal and Brazil. In: Biennial conference international biogeography society, 9, Málaga–Spain, Book of abstracts

Oliveira-Costa JLP, Figueiredo RF, Pivello VR (2020) A Invasora Acacia longifolia Andrews (Willd.) em Portugal e no Brasil: Investigação Preliminar Integrando Registros GBIF com Dados da Literatura e Condicionantes Ambientais e Regionais. Espaço em Revista 22:165–202

Oliveira-Costa JLP, Figueiredo RF, Pivello VR (2021a) Distribuição Atual e Susceptibilidade à Invasão por Acacia longifolia em Ambientes Litorais: Bases Teóricas e Metodológicas de um Projeto de Pesquisa sobre as Invasões por Acacia no Litoral Português e no Litoral Sul do Brasil. Geografia (UFPI) 1:1–50

Oliveira-Costa JLP, Figueiredo RF, Pivello VR (2021b) What is the role of coastlines in plant invasions dynamics? Three steps to understand the invasibility of coastal systems by Acacia longifolia: a data paper synthesizing geographic, climate and taxonomic informations from global to regional scale. Rendezvous. J Interfaces Arts Culture 1:1–50

Orians G, Milewski A (2007) Ecology of Australia: the effects of nutrient-poor soils and intense fires. Biol Rev 82:393–423

Pedley L (1978) A revision of Acacia Mill. in Queensland. Austrobaileya 1:75–234

Pieterse PJ (1987) Acacia longifolia and its control: to burn or not to burn? Veld Flora 73(2):67

Pieterse PJ, Cairns ALP (1986) The effect of fire on an Acacia longifolia seed bank in the south-western cape. S Afr J Bot 52(3):233–236

Pimentel D et al (2001) Economic and environmental threats of alien plant, animal, and microbe invasions. Agric Ecosyst Environ 84:1–20

Pohlman CL, Nicotra AB, Murray BR (2005) Geographic range size, seedling ecophysiology and phenotypic plasticity in Australian Acacia species. J Biogeogr 32(2):341–351

Pyšek P (2001) Past and future of predictions in plant invasions: a field test by time. Divers Distrib 7:145–151

Pyšek P, Prach K, Rejmànek M, Wade M (1995) Plant invasions: general aspects and special problems. SPB Academic Publishing, Amsterdam

Rascher K et al (2011) Acacia longifolia invasion impacts vegetation structure and regeneration dynamics in open dunes and pine forests. Biol Invasions 13(5):1099–1113

Rejmànek M (2011) Invasiveness. In: Encyclopedia of biological invasions. University of California Press, Berkeley and Los Angeles, CA

Rejmànek M, Richardson D (2013) Trees and shrubs as invasive alien species—update of the global database. Divers Distrib n.19:1903–1904

Rejmànek M, Richardson D, Pyšek P (2005) Plant invasions and invasibility of plant communities. Vegetat Ecol 13:332–355

Richards C et al (2006) Jack of all trades, master of some? On the role of phenotypic plasticity in plant invasions. Ecol Lett 9(8):981–999

Richardson DM (2004) Plant invasion ecology—dispatches from the front line. Divers Distrib 10(5–6):315–319

Richardson DM (2011) Fifty years of invasion ecology. The legacy of Charles Elton. Wiley-Blackwell, Oxford

Richardson D, Pyšek P (2006) Plant invasions: merging the concepts of species invasiveness and community invasibility. Prog Phys Geogr 30(409):409

Richardson DM et al (2011) Human mediated introductions of Australian Acacias—a global experiment in biogeography. Divers Distrib 17(5):771–787

Robinson J, Harris SA (2000) A plastid DNA phylogeny of the genus Acacia Miller (Acacieae, Leguminoseae). Bot J Linn Soc 132:195–222

Rodríguez-Echeverría S (2010) Rhizobial hitchhickers from down under: invasional meltdown in a plant-bacteria mutualism? J Biogeogr 37:1611–1622

Rodríguez-Echeverría S, Le Roux JJ, Crisóstomo JA, Ndlovu J (2011) Jack-of-all-trades and master of many? How does associated rhizobial diversity influence the colonization success of Australian Acacia species? Divers Distrib 17(5):946–957

Seebens H et al (2017) No saturation in the accumulation of alien species worldwide. Nat Commun 8:14435

Seebens H, Blackburn TM, Hulme PE, Van Kleunen M, Liebhold AM, Orlova-Bienkowskaja M et al (2021b) Around the world in 500 years: inter regional spread of alien species over recent centuries. Glob Ecol Biogeogr 30(8):1621–1632

Simberloff D, Rejmànek M (2011) Encyclopedia of biological invasions. University of California Press, Berkeley and Los Angeles, CA
Stellatelli OA, Vega LE, Block C, Cruz FB (2013) Effects on the thermoregulatory efficiency of two native lizards as a consequence of the habitat modification by the introduction of the exotic tree Acacia longifolia. J Therm Biol 38(3):135–142
Theoharides K, Dukes J (2007) Plant invasion across space and time: factors affecting nonindigenous species success during four stages of invasion. New Phytol 176:256–273
Turnbull J (1987) Australian Acacias in developing countries. In: Proceedings of an international workshop held at the forestry training Centre, Gympie, QLD, Australia. ACIAR proceedings no. 16, 196 p
Van Kleunen M, Dawson W, Essl F, Pergl J, Winter M, Weber E et al (2015) Global exchange and accumulation of non-native plants. Nature 525(7567):100–103
Vavilov NI (1992) Origin and geography of cultivated plants. Cambridge University Press, Cambridge
Vicente S et al (2018) Genetic diversity and differentiation of invasive Acacia longifolia in Portugal. Web Ecol 18(1):91–103
Vilà M et al (2011) Ecological impacts of invasive alien plants: a meta-analysis of their effects on species, communities and ecosystems. Ecol Lett 14(7):702–708
Vilà M, Dunn AM, Essl F, Gómez-Díaz E, Hulme PE, Jeschke JM et al (2021) Viewing emerging human infectious epidemics through the lens of invasion biology. Bioscience 71(7):722–740
Vitousek P (1990) Biological invasions and ecosystem processes: towards an integration of population biology and ecosystem studies. Oikos 57:7–13
Welgama A, Florentine S, Marchante H, Javaid MM, Turville C (2019) The germination success of Acacia longifolia subsp. longifolia (Fabaceae): a comparison between its native and exotic ranges. Aust J Bot 67(5):414–424
Williamson M (1996) Biological invasions. Chapman & Hall Publishing, London, p 244p
Zalba SM, Villamil CB (2002) Woody plant invasion in relictual grasslands. Biol Invasions 4:55–72
Zalba SM, Sanhueza C, Cuevas Y, Wong LJ, Pagad S (2020) Global register of introduced and invasive species—Argentina. Version 1.3. Invasive Species Specialist Group ISSG. Checklist dataset
Zenni R, Ziller S (2011) An overview of invasive plants in Brazil. Rev Bras Bot 34(3):431–446

Further Reading

Ahmed SE, Mcinerny G, O'Hara K, Harper R, Salido L, Emmott S, Joppa LN (2015) Scientists and software–surveying the species distribution modelling community. Divers Distrib 21(3): 258–267
Austin M (2007) Species distribution models and ecological theory: a critical assessment and some possible new approaches. Ecol Model 200(1–2):1–19
Barrett LG, Bever JD, Bissett A, Thrall PH (2015) Partner diversity and identity impacts on plant productivity in Acacia–rhizobial interactions. J Ecol 103(1):130–142
Booth TH (2018) Species distribution modelling tools and databases to assist managing forests under climate change. For Ecol Manag 430:196–203
Booth TH, Nix HA, Busby JR, Hutchinson MF (2014) BIOCLIM: the first species distribution modelling package, its early applications and relevance to most current MAXENT studies. Divers Distrib 20(1):1–9
Broennimann O et al (2007) Evidence of climatic niche shift during biological invasion. Ecol Lett 10:701–709
Camenen B, Larroudé P (2003) Comparison of sediment transport formulae for the coastal environment. Coast Eng 48(2):111–132

Capistrano D, Samper C, Lee MJ, Raudsepp-Hearne C (2005) Ecosystems and human well-being: multiscale assessments: findings of the sub-global assessments working group of the millennium ecosystem assessment. Island Press
Carter RWG (1988) Coastal environments: an introduction to the physical, ecological, and cultural systems of coastlines. Academic Press
Catford JA, Jansson R, Nilsson C (2009) Reducing redundancy in invasion ecology by integrating hypotheses into a single theoretical framework. Divers Distrib 15(1):22–40
Catford J et al (2012) Quantifying levels of biological invasion: towards the objective classification of invaded and invasible ecosystems. Glob Chang Biol 18:44–62
Darwin C (1859) The origin of species by means of natural selection, or, the preservation of favoured races in the struggle for life. Books, Incorporated, Pub.
Dewitt TJ, Sih A, Wilson DS (1998) Costs and limits of phenotypic plasticity. Trends Ecol Evol 13(2):77–81
Di Castri F (1990) On invading species and invaded ecosystems: the interplay of historical chance and biological necessity. In: Biological invasions in Europe and the Mediterranean Basin. Kluwer Academic Publishers, Dordrecht
Di Castri F, Mooney H (1973) Mediterranean type ecosystems: origin and structure. Springer Verlag, Berlin
Díaz, S. M., Settele, J., Brondízio, E., Ngo, H., Guèze, M., Agard, J., et al. (2019). The global assessment report on biodiversity and ecosystem services: summary for policy makers
Donaldson-Matasci MC, Lachmann M, Bergstrom CT (2008) Phenotypic diversity as an adaptation to environmental uncertainty. Evol Ecol Res 10(4):493–515
Drake JA et al (1989) Biological invasions: a global perspective. SCOPE
Elith J, Leathwick JR (2009) Species distribution models: ecological explanation and prediction across space and time. Annu Rev Ecol Evol Syst 40(1):677–697
Figueiredo AM, Leal I, Meireles R (2010) A territorialização de políticas públicas em Portugal. IFDR-Instituto Financeiro para o Desenvolvimento Regional, Lisboa
Forsman A (2014) Effects of genotypic and phenotypic variation on establishment are important for conservation, invasion, and infection biology. Proc Natl Acad Sci 111(1):302–307
Fridley JD (2011) Invasibility of communities and ecosystems. In: Encyclopedia of biological invasions. University of California Press, London
Gallardo B, Aldridge DC, González-Moreno P, Pergl J, Pizarro M, Pyšek P et al (2017) Protected areas offer refuge from invasive species spreading under climate change. Glob Chang Biol 23(12):5331–5343
Galvani F, Gaertner E (2006) Adequação da metodologia Kjeldahl para determinação de Nitrogênio Total e Proteína Bruta. Circular Técnica, 1–9 p
González-Moreno P et al (2013a) Quantifying the landscape influence on plant invasions in Mediterranean coastal habitats. Landscape Ecol 28:891–903
González-Moreno P et al (2015) Una visión a escala de paisaje de las invasiones biológicas. Ecosistemas 24(1):84–92
González-Moreno P, Diez JM, Ibáñez I, Font X, Vilà M (2014) Plant invasions are context dependent: multiscale effects of climate, human activity and habitat. Divers Distrib 20(6): 720–731
González-Moreno P, Pino J, Gassó N, Vila M (2013b) Landscape context modulates alien plant invasion in Mediterranean forest edges. Biol Invasions 15(3):547–557
González-Moreno P, Pino J, Cózar A, García-De-Lomas J, Vilà M (2017) The effects of landscape history and time-lags on plant invasion in Mediterranean coastal habitats. Biol Invasions 19(2): 549–561
Groves RH, Di Castri F (1991) Biogeography of Mediterranean invasions. Springer, London
Haggett P (1994) Prediction and predictability in geographical systems. Trans Inst Br Geogr 1(1):6
Henderson S et al (2006) Progress in invasive plants research. Prog Phys Geogr 30:25–46
Heywood J (1991) Spatial analysis of genetic variation in plant populations. Annu Rev Ecol Syst 22:335–355

Hijmans R et al (2005) Very high-resolution interpolated climate surfaces for global land areas. Int J Climatol 25:1965–1978
Holdridge LR (1947) Determination of world plant formations from simple climatic data. Science 105:367–368
Holmes PM, Esler KJ, Gaertner M, Geerts S, Hall SA, Nsikani MM et al (2020) Biological invasions and ecological restoration in South Africa. In: Biological invasions in South Africa. Springer, pp 665–700
Huggett R (1987) Systems analysis in geography. Oxford Clarendon Press
IUCN [International Union for the Conservation of Nature] (1999) Guidelines for the prevention of biodiversity loss due to biological invasion. Species 31(32):28–42p
Klink H (1974) Geoecology and natural regionalization—bases for environmental research. Institute for Scientific Cooperation, Applied Sciences and Development, Tübingen
Koike F et al (2008) Assessment and control of biological invasion risks. Shoukadoh Book Sellers, IUCN, Kyoto, Gland
Kolar C, Lodge D (2001) Progress in invasion biology: predicting invaders. Trends Ecol Evol 16(4):199–204
Kowaric I (1995) Time lags in biological invasions with regard to the success and failure of alien species. In: Plant invasions—general aspects and special problems. Academic Publishing, Amsterdam
Kussell E, Leibler S (2005) Phenotypic diversity, population growth, and information in fluctuating environments. Science 309(5743):2075–2078
Lee C (2002) Evolutionary genetics of invasive species. Trends Ecol Evol 17:386–339
Liu C, Wolter C, Xian W, Jeschke JM (2020) Species distribution models have limited spatial transferability for invasive species. Ecol Lett 23(11):1682–1692
Lockwood JL et al (2007) Invasion ecology. Blackwell Publishing
Lonsdale W (1999) Global patterns of plant invasions and the concept of invasibility. Ecology 80(5):1522–1536
Macarthur RH, Wilson EO (1967) The theory of Island biogeography. Princeton University Press, Princeton, NJ, p 203
MacDonald G (2003) Biogeography: introduction to space, time, and life. John Wiley & Sons Inc.
Malavasi M et al (2014) Landscape fragmentation, land-use legacy and propagule pressure promote plant invasion on coastal dunes: a patch-based approach. Landsc Ecol 29:1541–1550
Mcnelly J (2001) Invasive species: a costly catastrophe for native biodiversity. Land Use Water Resour 1:1–10
Meiners S et al (2001) Effects of plant invasions on the species richness of abandoned agricultural land. Ecography 24(6):633–644
Meneses P (2001) Sensoriamento remoto: reflectância dos alvos naturais. Brasília, Editora UnB, 262 p
Miner BG, Sultan SE, Morgan SG, Padilla DK, Relyea RA (2005) Ecological consequences of phenotypic plasticity. Trends Ecol Evol 20(12):685–692
Myers J, Bazely D (2010) Ecology and control of introduced plants. University Press, Cambridge
Naimi B, Araújo MB (2016) SDM: a reproducible and extensible R platform for species distribution modelling. Ecography 39(4):368–375
Norberg J, Swaney DP, Dushoff J, Lin J, Casagrandi R, Levin SA (2001) Phenotypic diversity and ecosystem functioning in changing environments: a theoretical framework. Proc Natl Acad Sci 98(20):11376–11381
O'Dowd D, Gill A (1986) Seed dispersal syndromes of Australian acacia. In: Seed dispersal. Academic Press, New York
Oliveira-Costa J et al (2016) Caracterizando as invasões por Acácias Australianas na Bacia do Rio Arouce desde 1960 usando a distribuição das espécies, a história do uso do solo e as condicionantes físicas—implicações à invasibilidade. Revista de Geografia e Ordenamento do Território 10:241–265

Pauchard A et al (2004) Comparing alien plant invasions among regions with similar climates: where to from here? Divers Distrib 10:371–375

Peacock L, Worner S (2008) Biological and ecological traits that assist establishment of alien invasive insects. N Z Plant Prot 61:1–7

Peterson AT (2003) Predicting the geography of species' invasions via ecological niche modeling. Q Rev Biol 78:419–443

Pigliucci M (2005) Evolution of phenotypic plasticity: where are we going now? Trends Ecol Evol 20(9):481–486

Pyšek P, Lambdon PW, Arianoutsou M, Kühn I, Pino J, Winter M (2009) Alien vascular plants of Europe. In: Handbook of alien species in Europe. Springer, Dordrecht, pp 43–61

Rejmànek M (1999) Invasive plants and invasible ecosystems. In: Invasive species and biodiversity management. Kluwer, Boston

Ricciardi A, Iacarella JC, Aldridge DC, Blackburn TM, Carlton JT, Catford JA et al (2021) Four priority areas to advance invasion science in the face of rapid environmental change. Environ Rev 29(2):119–141

Richling A (1981) Typology and regionalization as two independent systems of grouping geocomplexes. International Geographical Union, Bratislava

Robinson TP, Van Klinken RD, Metternicht G (2010) Comparison of alternative strategies for invasive species distribution modeling. Ecol Model 221(19):2261–2269

Schmidt IB, Sampaio AB, Borghetti F (2005) Efeitos da época de queima sobre a reprodução sexuada e estrutura populacional de Heteropterys pteropetala (Adr. Juss.), Malpighiaceae, em áreas de Cerrado sensu stricto submetidas a queimas bienais. Acta Bot Bras 19:927–934

Seebens H, Bacher S, Blackburn TM, Capinha C, Dawson W, Dullinger S et al (2021a) Projecting the continental accumulation of alien species through to 2050. Glob Chang Biol 27(5):970–982

Shigesada N, Kawasaki K (1997) Biological invasions: theory and practice, Oxford series in ecology and evolution. Oxford University Press

Smith TM, Shugart HH, Woodward FI (1997) Plant functional types: their relevance to ecosystem properties and global change. Cambridge University Press, Cambridge

Srivastava V, Lafond V, Griess VC (2019) Species distribution models (SDM): applications, benefits and challenges in invasive species management. CABI Rev 2019:1–13

Tivy J (1971) Biogeography: a study of plants in the ecosphere. Oliver & Boyd Inc.

Václavík T, Meentemeyer RK (2009) Invasive species distribution modeling (iSDM): are absence data and dispersal constraints needed to predict actual distributions? Ecol Model 220(23): 3248–3258

Vilà M, García-Berthou E (2010) Monitoring biological invasions in freshwater habitats. In: Conservation monitoring in freshwater habitats. Springer, Dordrecht, pp 91–100

Vilà M, Hulme P (2017) Impact of biological invasions on ecosystem services, Springer series in invasion ecology. Springer

Vila M, Ibáñez I (2011) Plant invasions in the landscape. Landsc Ecol 26(4):461–472

Watts D (1971) Principles of biogeography: an introduction to the functional mechanisms of ecosystems. McGraw-Hill Publishing Co.

Whitman DW, Agrawal AA (2009) What is phenotypic plasticity and why is it important. In: Phenotypic plasticity of insects: mechanisms and consequences. Science Publishers, pp 1–63

Wilson J et al (2016) Biological invasions and natural colonisations are different—the need for invasion science. NeoBiota 31:87–98

Wolfgang N (2016) Biological invasions (Ecological studies). Springer

Invasive Plants in India: Their Adaptability, Impact, and Response to Changing Climate

8

Sonia Rathee, Mustaqeem Ahmad, Padma Sharma, Daizy R. Batish, and Harminder Pal Singh

Abstract

In the twentieth century, plant invasions have become more frequent and intense, especially with changing climatic conditions. It is challenging to comprehend and forecast the patterns of plant invasion as these patterns are often specific to a place, spatial scale, time, and species under consideration. Invasive plants have been rapidly spreading, posing both economic and ecological harm. The impact on the natural ecosystem dynamics is mediated through alterations in the native floral diversity, composition, and soil environments. This chapter compiles 242 invasive plant species in India, together with information on their taxonomy, native range, degree of climatic adaptation, and other relevant factors. Most of the invasive flora belonged to 58 families, with Asteraceae (19%) and Fabaceae (11%) representing the most number of species. Of all the plants, 45% were annuals and 37% were perennials. Additionally, most of the invasive flora of India was observed to have originated from south America (51%), Europe (12%), and north America (9%). In India, the biodiversity and invasion hotspots are observed to coincide with each other. The invasion hotspots of India also coincide with the diverse ecosystems such as coastal forests, forest reserves, mangrove ecosystems, islands, and mountain ranges. Rising temperature and variability in

S. Rathee · D. R. Batish
Department of Botany, Panjab University, Chandigarh, India
e-mail: daizybatish@pu.ac.in

M. Ahmad
Laboratory for the Conservation of Endangered Species (LaCONES), CSIR-CCMB, Hyderabad, Telangana, India

P. Sharma · H. P. Singh (✉)
Department of Environment Studies, Panjab University, Chandigarh, India
e-mail: hpsingh_01@pu.ac.in

S. Tripathi et al. (eds.), *Plant Invasions and Global Climate Change*,
https://doi.org/10.1007/978-981-99-5910-5_8

precipitation associated with climate change would further promote invasive species spread to uninvaded regions and ecosystems. With changing climate, the areas under high suitability of several invasion species such as *Ageratina adenophora*, *Chromolaena odorata*, and *Lantana camara* are expected to increase further. Various factors contribute towards the expansion of these highly invasive species in India; however, phenotypic plasticity, local adaptation, and genetic diversity can be considered as the major reasons aiding the adaptation of invasive species in introduced regions. The need of the hour is to understand the mechanism of adaptation of invasive species in India, their impacts and subsequently devise appropriate management strategies.

Keywords

Climate change · Ecological impact · Genetic diversity · Invasive flora · Invasion hotspot · Phenotypic plasticity

8.1 Introduction

The introduction of alien flora into a novel environment beyond its native reach is drastically causing homogenisation of the flora of introduced regions (García et al. 2018). A significant rise in frequency and intensity of plant invasions in the twentieth century (Seebens et al. 2021) has invoked a keen interest in researchers for describing and anticipating the patterns, and reasons for these invasions (Davis 2006). Invasive species are either intentionally or unintentionally introduced into the new environment, and human intervention has been unarguably designated as one of the major causes of such invasions (Crooks et al. 2011). Generally, only those exotic species that proliferate and spread at a rate that poses negative consequences on the local ecosystem are termed as invasive species (Richardson and Pyšek 2008). An invasive species when entering an exotic range has to transit over four basic spatio-temporal stages: transport, colonisation, establishment, and spread (Colautti and MacIsaac 2004). Out of all alien species introduced in a region, only 10% are able to transit over to the next stage and establish themselves, and only ~10% of those become invasive (Keller et al. 2011). Rejmánek and Randall (1994) estimated that ≥20% of world's flora to be presently comprised of exotic species. The overall impact of invasive plant species is considered second only to habitat fragmentation, thereby making it a global concern (Sharma and Batish 2022). Invasive plants tend to destroy biogeographic regions and threaten ecosystems by altering their structure, function, processes, and stability, and therefore lead to biodiversity loss (Capinha et al. 2015; Pyšek et al. 2020). Moreover, the rising economic costs associated with invading species have also become a cause of worry (Bonanno 2016). Additionally, invasive species disrupt the phylogenetic and functional diversities and alter the abundance of species and composition of invaded plant communities, thus leading to species endangerment and extinction (Ricciardi et al. 2013; Bellard et al. 2016).

Interaction between plant invasions and changing climate may have extremely damaging environmental consequences (Crowl et al. 2008). Rapid changes in climate faced by the alien species in introduced areas, like the increase in CO_2 have shown a positive association with invasion spread, while others such as rising temperature and varying precipitation may have positive or negative impacts (Bradley et al. 2010; Demertzis and Iliadis 2018). The patterns of invasions are, thus, specific to a place, spatial scale, time, and species under consideration, thus making the prediction of future invasion very difficult based on data collected at regional scales (Theoharides and Dukes 2007). In this chapter, invasive flora of India, climate suitability and invasion hotspots, introduction pathways, role of climate change in plant invasion, adaptations in invasive plants with respect to changing climate, and socio-economic and socio-ecological impacts of plant invasion in India have been discussed.

8.2 Plant Diversity and Invasive Flora of India

India is one of the most diversity rich country of the world, with four biodiversity hotspots— Indo-Burma, Himalaya, Western Ghats-Sri Lanka, and Sundaland (Ahmad et al. 2022). India is home to diversified habitat, geology, geomorphology, climate, biological richness, and culture that encompasses ~2.4% of the world's landmass, and supports ~7–8% of the global biodiversity (Rathee and Kaur 2022). India has reportedly around 47,513 plant species (Kamble and Yele 2020) representing ~11% of the flora worldwide. Additionally, ~28% of the Indian flora is described to be endemic (Kamble and Yele 2020).

However, the invasive flora accumulation is also rising in India and the number has been increasing rapidly (Bhatt et al. 2011). Although numerous nonindigenous species have been introduced in India (Sankaran et al. 2021), the Botanical Survey of India (BSI 2020) reports the presence of 173 invasive alien plants in India. Increasing population, high trade rate, and fast economic growth of the country might be behind the increasing instances of invasion of alien species. According to Nayar (1977), ~18% of Indian floral diversity is constituted of exotic plants. In the Third National Report submitted to CBD (Convention on Biological Diversity), it was noted that ~40% of India's flora is exotic in origin, with ~21% of it being invasive (Khuroo et al. 2012). How the report cannot be considered reliable as the estimates were not based on any compiled inventory on alien species of India. Some publications have thereafter strived forward in this direction, such as Reddy et al. (2008), Khuroo et al. (2012), and Khuroo et al. (2021) who reported 173 invasive, 134 naturalised or invasive, and 145 invasive species, respectively, in India.

On the basis of published literature and online databases, we have compiled a list of 242 plants species reported to be invasive in India (Table 8.1). These species belonged to 58 families with Asteraceae (19%), Fabaceae (11%), Poaceae (7%), Amaranthaceae (6%), Solanaceae (5%), Euphorbiaceae (5%), Malvaceae (4%), and Convolvulaceae (4%) representing the majority of the invasive species in India (Fig. 8.1). All the other plant families such as Acanthaceae, Apocynaceae,

Table 8.1 A list of invasive flora (242 species) of India showing their lifespan, native region, and purpose of introduction (source: Khuroo et al. 2012; Jaryan et al. 2013; Khuroo et al. 2021; ENVIS 2023)

	Species name	Family name	Lifespan	Native region	Purpose of introduction
Climbers and vines					
1.	*Antigonon leptopus* Hook. & Arn.	Polygonaceae	P	South America	–
2.	*Convolvulus arvensis* L.	Convolvulaceae	P	Europe	Unintentional
3.	*Cryptostegia grandiflora* Roxb. ex R.Br.	Apocynaceae	P	Africa	–
4.	*Ipomoea indica* (Burm.) Merr.	Convolvulaceae	P	Europe/ South America	Ornamental
5.	*Ipomoea quamoclit* L.	Convolvulaceae	A/P	South America	Ornamental
6.	*Merremia aegyptia* (L.) Urb.	Convolvulaceae	P	South America	Unintentional
7.	*Mikania micrantha* Kunth	Asteraceae	P	South America	–
8.	*Passiflora foetida* L.	Passifloraceae	A/P	South America	–
Herbs					
1.	*Acanthospermum hispidum* DC.	Asteraceae	A	South America	Unintentional
2.	*Achillea millefolium* L.	Asteraceae	P	Europe	Medicinal
3.	*Adenostemma lavenia* (L.) Kuntze.	Asteraceae	A	South America	–
4.	*Aeschynomene americana* L.	Fabaceae	A	South America	–
5.	*Ageratina adenophora* (Spreng.) R.M.King & H. Rob.	Asteraceae	P	South America	Unintentional
6.	*Ageratum conyzoides* (L.) L.	Asteraceae	A	South America	Ornamental
7.	*Ageratum houstonianum* Mill.	Asteraceae	A	North America	–
8.	*Agrostis stolonifera* L.	Poaceae	A	North America	Fodder
9.	*Alisma plantago-aquatica* L.	Alismataceae	P	North America	Unintentional
10.	*Alternanthera ficoidea* (L.) Sm.	Amaranthaceae	P	South America	–
11.	*Alternanthera paronychioides* A.St.-Hil.	Amaranthaceae	P	South America	Unintentional
12.	*Alternanthera philoxeroides* (Mart.) Griseb.	Amaranthaceae	P	South America	Unintentional

(continued)

Table 8.1 (continued)

	Species name	Family name	Lifespan	Native region	Purpose of introduction
13.	*Alternanthera pungens* Kunth	Amaranthaceae	A/P	South America	–
14.	*Alternanthera spinosa* (Hornem.) Schult.	Amaranthaceae	A	North America/ South America	–
15.	*Amaranthus caudatus* L.	Amaranthaceae	A	South America	Food
16.	*Amaranthus spinosus* L.	Amaranthaceae	A	South America	Unintentional
17.	*Ambrosia artemisiifolia* L.	Asteraceae	A	North America	–
18.	*Anagallis arvensis* L.	Primulaceae	P	Europe	Unintentional
19.	*Anthemis cotula* L.	Asteraceae	A	Europe	Unintentional
20.	*Arenaria serpyllifolia* L.	Caryophyllaceae	A/P	Asia-Europe	Unintentional
21.	*Argemone mexicana* L.	Papaveraceae	A	North America	Medicinal
22.	*Argemone ochroleuca* Sweet	Papaveraceae	P	North America	Unintentional
23.	*Artemisia absinthium* L.	Asteraceae	P	Europe	Medicinal
24.	*Asclepias curassavica* L.	Apocynaceae	P	South America	–
25.	*Asphodelus tenuifolius* Cav.	Xanthorrhoeaceae	A	South America	Unintentional
26.	*Bidens biternata* (Lour.) Merr. & Sherff	Asteraceae	A	North America/ South America	–
27.	*Bidens pilosa* L.	Asteraceae	A	South America	Unintentional
28.	*Blainvillea acmella* (L.) Philipson	Asteraceae	A	South America	Unintentional
29.	*Blumea lacera* (Burm. f.) DC.	Asteraceae	A	South America	Unintentional
30.	*Blumea obliqua* (L.) Druce	Asteraceae	A	South America	Unintentional
31.	*Bothriochloa ischaemum* (L.) Keng.	Poaceae	P	Africa	Unintentional
32.	*Bromus inermis* Leyss.	Poaceae	A	Europe	Fodder
33.	*Bromus japonicus* Thunb.	Poaceae	A	Europe	Fodder
34.	*Calceolaria mexicana* Benth.	Calceolariaceae	A	North America	–
35.	*Calyptocarpus vialis* Less.	Asteraceae	P	South America	Unintentional

(continued)

Table 8.1 (continued)

	Species name	Family name	Lifespan	Native region	Purpose of introduction
36.	*Cannabis sativa* L.	Cannabaceae	A	Asia	Unintentional
37.	*Capsella bursa-pastoris* (L.) Medik.	Brassicaceae	A/B	Europe	Unintentional
38.	*Cardamine trichocarpa* Hochst. ex A. Rich.	Brassicaceae	A	Asia-temperate/ Europe	–
39.	*Carex notha* Kunth	Cyperaceae	A	Asia	Unintentional
40.	*Catharanthus pusillus* (Murray) G. Don	Apocynaceae	A	South America	Ornamental
41.	*Celosia argentea* L.	Amaranthaceae	A	South America	–
42.	*Ceratophyllum demersum* L.	Ceratophyllaceae	A/P	North America	Unintentional
43.	*Chamaecrista absus* (L.) H.S.Irwin & Barneby	Fabaceae	A	South America	Unintentional
44.	*Chamaecrista rotundifolia* (Pers.) Greene	Fabaceae	P	South America	–
45.	*Chenopodium album* L.	Amaranthaceae	A	Europe	Food
46.	*Chenopodium foliosum* Asch.	Amaranthaceae	A	Asia-Europe	Unintentional
47.	*Chenopodium hybridum* L.	Amaranthaceae	A	Asia-Europe	Food
48.	*Corchorus aestuans* L.	Malvaceae	A	South America	–
49.	*Chloris barbata* Sw.	Poaceae	P	Africa/ Southern America	–
50.	*Cirsium arvense* (L.) Scop.	Asteraceae	P	Asia	Unintentional
51.	*Cissampelos pareira* L.	Menispermaceae	A	South America	–
52.	*Cleome rutidosperma* DC.	Cleomaceae	A	Africa	–
53.	*Cleome viscosa* L.	Cleomaceae	A	South America	–
54.	*Corchorus aestuans* L.	Malvaceae	A/P	South America	–
55.	*Corchorus tridens* L.	Malvaceae	A	Africa	Unintentional
56.	*Corchorus trilocularis* L.	Malvaceae	A	Africa	Unintentional
57.	*Cosmos bipinnatus* Cav.	Asteraceae	A	South America	Ornamental
58.	*Crepis sancta* (L.) Bornm.	Asteraceae	A	Asia	Unintentional
59.	*Croton bonplandianus* Baill.	Euphorbiaceae	B/P	South America	Unintentional

(continued)

Table 8.1 (continued)

	Species name	Family name	Lifespan	Native region	Purpose of introduction
60.	*Crassocephalum crepidioides* S.Moore	Asteraceae	A	Africa	–
61.	*Cyclospermum leptophyllum* (Pers.) Sprague	Apiaceae	A	South America	–
62.	*Cyperus difformis* L.	Cyperaceae	A	Africa/ Europe	Unintentional
63.	*Cyperus rotundus* L.	Cyperaceae	P	Europe	Unintentional
64.	*Cytisus scoparius* (L.) Link.	Fabaceae	P	Europe	Unintentional
65.	*Dactylis glomerata* L.	Poaceae	P	Asia	Fodder
66.	*Datura innoxia* Mill.	Solanaceae	A	South America	–
67.	*Datura metel* L.	Solanaceae	P	South America	–
68.	*Datura stramonium* L.	Solanaceae	A	North America	Unintentional
69.	*Digera muricata* (L.) Mart.	Amaranthaceae	A/P	North America	Unintentional
70.	*Dysphania ambrosioides* (L.) Mosyakin & Clemants	Amaranthaceae	B	South America	–
71.	*Echinochloa colona* Link.	Poaceae	A	Africa/ Asia-tropical	–
72.	*Echinochloa crus-galli* (L.) P.Beauv.	Poaceae	A	Africa/ Asia-tropical	–
73.	*Eclipta prostrata* (L.) L.	Asteraceae	A	South America	Unintentional
74.	*Eichhornia crassipes* (Mart.) Solms	Pontederiaceae	P	South America	Ornamental
75.	*Emilia sonchifolia* (L.) DC. ex DC.	Asteraceae	A	South America	Unintentional
76.	*Epilobium hirsutum* L.	Onagraceae	A/P	Africa/ Europe	Unintentional
77.	*Eragrostis pilosa* (L.) P. Beauv.	Poaceae	A	Africa	Unintentional
78.	*Erigeron bonariensis* L.	Asteraceae	A	Asia-temperate/ Europe	–
79.	*Erigeron canadensis* L.	Asteraceae	A	North America	–
80.	*Erigeron karvinskianus* DC.	Asteraceae	P	South America	Unintentional

(continued)

Table 8.1 (continued)

	Species name	Family name	Lifespan	Native region	Purpose of introduction
81.	*Euphorbia heterophylla* L.	Euphorbiaceae	A	South America	Unintentional
82.	*Euphorbia hirta* L.	Euphorbiaceae	A	South America	Unintentional
83.	*Euphorbia hyssopifolia L.*	Euphorbiaceae	A	South America	–
84.	*Euphorbia prostrata* Aiton	Euphorbiaceae	A	South America	Unintentional
85.	*Euphorbia thymifolia* L.	Euphorbiaceae	A/P	South America	Unintentional
86.	*Evolvulus nummularius* (L.) L.	Convolvulaceae	P	South America	–
87.	*Flaveria trinervia* C. Mohar	Asteraceae	A	South America	–
88.	*Fumaria indica* (Hausskn.) Pugsley	Papaveraceae	A/P	Asia	Unintentional
89.	*Galinsoga parviflora* Cav.	Asteraceae	A	South America	Unintentional
90.	*Galinsoga quadriradiata* Ruiz & Pav.	Asteraceae	A	North America	–
91.	*Gamochaeta purpurea* (L.) Cabrera	Asteraceae	A	North America	Unintentional
92.	*Gnaphalium coarctatum* Willd.	Asteraceae	A	South America	–
93.	*Gnaphalium polycaulon* Pers.	Asteraceae	A	South America	Unintentional
94.	*Gnaphalium purpureum* L.	Asteraceae	A	South America	–
95.	*Gomphrena serrata* L.	Amaranthaceae	A/P	South America	–
96.	*Grangea maderaspatana* (L.) Desf.	Asteraceae	P	South America	Unintentional
97.	*Heliotropium indicum L.*	Boraginaceae	A	South America	–
98.	*Hyptis suaveolens* (L.) Poit.	Lamiaceae	A	South America	–
99.	*Ipomoea hederifolia* L.	Convolvulaceae	A	South America	Unintentional
100.	*Ipomoea obscura* (L.) Ker Gawl.	Convolvulaceae	A/P	Africa	Unintentional
101.	*Ipomoea pes-tigridis* L.	Convolvulaceae	A/P	Africa	Unintentional
102.	*Juncus articulatus* L.	Juncaceae	P	Asia-Europe	Unintentional
103.	*Lagascea mollis* Cav.	Asteraceae	A	North America	–

(continued)

Table 8.1 (continued)

	Species name	Family name	Lifespan	Native region	Purpose of introduction
104.	*Lemna minor* L.	Araceae	P	Asia/Africa	Unintentional
105.	*Leonotis nepetifolia* (L.) R.Br.	Lamiaceae	A/P	South America	–
106.	*Lepidium didymum* L.	Brassicaceae	A/P	South America	Food
107.	*Lithospermum arvense* L.	Boraginaceae	A	Asia-Europe	Unintentional
108.	*Lolium temulentum* L.	Poaceae	A	Europe	Fodder
109.	*Macroptilium atropurpureum* (DC.) Urb.	Fabaceae	P	South America	–
110.	*Macroptilium lathyroides* (L.) Urb.	Fabaceae	A/B	Europe	–
111.	*Malachra capitata L.*	Malvaceae	A/P	South America	–
112.	*Martynia annua* L.	Pedaliaceae	B/P	North America	Ornamental
113.	*Mecardonia procumbens* (Mill.) Small	Plantaginaceae	A	South America	–
114.	*Melilotus officinalis* subsp. *alba* (Medik.) H. Ohashi & Tateishi	Fabaceae	A	North America	Fodder
115.	*Mentha longifolia* (L.) L.	Lamiaceae	P	Africa/ Europe	Unintentional
116.	*Mimosa pudica* L.	Fabaceae	A/P	South America	–
117.	*Mirabilis jalapa* L.	Nyctaginaceae	A/P	South America	–
118.	*Monochoria vaginalis* (Burm. f.) C. Presl	Pontederiaceae	P	South America	Unintentional
119.	*Narcissus tazetta* L.	Amaryllidaceae	P	Europe	Ornamental
120.	*Nicandra physalodes* (L.) Gaertn.	Solanaceae	B	South America	–
121.	*Nicotiana plumbaginifolia* Viv.	Solanaceae	A	South America	Unintentional
122.	*Nymphoides peltatum* (S.G. Gmel.) Britten & Rendle	Menyanthaceae	P	Asia-Europe	Food
123.	*Ocimum americanum* L.	Lamiaceae	A	South America	Unintentional
124.	*Oenothera rosea* L'Hér. ex Aiton	Onagraceae	P	South America	Unintentional
125.	*Opuntia elatior* Mill.	Cactaceae	P	South America	Ornamental
126.	*Oxalis corniculata* L.	Oxalidaceae	P	Europe	Unintentional

(continued)

Table 8.1 (continued)

	Species name	Family name	Lifespan	Native region	Purpose of introduction
127.	*Oxalis debilis* var. *corymbosa* (DC.) Lourteig	Oxalidaceae	A	South America	Unintentional
128.	*Oxalis latifolia* H. B. & K.	Oxalidaceae	A	South America	–
129.	*Oxalis pes-caprae* Linn.	Oxalidaceae	A	Africa	–
130.	*Parthenium hysterophorus* L.	Asteraceae	A	South America	Unintentional
131.	*Pennisetum purpureum* Schumach.	Poaceae	A/P	Africa	–
132.	*Peperomia pellucida* (L.) Kunth	Piperaceae	A	South America	Unintentional
133.	*Persicaria hydropiper* (L.) Delarbre	Polygonaceae	A/P	Europe	Food
134.	*Phalaris arundinacea* L.	Poaceae	P	North America	–
135.	*Phyllanthus tenellus* Roxb.	Phyllanthaceae	A	Africa	–
136.	*Physalis lagascae* Roem. & Schult.	Solanaceae	A	South America	–
137.	*Physalis peruviana* L.	Solanaceae	P	South America	Horticultural
138.	*Physalis pruinosa L.*	Solanaceae	A	North America	–
139.	*Pilea microphylla* (L.) Liebm.	Urticaceae	A/B	South America	–
140.	*Plantago lanceolata* L.	Plantaginaceae	P	Africa/ Europe	Unintentional
141.	*Plantago major* L.	Plantaginaceae	A/P	Europe	Unintentional
142.	*Poa annua* L.	Poaceae	A	Europe	Unintentional
143.	*Polygonum aviculare* L.	Polygonaceae	A/P	Europe	Unintentional
144.	*Portulaca oleracea* L.	Portulacaceae	A	South America	Food
145.	*Portulaca pilosa* L.	Portulacaceae	A/P	South America	Ornamental
146.	*Portulaca quadrifida* L.	Portulacaceae	A	South America	Unintentional
147.	*Ranunculus arvensis* L.	Ranunculaceae	A/B	Africa/ Europe	Unintentional
148.	*Ranunculus distans* Royle	Ranunculaceae	P	Europe	Unintentional
149.	*Ranunculus muricatus* L.	Ranunculaceae	A	Africa/ Europe	Unintentional
150.	*Rivina humilis* L.	Phytolaccaceae	P	South America	Unintentional

(continued)

Table 8.1 (continued)

	Species name	Family name	Lifespan	Native region	Purpose of introduction
151.	*Rorippa dubia* (Pers.) H. Hara	Brassicaceae	A	South America	Unintentional
152.	*Ruellia tuberosa* L.	Acanthaceae	A	South America	–
153.	*Rumex hastatus* D. Don	Polygonaceae	P	Asia	Medicinal
154.	*Saccharum spontaneum* L.	Poaceae	P	Asia	Unintentional
155.	*Sagina saginoides* (L.) H. Karst.	Caryophyllaceae	A/P	Europe	Unintentional
156.	*Salvinia adnata* Desv.	Salvinaceae	A	South America	–
157.	*Sambucus wightiana* Wall. ex Wight & Arn.	Adoxaceae	P	Asia/Africa	Unintentional
158.	*Scoparia dulcis* L.	Plantaginaceae	A/P	South America	Medicinal
159.	*Senna obtusifolia* (L.) H. S.Irwin & Barneby	Fabaceae	A/P	South America	Unintentional
160.	*Senna occidentalis* (L.) Link	Fabaceae	A/P	South America	Unintentional
161.	*Senna tora* (L.) Roxb.	Fabaceae	A	South America	Unintentional
162.	*Senna uniflora* (Mill.) H. S.Irwin & Barneby	Fabaceae	A	South America	–
163.	*Sesbania bispinosa* (Jacq.) W.Wight	Fabaceae	A/P	South America	Unintentional
164.	*Setaria viridis* (L.) P. Beauv.	Poaceae	A	Asia/Africa	Fodder
165.	*Siegesbeckia orientalis* L.	Asteraceae	A	Africa	Unintentional
166.	*Sisymbrium loeselii* L.	Brassicaceae	A	Africa/ Europe	Unintentional
167.	*Solanum nigrum* L.	Solanaceae	A	South America	Unintentional
168.	*Solanum seaforthianum* Andrews	Solanaceae	P	South America	–
169.	*Solanum viarum* Dunal	Solanaceae	P	South America	Unintentional
170.	*Soliva anthemifolia* (Juss.) R.Br. ex Less.	Asteraceae	A	Australia	Unintentional
171.	*Sonchus arvensis* L.	Asteraceae	P	Asia/Africa	Unintentional
172.	*Sonchus asper* (L.) Hill.	Asteraceae	A	Asia	Unintentional
173.	*Sonchus oleraceus* (L.) L.	Asteraceae	A	Asia	Unintentional
174.	*Sorghum halepense* (L.) Pers.	Poaceae	P	Europe	Fodder
175.	*Spergula arvensis* L.	Caryophyllaceae	A	Asia/Africa	Unintentional

(continued)

Table 8.1 (continued)

	Species name	Family name	Lifespan	Native region	Purpose of introduction
176.	*Spermacoce alata* Aubl.	Rubiaceae	A	South America	–
177.	*Spirodela polyrhiza* (L.) Schleid.	Araceae	A	Asia/Africa	Unintentional
178.	*Stachytarpheta cayennensis* (Rich.) Vahl	Verbenaceae	P	South America	–
179.	*Stellaria media* (L.) Vill.	Caryophyllaceae	A	Europe	Unintentional
180.	*Stylosanthes hamata* (L.) Taub.	Fabaceae	P	South America	–
181.	*Synedrella nodiflora* (L.) Gaertn.	Asteraceae	A	South America	–
182.	*Tagetes minuta* L.	Asteraceae	A	South America	Unintentional
183.	*Themeda anathera* (Nees ex Steud.) Hack.	Poaceae	A/P	Asia	Fodder
184.	*Torenia fournieri* Linden ex Fourn	Linderniaceae	A/P	Asia-temperate/Europe	–
185.	*Trapa natans* L.	Lythraceae	A	Europe	Food
186.	*Tribulus terrestris* L.	Zygophyllaceae	A	South America	Unintentional
187.	*Tridax procumbens* (L.) L.	Asteraceae	A/P	South America	Unintentional
188.	*Triumfetta rhomboidea* Jacq.	Malvaceae	P	South America	Unintentional
189.	*Turnera ulmifolia* L.	Passifloraceae	P	South America	–
190.	*Typha domingensis* Pers	Typhaceae	P	South America	Unintentional
191.	*Urtica dioica* L.	Urticaceae	P	Africa/Europe	Unintentional
192.	*Vaccaria hispanica* (Mill.) Rauschert	Caryophyllaceae	A	Europe	Unintentional
193.	*Verbascum thapsus* L.	Scrophulariaceae	A	Europe	Unintentional
194.	*Verbesina encelioides* (Cav.) Benth. & Hook.f. ex A.Gray	Asteraceae	A	South America	–
195.	*Veronica persica* Poir.	Plantaginaceae	A	Asia	Unintentional
196.	*Vulpia myuros* (L.) C.C. Gmel.	Poaceae	A	Europe	Fodder
197.	*Youngia japonica* (L.) DC.	Asteraceae	A	South America	Unintentional
Shrubs					
1.	*Calotropis gigantea* (L.) Dryand.	Apocynaceae	P	Africa	Unintentional

(continued)

Table 8.1 (continued)

	Species name	Family name	Lifespan	Native region	Purpose of introduction
2.	*Calotropis procera* (Aiton) Dryand.	Apocynaceae	P	Africa	Unintentional
3.	*Clidemia hirta* (L.) D.Don	Melastomataceae	P	South America	–
4.	*Euphorbia pulcherrima* Willd. ex Klotzsch	Euphorbiaceae	P	North America	Ornamental
5.	*Euphorbia umbellata* (Pax) Bruyns	Euphorbiaceae	P	Africa	–
6.	*Hibiscus cannabinus* L.	Malvaceae	A/B	South America	Commercial
7.	*Indigofera linnaei* Ali	Fabaceae	A/P	Africa	Unintentional
8.	*Ipomoea carnea* Jacq.	Convolvulaceae	P	South America	Unintentional
9.	*Lantana camara* L.	Verbenaceae	P	South America	Ornamental
10.	*Lysiloma latisiliquum* (L.) Benth.	Fabaceae	P	South America	–
11.	*Opuntia dillenii* Haw.	Cactaceae	P	North America	–
12.	*Opuntia stricta* Haw.	Cactaceae	P	North America	–
13.	*Ricinus communis* L.	Euphorbiaceae	P	Africa	Food
14.	*Rubus ulmifolius* Schott	Rosaceae	P	Europe	Landscaping
15.	*Senna alata* (L.) Roxb.	Fabaceae	P	South America	–
16.	*Solanum rudepannum* Dunal	Solanaceae	P	South America	–
17.	*Ulex europaeus* L.	Fabaceae	P	Europe	–
18.	*Xanthium strumarium* L.	Asteraceae	A	South America	Unintentional
Subshrubs					
1.	*Desmanthus virgatus* (L.) Willd.	Fabaceae	P	South America	–
2.	*Euphorbia cyathophora* Murray	Euphorbiaceae	A/P	South America	–
3.	*Malvastrum coromandelianum* (L.) Garcke	Malvaceae	A/B	South America	Unintentional
4.	*Senna hirsuta* (L.) H.S. Irwin & Barneby	Fabaceae	P	South America	–
5.	*Stachytarpheta jamaicensis* (L.) Vahl	Verbenaceae	P	South America	–
6.	*Turnera subulata* J. E.	Passifloraceae	P	South America	–
7.	*Waltheria indica* L.	Malvaceae	P	South America	–

(continued)

Table 8.1 (continued)

	Species name	Family name	Lifespan	Native region	Purpose of introduction
Trees					
1.	*Acacia farnesiana* (L.) Willd.	Fabaceae	P	South America	Unintentional
2.	*Acacia mearnsii* De Wild	Fabaceae	P	Australasia	–
3.	*Ailanthus altissima* (Mill.) Swingle	Simaroubaceae	P	Asia	Plantation
4.	*Broussonetia papyrifera* (L.) L'Hér. ex Vent.	Moraceae	P	Asia	Unintentional
5.	*Casuarina equisetifolia* L.	Casuarinaceae	P	Asia-tropical	–
6.	*Leucaena leucocephala* (Lam.) de Wit	Fabaceae	P	South America	Ornamental
7.	*Parkinsonia aculeata* L.	Fabaceae	P	North America	Ornamental
8.	*Physalis angulata* L.	Solanaceae	P	North America	Medicinal
9.	*Pithecellobium dulce* (Roxb.) Benth.	Fabaceae	P	South America	–
10.	*Prosopis juliflora* (Sw.) DC.	Fabaceae	P	South America	Plantation
11.	*Robinia pseudoacacia* L.	Fabaceae	P	North America	Plantation
12.	*Triadica sebifera* (L.) Small	Euphorbiaceae	P	Asia	Commercial

A annual, *B* biennial, *P* perennial

Ranunculaceae, Portulacaceae, Onagraceae, Cyperaceae, Boraginaceae, Cactaceae, etc., represented <2% of the invasive plant species of India (Table 8.1). Annual plants (45%) represented most of the invasive plants, followed by perennial (37%) and annual/perennial (14%) species (Fig. 8.2a). Most of the invasive species listed were herbaceous (81%), and shrubs and trees formed only 7% and 5% of the invasive flora, respectively (Fig. 8.2b). In addition, most of the invasive flora of India was observed to have originated from south America (51%), Europe (12%), and north America (9%) (Fig. 8.3). Some of the invasive species had mixed records of origin in secondary literature, such as North America/South America, Europe/South America, Africa/Asia-Tropical, etc. (Fig. 8.3).

8.3 Climate Suitability and Invasion Hotspots in India

Different parts of the country experience different climatic conditions, and therefore, have different invasive plant diversity. Adhikari et al. (2015) reported that in India, the biodiversity and invasion hotspots coincided with each other. Invasion hotspots

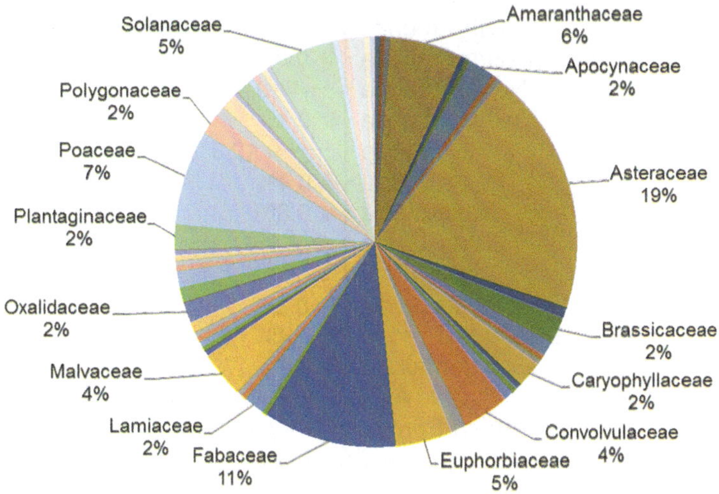

Fig. 8.1 Percentage distribution of invasive plant species in India belonging to different families. Unmarked components (families) in the pie chart represent ≤1% contribution of the invasive plants in India

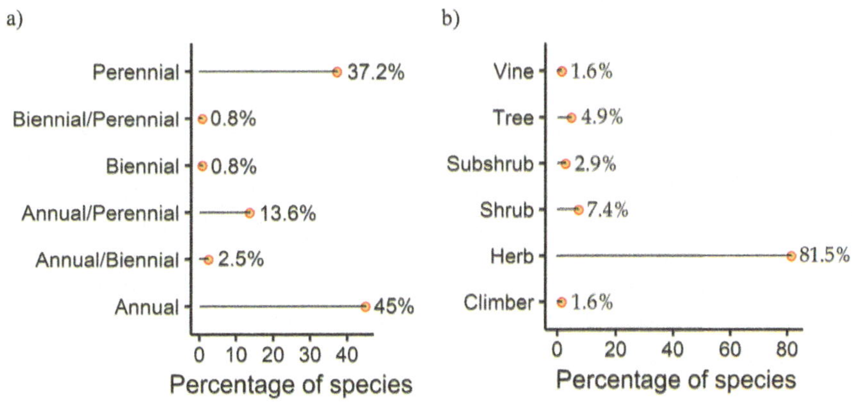

Fig. 8.2 Percentage of invasive plants in India belonging to different (**a**) lifespan and (**b**) life forms

were considered regions with 50% of its area showing high anthropogenic disturbance and climatic suitability for invasive alien plants. The reason could be the similar environmental requirements of invasive species with their native counterparts (Levine and D'Antonio 1999). According to Adhikari et al. (2015), 'high' suitability of the eastern coasts of the Peninsular region, and north-eastern region was predicted with projections of Australia and Africa, 'high' suitability of the western Himalaya was predicted with projections of South America, while 'high' suitability of the Aravalli range, western Himalaya, Hills in eastern Ghats, and Naga

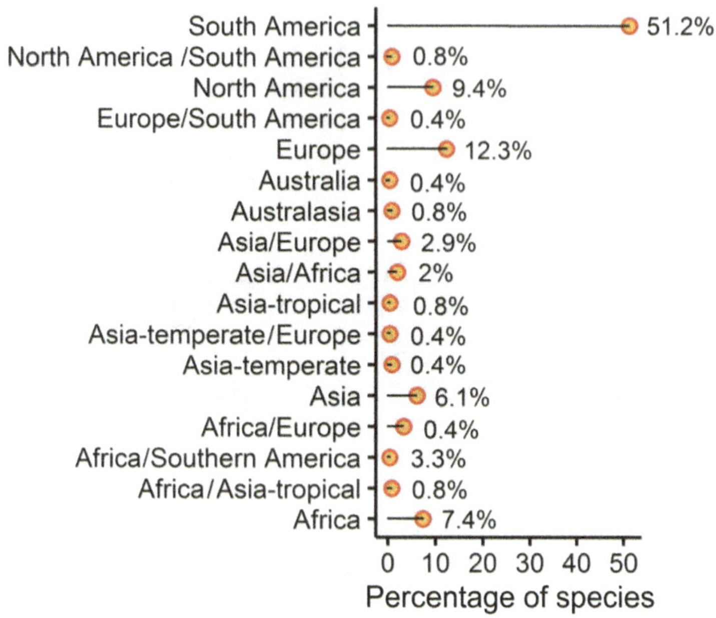

Fig. 8.3 Percentage of invasive plants in India belonging to different native regions

and Khasi hills of north-eastern region with projections of Europe. Overall, the projections of all five continents predicted north-eastern region, western Himalaya, and coastal areas to have high climatic suitability to invasive alien plants. In particular, eco-regions such as Brahmaputra valley semi-evergreen forests, Andaman islands rain forest, East Deccan dry-evergreen forests, Mizoram-Manipur-Kachin rainforests, Orissa semi-evergreen forests, Godavari-Krishna mangroves, Sunderbans mangroves, and freshwater swamp forests, etc., have 90% of their areas showing high climatic suitability for invasion. In addition, the Indian invasion hotspots also coincide with the diverse ecosystems such as coastal forests, forest reserves, mangrove ecosystems, and islands.

8.4 Introduction Pathways

Six major pathways of the introduction of alien species have been categorised: intentional release, escape from captivity, contamination of commodities, stowaways on transport vectors, anthropogenic corridors, and uncontrolled introduction from other invaded areas. However, these can be broadly classified into two categories: intentional and unintentional. Many flora and fauna have historically been planted and released on purpose for monetary, recreational, or aesthetic reasons (Pyšek et al. 2020). In India, some exotic plant species were introduced for their ornamental value, e.g., *Ageratum conyzoides*, *Asclepias curassavica*, *Cryptostegia*

Table 8.2 Purpose of introduction of some major invasive plant species in India

S. no.	Invasive species	Purpose of introduction
1.	*Ageratina adenophora*	Unintentional
2.	*Ageratum conyzoides*	Ornamental
3.	*Argemone mexicana*	Medicinal
4.	*Bidens pilosa*	Unintentional
5.	*Chromolaena odorata*	Ornamental
6.	*Eichhornia crassipes*	Ornamental
7.	*Lantana camara*	Ornamental
8.	*Parthenium hysterophorus*	Unintentional
9.	*Ricinus communis*	Food
10.	*Xanthium strumarium*	Unintentional

grandiflora, *Cytisus scoparius*, *Eichhornia crassipes*, *L. camara*, etc. and were introduced intentionally (Reddy et al. 2008). Another major factor is the accidental transport of alien flora propagules through ballast water, ocean rafting, etc. (Carlton and Geller 1993). Some major invasive species of India with their mode/reason of introduction are listed in Table 8.2.

8.5 Role of Changing Climate in Plant Invasion

The assessment of how invasive plant species are influenced by changing climate is filled with uncertainties, making the prediction of future trends more and more difficult. Studies have indicated an increase in the magnitude and severity of invasive plant spread in India under conditions of climate change (Kohli et al. 2012; Panda et al. 2018; Thapa et al. 2018; Mungi et al. 2020). It has been reported that the spread of invasive species to uninvaded regions and ecosystems would increase if the temperature and precipitation-related limitations are uplifted with changing climate (Parmesan 2006). Spatial shifting in species creates vacant niche spaces for opportunistic alien plant species to exploit, replacing the native and endemic flora in the process. Furthermore, due to increasing disturbance, nitrogen deposition, temperature, and CO_2 availability in the environment by automobiles and agricultural practices, more resources are available for invasive species to even invade resource-limited environments such as arid and alpine ecosystems (Dukes and Mooney 1999; Westerband et al. 2021). According to Weltzin et al. (2003), invasive species respond more rapidly to elevated atmospheric CO_2 levels. Moreover, with changing climate, greater reproduction rates, higher propagules generation, and enhanced tolerances to environmental stresses, the invasive species are displaying ability to quickly invade, colonise, and establish in a region, especially the disturbed and fragmented landscapes (Finch et al. 2021). Enhanced transportation, trade, horticulture, and frequent and intense disturbances associated with modernisation are also aiding the invasive species to show strong colonisation behaviour (Hulme 2009; Kühn et al. 2017). Changing climate, land-use pattern, pollution (Crooks et al. 2011), and the facilitative effect of other alien plant species can, therefore, be deemed to be the major reason behind the rising problem of plant invasion (Redding et al. 2019; Pyšek et al. 2020).

As for the Indian scenario, many studies have predicted the spread and distribution pattern of invasive species under climate change. Padalia et al. (2015) investigated the distribution potential of *Hyptis suaveolens* under A2a and B2a scenarios for climate change for 2050. The study indicated the formation of new niches and shift in the location of areas highly suitable for *H. suaveolens* within the existing distribution ranges, where the area currently less suitable for the species can turn into moderately or highly suitable areas. Furthermore, a general decrease in the global invasion range of *H. suaveolens* was predicted under both A2a and B2a scenarios. In India's case, the area under high suitability of *H. suaveolens* invasion might decline by 2050. In the Indian Himalayan region, Lamsal et al. (2018) investigated the current and future habitat of five major invasive alien plants: *A. adenophora*, *C. odorata*, *L. camara*, *A. conyzoides*, and *Parthenium hysterophorus*. The study revealed that *A. adenophora*, *C. odorata*, and *L. camara*, will observe an increase in suitable areas under the scenario of IPCC's (Intergovernmental Panel on Climate Change) $RCP_{4.5}$ and $RCP_{8.5}$ (Representative Concentration Pathway) by 2070, whereas, *P. hysterophorus* and *A. conyzoides* will observe a decrease in suitability. According to Lamsal et al. (2018), in the future, maximum suitability would be in the case of *A. adenophora*, while the minimum will be for *A. conyzoides*, as compared to the current scenario. *Lantana camara* on the other hand, would reportedly have least reduction in suitable area. Similarly, Sharma et al. (2022) reported that ~24% areas in India are currently susceptible to *C. odorata* invasion, and in the future the invasion would likely increase by ~1.53% in the event of climate change.

8.6 Adaptations in Invasive Plants

Various factors contribute to the expansion of these highly invasive species in India; however, phenotypic plasticity, local adaptation, and genetic diversity may be considered as the major reasons that aid in invasive species adaptation with changing climate in introduced regions.

8.6.1 Phenotypic Plasticity

Phenotypic plasticity is considered as the capacity of a species to produce phenotypes showing variability in their traits under heterogeneous environmental conditions (Rathee et al. 2021). Some examples of invasive plants that display phenotypic plasticity for adapting to different regions in India are described below.

Ageratina adenophora (Spreng.) R.M.King & H.Rob.—*Ageratina adenophora* exhibits plasticity in traits such as root length, plant height, above- and below-ground biomass, number of branches, leaves and capitula, root-shoot fraction, root-weight fraction, stem-weight fraction, leaf-weight fraction, inflorescence-weight fraction, and seed production for adaptation in diverse habitats along an altitudinal gradient (Khatri et al. 2022). The plant performance was reported to be the highest in mid-altitudinal areas (1000–2000 m a.s.l.), and in habitats such as wasteland and roadsides, compared to forest and agricultural land. Similarly, Datta et al. (2017) demonstrated that *A. adenophora* exhibits greater extent of phenotypic plasticity in functional traits, and that seed germination and winter mortality were the major drivers of the lower and upper range limits of the plants in the western Himalaya.

Anthemis cotula L.—The plant species exhibit season-specific phenotypic variations in various life-history traits that have a significant effect on the resource-allocation patterns of its pre- and post-winter populations. High dry biomass, relative growth rate, and per plant seed production were evident in the pre-winter population, indicating greater fecundity than the post-winter population. In disturbed habitats, the species allocated more biomass to above-ground parts, compared to below-ground parts (Allaie et al. 2005). Therefore, *A. cotula* due to its high fecundity, phenotypic variations, trade-off between life-history strategies, and early seedling emergence could establish and spread in the Kashmir Himalaya (Allaie et al. 2005).

Chenopodium murale L.—Adaptation of *C. murale* to environmentally diverse habitats has been attributed to its phenotypic plasticity and enhanced reproductive output (Shachi and Rup 2012), making the species an aggressive invader of dry tropical regions of India. The species exhibited increased biomass allocation to leaves and roots in the early life developmental stages for enhancing the uptake of nutrients and capturing light for increasing photosynthetic efficiency. Overall, the species displayed greater plasticity in leaf, stem, and reproductive part fraction for optimising growth during colonisation and establishment phase in novel environments.

Hyptis suaveolens (L.) Poit.—Sharma and Raghubanshi (2009) reported *H. suaveolens* to display greater plasticity in growth and reproductive traits for successful establishment and spread in diverse habitats, making the species a successful invader. Along roadsides the plant displayed greater height but produced lighter seeds, while in unfavourable habitats, the plants were shorter and produced heavier seeds. The study further explained that soil properties and neighbouring plant species were major drivers responsible for the plastic behaviour of *H. suaveolens*. The species also shows seed size dimorphism and produces dimorphic seeds (large- and small-sized seeds) that show significant differences in their germination behaviour, and may aid in expanding the species' germination niche and establishment in heterogeneous environments (Rathee et al. 2022).

Lantana camara L.—*Lantana camara* depicted higher phenotypic plasticity in the invaded areas through enhanced tolerance to high temperature, low nutrient, and shade conditions, compared to its native range (Mungi et al. 2020). The adaptive strategy of displaying phenotypic plasticity helped the species increase its spread to other areas and cause greater ecological and economic harm.

Mikania micrantha (Hieron.) B.L.Rob.—*Mikania micrantha* displays plasticity in its non-reproductive and reproductive traits for adaptation in disturbed urban environments. The species produced greater reproductive biomass in roadside habitats, and increased height and germination duration in disturbed habitats for maintaining its population size and adapting to novel environmental conditions. The study further indicated the potential of *M. micrantha* in occupying uninvaded areas in the vicinity of invaded areas and road networks. Apart from roadsides and disturbed areas, the species is also capable of invading the littoral habitats, i.e., margins of water bodies. This indicated that the plant species displayed high adaptive capacity in different habitats (Banerjee and Dewanji 2017). Additionally, the species is capable of generating season-specific modulation in its leaf traits, with an increase in traits such as Laminar N/area, LDMC (leaf dry matter content) during monsoons, and an increase in leaf area and thickness in summer and winter, respectively, to grow and propagate in both littoral and terrestrial habitats (Banerjee et al. 2017).

Parthenium hysterophorus L.—*Parthenium hysterophorus* exhibits phenotypic plasticity and variations in biomass allocation for adapting to stressful environments of high elevational areas in mountain environments (Rathee et al. 2021). The study reported that the species invested more resources in reproductive parts with increasing elevation, as compared to non-reproductive parts. The biomass of above-ground parts increased with elevation, whereas the below-ground biomass decreased. In terms of reproductive parts, total capitula count, seed thickness, and seed mass increased with elevation, whereas seed size decreased. Increased allocation of biomass to reproductive parts with rising elevation would aid the future populations of the plant to adapt and colonise high elevational areas in the western Himalaya.

Ricinus communis L.—Goyal et al. (2014) reported that the species displayed variations in vegetative, reproductive, and physiological traits for enhancing its invasiveness. *Ricinus communis* growing in urban environments (with high disturbance) displayed plasticity in reproductive traits, growth, and leaf proline content, which helped in the expansion of its invaded range (Goyal et al. 2014). This modulation of growth strategies helped *R. communis* to colonise and proliferate in the disturbed habitats.

8.6.2 Local Adaptations

Research has suggested that local adaptations may play a critical role in proliferation of invasive plant species in the introduced regions (Colautti and Barrett 2013). When plant populations are spatially separated with different environmental conditions affecting them, then natural selection acting on the populations might result in local adaptation (Kawecki and Ebert 2004). Local adaptation in a plant population can be investigated through common garden and reciprocal transplant experiments (Ebeling et al. 2011). A locally adapted plant population produces plant phenotypes that are better performing than other individuals of the same species that were transplanted into the habitat (Oduor et al. 2016). Datta et al. (2017) reported that *A. adenophora* showed lack of local adaptation in populations of low-, mid-, and high-elevations in western Himalaya (Datta et al. 2017).

8.6.3 Genetic Diversity

Apart from phenotypic plasticity and local adaptation, increased genetic diversity can also result in phenotypic variation among and within natural populations of an invasive plant species. Genetic studies conducted on *L. camara* reported high genetic diversity in the species, probably due to multiple introduction and probable hybridisation witnessed for the species in India (Kannan et al. 2013). Moreover, another study by Ray and Ray (2014) reported the presence of multiple genetic clusters of *Lantana* in India that were responsible for the local adaptation and invasiveness of the species. In the case of another aggressive invader species *P. hysterophorus*, the genetic diversity was evaluated from different regions of Jammu and Kashmir using ISSR markers (Dar et al. 2020). The study, however, observed low genetic diversity in *P. hysterophorus*, which was attributed to the recent invasion, frequent founder effects, and the bottleneck effect. Moreover, the study reported decreasing genetic diversity in *P. hysterophorus* populations with increase in elevation, which might be due to poor gene flow and harsh environment of high elevational areas. The study further revealed almost similar levels of within- and between-population genetic diversity in *P. hysterophorus* due to low genetic diversity.

8.7 Socio-economic Impacts of Invasive Species

Plant invasion in India is a huge environmental issue that not only causes ecological harm but also economic harm, and the situation is worsening with changing climate. Invasive plants have both direct and indirect socio-economic implications through their influence on the ecology of an ecosystem and the services provided by these ecosystems (Dogra et al. 2009). According to Babu et al. (2009), ~$18,700/ km^2 cost is incurred in the management of *L. camara* in India, reaching an estimated cost exceeding $5.5 billion (Mungi et al. 2020). The current climate change scenario would further increase the financial cost associated with the management of the species. According to Pimentel et al. (2001), the potential annual financial burden of invasive plants in India is ~$116 billion. However, Bang et al. (2022) estimated the total cost to be $182.6 billion from 1920 to 2020. Moreover, the study attributed the recorded cost of $616 million to west India, $64.2 million to south India, and $23.9 million to north India, while no costs were attributed to eastern, northern, north-eastern, and central India. As of 2020, the total invasion-related cost for 3.287 million km^2 landmass is ~$38,727/km^2. According to the study, India followed after the USA in per unit area costs related to plant invasion. The major costs related to invasive plants were attributed mainly to *Phalaris minor*, *M. micrantha*,

P. hysterophorus, and *L. camara*, indicating that only 2%, i.e., 4 out of 173 invasive plant species recorded in India were responsible for the majority of the species-specific invasion costs (Bang et al. 2022).

8.8 Socio-ecological Impacts of Invasive Species

Invasive alien plants are capable of inducing the invaded environments in various ways. Invasive species cause loss of biodiversity and negatively impacts on human, plant, and animal health alike (Sharma and Batish 2022). Rai and Singh (2020) emphasised the development of integrated eco-restoration strategies for understanding the impact of invasive plants on human health and environment. Invasive species impacts the natural ecosystem dynamics by altering the native floral diversity and composition (Dogra et al. 2009). Furthermore, lack of social awareness and invasive plant-related information has aggravated the issue of plant invasion in India. Most often, a drastic reduction in species diversity, richness, and evenness has been associated with the invasion of alien species in any ecosystem. In accordance, *L. camara* and *A. adenophora* reduce the diversity, density, richness, evenness, and biomass of native species, indicating altering of structure and composition of forest communities (Singh et al. 2014; Kumar et al. 2020). Additionally, due to its allelopathic properties, *L. camara* negatively impacted the seedling growth due to the release of organic inhibitors in the infested soil (Singh et al. 2014). Invasive species such as *L. camara* and *A. adenophora* also alter the soil-nutrient dynamics and further enhance the vulnerability of a forest ecosystem to invasion (Kumar et al. 2021). In contrast, Kumar and Mathur (2014) reported that *Prosopis juliflora* invaded areas showed an increase in species diversity, evenness, and richness, which was attributed to the presence of weedy flora along with *P. juliflora*. The study also reported that *P. juliflora* reduced the dominance of late-successional species and endangered species such as *Commiphora wightii*, in arid grazing lands of India (Kumar and Mathur 2014). Another invasive species, *P. hysterophorus*, also altered the species diversity, community composition, and structure in grasslands of Indo-Gangetic plains (Srivastava and Raghubanshi 2021). The negative impact of *P. hysterophorus* increased with time and caused significant variations in soil-nitrogen dynamics, increasing available nitrogen content, microbial biomass N, and N-mineralisation in the invaded grasslands. In rural India, displacement of native vegetation and scarcity in fodder availability were also associated with the spread of *P. hysterophorus* (Kohli et al. 2006). Moreover, the invasive species is further responsible for rendering the fodder unpalatable and inducing toxicity in animals. *Ageratum conyzoides* also interfere with crop species in agricultural fields and cause major yield losses (Kohli et al. 2006). Sharma et al. (2017) reported a decline of 46–52% in species richness in areas invaded with *H. suaveolens* in peri-urban areas of Chandigarh. Moreover, the invaded peri-urban areas had a higher number of exotic species and a low proportion of native species, including some economically important species such as *Murraya koenigii*, *Justicia adhatoda*, *Carissa carandas*, *Paspalidium flavidum*, *Dioscorea deltoidea*, etc.

8.9 Conclusions and Future Directions

India is facing the conundrum of balancing rising plant invasion and environmental conservation goals. It is evident that invasive plants have substantial ecological and socio-economic implications. Elton (1958) also indicated that "piling up of new human difficulties" would occur if the invasive species are not adequately managed and their impacts minimised. It is, therefore, extremely important to recognise and fill the gaps in understanding the ecological and economic impacts, and adaptation strategies of invasive plants in India, especially with climate change. Future research should focus on utilising inter-disciplinary approaches like people's perception studies, remote sensing technology, and molecular diagnostic techniques for identifying invasive plant species and developing management frameworks and policies.

References

Adhikari D, Tiwary R, Barik SK (2015) Modelling hotspots for invasive alien plants in India. PLoS One 10(7):e0134665

Ahmad M, Batish DR, Singh HP (2022) Biodiversity status, issues, and challenges: trees and shrubs. In: Biodiversity in India: status, issues and challenges. Springer Nature, Singapore, pp 45–59

Allaie RR, Reshi ZA, Wafai BA (2005) Demographic plasticity in relation to growth and resource allocation pattern in *Anthemis cotula*—an alien invasive species in Kashmir Himalaya, India. Appl Ecol Environ Res 4(1):63–74

Babu S, Love A, Babu CR (2009) Ecological restoration of *Lantana*-invaded landscapes in Corbett Tiger Reserve, India. Ecol Restor 27(4):467–477

Banerjee AK, Dewanji A (2017) Role of intraspecific trait plasticity in *Mikania micrantha* Kunth growth and impact of its abundance on community composition. J Asia-Pacific Biodiver 10(2): 237–249

Banerjee AK, Ghosh S, Dewanji A (2017) Do plants have a choice of traits to be modulated? Evidence from an invasive plant *Mikania micrantha* Kunth in different urban environments. Am J Plant Sci 8:835–855

Bang A, Cuthbert RN, Haubrock PJ, Fernandez RD, Moodley D, Diagne C, Turbelin AJ, Renault D, Dalu T, Courchamp F (2022) Massive economic costs of biological invasions despite widespread knowledge gaps: a dual setback for India. Biol Invasions 24:2017–2039

Bellard C, Cassey P, Blackburn TM (2016) Alien species as a driver of recent extinctions. Biol Lett 12(2):20150623

Bhatt JR, Singh JS, Singh SP, Tripathi RS, Kohli RK (2011) Invasive alien plants an ecological appraisal for the Indian subcontinent, vol 1. CABI

Bonanno G (2016) Alien species: to remove or not to remove? That is the question. Environ Sci Pol 59:67–73

Bradley BA, Blumenthal DM, Wilcove DS, Ziska LH (2010) Predicting plant invasions in an era of global change. Trends Ecol Evol 25:310–318

BSI (2020). http://www.bsienvis.nic.in/database/invasive_alien_species_15896.aspx. Accessed on 25 Mar 2023

Capinha C, Essl F, Seebens H, Moser D, Pereira HM (2015) The dispersal of alien species redefines biogeography in the anthropocene. Science 348:1248–1251

Carlton JT, Geller JG (1993) Ecological roulette: the global transport of nonindigenous marine organisms. Science 261:78–82

Colautti RI, Barrett SC (2013) Rapid adaptation to climate facilitates range expansion of an invasive plant. Science 342(6156):364–366

Colautti RI, MacIsaac HJ (2004) A neutral terminology to define 'invasive' species. Divers Distrib 10(2):135–141

Crooks JA, Chang AL, Ruiz GM (2011) Aquatic pollution increases the relative success of invasive species. Biol Invasions 13:165–176

Crowl TA, Crist TO, Parmenter RR, Belovsky G, Lugo AE (2008) The spread of invasive species and infectious disease as drivers of ecosystem change. Front Ecol Environ 6:238–246

Dar TUH, Bhat BA, Khuroo AA, Verma S, Islam SU (2020) Genetic diversity and population structure of an invasive plant species differ in two non-native regions with differing climate and invasion success. Nord J Bot 38(8):1

Datta A, Kühn I, Ahmad M, Michalski S, Auge H (2017) Processes affecting altitudinal distribution of invasive *Ageratina adenophora* in western Himalaya: the role of local adaptation and the importance of different life-cycle stages. PLoS One 12(11):e0187708

Davis MA (2006) Invasion biology 1958–2005: the pursuit of science and conservation. In: Cadotte MW, Mcmahon SM, Fukami T (eds) Conceptual ecology and invasion biology: reciprocal approaches to nature. Springer, Dordrecht, pp 35–64

Demertzis K, Iliadis L (2018) The impact of climate change on biodiversity: the ecological consequences of invasive species in Greece. In: Leal Filho W, Manolas E, Azul A, Azeiteiro U, McGhie H (eds) Handbook of climate change communication, Climate change management, vol 1. Springer, Cham

Dogra KS, Kohli RK, Sood SK (2009) An assessment and impact of three invasive species in the Shivalik hills of Himachal Pradesh, India. Int J Biodiver Conserv 1(1):4–10

Dukes JS, Mooney HA (1999) Does global change increase the success of biological invaders? Trends Ecol Evol 14(4):135–139

Ebeling SK, Stöcklin J, Hensen I, Auge H (2011) Multiple common garden experiments suggest lack of local adaptation in an invasive ornamental plant. J Plant Ecol 4(4):209–220

Elton CS (1958) The ecology of invasions by animals and plants. Metheun, London

ENVIS (2023) Invasive aline species. ENVIS Resource Partner on Biodiversity. Available online at http://www.bsienvis.nic.in/. Accessed 14 Jan 2023

Finch DM, Butler JL, Runyon JB, Fettig CJ, Kilkenny FF, Jose S, Frankel SJ, Cushman SA, Cobb RC, Dukes JS, Hicke JA (2021) Effects of climate change on invasive species. In: Invasive species in forests and rangelands of the United States: a comprehensive science synthesis for the United States forest sector. Springer, pp 57–83

García RA, Franzese J, Policelli N, Sasal Y, Zenni RD, Nuñez MA, Taylor K, Pauchard A (2018) Non-native pines are homogenizing the ecosystems of South America. In: Rozzi R, May RH, Chapin FS, Massardo F, Gavin MC, Klaver IJ, Pauchard A, Nuñez MA, Simberloff D (eds) From biocultural homogenization to biocultural conservation. Springer, Cham, pp 245–263

Goyal N, Pardha-Saradhi P, Sharma GP (2014) Can adaptive modulation of traits to urban environments facilitate Ricinus communis L. invasiveness? Environ Monit Assess 186(11): 7941–7948

Hulme PE (2009) Trade, transport and trouble: managing invasive species pathways in an era of globalization. J Appl Ecol 46(1):10–18

Jaryan V, Uniyal SK, Gupta RC, Singh RD (2013) Alien flora of Indian Himalayan state of Himachal Pradesh. Environ Monit Assess 185:6129–6153

Kamble VV, Yele RB (2020) Floristic survey of monocotyledonous plants from Man Tehsil of Satara District (Maharashtra) India. J Glob Biosci 9:7149–7159

Kannan R, Shackleton CM, Uma Shaanker R (2013) Reconstructing the history of introduction and spread of the invasive species, *Lantana*, at three spatial scales in India. Biol Invasions 15:1287–1302

Kawecki TJ, Ebert D (2004) Conceptual issues in local adaptation. Ecol Lett 7:1225–1241

Keller RP, Geist J, Jeschke JM, Kühn I (2011) Invasive species in Europe: ecology, status, and policy. Environ Sci Eur 23:23

Khatri K, Negi B, Bargali K, Bargali SS (2022) Phenotypic variation in morphology and associated functional traits in *Ageratina adenophora* along an altitudinal gradient in Kumaun Himalaya, India. Biologia 78:1–15

Khuroo AA, Reshi ZA, Malik AH, Weber E, Rashid I, Dar GH (2012) Alien flora of India: taxonomic composition, invasion status and biogeographic affiliations. Biol Invasions 14:99–113

Khuroo AA, Ahmad R, Hamid M, Rather ZA, Malik AH, Rashid I (2021) An annotated inventory of invasive alien flora of India. In: Invasive alien species: observations and issues from around the world, vol 2. Wiley Blackwell, pp 16–37

Kohli RK, Batish DR, Singh HP, Dogra KS (2006) Status, invasiveness and environmental threats of three tropical American invasive weeds (*Parthenium hysterophorus* L., *Ageratum conyzoides* L., *Lantana camara* L.) in India. Biol Invasions 8(7):1501–1510

Kohli RK, Batish DR, Singh JS, Singh HP, Bhatt JR (2012) Plant invasion in India: an overview. In: Invasive alien plants: an ecological appraisal for the Indian subcontinent. CABI, pp 1–9

Kühn I, Wolf J, Schneider A (2017) Is there an urban effect in alien plant invasions? Biol Invasions 19:3505–3513

Kumar S, Mathur M (2014) Impact of invasion by *Prosopis juliflora* on plant communities in arid grazing lands. Trop Ecol 55(1):33–46

Kumar M, Verma AK, Garkoti SC (2020) *Lantana camara* and *Ageratina adenophora* invasion alter the understory species composition and diversity of chir pine forest in central Himalaya, India. Acta Oecol 109:103642

Kumar M, Kumar S, Verma AK, Joshi RK, Garkoti SC (2021) Invasion of *Lantana camara* and *Ageratina adenophora* alters the soil physico-chemical characteristics and microbial biomass of chir pine forests in the central Himalaya, India. Catena 207:105624

Lamsal P, Kumar L, Aryal A, Atreya K (2018) Invasive alien plant species dynamics in the Himalayan region under climate change. Ambio 47(6):697–710

Levine JM, D'Antonio CM (1999) Elton revisited: a review of evidence linking diversity and invasibility. Oikos 87:15–26

Mungi NA, Qureshi Q, Jhala YV (2020) Expanding niche and degrading forests: key to the successful global invasion of *Lantana camara* (sensu lato). Glob Ecol Conserv 23:e01080

Nayar MP (1977) Changing patterns of the Indian flora. Nelumbo 19(1–4):145–155

Oduor AM, Leimu R, van Kleunen M (2016) Invasive plant species are locally adapted just as frequently and at least as strongly as native plant species. J Ecol 104(4):957–968

Padalia H, Srivastava V, Kushwaha SPS (2015) How climate change might influence the potential distribution of weed, bushmint (*Hyptis suaveolens*)? Environ Monit Assess 187:1–14

Panda RM, Behera MD, Roy PS (2018) Assessing distributions of two invasive species of contrasting habits in future climate. J Environ Manag 213:478–488

Parmesan C (2006) Ecological and evolutionary responses to recent climate change. Annu Rev Ecol Evol Syst 37:637–669

Pimentel D, McNair S, Janecka J, Wightman J, Simmonds C, O'Connell C, Wong E, Russel L, Zern J, Aquino T, Tsomondo T (2001) Economic and environmental threats of alien plant, animal, and microbe invasions. Agric Ecosyst Environ 84(1):1–20

Pyšek P, Hulme PE, Simberloff D, Bacher S, Blackburn TM, Carlton JT, Dawson W, Essl F, Foxcroft LC, Genovesi P, Jeschke JM (2020) Scientists' warning on invasive alien species. Biol Rev 95(6):1511–1534

Rai PK, Singh JS (2020) Invasive alien plant species: their impact on environment, ecosystem services and human health. Ecol Indic 111:106020

Rathee S, Kaur S (2022) Conservation of national biodiversity: efforts of the Indian government. In: Biodiversity in India: status, issues and challenges. Springer Nature, Singapore, pp 571–588

Rathee S, Ahmad M, Sharma P, Singh HP, Batish DR, Kaur S, Kaur A, Yadav SS, Kohli RK (2021) Biomass allocation and phenotypic plasticity are key elements of successful invasion of *Parthenium hysterophorus* at high elevation. Environ Exp Bot 184:104392

Rathee S, Sharma P, Ahmad M, Batish DR, Singh HP, Kaur S, Yadav SS (2022) Seed size dimorphism in *Hyptis suaveolens* aids in differentiation of the germination niche. Plant Biol 24(3):464–472

Ray A, Ray R (2014) Rapid divergence of ecotypes of an invasive plant. AoB Plants 6:plu052

Redding DW, Pigot AL, Dyer EE, Sekerçioglu CH, Kark S, Blackburn TM (2019) Location-level processes drive the establishment of alien bird populations worldwide. Nature 571:103–106

Reddy CS, Bagyanarayana G, Reddy KN, Raju VS (2008) Invasive alien flora of India. National Biological Information Infrastructure, USGS, pp 2–34

Rejmánek M, Randall JM (1994) Invasive alien plants in California: 1993 summary and comparison with other areas in North America. Madrono 41:161–177

Ricciardi A, Hoopes MF, Marchetti MP, Lockwood JL (2013) Progress toward understanding the ecological impacts of nonnative species. Ecol Monogr 83:263–282

Richardson DM, Pyšek P (2008) Fifty years of invasion ecology–the legacy of Charles Elton. Divers Distrib 14:161–168

Sankaran KV, Khuroo AA, Raghavan R, Molur S, Kumar B, Wong LJ, Pagad S (2021) Global register of introduced and invasive species—India. Version 1.5. Invasive Species Specialist Group ISSG. Checklist dataset. https://doi.org/10.15468/uvnf8m. Accessed via GBIF.org on 17 Mar 2021

Seebens H, Bacher S, Blackburn TM, Capinha C, Dawson W et al (2021) Projecting the continental accumulation of alien species through to 2050. Glob Chang Biol 27(5):970–982

Shachi G, Rup N (2012) Phenotypic plasticity of *Chenopodium murale* across contrasting habitat conditions in peri-urban areas in Indian dry tropics: is it indicative of its invasiveness? Plant Ecol 213(3):493–503

Sharma P, Batish DR (2022) Reasons of biodiversity loss in India. In: Biodiversity in India: status, issues and challenges. Springer Nature, Singapore, pp 555–567

Sharma GP, Raghubanshi AS (2009) Plastic responses to different habitat type contribute to *Hyptis suaveolens* Poit. invasiveness in the dry deciduous forest of India. AMBIO 38:342–344

Sharma A, Batish DR, Singh HP, Jaryan V, Kohli RK (2017) The impact of invasive *Hyptis suaveolens* on the floristic composition of the periurban ecosystems of Chandigarh, northwestern India. Flora 233:156–162

Sharma J, Singh R, Garai S, Rahaman SM, Khatun M, Ranjan A, Tiwari S (2022) Climate change and dispersion dynamics of the invasive plant species *Chromolaena odorata* and *Lantana camara* in parts of the central and eastern India. Eco Inform 72:101824

Singh HP, Batish DR, Dogra KS, Kaur S, Kohli RK, Negi A (2014) Negative effect of litter of invasive weed *Lantana camara* on structure and composition of vegetation in the lower Siwalik Hills, northern India. Environ Monit Assess 186(6):3379–3389

Srivastava P, Raghubanshi AS (2021) Impact of *Parthenium hysterophorus* L. invasion on soil nitrogen dynamics of grassland vegetation of Indo-Gangetic plains, India. Environ Monit Assess 193(5):1–16

Thapa S, Chitale V, Rijal SJ, Bisht N, Shrestha BB (2018) Understanding the dynamics in distribution of invasive alien plant species under predicted climate change in Western Himalaya. PLoS One 13(4):e0195752

Theoharides KA, Dukes JS (2007) Plant invasion across space and time: factors affecting nonindigenous species success during four stages of invasion. New Phytol 176(2):256–273

Weltzin JF, Belote RT, Sanders NJ (2003) Biological invaders in a greenhouse world: will elevated CO_2 fuel plant invasions? Front Ecol Environ 1(3):146–153

Westerband AC, Knight TM, Barton KE (2021) Intraspecific trait variation and reversals of trait strategies across key climate gradients in native Hawaiian plants and non-native invaders. Ann Bot 127(4):553–564

Plant Invasion in an Aquatic Ecosystem: A New Frontier Under Climate Change

9

Reemа Mishra, Renu Soni, Garvita Singh, Pritam Kaur, and Preeti Agarwal

Abstract

Climate change and invasive species impose severe threats to biodiversity, ecosystem, and economy; however, the impact on human well-being and livelihood is not much known. The interaction between these is complex and intensifying, and there is increasing evidence that climate change is amplifying the deleterious effects caused by invasive species. Worldwide, the damage resulting from invasive species accounts for 5% of the global economy and has an impact on a large number of sectors such as forestry, agriculture, aquaculture, trade, recreation, etc. Variations in climatic conditions are more likely to interrupt the existing populations of native as well as aquatic invasive species and also increase the susceptibility of the aquatic ecosystem by creating favourable conditions for invasive species as they are more adaptable to disturbances and varied environmental conditions. Climate change is anticipated to cause warmer water temperatures, minimize ice cover, change the pattern of streamflow, increase salinization, etc., which would modify the pathways through which invasive species infiltrate the aquatic bodies. In addition, climate change will transform the ecological effects of invasive species by increasing their predatory and competitive effect on indigenous species and by enhancing the harmfulness of certain diseases. The impact of invasive species is anticipated to be more deleterious as they proliferate both in numbers and degree; can considerably change the composition, chemistry, structure, and function of aquatic systems. However, a clear insight into how climate change upsets invasive species growth

All authors have contributed equally to this work.

R. Mishra · R. Soni · G. Singh · P. Kaur · P. Agarwal (✉)
Department of Botany, Gargi College, University of Delhi, Delhi, India
e-mail: reema.mishra@gargi.du.ac.in; renu.soni@gargi.du.ac.in; Garvita.singh@gargi.du.ac.in; pritam.kaur@gargi.du.ac.in

S. Tripathi et al. (eds.), *Plant Invasions and Global Climate Change*,
https://doi.org/10.1007/978-981-99-5910-5_9

and a study of their combined effects on the ecosystems is still required. Further to minimize the compounding impact of climate change on the devastating effect of invasive species, various preventive and control measures are required to regulate the invasive species that presently possess moderate effects and are restricted by seasonally adverse conditions. The present chapter focuses on how climate change affects plant invasion in the aquatic system and their complex interactions. This chapter also discusses various methods used for the management and restoration of the invaded ecosystem.

Keywords

Aquatic · Climate change · Ecosystem · Invasive aquatic species · Invasive plant species · Management · Restoration

Abbreviations

AIP	Aquatic invasive plants
AIS	Aquatic invasive species
IAAPS	Invasive aquatic alien plants species
IAPS	Invasive alien plants species
IAS	Invasive alien species
IPS	Invasive plant species

9.1 Introduction

An invasive species is a non-native species that enters a new area, becomes overpopulated, and alters the ecosystems that it colonizes. It is also termed as an alien, introduced, or exotic species. They impose a severe threat to the health, productivity, and sustainability of native ecosystems and cause huge economic loss. They exhibit a high dispersal rate, fast growth, a small lifespan, and increased tolerance to a wide range of environmental conditions that helps them to acclimatize to the new environment (Pimentel et al. 2005; Rai and Singh 2020). The impacts of invasive species are more severe as they flourish both in numbers and in degree. They extensively modify the structure and function of native aquatic systems through direct and indirect interactions (Wootton and Emmerson 2005; Burgiel and Muir 2010; Poland et al. 2021). The estimated damage from invasive species accounts for 5% of the world economy affecting various sectors such as forestry, agriculture, aquaculture, terrestrial habitat, waterways, trade, and recreation (Pimentel et al. 2001).

Freshwater habitats are more vulnerable to invasive species than terrestrial habitats (Moorhouse and Macdonald 2015). The susceptibility of aquatic bodies to invasion depends on various physical and chemical properties like their trophic state, depth, sediment, and flow rate. Thus, the degree and extent of destruction by invasive plants can be successfully controlled and it depends on various parameters like conditions of the site, recognition and response times, and management selection. Examples of submerged exotic aquatic plants, including *Brachiaria brizantha, Brachiaria mutica, Hydrocotyle vulgaris, Hydrilla verticillata, Myriophyllum aquaticum, Myriophyllum heterophyllum, Nitellopsis obtusa, Potamogeton crispus, Spartina alterniflora, Trapa natans*, etc.

Climate change intensifies the deleterious effect of invasive species. Both drivers (climate change and invasive species) are linked together in various manners (Walther et al. 2009; Smith et al. 2012). It increases the susceptibility of the aquatic ecosystem by creating conditions favourable for the invasive species.

Climate change effects like increased global temperature and CO_2 levels, severe weather events, changes in precipitation patterns and stream flow, increase in water temperature and salinization, decreased ice cover, etc. will result in transformation of pathways through which invasive species penetrate the aquatic systems. They favour these invasive species by increasing their chances to cross geographic barriers, spreading and establishing in new areas as they exhibit high adaptability to varied conditions (Walther et al. 2009; Burgiel and Muir 2010; Dai et al. 2022).

Detection at primary stages and eradication are regarded as the most efficient and cost-effective way to evade and regulate the introduction and establishment of invasive species. This also ensures long-term success in comparison to maintenance at post-entry stages. The outcome of invasive species is anticipated to further intensify with the change in climatic conditions; however, a clear insight into how climate change affects the growth of invasive species and their combined effects on the ecosystems still needs to be investigated. This chapter focuses on the impact of climate change on aquatic invasive species (AIS), how climate change affects plant invasion in the aquatic system and their complex interactions. This chapter also highlights various approaches used for the management and restoration of the invaded ecosystem.

9.2 Impact of Climate Change on Aquatic Ecosystem and Aquatic Invasive Plants

Rapidly increasing aquatic invaders pose a great risk to aquatic ecosystems. They can thrive in new surroundings and harm local ecosystems. Invasive species displaces native species, reduces ecological services, and also causes economic loss. Non-native species invasion is the primary source of biodiversity loss globally, especially in freshwater systems, which have more number of species in comparison to any ecosystem (Ricciardi and MacIsaac 2011; Thomaz et al. 2012). In freshwater ecosystems, invasion causes considerable harm by affecting the functional and structural integrity. The loss of species is more than that in terrestrial and marine habitats. These species spread to new locations through a variety of channels (Olden et al. 2006; Strecker et al. 2011). Human activity related to global trade has accelerated the spread of species to new locations and is the primary cause of most recent invasions (Levine and Antonio 2003). Freshwater systems, especially lakes, are vulnerable to invasion due to trophic linkages (Gallardo et al. 2016). Aquatic incursions influence ecosystem populations, communities, and processes (Ehrenfeld 2010). Once an invasion establishes itself, the species completely takes the place of the native species, consequently resulting in their elimination (Getsinger et al. 2014; Brundu 2015).

Effects of invasive species include shifts in the structure, composition, and even function of ecosystems (Lloret et al. 2004; Bobeldyk et al. 2015). It is well known that invasive species may alter the food webs of freshwater ecosystems (Vander Zanden et al. 1999). Invasive plant species (IPS) have negative societal effects as well. The IPS provides a lower-quality food supply for macroinvertebrates as well as higher-level consumers (Madsen et al. 1991).

Species abundance and richness, food web structure (Villamagna and Murphy 2010; Stiers et al. 2011), macrophyte composition (Hussner 2014), and even oxygen levels are all impacted by aquatic invasions (Shillinglaw 1981). IPS has the ability to reproduce clonally and spread quickly. Since clonal integration and invasion of alien plants are strongly connected, clonal plants reproduce rapidly and disperse to new areas (Maurer and Zedler 2002). Thus, due to their rapid proliferation, AIS poses a great danger to ecosystems and displays adverse effects on the environment as well as the economy (Brundu 2015).

Non-native plants proliferate in excess and create monospecific stands that block water flow. This affects water quality by reducing oxygen levels and odour. The extensive growth of aquatic weeds can impede water flow and block inlet pathways, which can result in floods (Hassan and Nawchoo 2020). The development, spread, and effects of IS may be exacerbated by increased nutrient levels, elimination of top predators, and altered flow regimes caused by increased overharvesting (Gherardi 2007). Floating aquatic plants may minimize freshwater extraction and navigation, fish harvesting, and water cycling and chemistry (MacDougall and Turkington 2005). Invasion effects are undoubtedly a reason for worry given the high level of biodiversity and susceptibility of freshwater ecosystems to biotic exchange (Sala et al. 2000). Invasive species affect ecosystems and the economy, which are responsible for several socio-ecological issues, and also impact people's health and livelihoods (Perrings et al. 2002). Management of foreign invasive species requires an understanding of invasive plant dispersion tactics, perpetuation time, and manner of invasion (Hassan and Nawchoo 2020).

9.2.1 Effect of Climatic Change on the Aquatic System

Global warming and climate change, which have forced ecological systems, biodiversity, and human existence to face the worst issue in history, have started to influence aquatic ecosystems, from plankton to mammals (Hoegh-Guldberg et al.

2019). Due to their size and diversity, oceans and seas are majorly impacted by the transformation brought on by global warming. In addition to the rising temperature of vast water bodies including oceans, seas, lakes, and ponds, an increase in atmospheric temperature also triggers hydrological processes that alter physical as well as chemical properties of water. Sea level rise, an increase in ocean temperature, and changes to current precipitation, wind, and water circulation patterns are all possible impacts of climate change (Scavia et al. 2002; Roessig et al. 2004).

Climatic changes are the most extreme component of global development. As a result of global warming, thermal stratification increases, glaciers melt, sea levels rise, coastal erosion increases, lakes evaporate more quickly, greenhouse effects are exacerbated, ocean acidity rises, carbonate concentration decreases, biological invasion increases, and biodiversity declines (Sivaramanan 2015). Climate change is not a national concern; it spans continents. The sudden spike in catastrophic climatic effects was caused by hydrologic shifts in worldwide water that migrated towards land. This makes aquatic species the most afflicted animals (Eissa and Zaki 2011).

The ongoing rise in sea level will, to some extent, put a large number of aquatic species in danger. Warming changes species ranges, fundamental metabolic processes, and the timing (or phenology) of critical biological events. Acidification limits the development of calcifying organisms and produces physiological stress in sensitive marine species (Waldbusser and Salisbury 2014; Asch 2015). Aquatic species distribution, range of aerobic conditions, and chances of survival can be affected by ocean deoxygenation and hypoxia conditions (Breitburg et al. 2018; Griffith and Gobler 2020). Many aquatic birds, including warblers, flamingos, aquatic swan geese, and pelicans as well as migratory fish species such as eels and mullet, other species like coral reefs, turtles, and some aquatic crustaceans are among those that are susceptible to such severe effects (Newson et al. 2009). According to Stocker et al. (2013), emissions of greenhouse gas by human activities have a significant role in climate change and ocean acidification, which has an effect on marine ecosystems and their products and services (Gattuso et al. 2015; Weatherdon et al. 2016). Climate change directly affects organisms' development, and their ability to reproduce. Thus indirectly, it results in a change in the structure, composition, function, and productivity of aquatic ecosystems (Ghosh et al. 2020).

9.2.2 Impact of Aquatic Invasive Plants (AIP) in Response to Global Climatic Change

In contrast to native species that cannot adapt to climate change, many alien species are anticipated to benefit from climate change and expand their range. IAS and climate change may progressively interact in a positive feedback loop, with the former creating new habitats for the latter and making ecosystems more vulnerable to the latter (McNeely 2000, 2001). According to the UN's Intergovernmental Platform for Biodiversity and Ecosystem Services (IPBES), biotic invaders threaten 1/5th of surface of the Earth, including biodiversity hotspots (IPBES 2019). Through several unusual physiological traits (such as large biomass, long roots, and increased transpiration), the IPS may enter aquatic systems and obstruct water flow, rendering it unfit for drinking and irrigation (Pejchar and Mooney 2009). Climate change and AIS pose a range of threats to ecosystems, biodiversity, human health, and socio-economic situations through a variety of methods (Bartz and Kowarik 2019; Rai and Singh 2020). In addition to having an impact on human health, invasive alien plant species (IAPS) also increase the frequency of floods by narrowing stream channels and changing the soil properties (such as decreasing its ability to retain water and increasing soil erosion) (Rai and Singh 2020). Ground and surface water supplies are also known to be impacted by IAPS (Shackleton et al. 2019). IAPS is known to interfere with water transportation regularly, which has a detrimental impact on recreation and tourism activities (Eiswerth 2005). Biologists who study invasions have recently concluded that not all invasions are harmful to ecosystems (Young and Larson 2011). Numerous IAPs are recognized for the positive effect they have on ecosystem services, which might include things like providing aesthetic value and entertainment, preserving cultural traditions, and enforcing laws and policies (Pejchar and Mooney 2009). It has been proposed that use of IAPS like *Phragmites* sp. and *Eichhornia crassipes* to create bioenergy might serve dual goals, i.e. to make renewable energy that won't run out and to get rid of weeds simultaneously (Rai et al. 2018; Stabenau et al. 2018). Effective phosphorus recycling by *Elodea nuttallii* may result in nutrient enrichment (eutrophication), which would be bad for aquatic habitats. Producing biogas and phosphorus-rich compost from this aquatic IAPS biomass is beneficial (Stabenau et al. 2018).

Several AIS have been discovered to have a detrimental effect on the Benthic Quality Index (BQI) in marine environments (Zaiko and Daunys 2015). Therefore, coastal invasive species may be used as a general indicator of the health of the marine ecosystem. Many foreign aquatic plants are intentionally introduced as they offer commercial, aesthetic, or environmental benefits; however, they also pose a negative impact on aquatic ecosystems by obstructing rivers, limiting aquatic life by lowering dissolved oxygen levels and reducing native biodiversity. They also offer a variety of ecosystem services like food, fodder, decorative use, ecological restoration, landscaping, and green manure (Wang et al. 2016). Aquatic alien plants, in particular, can induce oxygen deprivation, decrease native biodiversity, degrade water quality, and even disrupt food web structures in freshwater habitats once they have effectively invaded (Hussner 2014). These ecological consequences, whether favourable or unfavourable, might be amplified by global warming. Inputs of phosphorus and nitrogen may potentially change the status of some alien organisms. Additionally, the relationships between alien aquatic plants and herbivores have changed as an outcome of change in climatic conditions, which will affect how far they spread in the future (Wu and Ding 2019). The species makeup of plant communities may vary due to global change, and further affecting the ecological and physiological characteristics of alien plants in water habitats (Henriksen et al. 2018). Tabular representation of aquatic invasive plants that have been reported to expand under changing climatic conditions has been provided in Table 9.1.

Table 9.1 Tabular representation of list of aquatic invasive plants and effect of changing climatic conditions on the spread and invasiveness of these plants

S. No.	Plant species	Factors	Effects	References
1.	*Hydrilla verticillata*	Increased water temperature and carbon dioxide	The plants are more adaptable to warmer temperature. Increased in CO_2 level enhances the biomass under precise conditions	Chen et al. (1994), McFarland and Barko (1999), Williams et al. (2005), EPPO (2008)
2.	*Mimosa pigra*	Flooding and Rainfall	In Australia, flooding and rainfall assisted in seed dispersal by flotation	Lonsdale (1993)
3.	*Phragmites australis*	Increase in ambient air temperature	It is abundant on the Atlantic Coast and is quickly expanding to westward and northward	Wilcox et al. (2003)
4.	*Ranunculus trichophyllus*	Decreased length of ice cover	It has spread to non-vegetated lakes in the Himalayas	Lacoul and Freedman (2006)
5.	*Eichhornia crassipes* and *Typha angustifolia*	Storm, in case of after Tsunami occurred in southeast Asia in 2004	It spread to lagoon and estuaries. Storms resulted in increased disturbance in habitats and thus favoured the establishment and expansion of already existing invasive species	Bambaradeniya et al. (2006)
6.	*Posidonia oceanica*	Warming of water temperature	Warming was found to induce flowering	Diaz-Almela et al. (2007)
7.	*Arundo donax*	Climatic warming	It is native to riparian habitats of eastern Asia. It was introduced to South Africa and has expanded to riparian habitats of rivers and streams. They can withstand broad range of environment conditions and are suitable to South Africa's climatic conditions. Rooting of stem fragments was found	Milton (2004), Nel et al. (2004), Mgidi (2004), Wijte et al. (2005), Quinn and Holt (2008)

(continued)

Table 9.1 (continued)

S. No.	Plant species	Factors	Effects	References
			to be 100% at temperature 17.5 °C or greater than it. Thus, has increased its likelihood to expand and invade under changing climatic conditions	
8.	*Thalia dealbata*	Climate warming	It is predominant in China and has spread to upper altitude as result of warming	Chen and Ding (2011)
9.	*Eichhornia crassipes*	Warming, extreme rainfall	It is native to South America and has spread to Lake Victoria (Kenya), Tanzania, and Uganda. It is presently established in regions of southern Europe but is likely to expand to remaining parts of Mediterranean Basin and further to northward into Europe due to warming. It was introduced in China but later turned into invasive and has spread across 16 provinces. In addition to China, it has also expanded to Central America, Central Africa, Western Africa, Southeast Asia and South-eastern United States. These plants overcome winter as they possess floating vegetative tissues. The warm temperature of water avoids the root and leaves from being destroyed from frost	EPPO (2008), You et al. (2014), Wu and Ding (2019)

(continued)

Table 9.1 (continued)

S. No.	Plant species	Factors	Effects	References
			condition during winter. Their vegetative biomass (overwintering) also responds fast to increase in temperature and thus enhances their invasiveness. Extreme rainfall supports the transport of propagules across China	
10.	*Pistia stratiotes, Azolla filiculoides, Cabomba caroliniana,* and *Egeria densa*	Climatic warming, elevated rainfall	It was introduced in China and then turned into invasive. *P. stratiotes* is widely distributed in China and is found in more than 9 provinces. It is also reported to have spread in Germany. *A. filiculoides* is introduced in Spain and China, *and C. caroliniana* in China, *E. densa* in United states. In China, warming has resulted in transformation of these plants into invaders and thus has led to their expansion to new areas particularly to upper latitudes. Warming induces overwintering and their invasiveness. Increased rainfall has enhanced the survival and adaptation of these plants. It also assisted in propagules transport of these free-floating plants across China. Created favourable	Santos et al. (2011), Hussner (2014), Espinar et al. (2015), Gao et al. (2015), Vojtkó et al. (2017), Wu and Ding (2019)

(continued)

Table 9.1 (continued)

S. No.	Plant species	Factors	Effects	References
			conditions by providing more appropriate aquatic environments that helped in their spread and establishment at greater latitudes	
11.	*Nymphaea rubra*	Warming	Warming increases adaption	Hussner and Lösch (2005), Vojtkó et al. (2017)
12.	*Myriophyllum aquaticum*	High rainfall or water level variations	It was introduced in China and then turned into invasive. It is widely distributed in China and is found in more than 9 provinces. High rainfall or water level variations enhance clonal integration, number of branches and length of stolon	Chen et al. (2016), Wu and Ding (2019)
13.	*Myriophyllum spicatum*	Climatic warming	It is native to Europe, Asia and has invaded to North America. Warming has extended its growing season and thus has increased its abundance in freshwater and also improved its carbon stock as well as biomass	Velthuis et al. (2018)
14.	*Alternanthera philoxeroides*	Warming, Increased precipitation and variation in water level	It was introduced in China and then turned into invasive. It has spread to higher latitudes in North China and South America. Reported to have spread across 18 provinces in China. The fluctuation in water level increases length of shoot length and reduces	Yu (2011), You et al. (2013a, b), Lu et al. (2015), Chen et al. (2016), Wu et al. (2017a), Wu et al. (2017b), Wu and Ding (2019)

(continued)

Table 9.1 (continued)

S. No.	Plant species	Factors	Effects	References
			intraspecific competition. Warming induces increase in net rate of photosynthesis and morphological plasticity	

9.3 Climate Change and Aquatic Invasive Species Interactions

Invasive plant species (IPS) typically have a higher degree of environmental tolerance, faster rates of growth and dissemination, and shorter generation times, which make them more resilient to abrupt climate changes. Species interactions play a vital role in configuring different communities and these interactions are majorly influenced by climate. Tylianakis et al. (2008) in their review analysed the probable effect of global climate change on the terrestrial ecosystem and proposed that climate change might influence almost every species interaction. It can weaken the positive interactions (mutualism), can affect the food web, richness of taxa, intensity of predation, etc. Aquatic ecosystems are similarly vulnerable to these changes. Climate change might alter the competitive species interactions due to which the native communities may become more or less vulnerable to novel invasions or it can also lead to the establishment of already existing invaders. Alternatively, climate change might reduce the competitive capacity of primary invaders to the point that they are no longer deemed as invasive and this could enhance the abundance of secondary invaders (Bellard et al. 2013; Pearson et al. 2016). Predicting the future dispersal and species interaction of IPS in response to changes in climatic conditions is a challenging endeavour since many variables affect the local and transient invasion trends (Mainali et al. 2015). The impact of climate change on AIS introduction, establishment, spread, and dispersal is discussed in the following section.

9.3.1 Altered Mechanism of Invasive Species Introductions

It is predicted that climate change can increase the temperature of the water, decrease the thickness of the ice, influence the pattern of stream flow, and enhance salinization. Such changes might alter the pathways of invasive species introduction, growth, their spread and their dispersal (Rahel and Olden 2008; Kariyawasam et al. 2021). Studies have shown that the melting of ice has facilitated the migration of aquatic birds and mammals among the Atlantic and Pacific Ocean basins (McKeon et al. 2016). Plants have long been introduced for decorative and agricultural purposes. The majority of newly introduced plants have physiological characteristics that enable them to thrive in a variety of climatic situations and hasten their establishment and expansion (Bradley et al. 2010). The quest for plants that can withstand a variety of stress and are resistant to abiotic stress may increase due to climate change (Bradley et al. 2012). Many invasive species majorly spread to new sites as contaminants via human-assisted transport like cargo ships and as contaminants of agricultural products (Hulme 2009). Climate changes could modify human travel and connect previously disconnected locations. Such travel alterations could indirectly affect the invasive species' ability to propagate and establish itself in newer aquatic regions (Hellmann et al. 2008). According to Corlett and Westcott (2013), native plants possibly will face 'Migration lag' due to climate variation and such place when invaded by invasive species might change the community structure (Bernard-Verdier and Hulme 2015). In a nutshell, fluctuations in climate have the ability to modify the entry points and growth conditions that are favourable for invasive species in aquatic systems.

9.3.2 Influence of Climate Change on Establishment of Aquatic Invasive Species (AIS)

The establishment of AIS could be influenced by climate change negatively or positively. For many invasive species, phenotypic plasticity is thought to be a key factor in determining their establishment and growth. Acquired genetic variations may also regulate germination which in turn is crucial for the establishment of invasive species (Richards et al. 2006). Davis et al. (2000) suggested that instabilities in aquatic habitats due to eutrophication and other stresses can enhance plant invasions by raising their 'invasibility'. Wainwright et al. (2012) predicted that climate change will favour the establishment of species that have germination flexibility under a wide range of environmental variations. Information about the germination phenology of native and invasive species are very important for foreseeing the identification of species that may establish efficiently under varied climatic conditions (Gioria et al. 2018). Orbán et al. (2021) through their experiments on four invasive species suggested that disturbance parameters should also be considered while assessing the consequence of climate change on the growth and establishment of invasive species. It can be hypothesized that invasive species with flexible germinations will be able to establish successfully under variable climatic conditions in aquatic habitats.

9.3.3 Influence on Spread and Distribution Change of AIS

Climate change can significantly regulate the distribution and spread of AIS. The most significant factors in determining the geographic range of invasive species are the temperature and precipitation (Finch et al. 2021). In addition to enhancing survivability, milder winters in temperate regions due to enhanced temperature would lengthen the growing season, which could enhance reproductive productivity

(Hellmann et al. 2008). Species that can quickly shift their ranges may have an edge over other species. Water hyacinth, also known as *E. crassipes*, is one of the most troublesome species of tropical aquatic plant, and has invaded a number of other nations. You et al. (2013a, b) analysed the effect of temperature on the growth of water hyacinth and observed enhanced growth with the increase in temperature. From their experiments, they concluded that climate warming may increase the invasiveness of water hyacinth by increasing its distribution and spread (You et al. 2013a, b). Adhikari et al. (2019) studied the possible repercussions of climate change on the spread of IPS in the Republic of Korea (ROK). From their study, they predicted that climate change can enhance the IS richness and dispersion in the northern and eastern provinces of ROK. According to the findings of their research, Kariyawasam et al. (2021) concluded that climate variability will lead to the growth of AIP in the locations (different regions of Sri Lanka) that they studied. The dispersal of species has also been considerably enhanced by humans (Havel et al. 2015). Most research on the impact of climate on invasive species has been piloted on terrestrial systems; however, such research can aid in the design of experiments for AIS.

9.4 Management of Aquatic Species Vulnerable to Climate Change

Aquatic invasive species (AIS) is a threat to biodiversity loss and species extinction and is difficult to control. Reasons behind the invasion of species that are non-native are many; however, the main reason could be climate change. Wetlands are also vulnerable to invasive species and their impacts on the present diversity of the region, therefore, pose a major global concern (Zedler and Kercher 2004; Shackleton et al. 2018; Bolpagni et al. 2020; Adams et al. 2021; Lázaro-Lobo and Ervin 2021). Many attempts and also many efforts are made to restore ecosystems after an invasion explosion (Kettenring and Adams 2011; Prior et al. 2018). India due to its diverse environmental and varied climatic conditions is highly prone towards biological invasion and favours both accidental and intentional entry of plant species (Kohli et al. 2011). Plants in aquatic ecosystems are critical invasive species, namely *Alternanthera philoxeroides*, *E. crassipes*, *Lemna perpusilla, Marsilea quadrifolia, M. aquaticum*, *Salvinia molesta,* and *Ipomoea* spp. (Raghubanshi et al. 2005). *Eichhornia crassipes, A. philoxeroides, S. molesta,* and *Ipomoea* sp. invade aquatic ecosystems and cause much harm to the biodiversity of aquatic ecosystems (Reddy 2008). IPS is widely known for their harmful effects, and many nations are implementing strategies such as preventing the invasion of alien species, preventing its spread, detecting the invasions rapidly, eradicating it wherever possible, reducing the impact of consequences of invasive species and restoration of damaged ecosystems. Here, we review a few approaches to dealing with IPS. A schematic illustration of different stages of invasion, the successful establishment of invasive species in a region, and various management schemes that can be implemented at each stage is depicted in Fig. 9.1.

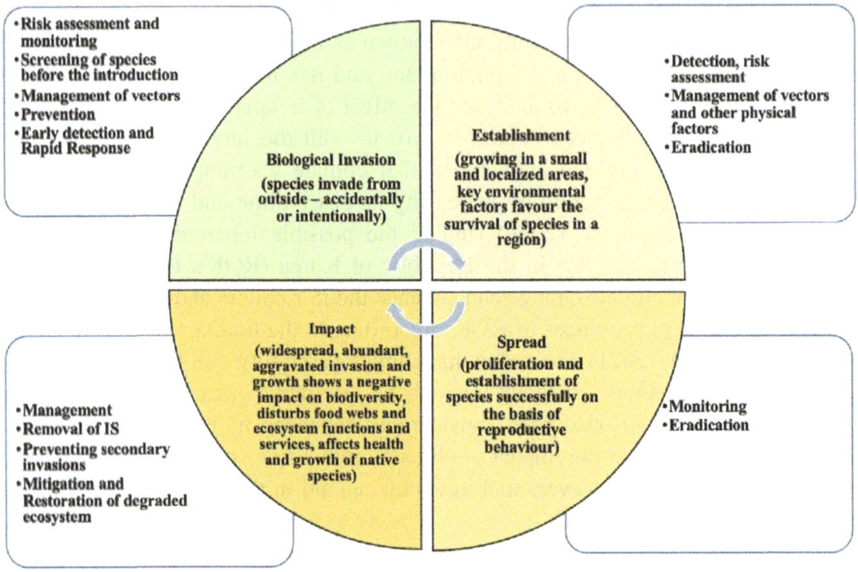

Fig. 9.1 Schematic representation of different stages of invasion and successful establishment of invasive species (IS) in a region and management strategies that are suggested to be implemented at each stage

9.4.1 Risk Assessment

It is a priority to assess the risk factors of establishment of alien species in aquatic systems and to consider the consequences that can arise upon the introduction. However, species are introduced for human welfare, and ornamental purposes and so humans are responsible for dispersal and establishment (Pyšek and Richardson 2010; Havel et al. 2015). The negative impact and consequences of invasion of aquatic alien plants can result in the change in the biodiversity of native species, aggravation of biological invasions, increases in non-target effects, disturbance in aquatic food webs, and accelerated water pollution, that change overall interspecific changes. Therefore, screening of species before introduction has to be done (Singh 2021). The history of species and the behaviour of growth and reproduction are crucial for screening. Also, weed risk assessment is significant for controlling high-risk species. Risk maps are to be created for determining invasive spread in fragmented areas and areas of higher risk, and therefore, remote sensing technology, computing, monitoring mechanisms, and modelling methods are being used nowadays (Bradley and Mustard 2006; Pyšek and Richardson 2010).

9.4.2 Management of Vectors

At various places, where climate change is the main problem, the management of vectors is necessary to reduce the invasion of alien species. In addition to the vectors and mechanisms of dispersal that are to be identified, other opportunities such as spread through garden escapes also make ornamental plant invasion and establishment easy. Therefore, there is an emergent need to find ways to control through the biological method, and measures of early detection of invasive species and alternatives to invasive species (Pyšek and Richardson 2010).

9.4.3 Early Detection and Rapid Response Strategy (EDRR)

Early detection and rapid response management strategy has a significant role in integrated techniques for the control of invasive species. Early detection of invaded species can aid in quick observation, thus, rapid responsiveness and safety regulation and control (Hulme et al. 2009). Sometimes, inconspicuous numbers and small sizes of invaders during the early stages of invasion escape early detection and mapping. Research and development are, therefore, focused on remote sensing (Koger et al. 2004) and mapping (Barnett et al. 2007). At places where species are introduced from many regions, their taxonomic identification can be difficult.

9.4.4 Eradication

Successful eradication of invasive species belonging to different taxons such as *Mytilopsis sallei* (marine mussel) from northern Australian harbour, *Caulerpa taxifolia* (seaweed) from a lagoon in California and *Bassia scoparia* (herb), and *Cenchrus echinatus* (grass) from a Hawaiian island, and Australia has been conducted and reported (Pyšek and Richardson 2010).

9.4.5 Difficulty in Controlling Key Environmental Factors

Degradation of ecosystems at accelerating rates due to multiple pressures of anthropogenic activities like urbanization, industrialization, and agriculture intensification leads to more frequent instances of species invasion (Kercher and Zedler 2004; Ervin et al. 2006). Though biological invasions also characterize degraded aquatic ecosystems. Therefore, an integrated approach of using effective control measures of preventing invasiveness and post-recovery mechanisms against various external factors and pressures is needed (Lavergne and Molofsky 2006). Botanists remain unaware of the spread and establishment of some invasive species, their mechanisms of propagation, and the dynamics of their growth and development, therefore, management is also tricky. Therefore, appropriate assessment of the risk of their potential invasiveness, early detection, forecasting and further rapid removal, education, raising awareness and legislation, and effective controls often require integrated long-term commitment techniques and approaches (Willby 2007).

9.4.6 Mitigation and Restoration

The strategies and approaches need to be focused on restoring ecosystems following degradation and their negative impacts. Also increasing incidences of 'secondary invasions', that is quick establishment of new invasive species in the place of earlier species in disturbed regions are reported that are favoured due to the various management strategies and interventions, control methods, and/or alteration of resources. Restoration involves the removal of invasive species. Though various control and restoration efforts were rather not appropriate, and therefore, exhibited consequences are not preferred in case to control the predator as this can cause further higher number of intermediate predators that affect trophic levels in food chains and food webs cascade through the ecosystem (Pyšek and Richardson 2010).

9.5 Restoration Methods for a Degraded Ecosystem

The methodology adopted for the restoration of aquatic systems is done through taking small steps towards stabilizing biodiversity with the constant increase in species count, using methods and approaches conserving habitats with their natural biodiversity and ecosystems. In general, habitat restoration can address the chemical properties of an ecosystem, such as re-oligotrophication or a decrease in the number of contaminants that are present in excess, as well as the rehabilitation of the physical-structural properties of an ecosystem, restoring connectivity, or any combination of these. In order to support ecosystem functioning, more emphasis is placed on the requirement to maintain habitat complexity and connectivity while focusing on biodiversity itself at the habitat, assemblage, or the individual species level (Dethier et al. 2003; Giller et al. 2004).

In order to create a balance in the ecosystem, the removal of IAS is frequently carried out via different restoration projects that have been approved to eradicate the alien species (Hobbs and Richardson 2011). A strong criticism was raised by ecologists due to the unrealistic methods of tackling with IAPS control (Richardson et al. 2004; Shaw et al. 2010). These studies utilized a restoration ecology approach that neglected the understanding of the basic cause of ecosystem damage. In order to improve restoration efforts, a common approach defining restoration ecology as well as invasion ecology together could bring clarity on the causes of invasion. This could further be supported via sharing and putting forward knowledge with supportive research, having application in the administration and restoration of the ecosystem. The main cause of the degradation of the ecosystem is competition because of IAPS and the most effective way is to eradicate them. However, the abrupt removal of invaders changes the natural habitat, which hinders the growth and re-establishment of native species or even results in the death of the native species that have been reintroduced into the ecosystem (Vila and Gimeno 2007; Beater et al. 2008; Bergstorm et al. 2009).

9.6 Ecological Restoration Practices

As per the Society of Ecological Restoration (SER), the main goal of restoration projects is to restore the ecosystem features that have been continuously destroyed, as a result of human interference (Ruiz-Jaen and Mitchell Aide 2005). According to reports by Benayas et al. (2009), ecological restoration benefits the recovery of native species and biodiversity. Based on meta-analyses research evaluating the impacts of restoration on various types of ecosystems across globe, ecological restoration projects raised the level of biodiversity present and also uplifted ecosystem benefits with 44% and 25%, respectively. This held true for additional ecological restoration meta-analyses carried out on more defined ecosystems, such as wetlands and forest reserves (Felton et al. 2010; Meli et al. 2014). Different passive or active strategies were used to implement ecological restoration for positive results. The removal of degrading elements is the first step in passive restoration, which is followed by the autogenic or natural regeneration of native species and their respective community. Active restoration (assisted regeneration) entails actions like adding desired plant species, amending the soil, and controlling fire regimes, which also drive secondary native succession (Holl and Aide 2011). It is difficult to reset the endpoint of ecological restoration, particularly for freshwater ecosystems, to that of the pre-invasion state because of changing environmental patterns such as climatic conditions, land use, and significant anthropogenic behaviour. As a result, the recovery of ecosystem processes and the regular operation of an ecosystem, which will produce ecosystem goods and services for society and wildlife, are the foundations for restoration success (Suding 2011). IAPS species management and restoration activities primarily use passive strategies in aquatic ecosystems, including herbicidal control, mechanical clearing, and the application of biological control measures (Coetzee et al. 2011; Stiers et al. 2011; Gaertner et al. 2012). In South Africa, passive restoration practices of alien invasive species resulted in the secondary invasion, according to Ruwanza et al. (2013). Their study noted following restoration management perspectives:

- Passive restoration alone is a slow and ineffective method that only permits the natural regeneration of native communities.
- Following catchment management strategies that constrained the discharge of nutrient-rich effluents, a freshwater lake in Scotland that was previously known to be highly eutrophic showed a significant reduction in its nutrient status. It demonstrated an autogenic recovery of the local species following a check on the lake's nutrient reduction (Carvalho et al. 2012).
- For successful ecosystem recovery, most eutrophic freshwater lake ecosystems require a combination of passive (reduction in nutrient input) and active restoration, using biological changes (Liu et al. 2018). This two-pronged approach to

restoration has enabled the recovery and re-establishment of native plant communities, followed by the distribution of related organisms, creating a balanced ecosystem with clear structure and function.

Even though active restoration can be expensive, it is justified for areas and regions with high conservation value, such as threatened or endangered biomes, biodiversity hotspots, and high-priority catchment areas for freshwater resources (Gaertner et al. 2012). In terms of the role that IAAP species invasion has played, biological control has been successful in reducing IAAP biomass and contributing to long-term benefits like water conservation and ecosystem recovery after control (Fraser et al. 2016; Martin et al. 2018). The South African Riparian invasion control proved to be an excellent example of dual restoration practice involving both passive and active methods. Implementation of a massive terrestrial and riparian invasive alien removal program leads to ecosystem balance/recovery studies showing complete establishment of introduced native species at those studied sites showing positive outcomes (Ruwanza et al. 2013; Nsikani et al. 2019).

Anthropogenic activities and landscape developments are the main reason behind the conversion of natural ecosystem to urban developments and agricultural space, which leads to natural habitat fragmentation, thus playing a significant role in compromising ecological recovery for freshwater ecosystems and limiting native gene pool flow (Kietzka et al. 2015). Elaborative studies and research practices on ecological restoration indicate the success and hardship in relation to the restoration of degraded ecosystems, with active long-term management studies providing evidence in order to develop knowledge and fulfill the bridge-gap of these approaches in understanding the complex variables. With regard to long-term post-IAAP species, management and restoration monitoring to give useful trajectories on restoration mechanisms within the aquatic environments was further supported by a number of researchers (Kettenring and Adams 2011; Suding 2011; Prior et al. 2018). These studies demonstrate the necessity and relevance of conducting additional IAAP species recovery studies following biological control, as the majority of meta-analyses and reports focus on restoration initiatives involving river channelization, urbanisation, deforestation, and mechanical removal of IAAP species (Miller et al. 2010; Kettenring and Adams 2011; Kail et al. 2015; Prior et al. 2018).

9.7 Conclusion and Future Prospects

Climate change and invasive species are two of the major threats to biodiversity and ecological services. There are evidence that invasive species has a greater impact on aquatic freshwaters in comparison to terrestrial ecosystems and is more susceptible to invasion. Moreover, climate change is intensifying the deleterious impact of invasive species. Global climate changes interfere with the population of native species and increase the vulnerability of the aquatic bodies to invasion by creating favourable conditions. Invasive species exerts a negative impact on the invaded habitat by modifying the structure and function of the native ecosystem via direct and indirect effects at various ecological levels. The primary intervention is a cost-effective method for controlling and managing invasive species. However, to ensure long-term success, restoration and rehabilitation should be aimed at attaining resilient ecosystem resistance to invasions. Further knowledge is required to:

- Understand as how and to what degree climate change is controlling the selection procedure on invasive species going through range extension that would aid in the effective management of invasive species. Insight into the relationship between climate change and genetic processes will be vital in predicting as how the invasive species adapts to climatic change.
- To gain insight into which species are more vulnerable including species that are tolerant to temperature and which systems are more susceptible to invasion in response to temperature change, water quality and quantity, nutrient availability, and changes in community compositions are required.
- The complexity created by the interaction between climatic variations and plant invasions can be resolved using a multidirectional approach. In order to apprehend the effect of biotic as well as abiotic interactions, transcriptomics along-with growth analyses are frequently utilized to locate and identify the genes involved in the IPS.
- By examining alien species at the population level in both native and invasive ranges and incorporating genomics and multi-omics approaches, we can learn more about the mechanisms underlying plant responses to climate change. Long-term experiments could help in gaining an in-depth understanding of how to target particular responsive genes by assessing the effects of environmental changes on invasions during each invasion stage.
- Need to find out the effect of mechanical, chemical, and biological controls under various climatic conditions and is also important to identify which control method is more robust, most adaptable, and healthy for the ecosystem.
- To develop integrated monitoring and information mechanism that syncs with new techniques for the management of aquatic IS.

References

Adams CR, Hovick SM, Anderson NO, Kettenring KM (2021) We can better manage ecosystems by connecting solutions to constraints: learning from wetland plant invasions. Front Environ Sci 349. https://doi.org/10.3389/fenvs.2021.715350

Adhikari P, Jeon JY, Kim HW, Shin MS, Adhikari P, Seo C (2019) Potential impact of climate change on plant invasion in the Republic of Korea. J Ecol Environ 43:1–12. https://doi.org/10.1186/s41610-019-0134-3

Asch RG (2015) Climate change and decadal shifts in the phenology of larval fishes in the California current ecosystem. PNAS 112:E4065–E4074. https://doi.org/10.1073/pnas.1421946112

Bambaradeniya CNB, Perera MSJ, Samarawickrema P (2006) A rapid assessment of post-tsunami environmental dynamics in relation to coastal zone rehabilitation and development activities in the Hambantota district of southern Sri Lanka. World Conservation Union (IUCN) Sri Lanka Country Office

Barnett DT, Stohlgren TJ, Jarnevich CS, Chong GW, Ericson JA et al (2007) The art and science of weed mapping. Environ Monit Assess 132:235–252. https://doi.org/10.1007/s10661-006-9530-0

Bartz R, Kowarik I (2019) Assessing the environmental impacts of invasive alien plants: a review of assessment approaches. NeoBiota 43:69–99. https://doi.org/10.3897/neobiota.43.30122

Beater MMT, Garner RD, Witkowski ETF (2008) Impacts of clearing invasive alien plants from 1995 to 2005 on vegetation structure, invasion intensity and ground cover in a temperate to subtropical riparian ecosystem. S Afr J Bot 74:495–507. https://doi.org/10.1016/j.sajb.2008.01.174

Bellard C, Thuiller W, Leroy B, Genovesi P, Bakkenes M, Courchamp F (2013) Will climate change promote future invasions? Glob Chang Biol 19:3740–3748. https://doi.org/10.1111/gcb.12344

Benayas JMR, Newton AC, Diaz A, Bullock JM (2009) Enhancement of biodiversity and ecosystem services by ecological restoration: a meta-analysis. Science 325:1121–1124. https://doi.org/10.1126/science.1172460

Bergstrom DM, Lucieer A, Kiefer K, Wasley J, Belbin L, Pedersen TK, Chown SL (2009) Indirect effects of invasive species removal devastate World Heritage Island. J Appl Ecol 46:3–81. https://doi.org/10.1111/j.1365-2664.2008.01601.x

Bernard-Verdier M, Hulme PE (2015) Alien and native plant species play different roles in plant community structure. J Ecol 103:143–152. https://doi.org/10.1111/1365-2745.12341

Bobeldyk AM, Rüegg J, Lamberti GA (2015) Freshwater hotspots of biological invasion are a function of species pathway interactions. Hydrobiologia 746:363–373. https://doi.org/10.1007/s10750-014-2009-z

Bolpagni R, Lastrucci L, Brundu G, Hussner A (2020) Multiple roles of alien Plants in aquatic ecosystems: from processes to modelling. Front Plant Sci 11:1299. https://doi.org/10.3389/fpls.2020.01299

Bradley BA, Mustard JF (2006) Characterizing the landscape dynamics of an invasive plant and risk of invasion using remote sensing. Ecol Appl 16:1132–1147. https://doi.org/10.1890/1051-0761(2006)016[1132:CTLDOA]2.0.CO;2

Bradley BA, Blumenthal DM, Wilcove DS, Ziska LH (2010) Predicting plant invasions in an era of global change. Trends Ecol Evol 25:310–318. https://doi.org/10.1016/j.tree.2009.12.003

Bradley BA, Blumenthal DM, Early R, Grosholz ED, Lawler JJ, Miller LP, Olden JD (2012) Global change, global trade, and the next wave of plant invasions. Front Ecol Environ 10:20–28. https://doi.org/10.1890/110145

Breitburg D, Levin LA, Oschlies A, Grégoire M, Chavez FP, Conley DJ, Garçon V, Gilbert D, Gutiérrez D, Isensee K, Jacinto GS (2018) Declining oxygen in the global ocean and coastal waters. Science 359:eaam7240. https://doi.org/10.1126/science.aam7240

Brundu G (2015) Plant invaders in European and Mediterranean inland waters: profiles, distribution, and threats. Hydrobiology 746:61–79. https://doi.org/10.1007/s10750-014-1910-9

Burgiel SW, Muir AA (2010) Invasive species, climate change and ecosystem-based adaptation: addressing multiple drivers of global change. IUCN: International Union for Conservation of Nature. Retrieved from https://policycommons.net/artifacts/1375221/invasive-species-climate-change-and-ecosystem-based-adaptation/1989482/ on 23 Nov 2022. CID: 20.500.12592/fbxd6m. https://doi.org/10.13140/2.1.1460.8161

Carvalho L, Miller C, Spears BM, Gunn ID, Bennion H, Kirika A, May L (2012) Water quality of Loch Leven: responses to enrichment, restoration and climate change. Hydrobiologia 681:35–47. https://doi.org/10.1007/s10750-011-0923

Chen S, Ding JQ (2011) Risk assessment and spread potential of alien wetland plant species *Thalia dealbata* in China. Plant Sci J 6:675–682. https://doi.org/10.3724/SP.J.1142.2011.60675

Chen DX, Coughenour MB, Eberts D, Thullen JS (1994) Interactive effects of CO_2 enrichment and temperature on the growth of dioecious *Hydrilla verticillata*. Environ Exp Bot 34:345–353. https://doi.org/10.1016/0098-8472(94)90016-7

Chen XW, Yu D, Liu CH (2016) Effect of water level fluctuation frequency on *Alternanthera philoxeroides*, *Myriophyllum aquaticum* and *Ludwigia adscendens* in autumn. Chin J Plant Ecol 40:493–501. https://doi.org/10.17521/cjpe.2015.0174

Coetzee JA, Hill MP, Byrne MJ, Bownes A (2011) A review of the biological control programmes on *Eichhornia crassipes* (C. mart.) solms (Pontederiaceae), *Salvinia molesta* DS Mitch. (Salviniaceae), *Pistia stratiotes* L.(Araceae), *Myriophyllum aquaticum* (vell.) verdc. (Haloragaceae) and *Azolla filiculoides* Lam.(Azollaceae) in South Africa. Afr Entomol 19: 451–468. https://hdl.handle.net/10520/EJC32900

Corlett RT, Westcott DA (2013) Will plant movements keep up with climate change? Trends Ecol Evol 28:482–488. https://doi.org/10.1016/j.tree.2013.04.003

Dai ZC, Zhu B, Wan JSH, Rutherford S (2022) Editorial: global changes and plant invasions. Front Ecol Evol 78. https://doi.org/10.3389/fevo.2022.845816

Davis MA, Grime JP, Thompson K (2000) Fluctuating resources in plant communities: a general theory of invasibility. J Ecol 88:528–534. https://doi.org/10.1046/j.1365-2745.2000.00473.x

Dethier MN, McDonald K, Strathmann RR (2003) Colonization and connectivity of habitat patches for coastal marine species distant from source populations. Conserv Biol 17:1024–1035. https://doi.org/10.1046/j.1523-1739.2003.01606.x

Diaz-Almela E, Marba N, Duarte CM (2007) Consequences of Mediterranean warming events in seagrass (*Posidonia oceanica*) flowering records. Glob Chang Biol 13:224–235. https://doi.org/10.1111/j.1365-2486.2006.01260.x

Espinar JL, Díaz-Delgado R, Bravo-Utrera MA, Vilà M (2015) Linking *Azolla filiculoides* invasion to increased winter temperatures in the Doñana marshland (SW Spain). Aquat Invasions 10:17–24. https://doi.org/10.3391/ai.2015.10.1.02

Ehrenfeld JG (2010) Ecosystem consequences of biological invasions. Annu Rev Ecol Evol Syst 41:59–80. https://doi.org/10.1146/annurev-ecolsys-102209-144650

Eissa AE, Zaki MM (2011) The impact of global climatic changes on the aquatic environment. Procedia Environ Sci 4:251–259. https://doi.org/10.1016/j.proenv.2011.03.030

Eiswerth ME (2005) Input-output modeling, outdoor recreation, and the economic impact of weeds. Weed Sci 53:130–137. https://doi.org/10.1614/WS-04-022R

EPPO (European and Mediterranean Plant Protection Organization) (2008) Report of a pest risk analysis: *Eichhornia crassipes* (08–14408). EPPO, Paris

Ervin G, Smothers M, Holly C, Anderson C, Linville J (2006) Relative importance of wetland type versus anthropogenic activities in determining site invisibility. Biol Invasions 8:1425–1432. https://doi.org/10.1007/s10530-006-0006-5

Felton A, Knight E, Wood J, Zammit C, Lindenmayer D (2010) A meta-analysis of fauna and flora species richness and abundance in plantations and pasture lands. Biol Conserv 143:545–554. https://doi.org/10.1016/j.biocon.2009.11.030

Finch DM, Butler JL, Runyon JB, Fettig CJ, Kilkenny FF, Jose S, Frankel SJ, Cushman SA, Cobb RC, Dukes JS, Hicke JA, Amelon SK (2021) Effects of climate change on invasive species. In: Invasive species in forests and Rangelands of the United States. Springer International Publishing, pp 57–83. https://doi.org/10.1007/978-3-030-45367-1_4

Fraser GC, Hill MP, Martin JA (2016) Economic evaluation of water loss saving due to the biological control of water hyacinth at New Year's Dam, Eastern Cape province, South Africa. Afr J Aquat Sci 41:227–234. https://doi.org/10.2989/16085914.2016.1151765

Gaertner M, Nottebrock H, Fourie H, Privett SD, Richardson DM (2012) Plant invasions, restoration, and economics: perspectives from South African fynbos. Perspect Plant Ecol Evol Syst 14: 341–353. https://doi.org/10.1016/j.ppees.2012.05.001

Gallardo B, Clavero M, Sánchez MI, Vilà M (2016) Global ecological impacts of invasive species in aquatic ecosystems. Glob Change Biol 22:151–163

Gao JQ, Yang XG, Dong ZY, Li KN (2015) Precipitation resource changed characteristics in arid and humid regions in Northern China with climate changes. Trans Chin Soc Agric Eng 31:99–110. https://doi.org/10.11975/j.issn.1002-6819.2015.12.014

Gattuso J-P, Magnan A, Billé R, Cheung WWL, Howes EL, Joos D, Allemand D, Bopp L, Cooley SR, Eakin CM, Hoegh-Guldberg O, Kelly RP, Portner HO, Rogers AD, Baxter JM, Laffoley D, Osborn D, Rankovic A, Rochette J, Sumaila UR, Treyer S, Turley C (2015) Contrasting futures for ocean and society from different anthropogenic CO2 emissions scenarios. Science 349: aac4722. https://doi.org/10.1126/science.aac4722

Getsinger KD, Dibble E, Rodgers JH, Spencer D (2014) Benefits of controlling nuisance aquatic plants and algae in the United States. Council of Agricultural Science and Technology (CAST) Commentary, QTA, Ames, IA, pp 1–12

Gherardi F (2007) Biological invasions in inland waters: an overview. In: Gherardi F (ed) Biological invaders in inland waters: profiles, distribution, and threats. Springer, Netherlands, pp 3–25. https://doi.org/10.1007/978-1-4020-609-8_1

Ghosh S, Chatterjee S, Prasad GS, Pal P (2020) Effect of climate change on aquatic ecosystem and production of fisheries. In: Devlin A, Pan J, Shah M (eds) Inland waters-dynamics and ecology. IntechOpen, London. https://doi.org/10.5772/intechopen.93784

Giller PS, Hillebrand H, Berninger UG, Gessner OM, Hawkins S, Inchausti P, Inglis C, Leslie H, Malmqvist B, Monaghan TM, Morin JP (2004) Biodiversity effects on ecosystem functioning: emerging issues and their experimental test in aquatic environments. Oikos 104:423–436. https://doi.org/10.1111/j.0030-1299.2004.13253.x

Gioria M, Pyšek P, Osborne BA (2018) Timing is everything: does early and late germination favor invasions by herbaceous alien plants? J Plant Ecol 11:4–16. https://doi.org/10.1093/jpe/rtw105

Griffith AW, Gobler CJ (2020) Harmful algal blooms: a climate change co-stressor in marine and freshwater ecosystems. Harmful Algae 91:101590. https://doi.org/10.1016/j.hal.2019.03.008

Hassan A, Nawchoo IA (2020) Impact of invasive plants in aquatic ecosystems. In: Hakeem KR, Bhat RA, Qadri H (eds) Bioremediation and biotechnology. Springer, Cham, pp 55–73. https://doi.org/10.1007/978-3-030-35691-0_3

Havel JE, Kovalenko KE, Thomaz SM, Amalfitano S, Kats LB (2015) Aquatic invasive species: challenges for the future. Hydrobiologia 750:147–170. https://doi.org/10.1007/s10750-014-2166-0

Hellmann JJ, Byers JE, Bierwagen BG, Dukes JS (2008) Five potential consequences of climate change for invasive species. Conserv Biol 22:534–543. https://doi.org/10.1111/j.1523-1739.2008.00951.x

Henriksen JW, Lim DS, Lu XM, Ding JQ, Siemann E (2018) Strong effects of hydrologic environment and weak effects of elevated CO2 on the invasive weed *Alternanthera philoxeroides* and the biocontrol beetle *Agasicles hygrophila*. Arthropod Plant Int 12:691–700. https://doi.org/10.1007/s11829-018-9614-0

Hobbs RJ, Richardson DM (2011) Invasion ecology and restoration ecology: Parallel evolution in two fields of endeavour. In: Richardson DM (ed) Fifty years of invasion ecology. The legacy of Charles Elton. Wiley-Blackwell, Oxford, pp 61–69. https://doi.org/10.1002/9781444329988.ch6

Hoegh-Guldberg O, Caldeira K, Chopin T, Gaines S, Haugan P, Hemer M, Tyedmers P (2019) The ocean as a solution to climate change: five opportunities for action. Report. World Resources Institute, Washington, DC. https://doi.org/10.1111/gcb.13004

Holl KD, Aide TM (2011) When and where to actively restore ecosystems? For Ecol Manag 261: 1558–1563. https://doi.org/10.1016/j.foreco.2010.07.004

Hulme PE (2009) Trade, transport and trouble: managing invasive species pathways in an era of globalization. J Appl Ecol 46:10–18. https://doi.org/10.1111/j.1365-2664.2008.01600.x

Hulme PE, Nentwig W, Pyšek P, Vilà M (2009) Common market, shared problems: time for a coordinated response to biological invasions in Europe? Neobiota 8:3–19

Hussner A (2014) Long-term macrophyte mapping documents a continuously shift from native to non-native aquatic plant dominance in the thermally abnormal River Erft (North Rhine-Westphalia, Germany). Limnol Ecol Manag Inland Waters 48:39–45. https://doi.org/10.1016/j.limno.2014.05.003

Hussner A, Lösch R (2005) Alien aquatic plants in a thermally abnormal river and their assembly to neophyte-dominated macrophyte stands (River Erft, Northrhine Westphalia). Limnologica 35: 18–30. https://doi.org/10.1016/j.limno.2005.01.001

IPBES (2019) Summary for policymakers of the global assessment report on biodiversity and ecosystem services of the Intergovernmental Science-Policy Platform on Biodiversity and Ecosystem Services. In: Díaz S, Settele J, Brondizio ES, Ngo HT, Guèze M, Agard J, Arneth A, Balvanera P, Brauman KA, Butchart SHM et al (eds) IPBES secretariat. IPBES, Bonn, Germany. (Accessed 27th September, 2019)

Kail J, Brabec K, Poppe M, Januschke K (2015) The effect of river restoration on fish, macroinvertebrates and aquatic macrophytes: a meta-analysis. Ecol Indic 58:311–321. https://doi.org/10.1016/j.ecolind.2015.06.011

Kariyawasam CS, Kumar L, Kogo BK, Ratnayake SS (2021) Long-term changes of aquatic invasive plants and implications for future distribution: a case study using a tank cascade system in Sri Lanka. Climate 9:31. https://doi.org/10.3390/cli9020031

Kercher SM, Zedler JB (2004) Multiple disturbances accelerate invasion of reed canary grass (*Phalaris arundinacea* L.) in a mesocosm study. Oecologia 138:455–464. https://doi.org/10.1007/s00442-003-1453-7

Kettenring KM, Adams CR (2011) Lessons learned from invasive plant control experiments: a systematic review and meta-analysis. J Appl Ecol 48:970–979. https://doi.org/10.1111/j.1365-2664.2011.01979.x

Kietzka GJ, Pryke JS, Samways MJ (2015) Landscape ecological networks are successful in supporting a diverse dragonfly assemblage. Insect Conserv Diver 8:229–237. https://doi.org/10.1111/icad.12099

Koger CH, Shaw DR, Reddy KN, Bruce LM (2004) Detection of pitted morning glory (*Ipomoea lacunosa*) with hyperspectral remote sensing. II. Effects of vegetation ground cover and reflectance properties. Weed Sci 52:230–235. https://doi.org/10.1614/WS-03-083R1

Kohli RK, Batish DR, Singh JS, Singh HP, Bhatt JR (2011) Plant invasion in India: an overview. In: Bhatt JR, Singh JS, Singh SP, Tripathi RS, Kohli RK (eds) Invasive alien plants: an ecological appraisal for the Indian Subcontinent. CABI, pp 1–9. ISBN:978-1-84593-907-6

Lacoul P, Freedman B (2006) Recent observation of a proliferation of *Ranunculus trichophyllus* Chaix. in high-altitude lakes of the Mount Everest region. Arct Antarct Alp Res 38:94–398

Lavergne S, Molofsky J (2006) Control strategies for the invasive reed canary grass (*Phalaris arundinacea* L.) in North American wetlands: the need for an integrated management plan. Nat Areas J 26:208–214. https://doi.org/10.3375/0885-8608(2006)26[208:CSFTIR]2.0.CO;2

Lázaro-Lobo A, Ervin GN (2021) Wetland invasion: a multi-faceted challenge during a time of rapid global change. Wetlands 41:64. https://doi.org/10.1007/s13157-021-01462-1

Levine JMD, Antonio CM (2003) Forecasting biological invasion with increasing international trade. Conserv Biol 17:322–326. https://doi.org/10.1046/j.1523-1739.2003.02038.x

Liu Z, Hu J, Zhong P, Zhang X, Ning J, Larsen SE, Chen D, Gao Y, He H, Jeppesen E (2018) Successful restoration of a tropical shallow eutrophic lake: strong bottom-up but weak top-down effects recorded. Water Res 146:88–97. https://doi.org/10.1016/j.watres.2018.09.007

Lloret F, Dail F, Brundu G, Hulme PE (2004) Local and regional abundance of exotic plant species on Mediterranean islands: are species traits important? Glob Ecol Biogeogr 13:37–45. https://doi.org/10.1111/j.1466-882X.2004.00064.x

Lonsdale WM (1993) Rates of spread of an invading species—*Mimosa pigra* in northern Australia. J Ecol 1:513–521. https://doi.org/10.2307/2261529

Lu XM, Siemann E, He MY, Wei H, Shao X, Ding JQ (2015) Climate warming increases biological control agent impact on a non-target species. Ecol Lett 18:48–56. https://doi.org/10.1111/ele.12391

MacDougall AS, Turkington R (2005) Are invasive species the drivers orpassengers of change in degraded ecosystems? Ecology 81:42–55. https://doi.org/10.1890/04-0669

Madsen JD, Sutherland JW, Bloomfield JA, Eichler LW, Boylen CW (1991) The decline of native vegetation under dense Eurasian watermilfoil canopies. J Aquat Plant Manag 29:94–99

Mainali KP, Warren DL, Dhileepan K, McConnachie A, Strathie L, Hassan G, Karki D, Shrestha BB, Parmesan C (2015) Projecting future expansion of invasive species: comparing and improving methodologies for species distribution modeling. Glob Chang Biol 21:4464–4480. https://doi.org/10.1111/gcb.13038

Martin GD, Coetzee JA, Weyl PS, Parkinson MC, Hill MP (2018) Biological control of *Salvinia molesta* in South Africa revisited. Biol Control 125:74–80. https://doi.org/10.1016/j.biocontrol.2018.06.011

Maurer DA, Zedler JB (2002) Differential invasion of a wetland grass explained by tests of nutrients and light availability on establishment and clonal growth. Oecologia 131:279–288. https://doi.org/10.1007/s00442-002-0886-8

McFarland DJ, Barko JW (1999) High-temperature effects on growth and propagule formation in hydrilla biotypes. J Aquat Plant Manage 37:17–35

McKeon CS, Weber MX, Alter SE, Seavy NE, Crandall ED, Barshis DJ, Oleson KL (2016) Melting barriers to faunal exchange across ocean basins. Glob Chang Biol 22:465–473. https://doi.org/10.1111/gcb.13116

McNeely JA (2000) The future of alien invasive species: changing social views. In: Mooney A, Hobbs RJ (eds) Invasive species in a changing world. Island Press, Washington, DC, pp 171–189

McNeely JA (2001) Invasive species: a costly catastrophe for native biodiversity. Land Use Water Resour Res 1:1–10. https://doi.org/10.22004/ag.econ.47850

Meli P, Rey Benayas JM, Balvanera P, Martínez Ramos M (2014) Restoration enhances wetland biodiversity and ecosystem service supply, but results are context-dependent: a meta-analysis. PLoS One 9:1–9. https://doi.org/10.1371/journal.pone.0093507

Mgidi T (2004) An assessment of invasion potential of invasive alien plant species in South Africa. CSIR-Environmentek, Pretoria, South Africa

Miller SW, Budy P, Schmidt JC (2010) Quantifying macroinvertebrate responses to in-stream habitat restoration: applications of meta-analysis to river restoration. Restor Ecol 18:8–19. https://doi.org/10.1111/j.1526-100X.2009.00605.x

Milton SJ (2004) Grasses as invasive alien plants in South Africa. S Afr J Sci 100:69–75

Moorhouse TP, Macdonald DW (2015) Are invasives worse in freshwater than terrestrial ecosystems? WIREs Water 2015:1–8. https://doi.org/10.1002/wat2.1059

Nel JL, Richardson DM, Rouget M, Mgidi TN, Mdzeke N, Le Maitre DC, Van Wilgen BW, Schonegevel L, Henderson L, Neser S (2004) A proposed classification of invasive alien plant species in South Africa: towards prioritising species and areas for management action. South Afr J Sci 100:53–64

Newson SE, Mendes S, Crick HQP, Dulvy NK, Houghton JDR, Hays GC, Hutson AM, Macleod CD, Pierce GC, Robinson RA (2009) Indicators of the impact of climate change on migratory species. Endang Species Res 7:101–113. https://doi.org/10.3354/esr00162

Nsikani MM, Gaertner M, Kritzinger-Klopper S, Ngubane NP, Esler KJ (2019) Secondary invasion after clearing invasive *Acacia saligna* in the South African fynbos. S Afr J Bot 125:280–289. https://doi.org/10.1016/j.sajb.2019.07.034

Olden JD, McCarthy JM, Maxted JT, Fetzer WW, Vander Zanden MJ (2006) The rapid spread of rusty crayfish (*Orconectes rusticus*) with observations on native crayfish declines in Wisconsin (U.S.A.) over the past 130 years. Biol Invasions 8:1621–1628. https://doi.org/10.1007/s10530-005-7854-2

Orbán I, Szitár K, Kalapos T, Körel-Dulay G (2021) The role of disturbance in invasive plant establishment in a changing climate: insights from a drought experiment. Biol Invasions 23:1877–1890. https://doi.org/10.1007/s10530-021-02478-8

Pearson DE, Ortega YK, Runyon JB, Butler JL (2016) Secondary invasion: the bane of weed management. Biol Conserv 197:8–17. https://doi.org/10.1016/j.biocon.2016.02.029

Pejchar L, Mooney HA (2009) Invasive species, ecosystem services and human well-being. Trends Ecol Evol 24:497–504. https://doi.org/10.1016/j.tree.2009.03.016

Perrings C, Williamson M, Barbier EB, Delfino D, Dalmazzone S, Shogren J, Simmons P, Watkinson A (2002) Biological invasions risks and the public good: an economic perspective. Conserv Ecol 6:1. https://doi.org/10.5751/ES-00396-060101

Pimentel D, McNair S, Janecka J, Wightman J, Simmonds C, O'connell C, Wong E, Russel L, Zern J, Aquino T, Tsomondo T (2001) Economic and environmental threats of alien plant, animal, and microbe invasions. Agric Ecosyst Environ 84:1–20. https://doi.org/10.1016/S0167-8809(00)00178-X

Pimentel D, Zuniga A, Morrison D (2005) Update on the environmental and economic costs associated with alien-invasive species in the United States. Ecol Econ 52:273–288. https://doi.org/10.1016/j.ecolecon.2004.10.002

Poland TM, Patel-Weynand T, Finch DM, Miniat CF, Hayes DC, Lopez VM (2021) Invasive species in forests and Rangelands of the United States: a comprehensive science synthesis for the United States Forest Sector. Springer Nature, p 455. https://doi.org/10.1007/978-3-030-45367-1

Prior KM, Adams DC, Klepzig KD, Hulcr J (2018) When does invasive species removal lead to ecological recovery? Implications for management success. Biol Invasions 20:267–283. https://doi.org/10.1007/s10530-017-1542-x

Pyšek P, Richardson DM (2010) Invasive species, environmental change and management, and health. Annu Rev Environ Resour 35:25–55. https://doi.org/10.1146/annurev-environ-033009-095548

Quinn LD, Holt JS (2008) Ecological correlates of invasion by *Arundo donax* in three southern California riparian habitats. Biol Invasions 10:591–601. https://doi.org/10.1007/s10530-007-9155-4

Raghubanshi AS, Rai LC, Gaur JP, Singh JS (2005) Invasive alien species and biodiversity in India. Curr Sci 88:539–540

Rahel FJ, Olden JD (2008) Assessing the effects of climate change on aquatic invasive species. Conserv Biol 22:521–533. https://doi.org/10.1111/j.1523-1739.2008.00950.x

Rai PK, Singh JS (2020) Invasive alien plant species: their impact on environment, ecosystem services and human health. Ecol Indic 111:06020. https://doi.org/10.1016/j.ecolind.2019.106020

Rai PK, Kumar V, Tsang YF, Naddem OY, Kim JH, Tsang YF (2018) Nanoparticle-plant interaction: implications in energy, the environment, and agriculture. Environ Int 119:1–19. https://doi.org/10.1016/j.envint.2018.06.012

Reddy CS (2008) Catalogue of invasive alien flora of India. Life Sci 5:8–89. ISSN: 1097–8135

Ricciardi A, MacIsaac HJ (2011) Impacts of biological invasions on freshwater ecosystems. In: Richardson D (ed) Fifty years of invasion ecology: the legacy of Charles Elton, 1st edn. Blackwell Publishing Ltd, West Sussex, pp 211–224

Richards CL, Bossdorf O, Muth NZ, Gurevitch J, Pigliucci M (2006) Jack of all trades, master of some? On the role of phenotypic plasticity in plant invasions. Ecol Lett 9:981–993. https://doi.org/10.1111/j.1461-0248.2006.00950.x

Richardson DM, Moran VC, Le Maitre DC, Rouget M, Foxcroft LC (2004) Recent developments in the science and management of invasive alien plants: connecting the dots of research knowledge, and linking disciplinary boxes: working for water. S Afr J Sci 100:26–128

Roessig JM, Woodley CM, Cech JJ, Hansen LJ (2004) Effects of global climate change on marine and estuarine fishes and fisheries. Rev Fish Biol Fisheries 14:251–275. https://doi.org/10.1007/s11160-004-6749-0

Ruiz-Jaen MC, Mitchell Aide T (2005) Restoration success: how is it being measured? Restor Ecol 13:569–577. https://doi.org/10.1111/j.1526-100X.2005.00072.x

Ruwanza S, Gaertner M, Esler KJ, Richardson DM (2013) The effectiveness of active and passive restoration on recovery of indigenous vegetation in riparian zones in the Western Cape, South Africa: a preliminary assessment. S Afr J Bot 88:132–141. https://doi.org/10.1016/j.sajb.2013.06.022

Sala OE, Chapin FS, Armesto JJ, Berlow E, Bloomfield J, Dirzo R, Huber-Sanwald E, Huenneke LF, Jackson RB, Kinzig A, Leemans R, Lodge DM, Mooney HA, Oesterheld M, Poff NL, Sykes MT, Walker BH, Walker M, Wall DH (2000) Global biodiversity scenarios for the year 2100. Science 287:1770–1774. https://doi.org/10.1126/science.287.5459.1770

Santos M, Anderson JLW, Ustin SL (2011) Effects of invasive species on plant communities: an example using submersed aquatic plants at the regional scale. Biol Invasions 13:443–457. https://doi.org/10.1007/s10530-010-9840-6

Scavia D, Field JC, Boesch DF, Buddemeier RW, Burkett V, Cayan DR, Fogarty M, Harwell M, Howarth RW, Mason C, Reed DJ, Royer T, Salinger AH, Titus JG (2002) Climate change impacts on U.S. Coastal and marine ecosystems. Estuaries 25:149–164. https://doi.org/10.1007/BF02691304

Shackleton RT, Biggs R, Richardson DM, Larson BMH (2018) Social-ecological drivers and impacts of invasion-related regime shifts: consequences for ecosystem services and human wellbeing. Environ Sci Pol 89:300–314. https://doi.org/10.1016/j.envsci.2018.08.005

Shackleton RT, Shackleton CM, Kull CA (2019) The role of invasive alien species in shaping local livelihoods and human well-being: a review. J Environ Manag 229:145–157. https://doi.org/10.1016/j.jenvman.2018.05.007

Shaw JD, Wilson JR, Richardson DM (2010) Initiating dialogue between scientists and managers of biological invasions. Biol Invasions 12:4077–4083. https://doi.org/10.1007/s10530-010-9821-9

Shillinglaw SN (1981) Dissolved oxygen depletion and nutrient uptake in an impoundment infested with *Eichhornia crassipes* (Mart.) Solms. J Limnol Soc S Afr 7:63–66. https://doi.org/10.1080/03779688.1981.9632945

Singh AK (2021) Management of alien aquatic invasive species: strategic guidelines and policy in India. Aquat Ecosyst Health Manag 24:86–95. https://doi.org/10.14321/aehm.024.02.12

Sivaramanan S (2015) Global warming and climate change, causes, impacts and mitigation, vol 2. Central Environmental Authority. https://doi.org/10.13140/RG.2.1.4889.7128

Smith AL, Hewitt N, Klenk N, Bazely DR, Yan N, Wood S, Henriques I, MacLellan JI, Lipsig-Mummé C (2012) Effects of climate change on the distribution of invasive alien species in Canada: a knowledge synthesis of range change projections in a warming world. Environ Rev 20:1–16. https://doi.org/10.1139/a11-020

Stabenau N, Zehnsdorf A, Rönicke H, Wedwitschka H, Moeller L, Ibrahim B, Stinner W (2018) A potential phosphorous fertilizer for organic farming: recovery of phosphorous resources in the course of bioenergy production through anaerobic digestion of aquatic macrophytes. Energy Sustain Soc 8:16. https://doi.org/10.1186/s13705-018-0155-2

Stiers I, Crohain N, Josens G, Triest L (2011) Impact of three aquatic invasive species on native plants and macroinvertebrates in temperate ponds. Biol Invas 13:2715–2726. https://doi.org/10.1007/s10530-011-9942-9

Stocker TF, Qin D, Plattner GK et al (2013) Climate change 2013: The physical science basis. Contribution of working group I to the fifth assessment report of the intergovernmental panel on climate change. Cambridge University Press, Cambridge

Strecker AL, Campbell PM, Olden JD (2011) The aquarium trade as an invasion pathway in the Pacific Northwest. Fisheries 36:74–85. https://doi.org/10.1577/03632415.2011.10389070

Suding KN (2011) Toward an era of restoration in ecology: successes, failures, and opportunities ahead. Annu Rev Ecol Evol Syst 42:465–487. https://doi.org/10.1146/annurev-ecolsys-102710-145115

Thomaz SM, Agostinho AA, Gomes LC, Silveira MJ, Rejmánek M, Aslan CE, Chow E (2012) Using space-for-time substitution and time sequence approaches in invasion ecology. Freshw Biol 57:2401–2410. https://doi.org/10.1111/fwb.12005

Tylianakis JM, Didham RK, Bascompte J, Wardle DA (2008) Global change and species interactions in terrestrial ecosystems. Ecol Lett 11:1351–1363. https://doi.org/10.1111/j.1461-0248.2008.01250.x

Vander Zanden MJ, Casselman JM, Rasmussen JB (1999) Stable isotope evidence for the food web consequences of species invasions in lakes. Nature 401:464–467

Velthuis M, Kosten S, Aben R, Kazanjian G, Hilt S, Peeters ET, van Donk E, Bakker ES (2018) Warming enhances sedimentation and decomposition of organic carbon in shallow macrophyte-dominated systems with zero net effect on carbon burial. Glob Change Biol 24:5231–5242. https://doi.org/10.1111/gcb.14387

Vilà M, Gimeno I (2007) Does invasion by an alien plant species affect the soil seed bank? J Veg Sci 18:423–430. https://doi.org/10.1111/j.1654-1103.2007.tb02554.x

Villamagna AM, Murphy BR (2010) Ecological and socio-economic impacts of invasive water hyacinth *E(ichhornia crassipes*): a review. Freshw Biol 55:282–298. https://doi.org/10.1111/j.1365-2427.2009.02294.x

Vojtkó AE, Mesterházy A, Süveges K, Valkó O, Lukács BA (2017) Changes in sediment seed-bank composition of invaded macrophyte communities in a thermal river. Freshw Biol 62:1024–1035. https://doi.org/10.1111/fwb.12922

Wainwright CE, Wolkovich EM, Cleland EE (2012) Seasonal priority effects: implications for invasion and restoration in a semi-arid system. J Appl Ecol 49:234–241. https://doi.org/10.1111/j.1365-2664.2011.02088.x

Waldbusser GG, Salisbury JE (2014) Ocean acidification in the coastal zone from an organism's perspective: multiple system parameters, frequency domains, and habitats. Annu Rev Mar Sci 6: 221–247. https://doi.org/10.1146/annurev-marine-121211-172238

Walther GR, Roques A, Hulme PE, Sykes MT, Pyšek P, Kühn I, Zobel M, Bacher S, Botta-Dukát Z, Bugmann H, Czúcz B (2009) Alien species in a warmer world: risks and opportunities. Trends Ecol Evol 24:686–693. https://doi.org/10.1016/j.tree.2009.06.008

Wang H, Wang Q, Bowler PA, Xiong W (2016) Invasive aquatic plants in China. Aquat Invas 11: 1–9. https://doi.org/10.3391/ai.2016.11.1.01

Weatherdon LV, Magnan AK, Rogers AD, Sumaila UR, Cheung WWL (2016) Observed and projected impacts of climate change on marine fisheries, aquaculture, coastal tourism, and human health: an update. Front Mar Sci 3:48. https://doi.org/10.3389/fmars.2016.00048

Wijte AH, Mizutani T, Motamed ER, Merryfield ML, Miller DE, Alexander DE (2005) Temperature and endogenous factors cause seasonal patterns in rooting by stem fragments of the invasive giant reed, *Arundo donax* (Poaceae). IJPS 166:507–517. https://doi.org/10.1086/428915

Wilcox KL, Petrie SA, Maynard LA, Meyer SW (2003) Historical distribution and abundance of *Phragmites australis* at Long Point, Lake Erie, Ontario. J Great Lakes Res 29:664–680. https://doi.org/10.1016/S0380-1330(03)70469-9

Willby NJ (2007) Managing invasive aquatic plants: problems and prospects. Aquat Conserv 17: 659–665. https://doi.org/10.1002/aqc.913

Williams AE, Duthie HC, Hecky RE (2005) Water hyacinth in Lake Victoria: Why did it vanish so quickly and will it return? Aquat Bot 81:300–314

Wootton JT, Emmerson M (2005) Measurement of interaction strength in nature. Annu Rev Ecol Evol Syst 6:419–444. https://doi.org/10.1146/annurev.ecolsys.36.091704.175535

Wu H, Ding J (2019) Global change sharpens the double-edged sword effect of aquatic alien plants in China and beyond. Front Plant Sci 10:787. https://doi.org/10.3389/fpls.2019.00787

Wu H, Carrillo J, Ding JQ (2017a) Species diversity and environmental determinants of aquatic and terrestrial communities invaded by *Alternanthera philoxeroides*. Sci Total Environ 581–582: 666–675. https://doi.org/10.1016/j.scitotenv.2016.12.177

Wu H, Ismail M, Ding JQ (2017b) Global warming increases the interspecific competitiveness of the invasive plant alligator weed, *Alternanthera philoxeroides*. Sci Total Environ 575:1415–1422. https://doi.org/10.1016/j.scitotenv.2016.09.226

You W, Yu D, Xie D, Yu L (2013a) Overwintering survival and regrowth of the invasive plant *Eichhornia crassipes* are enhanced by experimental warming in winter. Aquat Biol 19:45–53. https://doi.org/10.3354/ab00519

You WH, Yu D, Liu CH, Xie D, Xiong W (2013b) Clonal integration facilitates invasiveness of the alien aquatic plant *Myriophyllum aquaticum* L. Under heterogeneous water availability. Hydrobiologia 718:27–39. https://doi.org/10.1007/s10750-013-1596-4

You W, Yu D, Xie D, Yu L, Xiong W, Han C (2014) Responses of the invasive aquatic plant water hyacinth to altered nutrient levels under experimental warming in China. Aquat Bot 119:51–56. https://doi.org/10.1016/j.aquabot.2014.06.004

Young AM, Larson BMH (2011) Clarifying debates in invasion biology: a survey of invasion biologists. Environ Res 111:893–898. https://doi.org/10.1016/j.envres.2011.06.006

Yu GL (2011) Effects of waterlogging on intraspecific interactions of the clonal herb *Alternanthera philoxeroides*. Chin J Plant Ecol 35:973–980. https://doi.org/10.3724/SP.J.1258.2011.00973

Zaiko A, Daunys D (2015) Invasive ecosystem engineers and biotic indices: giving a wrong impression of water quality improvement? Ecol Indic 52:292–299. https://doi.org/10.1016/j.ecolind.2014.12.023

Zedler JB, Kercher S (2004) Causes and consequences of invasive plants in wetlands: opportunities, opportunists, and outcomes. Crit Rev Plant Sci 23:431–452. https://doi.org/10.1080/07352680490514673

Plant Invasion and Soil Processes: A Mechanistic Understanding

10

Talat Afreen, Sweety Kumari, Rahul Bhadouria, Rajkumari S. Devi, Swati Singh, and Sachchidanand Tripathi

Abstract

In current scenario of changing climate, invasive plants are affecting the global environment by interfering with the biological community structure and composition and soil processes unprecedently. Plant invasions have dramatically threatened native flora and are responsible for loss in biodiversity and ecosystem functioning across the globe. Invasive plants are equipped with rapid growth, high self-regeneration capability and competitive edge in resource acquisition, higher reproduction capacity, multi-resource consumption, and adaptability to diverse niches. It is reported that invasive plants negatively impact agricultural crops, soil nutrients and microbial communities, and consequently overall soil health. They can further influence soil nutrient cycling by altering the soil microbial population through allelopathy and extension. Studies indicate that presence of invasive plants has tendency to promote the activity of N and P metabolizing enzymes. Moreover, plant invasion substantially supports the bacterial and fungal diversity and hence have determining impact on soil processes. Exhibition of variation in their functional traits provides invasive plants a

T. Afreen · S. Kumari
PG Centre of Botany, Gaya College (A Constituent Unit of Magadh University), Bodhgaya, Bihar, India

R. Bhadouria
Department of Environmental Studies, Delhi College of Arts and Commerce, University of Delhi, New Delhi, India

R. S. Devi · S. Tripathi (✉)
Department of Botany, Deen Dayal Upadhyaya College (University of Delhi), New Delhi, India
e-mail: rajkumarisana@ddu.du.ac.in; drsachchidanandtripathi@ddu.du.ac.in

S. Singh
Central Pollution Control Board, Delhi, India

S. Tripathi et al. (eds.), *Plant Invasions and Global Climate Change*,
https://doi.org/10.1007/978-981-99-5910-5_10

competitive edge against the native plants. Investigations have been done to understand the successful mechanisms of plant invasions and to comprehend why several plant species have more invasive potential than others. This chapter focuses on the impact of invasive plants on soil nutrient profile, microbial activities, and phytodiversity.

Keywords

Climate change · Functional traits · Invasive plants · Microbial community · Nutrient cycling · Plant diversity · Soil nutrient

10.1 Introduction

In the era of climate change, plant invasion is affecting the global environment by interfering with the biological community structure and composition and soil processes unprecedently (Guido and Pillar 2017; Sardans et al. 2017; Zhou and Staver 2019; McLeod et al. 2021; Fig. 10.1). Further, they have threatened native flora and are responsible for loss in biodiversity and ecosystem functioning across the globe (Gaertner et al. 2009; Divíšek et al. 2018; McLeod et al. 2021). Invasive plants are equipped with rapid growth, high self-regeneration capability and competitive edge in resource acquisition, higher reproduction capacity, multi-resource consumption, and adaptability to diverse niches (Rai 2015; Sardans et al. 2017; McLeod et al. 2021; Sperry et al. 2021).

Studies indicate that invasive plants negatively impact agricultural crops, soil nutrients and microbial communities, and consequently overall soil health

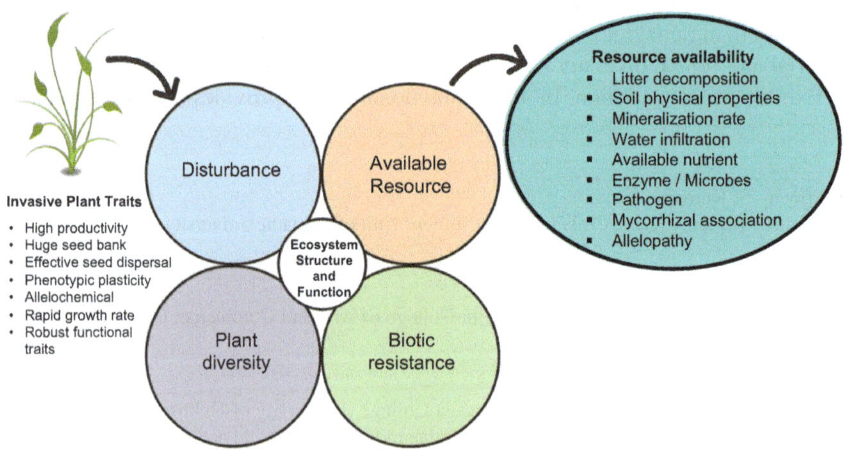

Fig. 10.1 Invasive species traits and ecosystem structure/function along with soil properties that facilitate invasion processes in recipient ecosystem

(Rodríguez-Caballero et al. 2017; Sardans et al. 2017; Zhou and Staver 2019; Jesse et al. 2020). Sufficient availability of soil nutrients and required microbial communities are critical for plant growth and productivity (Tilman 1990; Vilà et al. 2011; Zhang et al. 2019a, b). Therefore, it has been observed that to acquire a competitive edge, the invasive plants enhance their resource use efficiency or can alter the soil nutrient profile by altering the quality and quantity of the nutrients (Ens et al. 2015; Zhou and Staver 2019). Further, the role of soil microbiota in maintaining the soil nutrient profile through nutrient cycling is well known (Zhou and Staver 2019). It has been reported that invasive plants can influence soil nutrient cycling by altering the soil microbial population through allelopathy and extension (Thorpe et al. 2009; Tian et al. 2021; Tamura and Tharayil 2014). It is found recently that plant invasion substantially supports the bacterial and fungal diversity and hence have determining impact on soil processes (Stefanowicz et al. 2016, 2017). Further, studies indicate that presence of invasive plants has tendency to promote the activity of N and P metabolizing enzymes as compared to the C acquiring enzyme (Saiya-Cork et al. 2002; Zhou and Staver 2019). This may be due to the diversity of enzyme function in soil nutrient cycling and litter decomposition.

It is a common fact that soil moisture and soil texture play an important role in plant growth and development, and soil processes; however, the impact of invasive plants on these soil properties need to be understood (Liao et al. 2008; Prescott and Zukswert 2016; Zhai et al. 2020). Therefore, studies focusing on the response of soil microbiota and soil physicochemical properties to plant invasion will assist in generating a comprehensive understanding on the alterations occurring in soil microbial activity and nutrient cycling and consequently the impact on phyto-diversity. Recently, investigations have been done to understand the successful mechanisms of plant invasions and to comprehend why several plant species have more invasive potential than others (Rai and Singh 2020; Kato-Noguchi and Kurniadie 2021; McLeod et al. 2021). This may be augmenting in mitigating the impacts of plant invasion on native plant communities and in the management of further spread of invasive species. It has been observed that some of the invasive plant species promote rhizospheric soil microbial activity and community structure, thereby strengthening the soil nutrient profile and facilitate further invasion (Prescott and Zukswert 2016; Sun et al. 2019; Zhang et al. 2019a, b; Li et al. 2022a, b). Moreover, an understanding of a functional and active rhizospheric microbial community structure of invasive plants may help in developing biocontrol measures for these invaders (Dawkins et al. 2022). Overall, majority of the belowground mechanisms adopted by invasive plants have a significant impact on soil microbiota, physicochemical and biogeochemical properties of the soil in the recently invaded community.

Further, invasive plants exhibit variation in their functional traits (van Kleunen et al. 2010; Mathakutha et al. 2019; Vasquez-Valderrama et al. 2020) which provides them a competitive edge against the native plants, and therefore, such characteristics have led to the development of diverse but interlinked hypothesis, such as the enemy release hypothesis (Keane and Crawley 2002) and the resource hypothesis (Davis et al. 2000). It is suggested that disturbance events like fire, logging, forest clearing, and ploughing for agricultural lands may provide a resource rich environment to the fast-growing plant invaders, thereby assisting them to acquire diverse niches (Blumenthal and Kray 2014; Sardans et al. 2017). This chapter focuses on the impact of invasive plant species on soil nutrient profile, microbial activities, and plant diversity.

10.2 Plant Invasion and Soil Physicochemical and Biological Properties

Plant invasion significantly affects the abundance, diversity, and structure of biological communities (Lenda et al. 2013; Afreen et al. 2018; Dar et al. 2023), along with considerable modification of soil's physicochemical and biological properties and ultimately altering the key ecosystem processes (Vitousek 1990; Osunkoya and Perrett 2011; Jandová et al. 2014; Stefanowicz et al. 2017, 2018; Carboni et al. 2021; Table 10.1). It has been observed by many studies that invasive species modify the soil nutrient pool by enhancing the soil nutrients (C, N, P) along with acceleration in soil processes and litter decomposition, and mineralization dynamics (Liao et al. 2008; Zhang et al. 2019a, b; Kumar and Garkoti 2022). This impetus provided to soil nutrient cycling, particularly in C and N cycle regulation by invasive plants may be due to robustness in their functional attributes, viz. physiological, leaf area and above ground allocation, and relative growth rate and net assimilation rate, size, and fitness (Liao et al. 2008; Afreen et al. 2018; Kaushik et al. 2022). However, the impact of invasive plants on soil processes are species-specific and may be affected by the interactive effect of habitat soil, duration of the invasion, precipitation pattern, and seasonality (Liao et al. 2008; Stefanowicz et al. 2016; Bradley et al. 2018; Spear et al. 2021). Stefanowicz et al. (2020) suggested that the response of an ecosystem to plant invasion may be site-specific and rely on the initial properties of the concerned ecosystem. These alterations in soil properties inflicted by invasive plants may further assist or accelerated the invasion process.

It has been observed that the changes brought about by invasive plants may exist even after its eradication and pose challenges for the recolonization of native flora. This process is called 'invasive plant legacy' and has negative implication for restoration and management of invaded sites (Corbin and D'Antonio 2012; Perkins and Hatfield 2014; Hess et al. 2019). Xu et al. (2021) suggested that nitrogen deposition scenario may facilitate invasive plants to establish soil legacies by hijacking nitrogen absorption that may suppress the advantages of soil microbes. In a similar study, Hawkes et al. (2005) reported that exotic grasses have enhanced gross nitrification rates by supporting the availability and altering the composition of ammonia-oxidizing soil bacteria. These alterations may alter the N budget of an ecosystem and its soil processes therein, promoting the plant invader's legacy. Gibbons et al. (2017) reported that each invasive plant species has its specific effect on soil physicochemical properties. They observed that plant invasion drives modulations in abundance of microbial diversity, while other belowground

Table 10.1 Globally recognized invasive species and their destructive effect on soil

Serial no.	Name of invasive species	Family	Affected soil properties	Reference
1.	*Ailanthus altissima* (P. Mill) Swingle	Simaroubaceae	Increased soil pH, total N, organic C, and decreased C: N ratio	Vilà et al. (2006)
2.	*Alliaria petiolata* (Garlic mustard)	Brassicaceae	Antifungal phytochemicals, supresses mycorrhizal fungi, dominance of plant pathogenic fungal and saprotrophs	Barto et al. (2011), Lankau (2011), Anthony et al. (2017), Anthony et al. (2020)
3.	*Anthemis cotula* L. (Stinking chamomile)	Asteraceae	Alter soil physiochemical properties: soil pH, water holding capacity, porosity and electrical conductivity, and increases soil nutrients, except soil P content	Dar et al. (2023)
4.	*Bromus tectorum* (Downy brome)	Poaceae	Increases soil nitrate, ammonium oxidizing bacteria, labile N concentration, excess root exudate rich in organic matter	McLeod et al. (2016), Morris et al. (2016)
5.	*Centaurea maculosa* Lam. (Spotted Knapweed)	Asteraceae	Root exudate produces both enantiomer of polyphenol (±)-catechin, inhibits growth of native species	Thorpe et al. (2009)
6.	*Conyza canadensis* (Horseweed)	Asteraceae	Alter available – N, organic C, increase catalase activity, affect soil microbiota related with bacteria and fungi	Zhang et al. (2020)
7.	*Hyptis suaveolens* (L.) Poit, (Bush Mint or pignut)	Lamiaceae	Affects biodiversity, diverting N -mineralization pathway especially nitrification, maintains high inorganic-N content	Afreen et al. (2018)
8.	*Lantana camara* L. (Shrub verbena)	Verbenaceae	Produce allelochemicals: sesquiterpenes, flavonoid, phenolic compound, and triterpenes present in its rhizosphere soil, residues, extract, and essential oil. Affects native species	Kato-Noguchi and Kurniadie (2021)
9.	*Mikania micrantha*	Asteraceae	Harbour bacterivore nematode in rhizosphere,	Sun et al. (2019)

(continued)

Table 10.1 (continued)

Serial no.	Name of invasive species	Family	Affected soil properties	Reference
	(L.) (American Rope)		which stimulate and feeds on potassium (K) solubilizing bacteria, significantly increases K concentration in soil	
10.	*Parthenium hysterophorus* (Santa-Maria, Famine weed)	Asteraceae	Increase soil pH, phosphorus, potassium, soil - N and organic C in invaded soil	Timsina et al. (2011)
11.	*Polygonum cuspidatum* (Japanese knotweed)	Polygonaceae	Litter rich in recalcitrant compound that slows the decomposition process, alters C-cycle, litters rich in lignin and polyphenols including tannins and flavonoids that hinder microbial growth, legacy effect	Tamura and Tharayil (2014), Tamura et al. (2017), Zhang et al. (2021a, b)
12.	*Solidago gigantea* (Early goldenrod)	Asteraceae	Affects phosphorus (P) turnover rate, increases labile P by mediating mineralization process via enhanced alkaline and acid phosphomonoesterase activities	Chapuis-Lardy et al. (2006)
13.	*Spartina alterniflora* (Smooth cord grass)	Poaceae	Increases P storage in biomass and soil, influence wetland ecosystem by affecting soil N:P ratio, increases fungi: bacteria ratio, increases organic C, stimulates soil denitrification activity and hence N_2O emission	Gao et al. (2019), Wang et al. (2019), Zhang et al. (2019a, b)
14.	*Triadica sebifera* (Chinese tallow)	Euphorbiaceae	Increase flavonoid quercetin in root exudate, facilitate growth of arbuscular mycorrhizal (AM) fungi, and enhance mutual association of AM with root	Tian et al. (2021)
15.	*Wedelia trilobata* (Yellow Creeping Daisy)	Asteraceae	Affects soil pH, soil calcium, significantly increases soil fungal community richness, affects microbes of N-cycling	Si et al. (2013)

community structure and function were constant. Further, they suggested that older invasion has more impact on soil abiotic parameters, hence indicate multi-layered succession.

In a case of wetland invasion by *Spartina alterniflora*, a direct impact on soil organic C (SOC) was reported (Yang et al. 2019) due to its effect on biomass production. It was further suggested that enhancement in SOC storage was due to interactive effect between *S. alterniflora* and soil biogeochemical attributes such as soil salinity, fine soil fraction, and bulk density, thereby directly contributing to increase in soil nutrient pool of the invaded lands (Zhou and Staver 2019; Stefanowicz et al. 2020). Zhou and Staver (2019) reported that invasive plants significantly affect the soil enzymes activity and there is a higher impact on the enzyme activities linked to N- and P-mineralization on invaded sites, as compared to C-decomposing enzymes. They also suggested that invaded soils have comparatively higher soil nutrient and microbial biomass pool than the non-invaded one, and therefore, creating a nutrient abundant niche that supports invaders and augment their invasiveness.

In a recent study, Sun et al. (2021) suggested that it is the soil depth which significantly determines the impact of invasive plants on litter decomposition and native plant species. While studying with the invasion of Moso bamboo, Li et al. (2019) reported that changes in N form during invasion modulate the soil microbial species. In a bibliometric study, Dawkins et al. (2022) found that, '*the impact on plant invasion and inability of the native plants to compete was due to specific microbial associations of the invasive plant or disruption of the soil microbial community*'. They further argued that the shift in microbial community relates to the alteration in physiochemical properties of the soil and consequent negative impact on the native flora.

10.3 Plant Invasion and Phytodiversity

In current scenario, invasive plants are dramatically altering the global environment by modifying biological community structure at an alarming rate (Strayer et al. 2006; Gaertner et al. 2014; Sardans et al. 2017). It is suggested that invasive plant species are a serious threat to the existence and diversity of native plant species (Gaertner et al. 2009), therefore, may accelerate decline in overall biodiversity and function of the ecosystem (Shabani et al. 2020). Several characteristics are rapid growth rate, greater self-regeneration capacity and stronger competitiveness for resources, and higher reproductive rate, flexibility in resource consumption, and affinity to different niches in the plant community (Sardans et al. 2017; McLeod et al. 2021; Sperry et al. 2021). Furthermore, it has also been suggested that successful colonization of invasive plants in new habitats is usually regulated by natural enemy evasion, mechanisms of plant adaptation, and allelopathy, thereby affecting nutrient cycling and native plant diversity (Blumenthal 2005, 2006; Kato-Noguchi and Kurniadie 2021; Zhang et al. 2021a, b). Moreover, the factors such as climate, soil environment, and interspecific interactions between native and invasive plants also determine the success of an invasive species in a habitat (Vasquez et al. 2008; Shen et al. 2019; Bell et al. 2020). However, it has been suggested by many studies that the mechanism of success of an invasive species in colonization of a new habitat is very complicated and species-specific, and therefore, it is critical to understand this mechanism to get an insight on the overall decline in plant diversity along with control and management of plant invasion (Mack et al. 2000; Thuiller et al. 2005; van Kleunen et al. 2010).

Sufficient availability of nutrients in soil supports invasive potential of an invasive species (Stefanowicz et al. 2019). Furthermore, during the invasion process, an invasive plant strive to achieve a competitive edge by acquiring a higher nutrient-use efficiency and modulating the soil nutrient cycling by increasing the quality and quantity of the soil nutrients (Sardans et al. 2017; Sun et al. 2019; Zhou and Staver 2019). Another important factor in successful establishment of an invasive plant species is the shift of microbial community in favour of invasive plants and thereby play an important role in alteration of phytodiversity of a plant community (Zhou and Staver 2019). It has been recently reported that invasive plants substantially enhance the diversity of bacterial and fungal communities which further assist in their successful colonization (Custer and van Diepen 2020).

It is suggested that invasive plants alter the soil moisture, particle composition and texture during the invasion process, which has impact on soil permeability, compactness, and cohesiveness of soil and ultimately on the soil nutrient cycling (Stefanowicz et al. 2017; Tian et al. 2018; Zhai et al. 2020). However, investigations are further needed to understand the mechanism of alteration of soil moisture and particle composition during colonization (Xu et al. 2022). There are, however, several studies which reported that soil C amendments may favour the reversal of niche ‘construction and legacy effects’ of invasive species and suggested that amendment of invaded soils with biochar may help in restoration of invaded lands (Zhang et al. 2021a, b). Linders et al. (2019), while studying on the invasion process of *Prosopis* suggested that indirect effect by the invader is more important than the direct individual effect in describing the impact of plant invasion on phytodiversity and ecosystem functions. They further indicated that a successful management of plant invaders also requires restoration of productive and diverse herb communities.

In a study, Castro-Díez et al. (2016) reported that a range of functional traits in invaded lands changed towards greater woodiness and ever greenness, along with decline in species richness, functional richness, divergence, dispersion, and redundancy. On the contrary, ecosystem properties were found to be least responsive to invasion. They further reported that *Carpobrotus* invasion supported functional homogenization within communities, and the functional organization was more affected by invasion than the ecosystem properties.

10.4 Terrestrial Plant Invasion and Microbes

Belowground microbial communities have a critical role in soil nutrient cycling and bioavailability of essential plant nutrients (Trognitz et al. 2016). Various studies have reported differential impact of plant invasion in affected habitats, along with a significant shift in community structure of soil microbes during different stages of invasion (Elgersma and Ehrenfeld 2011; Huangfu et al. 2015). Waller et al. (2020) observed that soil microorganisms and herbivores interactively drive the effects of plant invader, however, traits related to nutrient acquisition and growth mainly determine invasive plant's interaction with local organisms and success thereafter. Si et al. (2013) observed that differential degree of *Wedelia trilobata* invasion supported fungal community compared to the bacterial, however, it has an explicit impact on bacterial community involved in soil N cycling. Further, it has been also observed that enhanced levels of invasion significantly affect community structure of soil microbes, including those involved in nutrient cycling (Zhang et al. 2020), and therefore, supports the establishment and spread of invasive species. Moreover, it is suggested that a successful establishment of invasive plant require positive plant-soil feedback mechanisms (Rodriguez-Echeverria et al. 2013). Such plant-soil feedback mechanisms are reported from various ecosystems and being directionally operated by plants to alter biotic and abiotic components of the soil impacting their growth directly, ultimately affecting the plant community composition of the inhabited ecosystem (Callaway et al. 2004; Day et al. 2015; Bennett and Kliromomos 2019). Furthermore, mutualistic interaction between invasive plants and various bacterial and fungal species are key drivers of plant-soil feedbacks (Mariotte et al. 2018), and in many cases, it has been observed that invasive plants have greater reliance on such mutualistic interactions to assert their invasiveness (Massenssini et al. 2014; Trognitz et al. 2016).

Allelopathic influence through root secretion or litter decomposition is also an important component of plant invasion processes that may influence microbial community structure, and to some of the specific microbial groups involve in metabolism and nutrient acquisition (Ehrenfeld et al. 2001; Kramer et al. 2020). The 'novel weapons hypothesis' propagates that successful invasion of plants relies on release of detrimental biochemicals, containing allelopathic roots exudate that actively suppress plant growth and soil microbes. Qu et al. (2021) observed that *Rhus typhina* root extracts inhibit both plant growth and soil microbial activity that may attributed to strong allelopathic effects.

Rodríguez-Caballero et al. (2020) observed a marked difference in composition of rhizospheric bacterial and fungal communities between native and invasive plant species. Jo et al. (2017) suggested linking of above and belowground processes to comprehend the impact of plant invader and soil nutrient cycling. Zhang et al. (2018) suggested that invasive plant species may promote decomposers more in rhizosphere to accelerate nutrient release and maximize nutrient uptake by establishing more mutualistic interaction with soil microbiota. They further indicated that changes in soil microbial community structure were strongly determined by soil parameters such as soil pH and soil nutrient availability. In a metanalysis, Torres et al. (2021)

suggested that there may be differential impact of plant invasion on the soil microbiota, and more understanding in terms of soil N and C cycles, litter composition, and other biochemicals involved in success of invasive plants. They further argue that a shift in bacterial diversity may alter soil nutrient cycle, enzymatic processes, soil mineralization rates, and C and N content. However, Custer and van Diepen (2020) concluded that invasive plants have highly heterogenous and minimal impact on the α-diversity of soil microbes. Moreover, Stefanowicz (2016) suggested that impact of invasive plants on soil microbial communities in terms of activity, biomass, and composition was species-specific.

10.5 Plant Invasion and Soil Hydraulic Properties

Though the impact of functional diversity of plants on soil hydraulic properties is understudied, a number of studies reported an increase in higher water consumption due to higher stomatal conductance, photosynthetic rate, and transpiration rate in invasive plants (Leishman et al. 2007; Cavaleri and Sack 2010; Vasquez-Valderrama et al. 2020). Further, a shift in runoff, evapotranspiration, and lower soil moisture has been reported in invaded landscape (Levine et al. 2003). Moreover, a decline in water infiltration in soil and enhanced soil organic matter has also been observed in invaded areas (Vanderhoeven et al. 2005; te Beest et al. 2015; Castro-Díez et al. 2019). However, an understating of functional mechanisms that determine ecosystem processes is still warranted to understand the impact of invasion process on soil hydraulic properties in many tropical ecosystems. In a study, Vasquez-Valderrama et al. (2020) indicated that clarity in terms of differences in functional traits is lacking between native and invasive species in tropical dry forests (TDFs). However, in TDFs, studies have reported robustness in plant functional traits, i.e. high wood density, low tree height, deep roots, and small leaf size in invasive plants which supports them under water stressed soils (Ratnam et al. 2019).

10.6 Plant Invasion and Functional Traits

Invasive plant species acquire the potential to alter the inhabiting ecosystem due to robustness in their functional traits as compared to the natives (Godoy et al. 2011; te Beest et al. 2015; Zhang et al. 2016). Some of the functional traits that provide them an edge over native community are high productivity (McLeod et al. 2021), tremendous seed bank (Sperry et al. 2021), effective dispersal of seed (Sperry et al. 2021), phenotypic plasticity (Rai 2015), and allelochemicals (Kato-Noguchi and Kurniadie 2021). The high productivity of the invasive plant is supported by the idea of enemy release hypothesis (Inderjit and Van der Putten 2010). Van Kleunen et al. (2010) suggested that invasive plants have higher physiological, leaf-area allocation, shoot allocation, growth rate, size and fitness values as compared to the native species. Scharfy et al. (2011) reported that superiority in functional traits such as leaf tissue density, leaf life span, litter decomposition rate, and N use efficiency provides a competitive advantage to invasive forbs against native graminoids. It has been indicated by many studies that the quantity and quality of plant biomass are keys for the invasion effect on the soil (Godoy et al. 2011; Stanek and Stefanowicz 2019; Stefanowicz et al. 2020).

10.7 Plant Invasion and Rhizospheric Response

Plant is anchored in soil through root, and root is constituently attached by different microbes such as bacteria, fungi, oomycetes, etc. (Baetz and Martinoia 2014). Living root of invasive plants provides tissue and exudates for soil microorganism (Baetz and Martinoia 2014). Thus, to defend itself root releases some defense compounds which are biologically active (Baetz and Martinoia 2014). These defense molecules may act as stimulant, signaling attractants, inhibitors, or repellents (Baetz and Martinoia 2014). These root exudates formed the basis of Novel Weapon Hypothesis (Inderjit and Van der Putten 2010). Invasive plant secretes chemical that inhibits the growth of other native species called allelochemical (Inderjit and Van der Putten 2010). These chemicals inhibit the growth of certain bacteria, herbivores, and predators (Thorpe et al. 2009). Nature of these chemicals varies from sesquiterpenes, phenolics to flavonoid, and enantiomer of catechin depending upon the invading species (Thorpe et al. 2009; Tian et al. 2021).

Tian et al. (2021) reported that invaded population of *Triadica sebifera* has enhanced flavonoid production in root exudate and arbuscular mycorrhizal (AM) fungi population in their rhizosphere, thus indicating that flavonoid may act as a signaling molecule to enhance AM fungal association with invasive species. Meta-analysis by Zhang et al. (2019a, b) found that invasive plant with allelopathic effect has greater suppressive control over bacterial population than non-allelopathic one. Another study by Thorpe et al. (2009) supported that root exudates are allelopathic in invaded range, however, non-allelopathic in native range. In a study *Centaurea maculosa*, native to Eurasia invades Western North America and produces (±) catechin as root exudates, which are more phytotoxic in invaded range than non-invaded.

10.8 Plant Invasion, Enzymes, and Soil Nutrient Cycling

An efficient nutrient cycling and subsequent availability of higher soil nutrient in any ecosystem will determine its suitability as a potential site of plant invasion (Funk et al. 2008; Allison et al. 2011). The plant invader must be equipped with traits such as higher nutrient-use efficiency that enables them to outcompete native species in resource stressed soils (Funk et al. 2008; Ens et al. 2015). The invasive plants must possess a rapid nutrient cycling capability to enhance the soil nutrient content that may guarantee their colonization in any habitat (Allison and Vitousek 2004; Lee et al. 2012, 2017; Zhang et al. 2019a, b). A number of studies indicated that invaded sites contain a higher quality and quantity of litter input along with higher soil N and

P compared to the non-invaded ones (Vilà et al. 2011; Tamura and Tharayil 2014; Sardans et al. 2017; Tamura et al. 2017; Zhang et al. 2021a, b). This further suggests soil microbes mediated enhanced rate of litter decomposition and accelerated nutrient cycling in plant invaded sites. Moreover, a number of studies reported the role of soil extracellular enzymes in release of nutrient in soil through degrading complex organic compounds (Chapuis-Lardy et al. 2006; Burns et al. 2013; Sardans et al. 2017; Zhou and Staver 2019; Hu et al. 2021). The enzymes those are studied more belong to the categories that involve breakdown of lignin and cellulose, hydrolysis of proteins and other complex organic compounds such as chitin and mineralization of compounds containing organic phosphorus (Sinsabaugh and Follstad Shah 2012). However, there were contradictory hypotheses regarding enzymes mediated cycling of nutrients in soils invaded by plants.

According to a hypothesis, since the enzyme production needs C and N, and because the invaded habitats are already rich in required nutrients, the enzyme production will be lowered (Schimel and Weintraub 2003; Allison et al. 2011). Another hypothesis suggests that since invasive plants have to compete with soil microbes for nutrient acquisition, they will produce enzymes despite the ample availability of soil nutrient at the invaded sites (Craig and Fraterrigo 2017; Min and Suseela 2020; Stanek et al. 2021). Invasive plant species have effective mechanism of soil nutrient utilization and nutrient-use efficiency as well as rate of nutrient acquisition, that is mediated via their association with fungal groups, i.e. arbuscular mycorrhizal fungi (Stinson et al. 2006; Majewska et al. 2017), apart from their robustness in specific root length and density of fine roots that further provides a greater surface area for nutrient scavenging (McCormack et al. 2014; Jo et al. 2017; Davidson et al. 2016; Li et al. 2023). Further, the third hypothesis suggests that the invasive plants owing to their fast growth and enhanced biomass increase N rich litter addition, thereby stimulating microbial activity and related enzymatic input (Kuzyakov 2010; Burns et al. 2013; Sardans et al. 2017). This indicates a trade-off between activities of nutrient-releasing and carbon-oxidizing enzymes (Jian et al. 2016).

10.9 Research Gaps and Future Perspectives

In the scenario of global climate change and emerging heterogenous niches, colonization of invasive plants may exhibit species-specific response and mechanisms may be highly complex, that need to be understood for their control and management (Jiménez-Ruiz et al. 2021). Further, invasive plants are a serious threat to local plant diversity and evidences are coming for the overall decline of the same, therefore, comprehending the invasion mechanism and possible causes of diversity decline may assist in effective control and dissemination of these invasives (North et al. 2021; Rai 2022). It has been suggested by many studies that responses of soil microorganisms and soil nutrients may be very complexed (Simberloff et al. 2021; Yang et al. 2021). Furthermore, the ongoing investigations only focus on certain environmental variables, leaving emerging factors such as N deposition and climate change. Studies suggested that N deposition, climate change and related drought may interact to pose challenging conditions for soil microorganisms and soil nutrient cycling, impacting plant diversity and promoting invasive species (Engelhardt et al. 2018, 2021; Yang et al. 2021). These challenges warrant investigation to be planned on above-mentioned interactive effects to understand their effect on plant diversity and emerging invasive plants. Climate change is further claimed to alter the precipitation across the globe and thereby going to influence soil moisture availability, therefore, under such scenario it is imperative to understand how invasive plants invade by affecting soil moisture and soil particle composition.

Since the connectivity across the globe has been tremendously increased, there are ample chances of dispersal of invasive plant's propagule to various geographical regions. This further indicates the urgency to plan studies including larger geographical regions. Therefore, future research on invasive plants and invasion process therein should be taken up at various geographical regions and larger regional scales.

References

Afreen T, Srivastava P, Singh H, Singh JS (2018) Effect of invasion by Hyptis suaveolens on plant diversity and selected soil properties of a constructed tropical grassland. J Plant Ecol 11(5): 751–760

Allison SD, Vitousek PM (2004) Rapid nutrient cycling in leaf litter from invasive plants in Hawai'i. Oecologia 141:612–619

Allison SD, Weintraub MN, Gartner TB, Waldrop MP (2011) Evolutionary-economic principles as regulators of soil enzyme production and ecosystem function. Soil Enzymol (1):229–243

Anthony MA, Frey SD, Stinson KA (2017) Fungal community homogenization, shift in dominant trophic guild, and appearance of novel taxa with biotic invasion. Ecosphere 8(9):e01951

Anthony MA, Stinson KA, Moore JA, Frey SD (2020) Plant invasion impacts on fungal community structure and function depend on soil warming and nitrogen enrichment. Oecologia 194:659–672

Baetz U, Martinoia E (2014) Root exudates: the hidden part of plant defense. Trends Plant Sci 19(2):90–98

Barto EK, Antunes PM, Stinson K, Koch AM, Klironomos JN, Cipollini D (2011) Differences in arbuscular mycorrhizal fungal communities associated with sugar maple seedlings in and outside of invaded garlic mustard forest patches. Biol Invasions 13:2755–2762

Bell JK, Siciliano SD, Lamb EG (2020) A survey of invasive plants on grassland soil microbial communities and ecosystem services. Scientific Data 7(1):86

Bennett JA, Klironomos J (2019) Mechanisms of plant–soil feedback: interactions among biotic and abiotic drivers. New Phytol 222(1):91–96

Blumenthal D (2005) Interrelated causes of plant invasion. Science 310(5746):243–244

Blumenthal DM (2006) Interactions between resource availability and enemy release in plant invasion. Ecol Lett 9(7):887–895

Blumenthal DM, Kray JA (2014) Climate change, plant traits and invasion in natural and agricultural ecosystems. In: Invasive species and global climate change. CABI, Wallingford, UK, pp 62–79

Bradley, B.A., Beaury, E., Fusco, E.J., Laginhas, B. and Pasquarella, V., 2018. Regional invasive species & climate change management challenge: preparing for sleeper species

Burns RG, DeForest JL, Marxsen J, Sinsabaugh RL, Stromberger ME, Wallenstein MD, Weintraub MN, Zoppini A (2013) Soil enzymes in a changing environment: current knowledge and future directions. Soil Biol Biochem 58:216–234

Callaway RM, Thelen GC, Rodriguez A, Holben WE (2004) Soil biota and exotic plant invasion. Nature 427(6976):731–733
Carboni M, Livingstone SW, Isaac ME, Cadotte MW (2021) Invasion drives plant diversity loss through competition and ecosystem modification. J Ecol 109(10):3587–3601
Castro-Díez P, Pauchard A, Traveset A, Vilà M (2016) Linking the impacts of plant invasion on community functional structure and ecosystem properties. J Veg Sci 27(6):1233–1242
Castro-Díez P, Vaz AS, Silva JS, Van Loo M, Alonso Á, Aponte C, Bayón Á, Bellingham PJ, Chiuffo MC, DiManno N, Julian K (2019) Global effects of non-native tree species on multiple ecosystem services. Biol Rev 94(4):1477–1501
Cavaleri MA, Sack L (2010) Comparative water use of native and invasive plants at multiple scales: a global meta-analysis. Ecology 91(9):2705–2715
Chapuis-Lardy L, Vanderhoeven SONIA, Dassonville N, Koutika LS, Meerts P (2006) Effect of the exotic invasive plant Solidago gigantea on soil phosphorus status. Biol Fertil Soils 42:481–489
Corbin JD, D'Antonio CM (2012) Gone but not forgotten? Invasive plants' legacies on community and ecosystem properties. Invas Plant Sci Manage 5(1):117–124
Craig ME, Fraterrigo JM (2017) Plant–microbial competition for nitrogen increases microbial activities and carbon loss in invaded soils. Oecologia 184:583–596
Custer GF, van Diepen LT (2020) Plant invasion has limited impact on soil microbial α-diversity: a meta-analysis. Diversity 12(3):112
Dar MA, Ahmad M, Singh R, Kohli RK, Singh HP, Batish DR (2023) Invasive plants alter soil properties and nutrient dynamics: a case study of Anthemis cotula invasion in Kashmir Himalaya. Catena 226:107069
Davidson BE, Novak SJ, Serpe MD (2016) Consequences of inoculation with native arbuscular mycorrhizal fungi for root colonization and survival of Artemisia tridentata ssp. wyomingensis seedlings after transplanting. Mycorrhiza 26:595–608
Davis MA, Grime JP, Thompson K (2000) Fluctuating resources in plant communities: a general theory of invasibility. J Ecol 88(3):528–534
Dawkins K, Mendonca J, Sutherland O, Esiobu N (2022) A systematic review of terrestrial plant invasion mechanisms mediated by microbes and restoration implications. Am J Plant Sci 13(2): 205–222
Day NJ, Dunfield KE, Antunes PM (2015) Temporal dynamics of plant–soil feedback and root-associated fungal communities over 100 years of invasion by a non-native plant. J Ecol 103(6): 1557–1569
Divíšek J, Chytrý M, Beckage B, Gotelli NJ, Lososová Z, Pyšek P, Richardson DM, Molofsky J (2018) Similarity of introduced plant species to native ones facilitates naturalization, but differences enhance invasion success. Nat Commun 9(1):4631
Ehrenfeld JG, Kourtev P, Huang W (2001) Changes in soil functions following invasions of exotic understory plants in deciduous forests. Ecol Appl 11(5):1287–1300
Elgersma KJ, Ehrenfeld JG (2011) Linear and non-linear impacts of a non-native plant invasion on soil microbial community structure and function. Biol Invas 13:757–768
Engelhardt IC, Welty A, Blazewicz SJ, Bru D, Rouard N, Breuil MC, Gessler A, Galiano L, Miranda JC, Spor A, Barnard RL (2018) Depth matters: effects of precipitation regime on soil microbial activity upon rewetting of a plant-soil system. ISME J 12(4):1061–1071
Engelhardt IC, Niklaus PA, Bizouard F, Breuil MC, Rouard N, Deau F, Philippot L, Barnard RL (2021) Precipitation patterns and N availability alter plant-soil microbial C and N dynamics. Plant Soil 466(1–2):151–163
Ens E, Hutley LB, Rossiter-Rachor NA, Douglas MM, Setterfield SA (2015) Resource-use efficiency explains grassy weed invasion in a low-resource savanna in north Australia. Front Plant Sci 6:560
Funk JL, Cleland EE, Suding KN, Zavaleta ES (2008) Restoration through reassembly: plant traits and invasion resistance. Trends Ecol Evol 23(12):695–703

Gaertner M, Den Breeyen A, Hui C, Richardson DM (2009) Impacts of alien plant invasions on species richness in Mediterranean-type ecosystems: a meta-analysis. Prog Phys Geogr 33(3): 319–338

Gaertner M, Biggs R, Te Beest M, Hui C, Molofsky J, Richardson DM (2014) Invasive plants as drivers of regime shifts: identifying high-priority invaders that alter feedback relationships. Divers Distrib 20(7):733–744

Gao GF, Li PF, Zhong JX, Shen ZJ, Chen J, Li YT, Isabwe A, Zhu XY, Ding QS, Zhang S, Gao CH (2019) Spartina alterniflora invasion alters soil bacterial communities and enhances soil N2O emissions by stimulating soil denitrification in mangrove wetland. Sci Total Environ 653:231–240

Gibbons SM, Lekberg Y, Mummey DL, Sangwan N, Ramsey PW, Gilbert JA (2017) Invasive plants rapidly reshape soil properties in a grassland ecosystem. MSystems 2(2):e00178–e00116

Godoy O, Valladares F, Castro-Díez P (2011) Multispecies comparison reveals that invasive and native plants differ in their traits but not in their plasticity. Funct Ecol 25(6):1248–1259

Guido A, Pillar VD (2017) Invasive plant removal: assessing community impact and recovery from invasion. J Appl Ecol 54(4):1230–1237

Hawkes CV, Wren IF, Herman DJ, Firestone MK (2005) Plant invasion alters nitrogen cycling by modifying the soil nitrifying community. Ecol Lett 8(9):976–985

Hess MC, Mesléard F, Buisson E (2019) Priority effects: emerging principles for invasive plant species management. Ecol Eng 127:48–57

Hu Z, Li J, Shi K, Ren G, Dai Z, Sun J, Zheng X, Zhou Y, Zhang J, Li G, Du D (2021) Effects of Canada goldenrod invasion on soil extracellular enzyme activities and ecoenzymatic stoichiometry. Sustainability 13(7):3768

Huangfu C, Li H, Chen X, Liu H, Yang D (2015) The effects of exotic weed Flaveria bidentis with different invasion stages on soil bacterial community structures. Afr J Biotechnol 14(35):2636

Inderjit, van der Putten WH (2010) Impacts of soil microbial communities on exotic plant invasions. Trends Ecol Evol 25(9):512–519. https://doi.org/10.1016/j.tree.2010.06.006

Jandová K, Klinerová T, Müllerová J, Pyšek P, Pergl J, Cajthaml T, Dostál P (2014) Long-term impact of Heracleum mantegazzianum invasion on soil chemical and biological characteristics. Soil Biol Biochem 68:270–278

Jesse WA, Molleman J, Franken O, Lammers M, Berg MP, Behm JE, Helmus MR, Ellers J (2020) Disentangling the effects of plant species invasion and urban development on arthropod community composition. Glob Chang Biol 26(6):3294–3306

Jian S, Li J, Chen JI, Wang G, Mayes MA, Dzantor KE, Hui D, Luo Y (2016) Soil extracellular enzyme activities, soil carbon and nitrogen storage under nitrogen fertilization: a meta-analysis. Soil Biol Biochem 101:32–43

Jiménez-Ruiz J, Hardion L, Del Monte JP, Vila B, Santín-Montanyá MI (2021) Monographs on invasive plants in Europe N° 4: Arundo donax L. Bot Lett 168(1):131–151

Jo I, Fridley JD, Frank DA (2017) Invasive plants accelerate nitrogen cycling: evidence from experimental woody monocultures. J Ecol 105(4):1105–1110

Kato-Noguchi H, Kurniadie D (2021) Allelopathy of Lantana camara as an invasive plant. Plan Theory 10(5):1028

Kaushik P, Pati PK, Khan ML, Khare PK (2022) Plant functional traits best explain invasive species' performance within a dynamic ecosystem: a review, vol 8. Trees Forests People, p 100260

Keane RM, Crawley MJ (2002) Exotic plant invasions and the enemy release hypothesis. Trends Ecol Evol 17(4):164–170

Kramer J, Özkaya Ö, Kümmerli R (2020) Bacterial siderophores in community and host interactions. Nat Rev Microbiol 18(3):152–163

Kumar S, Garkoti SC (2022) Rhizosphere influence on soil microbial biomass and enzyme activity in banj oak, chir pine and banj oak regeneration forests in the central Himalaya. Geoderma 409: 115626

Kuzyakov Y (2010) Priming effects: interactions between living and dead organic matter. Soil Biol Biochem 42(9):1363–1371

Lankau RA (2011) Resistance and recovery of soil microbial communities in the face of Alliaria petiolata invasions. New Phytol 189(2):536–548

Lee MR, Flory SL, Phillips RP (2012) Positive feedbacks to growth of an invasive grass through alteration of nitrogen cycling. Oecologia 170:457–465

Lee MR, Bernhardt ES, van Bodegom PM, Cornelissen JHC, Kattge J, Laughlin DC, Niinemets Ü, Peñuelas J, Reich PB, Yguel B, Wright JP (2017) Invasive species' leaf traits and dissimilarity from natives shape their impact on nitrogen cycling: a meta-analysis. New Phytol 213(1): 128–139

Leishman MR, Haslehurst T, Ares A, Baruch Z (2007) Leaf trait relationships of native and invasive plants: community-and global-scale comparisons. New Phytol 176(3):635–643

Lenda M, Witek M, Skórka P, Moroń D, Woyciechowski M (2013) Invasive alien plants affect grassland ant communities, colony size and foraging behaviour. Biol Invasions 15:2403–2414

Levine JM, Vilà M, Antonio CM, Dukes JS, Grigulis K, Lavorel S (2003) Mechanisms underlying the impacts of exotic plant invasions. Proc Royal Soc B: Biol Sci 270(1517):775–781

Li Q, Song X, Chang SX, Peng C, Xiao W, Zhang J, Xiang W, Li Y, Wang W (2019) Nitrogen depositions increase soil respiration and decrease temperature sensitivity in a Moso bamboo forest. Agric For Meteorol 268:48–54

Li SP, Jia P, Fan SY, Wu Y, Liu X, Meng Y, Li Y, Shu WS, Li JT, Jiang L (2022a) Functional traits explain the consistent resistance of biodiversity to plant invasion under nitrogen enrichment. Ecol Lett 25(4):778–789

Li S, Xie D, Ge X, Dong W, Luan J (2022b) Altered diversity and functioning of soil and root-associated microbiomes by an invasive native plant. Plant Soil 473(1–2):235–249

Li W, Wang L, Qian S, He M, Cai X, Ding J (2023) Root characteristics explain greater water use efficiency and drought tolerance in invasive Compositae plants. Plant Soil 483(1–2):209–223

Liao C, Peng R, Luo Y, Zhou X, Wu X, Fang C, Chen J, Li B (2008) Altered ecosystem carbon and nitrogen cycles by plant invasion: a meta-analysis. New Phytol 177(3):706–714

Linders TEW, Schaffner U, Eschen R, Abebe A, Choge SK, Nigatu L, Mbaabu PR, Shiferaw H, Allan E (2019) Direct and indirect effects of invasive species: biodiversity loss is a major mechanism by which an invasive tree affects ecosystem functioning. J Ecol 107(6):2660–2672

Mack RN, Simberloff D, Mark Lonsdale W, Harry E, Michael C, Fakhri FA (2000) Biotic invasions: causes, epidemiology, global consequences, and control. Ecol Appl 10:689–710

Majewska ML, Rola K, Zubek S (2017) The growth and phosphorus acquisition of invasive plants Rudbeckia laciniata and Solidago gigantea are enhanced by arbuscular mycorrhizal fungi. Mycorrhiza 27:83–94

Mariotte P, Mehrabi Z, Bezemer TM, De Deyn GB, Kulmatiski A, Drigo B, Veen GC, Van der Heijden MG, Kardol P (2018) Plant–soil feedback: bridging natural and agricultural sciences. Trends Ecol Evol 33(2):129–142

Massenssini AM, Bonduki VHA, Melo CAD, Tótola MR, Ferreira FA, Costa MD (2014) Soil microorganisms and their role in the interactions between weeds and crops. Planta Daninha 32: 873–884

Mathakutha R, Steyn C, le Roux PC, Blom IJ, Chown SL, Daru BH, Ripley BS, Louw A, Greve M (2019) Invasive species differ in key functional traits from native and non-invasive alien plant species. J Veg Sci 30(5):994–1006

McCormack ML, Lavely E, Ma Z (2014) Fine-root and mycorrhizal traits help explain ecosystem processes and responses to global change. New Phytol 204(3):455–458

McLeod ML, Cleveland CC, Lekberg Y, Maron JL, Philippot L, Bru D, Callaway RM (2016) Exotic invasive plants increase productivity, abundance of ammonia-oxidizing bacteria and nitrogen availability in intermountain grasslands. J Ecol 104(4):994–1002

McLeod ML, Bullington L, Cleveland CC, Rousk J, Lekberg Y (2021) Invasive plant-derived dissolved organic matter alters microbial communities and carbon cycling in soils. Soil Biol Biochem 156:108191

Min K, Suseela V (2020) Plant invasion alters the Michaelis–Menten kinetics of microbial extracellular enzymes and soil organic matter chemistry along soil depth. Biogeochemistry 150(2):181–196
Morris KA, Stark JM, Bugbee B, Norton JM (2016) The invasive annual cheatgrass releases more nitrogen than crested wheatgrass through root exudation and senescence. Oecologia 181:971–983
North HL, McGaughran A, Jiggins CD (2021) Insights into invasive species from whole-genome resequencing. Mol Ecol 30(23):6289–6308
Osunkoya OO, Perrett C (2011) Lantana camara L.(Verbenaceae) invasion effects on soil physico-chemical properties. Biol Fertil Soils 47:349–355
Perkins LB, Hatfield G (2014) Competition, legacy, and priority and the success of three invasive species. Biol Invas 16:2543–2550
Prescott CE, Zukswert JM (2016) Invasive plant species and litter decomposition: time to challenge assumptions. New Phytol 209(1):5–7
Qu T, Du X, Peng Y, Guo W, Zhao C, Losapio G (2021) Invasive species allelopathy decreases plant growth and soil microbial activity. PLoS One 16(2):e0246685
Rai PK (2015) What makes the plant invasion possible? Paradigm of invasion mechanisms, theories and attributes. Environ Skept Crit 4(2):36
Rai PK (2022) Environmental degradation by invasive alien plants in the anthropocene: challenges and prospects for sustainable restoration. Anthropocene Sci 1(1):5–28
Rai PK, Singh JS (2020) Invasive alien plant species: Their impact on environment, ecosystem services and human health. Ecol Indic 111:106020
Ratnam J, Chengappa SK, Machado SJ, Nataraj N, Osuri AM, Sankaran M (2019) Functional traits of trees from dry deciduous "forests" of southern India suggest seasonal drought and fire are important drivers. Front Ecol Evol 7:8
Rodríguez-Caballero G, Caravaca F, Alguacil MM, Fernández-López M, Fernández-González AJ, Roldán A (2017) Striking alterations in the soil bacterial community structure and functioning of the biological N cycle induced by Pennisetum setaceum invasion in a semiarid environment. Soil Biol Biochem 109:176–187
Rodríguez-Caballero G, Roldán A, Caravaca F (2020) Invasive Nicotiana glauca shifts the soil microbial community composition and functioning of harsh and disturbed semiarid Mediterranean environments. Biol Invasions 22:2923–2940
Rodríguez-Echeverría S, Afonso C, Correia M, Lorenzo P, Roiloa SR (2013) The effect of soil legacy on competition and invasion by Acacia dealbata Link. Plant Ecol 214:1139–1146
Saiya-Cork KR, Sinsabaugh RL, Zak DR (2002) The effects of long term nitrogen deposition on extracellular enzyme activity in an Acer saccharum forest soil. Soil Biol Biochem 34(9): 1309–1315
Sardans J, Bartrons M, Margalef O, Gargallo-Garriga A, Janssens IA, Ciais P, Obersteiner M, Sigurdsson BD, Chen HY, Peñuelas J (2017) Plant invasion is associated with higher plant–soil nutrient concentrations in nutrient-poor environments. Glob Chang Biol 23(3):1282–1291
Scharfy D, Funk A, Olde Venterink H, Güsewell S (2011) Invasive forbs differ functionally from native graminoids, but are similar to native forbs. New Phytol 189(3):818–828
Schimel JP, Weintraub MN (2003) The implications of exoenzyme activity on microbial carbon and nitrogen limitation in soil: a theoretical model. Soil Biol Biochem 35(4):549–563
Shabani F, Ahmadi M, Kumar L, Solhjouy-fard S, Tehrany MS, Shabani F, Kalantar B, Esmaeili A (2020) Invasive weed species' threats to global biodiversity: future scenarios of changes in the number of invasive species in a changing climate. Ecol Indic 116:106436
Shen N, Yu H, Yu S, Yu D, Liu C (2019) Does soil nutrient heterogeneity improve the growth performance and intraspecific competition of the invasive plant Myriophyllum aquaticum? Front Plant Sci 10:723
Si C, Liu X, Wang C, Wang L, Dai Z, Qi S, Du D (2013) Different degrees of plant invasion significantly affect the richness of the soil fungal community. PLoS One 8(12):e85490

Simberloff D, Kaur H, Kalisz S, Bezemer TM (2021) Novel chemicals engender myriad invasion mechanisms. New Phytol 232(3):1184–1200
Sinsabaugh RL, Follstad Shah JJ (2012) Ecoenzymatic stoichiometry and ecological theory. Annu Rev Ecol Evol Syst 43:313–343
Spear MJ, Walsh JR, Ricciardi A, Zanden MJV (2021) The invasion ecology of sleeper populations: prevalence, persistence, and abrupt shifts. Bioscience 71(4):357–369
Sperry JH, O'Hearn D, Drake DR, Hruska AM, Case SB, Vizentin-Bugoni J, Arnett C, Chambers T, Tarwater CE (2021) Fruit and seed traits of native and invasive plant species in Hawai 'i: implications for seed dispersal by non-native birds. Biol Invasions 23:1819–1835
Stanek M, Stefanowicz AM (2019) Invasive Quercus rubra negatively affected soil microbial communities relative to native Quercus robur in a semi-natural forest. Sci Total Environ 696: 133977
Stanek M, Zubek S, Stefanowicz AM (2021) Differences in phenolics produced by invasive Quercus rubra and native plant communities induced changes in soil microbial properties and enzymatic activity. For Ecol Manag 482:118901
Stefanowicz AM, Stanek M, Nobis M, Zubek S (2016) Species-specific effects of plant invasions on activity, biomass, and composition of soil microbial communities. Biol Fertil Soils 52:841–852
Stefanowicz AM, Stanek M, Nobis M, Zubek S (2017) Few effects of invasive plants Reynoutria japonica, Rudbeckia laciniata and Solidago gigantea on soil physical and chemical properties. Sci Total Environ 574:938–946
Stefanowicz AM, Majewska ML, Stanek M, Nobis M, Zubek S (2018) Differential influence of four invasive plant species on soil physicochemical properties in a pot experiment. J Soils Sediments 18:1409–1423
Stefanowicz AM, Stanek M, Majewska ML, Nobis M, Zubek S (2019) Invasive plant species identity affects soil microbial communities in a mesocosm experiment. Appl Soil Ecol 136:168–177
Stefanowicz AM, Banaś A, Stanek M, Woch MW, Zubek S (2020) Large differences in biomass quantity and quality between invasive Reynoutria japonica and resident vegetation are not reflected in topsoil physicochemical properties. Geoderma 368:114307
Stinson KA, Campbell SA, Powell JR, Wolfe BE, Callaway RM, Thelen GC, Hallett SG, Prati D, Klironomos JN (2006) Invasive plant suppresses the growth of native tree seedlings by disrupting belowground mutualisms. PLoS Biol 4(5):e140
Strayer DL, Eviner VT, Jeschke JM, Pace ML (2006) Understanding the long-term effects of species invasions. Trends Ecol Evol 21(11):645–651
Sun F, Ou Q, Yu H, Li N, Peng C (2019) The invasive plant Mikania micrantha affects the soil food web and plant-soil nutrient contents in orchards. Soil Biol Biochem 139:107630
Sun T, Wang Y, Lucas-Borja ME, Jing X, Feng W (2021) Divergent vertical distributions of microbial biomass with soil depth among groups and land uses. J Environ Manag 292:112755
Tamura M, Tharayil N (2014) Plant litter chemistry and microbial priming regulate the accrual, composition and stability of soil carbon in invaded ecosystems. New Phytol 203(1):110–124
Tamura M, Suseela V, Simpson M, Powell B, Tharayil N (2017) Plant litter chemistry alters the content and composition of organic carbon associated with soil mineral and aggregate fractions in invaded ecosystems. Glob Change Biol 23(10):4002–4018
Te Beest M, Esler KJ, Richardson DM (2015) Linking functional traits to impacts of invasive plant species: a case study. Plant Ecol 216:293–305
Thorpe AS, Thelen GC, Diaconu A, Callaway RM (2009) Root exudate is allelopathic in invaded community but not in native community: field evidence for the novel weapons hypothesis. J Ecol 97(4):641–645
Thuiller W, Richardson DM, Pyšek P, Midgley GF, Hughes GO, Rouget M (2005) Niche-based modelling as a tool for predicting the risk of alien plant invasions at a global scale. Glob Chang Biol 11(12):2234–2250

Tian L, Zhao L, Wu X, Fang H, Zhao Y, Hu G, Yue G, Sheng Y, Wu J, Chen J, Wang Z (2018) Soil moisture and texture primarily control the soil nutrient stoichiometry across the Tibetan grassland. Sci Total Environ 622:192–202

Tian B, Pei Y, Huang W, Ding J, Siemann E (2021) Increasing flavonoid concentrations in root exudates enhance associations between arbuscular mycorrhizal fungi and an invasive plant. ISME J 15(7):1919–1930

Tilman D (1990) Mechanisms of plant competition for nutrients: the elements of a predictive theory of competition. In: Grace, Tilman D (eds) Perspectives on plant competition, Academic Press, New York. pp. 117–141

Timsina B, Shrestha BB, Rokaya MB, Münzbergová Z (2011) Impact of Parthenium hysterophorus L. invasion on plant species composition and soil properties of grassland communities in Nepal. Flora-Morphol Distrib Funct Ecol Plants 206(3):233–240

Torres N, Herrera I, Fajardo L, Bustamante RO (2021) Meta-analysis of the impact of plant invasions on soil microbial communities. BMC Ecol Evol 21:1–8

Trognitz F, Hackl E, Widhalm S, Sessitsch A (2016) The role of plant–microbiome interactions in weed establishment and control. FEMS Microbiol Ecol 92(10):fiw138

Van Kleunen M, Weber E, Fischer M (2010) A meta-analysis of trait differences between invasive and non-invasive plant species. Ecol Lett 13(2):235–245

Vanderhoeven S, Dassonville N, Meerts P (2005) Increased topsoil mineral nutrient concentrations under exotic invasive plants in Belgium. Plant Soil 275:169–179

Vasquez E, Sheley R, Svejcar T (2008) Nitrogen enhances the competitive ability of cheatgrass (Bromus tectorum) relative to native grasses. Invas Plant Sci and Manage 1(3):287–295

Vasquez-Valderrama M, González-M R, López-Camacho R, Baptiste MP, Salgado-Negret B (2020) Impact of invasive species on soil hydraulic properties: importance of functional traits. Biol Invasions 22(6):1849–1863

Vilà M, Tessier M, Suehs CM, Brundu G, Carta L, Galanidis A, Lambdon P, Manca M, Médail F, Moragues E, Traveset A (2006) Local and regional assessments of the impacts of plant invaders on vegetation structure and soil properties of Mediterranean islands. J Biogeogr 33(5):853–861

Vilà M, Espinar JL, Hejda M, Hulme PE, Jarošík V, Maron JL, Pergl J, Schaffner U, Sun Y, Pyšek P (2011) Ecological impacts of invasive alien plants: a meta-analysis of their effects on species, communities and ecosystems. Ecol Lett 14(7):702–708

Vitousek PM (1990) Biological invasions and ecosystem processes: towards an integration of population biology and ecosystem studies. Oikos (1):7–13

Waller LP, Allen WJ, Barratt BIP, Condron LM, França FM, Hunt JE, Koele N, Orwin KH, Steel GS, Tylianakis JM, Wakelin SA (2020) Biotic interactions drive ecosystem responses to exotic plant invaders. Science 368(6494):967–972

Wang W, Sardans J, Wang C, Zeng C, Tong C, Chen G, Huang J, Pan H, Peguero G, Vallicrosa H, Peñuelas J (2019) The response of stocks of C, N, and P to plant invasion in the coastal wetlands of China. Glob Chang Biol 25(2):733–743

Xu Z, Guo X, Caplan JS, Li M, Guo W (2021) Novel plant-soil feedbacks drive adaption of invasive plants to soil legacies of native plants under nitrogen deposition. Plant Soil 467:47–65

Xu H, Liu Q, Wang S, Yang G, Xue S (2022) A global meta-analysis of the impacts of exotic plant species invasion on plant diversity and soil properties. Sci Total Environ 810:152286

Yang Y, Tilman D, Furey G, Lehman C (2019) Soil carbon sequestration accelerated by restoration of grassland biodiversity. Nat Commun 10(1):718

Yang Y, Li T, Wang Y, Cheng H, Chang SX, Liang C, An S (2021) Negative effects of multiple global change factors on soil microbial diversity. Soil Biol Biochem 156:108229

Zhai J, Song Y, Entemake W, Xu H, Wu Y, Qu Q, Xue S (2020) Change in soil particle size distribution and erodibility with latitude and vegetation restoration chronosequence on the Loess Plateau, China. Int J Environ Res Public Health 17(3):822

Zhang L, Xiaochi MA, Hong WA, Shuwei LI, Siemann E, Jianwen ZO (2016) Soil respiration and litter decomposition increased following perennial forb invasion into an annual grassland. Pedosphere 26(4):567–576

Zhang P, Neher DA, Li B, Wu J (2018) The impacts of above-and belowground plant input on soil microbiota: invasive Spartina alterniflora versus native Phragmites australis. Ecosystems 21: 469–481

Zhang G, Bai J, Jia J, Wang W, Wang X, Zhao Q, Lu Q (2019a) Shifts of soil microbial community composition along a short-term invasion chronosequence of Spartina alterniflora in a Chinese estuary. Sci Total Environ 657:222–233

Zhang P, Li B, Wu J, Hu S (2019b) Invasive plants differentially affect soil biota through litter and rhizosphere pathways: a meta-analysis. Ecol Lett 22(1):200–210

Zhang HY, Goncalves P, Copeland E, Qi SS, Dai ZC, Li GL, Wang CY, Du DL, Thomas T (2020) Invasion by the weed Conyza canadensis alters soil nutrient supply and shifts microbiota structure. Soil Biol Biochem 143:107739

Zhang Z, Bhowmik PC, Suseela V (2021a) Effect of soil carbon amendments in reversing the legacy effect of plant invasion. J Appl Ecol 58(1):181–191

Zhang Z, Liu Y, Yuan L, Weber E, van Kleunen M (2021b) Effect of allelopathy on plant performance: a meta-analysis. Ecol Lett 24(2):348–362

Zhou Y, Staver AC (2019) Enhanced activity of soil nutrient-releasing enzymes after plant invasion: a meta-analysis. Ecology 100(11):e02830

Plant Invasion Dynamics in Mountain Ecosystems Under Changing Climate Scenario

11

Mushtaq Ahmad Dar, Rigzin Chuskit, Rishikesh Singh, Shalinder Kaur, Sachchidanand Tripathi, and Daizy R. Batish

Abstract

Under changing climate scenario, it is anticipated that the distribution of various native and invasive species will be shifting to different novel regions. The possibility of biological invasion has amplified nowadays, because climate change has increased the range of geographical regions where invasive alien species can flourish. Most of the studies on invasive species generally concentrated on the lowland regions that have undergone substantial changes. However, less concern has been provided to the higher elevation ranges that have shown comparatively less disruption. The mountainous ecosystems provide several ecosystem services and have high conservation values. Since most of the invasive species are well adapted to the lowland conditions, they generally have limited scope to spread in the harsher conditions of mountain ecosystems. Thus, they have comparatively lesser negative impact on mountainous landscapes than other lowland ecosystems. However, because of ongoing climate change and increased anthropogenic interventions, invasive species might easily go uphill and affect mountain ecosystems at mid- and then high-elevation ranges. There is increasing evidence of plant invasions in these regions nowadays. Several exotic species have been reported to become established with the passage of time in high-elevation areas all around the world. Most of these species are not aggressive, but a few could pose a significant threat to the surrounding mountain

M. A. Dar · R. Chuskit · R. Singh · S. Kaur (✉) · D. R. Batish
Department of Botany, Panjab University, Chandigarh, India
e-mail: darmushtaq199@gmail.com; rigzinshel@gmail.com; rishikesh.iesd@gmail.com; shalinder@pu.ac.in; daizybatish@yahoo.com

S. Tripathi
Department of Botany, Deen Dayal Upadhyaya College (University of Delhi), New Delhi, India
e-mail: drsachchidanandtripathi@ddu.du.ac.in

S. Tripathi et al. (eds.), *Plant Invasions and Global Climate Change*,
https://doi.org/10.1007/978-981-99-5910-5_11

ecosystems. In this chapter, emphasis has been given to highlight the role of mountains in studying plant invasion dynamics, mechanism of plant invasion, examples of a few major invasive species in the Indian Himalayan Region (IHR), impacts of invasive plants, followed by the suggestions for managing invasive plant species. Overall, the present chapter provides an overview on the plant invasion dynamics in the mountain ecosystems.

Keywords

Biological invasion · Climate change · Invasive species · Mountain ecosystems · Plant dynamics · Soil conditions

11.1 Introduction

Habitat destruction, biological invasion and climate change are the major trinities considered as primary threats to the biodiversity (Bellard et al. 2021; Fuentes-Lillo et al. 2021). In addition, rapid expansion of global trade has resulted in a variety of exchanges increasing the number of introduced species to the novel ecosystems (Catford et al. 2018; Pyšek et al. 2020; Hulme 2021). Due to their widespread economic and ecological harms, particularly to the structure, functions, and services provided by native ecosystems, biological invasions are drawing increasing public attention, globally (Esler et al. 2010; Diagne et al. 2020; Shah and Sharma 2022). The key components of biological invasions are the spread and effects of invasive species on the surrounding ecosystems (McGeoch and Jetz 2019). Depending on the receiving environments' randomness and the species inherent demography, invasive species can spread widely (Hui et al. 2016; Lustig et al. 2017). It is well known that demographic parameters, such as population size, density, and intraspecific competition play a major role in predicting how successfully non-native species can colonise new areas and establish themselves (Goyal et al. 2018; Shah and Sharma 2022). The species composition of a natural ecosystem can be altered by invasive species which can do so by driving out native species through competition, predation, and altered nutrient cycling (Foster et al. 2021; Lázaro-Lobo and Ervin 2021; Carvalho et al. 2022).

The climate plays a major role in determining worldwide vegetation distribution patterns and the overall forest ecology (Wu et al. 2015). Climate change intensifies the ecosystem degradation and threat to the biodiversity through a variety of factors including the elimination of major climatic barriers and increasing the spread of invasive species (Adhikari et al. 2019; Shrestha and Shrestha 2019). Therefore, global climate change is considered as one of the key factors contributing to the spread of invasive species (Bogale and Tolossa 2021). Several reports claim that invasive alien species are more likely to spread as a result of climate change (Pathak et al. 2019; Kariyawasam et al. 2021; Tripathi et al. 2022). It is believed that invasive species and climate change together will cause a significant and irreversible loss of the biodiversity (Sintayehu 2018; Bellard et al. 2021). With respect to vegetative reproduction, invasive alien plants frequently multiply more quickly and are typically more susceptible to rising atmospheric carbon dioxide (CO_2) concentrations than other species (Vilà et al. 2007; Gazoulis et al. 2022). It is advantageous for invading alien species to be able to outcompete native species when they migrate to higher altitudes in the presence of rising CO_2 levels (Ziska et al. 2019). These facts have led to the claims that a number of alien biota, which is often confined to the lower altitudes, are likely to migrate to the higher altitudes via growing human interferences and global climate change (Pathak et al. 2019; Negi et al. 2022).

According to Fuentes-Lillo et al. (2021), climate and habitat characteristics both have an impact on the altitudinal distribution of non-native plants. Non-native invasive species are generally found at lower and mid-elevation levels in the mountain ecosystems. However, due to ongoing anthropogenic interferences and climate change, it is projected that they will spread and take control at higher elevations as well (Alexander et al. 2016; Tito et al. 2020). For example, global mean temperature has increased by 0.78 °C (in twentieth century), which is further predicted to rise by 2.6–4.8 °C by the end of twenty-first century (IPCC 2014; Reisinger and Clark 2018). In the mountain ecosystems, a possible invasion of alien plant species could happen if population growth at higher elevations is sustained under a warmer climatic conditions (Kueffer et al. 2013; Lamsal et al. 2018). Climate change may, therefore, have an effect on the invasive plants' ability to survive in these terrains, either directly or indirectly. Given the low propagule pressure, energy constraints, and disturbances present in the mountain ecosystems, it is also known that high-elevation regions are comparatively less prone to plant invasions than the low- and mid-elevation regions, even when projected temperature increases are taken into account (Marini et al. 2009). However, recently it is recognised that in mountain ecosystems, climatic and inherent ecological conditions of the site regulate the altitudinal spread of alien species (Haider et al. 2010; Petitpierre et al. 2016). Therefore, it appears that in the upcoming decades, high-elevation ecosystems will see a considerable increase in plant invasion by the lowland alien species (Hulme 2017; Lamsal et al. 2018). Knowing an alien species distribution pattern is necessary to predict its likely expansion under various climate change scenarios. For example, some of the modelling studies conducted to envisage the spread of invasive species under different climate change scenarios have predicted that *Ageratum conyzoides* L. and *Parthenium hysterophorus* L. will lose the majority of their suitable habitat from the Himalayas by the year 2070, whereas *Ageratina adenophora* (Spreng.) R.M. King & H. Rob, *Chromolaena odorata* L. R.M King & H., and *Lantana camara* L. are predicted to spread in more suitable locations in the Himalayas (Lamsal et al. 2018; Pathak et al. 2019; Shrestha and Shrestha 2021).

The Indian Himalayan Region (IHR) with an area of more than 5.3 lakh km^2 is made up of the gigantic mountain ranges extending over 2500 km between the Indus and Brahmaputra River basins (Nandi and Rawat 2019; Singh et al. 2021). These mountain ranges are more than 8000 metres above sea level (m asl) and climb vertically from the low-lying lowlands. As a young mountain that is still rising (between 30 and 40 million years old), the Himalayas is prone to landslides, landslips, and accumulations of debris from the broken, cracked, and crushed rocks (Pradhan and Siddique 2020). The IHR is spread across 13 Indian states/ Union territories: Arunachal Pradesh, Himachal Pradesh, Jammu and Kashmir, Ladakh, Manipur, Meghalaya, Mizoram, Nagaland, Sikkim, Tripura, and Uttarakhand, whereas Assam and West Bengal (Darjeeling) Hills make up a section of the IHR (Palni and Rawal 2010; Rawal et al. 2013; NITI Aayog 2023). The regions' biophysical richness is further reflected in its biogeographical divisions. The region is consisting of three of the 10 biogeographic zones, viz. Trans Himalaya, Himalaya, and North-East India, together with nine of the India's 27 biogeographic provinces (Palni and Rawal 2010). In this chapter, a brief insight has been given on the process of plant invasion, followed by the examples of a few major invasive species and their invasion dynamics, impact of invasive species, and strategies for managing plant invasion with a particular reference to the mountain ecosystems.

11.2 The Way How Plants Invade in an Ecosystem

Over the past 1000 years, humans have intentionally and unintentionally spread plants far beyond their original ranges through transportation and commerce (Hulme 2021; Wu et al. 2022). The success of the invasion by any alien plant species is determined by a number of variables, including the biology of reproduction, the availability of resources (light, water, nutrients, etc.), competition, and the obscurity of the receiving habitat that modulates the spread and establishment of invasive species to become naturalised (Pathak et al. 2019; Hakim et al. 2023). Figure 11.1 shows the schematic representation of the process of arrival, establishment, and dominance of alien species in an ecosystem. However, this does not necessarily apply to all the introduced species. In the IHR, a variety of plant invasion mechanisms have been investigated, and majority of them are found to be species-specific (Jaryan et al. 2007). For example, the majority of research has been conducted on the characterisation, distribution patterns, and exploring the mechanisms of plant invasion in the Kashmir Himalaya (Khuroo et al. 2011; Manish 2021).

In general, disturbance events facilitate the biological invasion, which may lead to the loss of native biodiversity (Hulme 2006; Diez et al. 2012). During the initial phases of forest disturbance and degradation, invasive species act as 'passengers' to visit these areas and spreading moderately, while later they overtake the entire disturbed land and act as "drivers" which interfere the forest regeneration processes (Lugo 2009; Pathak et al. 2021). The range of present invaders is anticipated to eventually spread into untouched areas due to a number of factors, including climate change and localised species adaptions (Pauchard et al. 2009; Petitpierre et al. 2016). Some of the key features and processes describing the mechanism of plant invasion in an ecosystem have been shown in Fig. 11.2.

Due to their resilience to disturbance, invasive alien species have a greater chance of expanding their distribution under the changing climate scenario (Pauchard et al. 2009; Clements and Ditommaso 2011). Although higher elevation mountain ecosystems have been thought to be less susceptible to the plant invasions compared

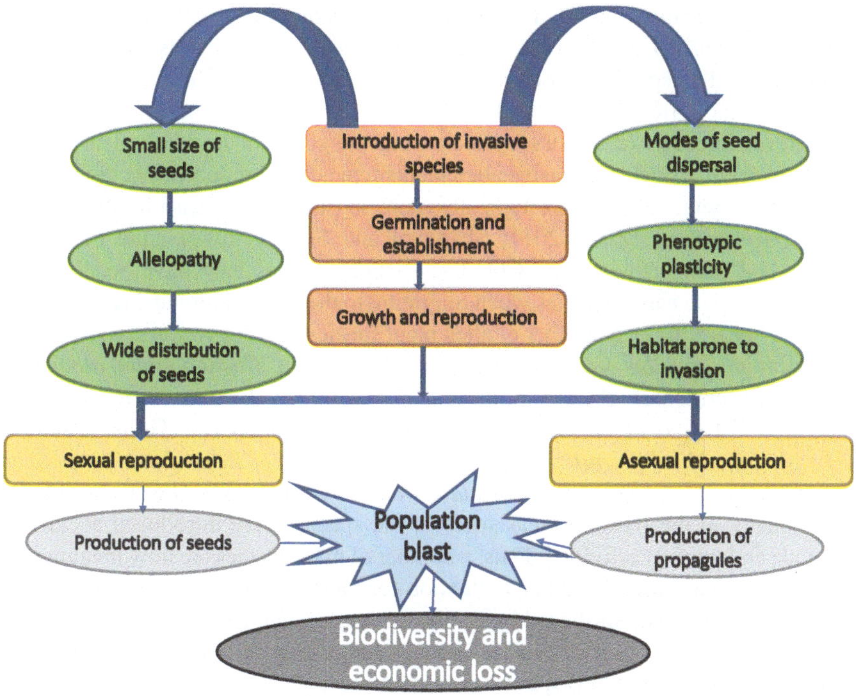

Fig. 11.1 A schematic representation of the process of an alien plant species arrival, establishment, and becoming invasive in an ecosystem

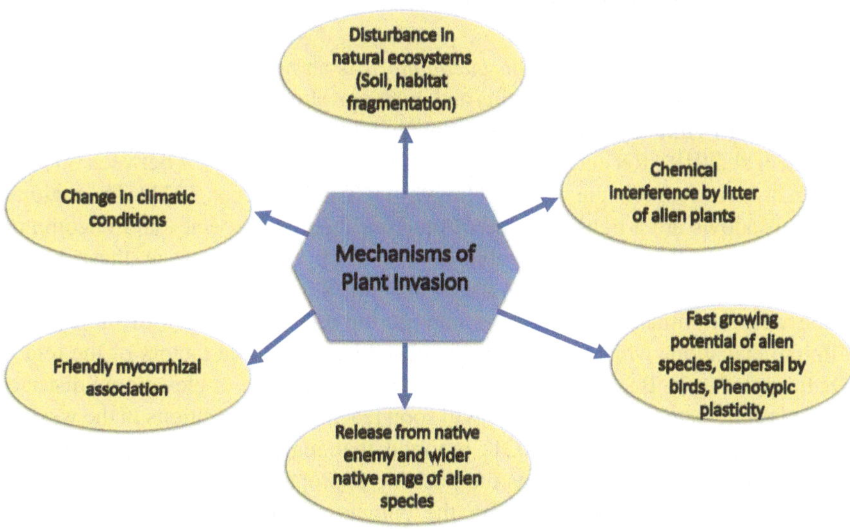

Fig. 11.2 Features of an invasive species and the surrounding ecological processes facilitating it in becoming invasive in an ecosystem

to the lower elevation mountain ecosystems (McDougall et al. 2011; Kueffer et al. 2013; Joshi et al. 2022), however, increased tourism in mountainous areas may help in spreading alien species, those are currently restricted to the mid-elevations and/or foothills of the IHR (Joshi et al. 2022). Climate change may contribute to a major and irreversible loss of species by increasing the cover of invasive plant species (Mittermeier et al. 2011). Invasive alien plants typically reproduce more quickly and propagate vegetatively and are typically more sensitive to increase in atmospheric CO_2 concentrations (Ziska 2011; Sheppard and Stanley 2014). Due to climate change and rising atmospheric CO_2 levels, these species may have benefits, and thus, outcompete native species when migrating to the higher altitudes. According to Chandra Sekar (2012), a number of alien plant species, e.g. *Ageratum conyzoides* L., *Catharanthus pusillus* G. Don., *Celosia argentea* L., *Chenopodium album* L., *Eichhornia crassipes* (Mart.) Solms-Laub., *Impatiens balsamina* L., *Ipomoea eriocarpa* R.Br., *Lantana camara* L., *Leucaena latisiliqua* (L.) Gillis, *Melilotus alba* Medik., *Mirabilis jalapa* L., *Passiflora foetida* L., *Pennisetum purpureum* (Schumach.) Morrone, and *Portulaca oleracea* L. have been introduced intentionally in the IHR, whereas some of the species were introduced accidently via trade and tourism activities. Thus, a number of plant species are now invading the mountainous landscapes and need attention for their judicious management so that the native biodiversity of these highly diverse regions can be sustained. In the next section, some of the major invasive species dominating the mountainous regions, particularly Indian Himalayas, have been described.

11.3 Examples of Major Plants Invading in the Mountain Ecosystems of India

Ageratina adenophora: *Ageratina adenophora* is an aggressive alien species that was formerly confined to the foothills and moderate altitudes in the IHR, particularly in the western Himalayan region. It is now exhibiting upward mobility in the IHR (Pathak et al. 2019). Recently a small population of *A. adenophora* between 300 and 2500 m asl was found in the western Himalaya (Chaudhary et al. 2021). The regions' alpine meadows are a treasure mine of precious medicinal species and uncommon plant elements, thus, the anticipated spread of *A. adenophora* at these higher elevations could be problematic in the near future. For its survival and propagation in the western Himalaya, the species may go through many stages of life cycle at different elevations which are mainly regulated by the abiotic factors of the region (Datta et al. 2017). Because of *A. adenophora*'s wide range of elevational distribution, it is anticipated that this species will soon reach higher elevations in the western Himalaya. Changejun et al. (2021) reported that regions with higher elevations (3000–3500 m) would be seen as the possible appropriate areas for the expansion of *A. adenophora*'s population with the climate change. Thus, due to its wide ecological amplitude and further risks of spread to the higher elevational ranges, there is a need of immediate management and planning considerations for this species (Pathak et al. 2019).

Anthemis cotula: *Anthemis cotula* L. (known as Mayweed or stinking chamomile) is an annual weed with an unpleasant aroma. It was first reported in the Mediterranean region (Kay 1971), and nowadays showed its spread throughout the world, particularly in dump soil in ditches, roadsides, and other disturbed agricultural lands (Adhikari et al. 2020). Its worldwide and local spread is thought to be anthropogenic, particularly through the movement of agricultural equipment and other vehicles which can contaminate seeds of the crops (Kay 1971; Shankar et al. 2012). However, the precise timeline and routes of global invasion of this species remain unknown. Despite its wider infestations as well as having risks of further spread in different new areas, the weed is having detrimental effects in the Mediterranean-like climates such as Kashmir valley in India and the Pacific Northwest region in the United States (Shah and Reshi 2007; Adhikari et al. 2020).

With its first occurrence reported in Kashmir Himalaya in 1972 (Stewart et al. 1972), this plant is currently thought to be one of the most damaging invaders in this region (Shah and Reshi 2007). It has a wide elevational distribution range (1600–2800 m asl) in the western Himalaya. *Anthemis cotula* favours soils that are N-rich, alkaline, rather dry, and warmer in nature (Ziada et al. 2014; Dar et al. 2023). It frequently occurs in disturbed areas as well as in agricultural and forest zones and has several negative ecological and economic impacts (Shah and Reshi 2007). Its prolonged recruitment pattern, which was facilitated by habitat disruption and large population size, has been linked to its invasion in disturbed and abandoned habitats in Kashmir Himalaya (Allaie et al. 2005). In addition, allelopathic activity (Allaie et al. 2006), coordinated germination, abundant achenes synthesis, and favourable environmental conditions facilitate *A. cotula*'s growth and survival even after seedling mortality (Rashid et al. 2007). Moreover, heavy litter production and faster decomposition, along with the Arbuscular Mycorrhizal fungi (AMF) association help this species to dominate the P-deficient regions of the Kashmir Himalayas (Shah and Reshi 2007; Dar et al. 2023). With the ongoing climate change, the species may shift its range to the newer regions of Kashmir Himalaya.

Lantana camara: *Lantana camara* L. is one amongst the top 100 worst invasive species in the world (Sharma et al. 2005; Dobhal et al. 2011). It is one of the dominant invaders in the wastelands, pastures and woodlands, and forests of the country—India (Sharma et al. 2005; Dobhal et al. 2011). Its distribution varies between 300 and 2500 m asl elevation ranges (Chaudhary et al. 2021). The plant has high capacity to flourish in various climatic conditions (Taylor et al. 2012), and produces numerous fruits which may help in its success in diverse environmental conditions (Kato-Noguchi and Kurniadie 2021). Moreover, the allelochemicals released by roots in the rhizospheric soil prevent the growth of local plants (Kong 2010; Hamad et al. 2022). The socioeconomic circumstances in locations where it has invaded can be changed by this species. Although *L. camara* is well recognised as a harmful invader of many terrestrial environments, little research has concentrated on the extent to which this weed affects the biodiversity of native ecosystems (Pyšek and Richardson 2010). Recent studies revealed that this plant is resulted in considerably lower species diversity, basal area, and biomass of other species in the Himalayan regions (Mandal and Joshi 2015). Fast growth rate and heavy litterfall of this species help in improved nutrient cycling, particularly organic C and N enrichment in the invaded area which further facilitate the plant invasion success (Kumar and Garkoti 2021). This species is posing serious threat to different Himalayan and Vindhyan highlands of India (Sharma et al. 2005; Kumar and Garkoti 2021).

Leucanthemum vulgare: *Leucanthemum vulgare* Lam. (known as Ox-eye daisy) is also one of worst invasive weed in the world. Outside of its native biogeographical region (i.e. Europe and Western Asia), *L. vulgare* swiftly expanded to Western Asia and Europe, and it eventually became naturalised globally, except Antarctica (Stutz et al. 2018). It is a well-known high-altitude invasive plant that prefers disturbed regions such as roadsides, heavily grazed pastures, open meadows, open forests, and woodlands for its spread and invasion (McDougall et al. 2018). *Leucanthemum vulgare* has the ability to colonise the higher elevation mountain ranges like alpine ecosystems where it poses a serious threat to the endemic plant diversity and essential ecosystem services supporting the life over there (Khuroo et al. 2009; McDougall et al. 2018). For example, *L. vulgare* invasion in Kashmir Himalaya, an ecologically fragile region, is a major problem (Khuroo et al. 2010). It was introduced as an ornamental plant in the regions during the British colonial era in India (Stewart et al. 1972), and nowadays has widely spread over different elevation ranges from 1300 to 2500 m asl (Khuroo et al. 2010). After being introduced, the plant spread into the surrounding natural environments at an alarming rate, eventually becoming a wild plant (Ahmad et al. 2019a). The rapid invasion in the higher elevation landscapes, particularly protected areas, is leading to a substantial decline in native species diversity and distribution (Khuroo et al. 2010). The invasion by *L. vulgare* has shown significant impact on soil physicochemical properties where it alters the nutrient dynamics for facilitating its own growth (Ahmad et al. 2019a). In a recent modelling-based study on the distribution (niche dynamics) of *L. vulgare* with changing climatic scenarios, Ahmad et al. (2019b) predicted a high risk of invasion of this species with a global increase in its habitat suitability. Oceania has been identified as the high-risk region for its invasion, and niche shifting was predicted for Asia, Africa, and South America continents (Ahmad et al. 2019b).

Parthenium hysterophorus: An annual herbaceous weed with Neotropical origins, *P. hysterophorus* L. has spread to many different parts of the world, including semi-arid tropical and subtropical, and warmer temperate zones (Rathee et al. 2021). The weed has largely been observed in India's tropical and subtropical climate zones but it has also been shown to be rapidly spreading at an elevation range of 1600–1700 m asl in the lower Himalayas (Khuroo et al. 2007). In the Nepal Himalayas, its maximum spread has been observed to reach up to 2000 m asl (Shrestha et al. 2019). Currently, the species has spread all over the country—India (Gnanavel and Natarajan 2013) and can be particularly observed flourishing in the riparian zones, seasonal floodplains, along the road edges, across rail-lines/roadsides, managed and unmanaged grasslands, pastures, open and/or disturbed forests, wastelands, and agricultural lands (Evans 1997; Srivastava and Raghubanshi 2021). As long as there is enough moisture content, *P. hysterophorus* can thrive throughout the year at temperatures ranging from 12 to 30 °C on average (Nguyen et al. 2017). *Parthenium hysterophorus* causes serious risks to endemic biodiversity, agricultural crops, and both human and animal health in the invaded areas (Shabbir and Bajwa 2006). Particularly notorious to the lowlands, the species is spreading rigorously in the higher elevation zones due to its wider phenotypic plasticity (Kaur et al. 2019; Rathee et al. 2021). Two morphotypes of the species help in adjusting it to different environmental conditions and may enhance the chances of its spread with the changing climatic conditions (Kaur et al. 2019).

Upto 30,000 seeds can be produced by the weed per plant (Evans 1997). Climate has a major role on how long *Parthenium* seeds can survive in soil seed banks (Nguyen et al. 2017). The development and physiological activity of *Parthenium* have been reported to benefit from increasing atmospheric CO_2 levels. According to Pandey et al. (2003), a high CO_2 concentration (upto 700 ppm) increased the *Parthenium* weed's net photosynthetic efficiency, photosynthetic rate, and water use efficiency while decreased the plant's need for light for net photosynthesis, stomatal conductance, and consequently transpiration rate. As a result, *Parthenium* weed is expected to exhibit a faster rate of development in an environment with more CO_2 and higher temperatures (Pandey et al. 2003). Climate change induced variabilities in temperature in the mountain regions (high elevations), which already have higher atmospheric CO_2 concentration, can be a potential habitat for this species in the near future. Therefore, there is a need to closely observe the invasion ecology of all these invasive species in the near future.

In addition to a few species elaborated above, a detailed list of plants invading in different Himalayan states of India is provided in Table 11.1. With the changing climatic scenario, there is a possibility that a number of invasive species can make room for their inclusion in the list of dominating invaders in the mountainous landscapes, if appropriate measures and policies are not taken into consideration.

11.4 Impact of Invasive Plants on Mountain Ecosystems

All around the world, biological invasions cause harm to natural ecosystems (Pyšek et al. 2020; Fantle-Lepczyk et al. 2022). The invasion of alien species has posed serious concerns to the local biodiversity that can have detrimental ecological and economic effects, and affects ecosystem services, environmental quality, and human health (Pyšek and Richardson 2010; Bartz and Kowarik 2019; Rai and Singh 2020). About 48 countries have been invaded by the highly invasive weed *Parthenium* (Chhogyel et al. 2021). As per the record, about 13,168 vascular plant species have become naturalised in 843 distinct areas of the globe (Van Kleunen et al. 2015). It accounts for around 3.9% of all the fauna (plants) that are present outside of their natural/native habitat range on the Earth (Van Kleunen et al. 2015). According to the Intergovernmental Platform for Biodiversity and Ecosystem Services (IPBES) of the United Nations (UN), biotic invaders threaten around one fifth of the Earth's surface, which also include the biodiversity hotspots of the world (Rai and Singh 2020). The increased trade and transportation operations in nations with high per capita incomes are to blame for this trend. It is broadly recognised as posing a significant risk to the

Table 11.1 Most aggressive invasive alien plant species (IAPS) reported in various Himalayan states of India

S. no.	States	Invasive alien species	References
1	Arunachal Pradesh	*Ageratum conyzoides* L., *Bidens pilosa* L., *Chromolaena odorata* L. R.M King and H., *Mikania micrantha* Kunth	Kosaka et al. (2010), Tripathi (2013)
2	Assam hills	*Ludwigia peruviana* (L.) Hara., *M. micrantha*	Barua et al. (2017)
3	Darjeeling (West Bengal) Hills	*Ageratina adenophora* (Spreng.) R.M. King & H. Rob, *Ageratum houstonianum* Mill.	Moktan and Das (2013)
4	Himachal Pradesh	*Ageratum conyzoides*, *Lantana camara* L., *Parthenium hysterophorus* L., *Sapium sebiferum* (L.) Roxb.	Kohli et al. (2004), Jaryan et al. (2013), Sekar et al. (2015), Rathee et al. (2021)
5	Jammu and Kashmir	*Ageratum conyzoides*, *Anthemis cotula* L., *Argemone mexicana* L., *Cassia tora* L., *L. camara*, *Leucanthemum vulgare* Lam., *P. hysterophorus*	Khuroo et al. (2007, 2012a, b), Shah and Reshi (2007), Singh and Dangwal (2014), Ahmad et al. (2019a, b), Dar et al. (2023)
6	Manipur	*Ageratina adenophora*, *A. conyzoides*, *A. houstonianum*, *M. micrantha*, *L. camara*	Tripathi (2013), Singh et al. (2015)
7	Meghalaya	*Ageratina adenophora*, *A. conyzoides*, *Artemisia nilagirica* L., *C. odorata*, *Imperata cylindrica* (L.) Raeuschel, *L. camara*, *Ligustrum robustum* (Roxb.) Blume	Singh et al. (2011a, b), Shankar et al. (2012), Tripathi (2013)
8	Mizoram	*Ageratum conyzoides*, *Eupatorium serotinum* Michx., *L. camara*, *M. micrantha*	State Action Plan on Climate Change-Mizoram (2012–2017), Rai (2015), Rai and Singh (2015), Rai and Singh (2021)
9	Nagaland	*Chromolaena odorata*, *L. camara*, *P. hysterophorus*	Naithani (1987), Tripathi (2013)
10	Sikkim	*Ageratina adenophora*, *C. odorata*, *Eupatorium riparium* Reg., *L. camara*, *M. micrantha*, *Rumex nepalensis* Spreng.	Tripathi and Yadav (1982), Tripathi et al. (2006), Sikkim Biodiversity Action Plan (2011)
11	Tripura	*Ageratum conyzoides*, *Alternanthera philoxeroides* (Mart.) Griseb., *C. odorata*, *Eclipta prostrata* (L.) L., *L. camara*, *M. micrantha*, *Mimosa pudica* L., *P. hysterophorus*	Debnath et al. (2015a, b), Debnath and Debnath (2017)
12	Uttarakhand	*Ageratina adenophora*, *A. conyzoides*, *L. camara*, *P. hysterophorus*, *Rubus niveus* Thunb.	Bughani and Rajwar (2005), Dobhal et al. (2011), Khanduri et al. (2017)

local biodiversity, ecological function, agricultural output, health, and socioeconomic stability (Bajwa et al. 2019). Once an invasive species has invaded an area, it has impact on all the components of the environment and changes the ecological conditions of the area (Gritti et al. 2006). For example, invasive plants affect the native plant species diversity, water availability, and soil nutrient quality (Charles and Dukes 2007). Moreover, invasive species affect the light availability and acquisition patterns, temperature levels, and solar radiation in the invaded areas which provide strong competition to the native species (Gritti et al. 2006). In addition, plant invasion has considerable effects on availability, distribution, and quality of resources, e.g. food, water, and building materials (Heckscher and Taylor 2014; Stewart et al. 2021).

There are several accounts of invasive alien species from different regions in India (Khuroo et al. 2012a, b; Niphadkar and Nagendra 2016). Invasive alien species such as *A. adenophora*, *A. conyzoides*, *L. camara*, *M. micrantha,* and *P. hysterophorus* have caused concern for native plant populations in India's plains as well as hills (Sharma et al. 2005; Dogra et al. 2009; Singh et al. 2014; Srivastava and Raghubanshi 2021). The emergence of new invasive alien plants in novel environments further arises the potential threats to the human and environmental health (Banerjee et al. 2021). For example, the introduction of invasive plants like *Opuntia stricta* (Haw.) Haw. or *L. camara* can result in inadequate feed, the sickness or even death of animals in the novel ecosystems/habitats (Reynolds et al. 2020). Moreover, introduction of invasive woody riparian species like Australian *Acacia* spp. has been identified as to reduce the water table as well as water quality by extracting the groundwater, which makes it more challenging for families and farmers to acquire freshwater (Chamier et al. 2012). IPBES, 2019's Global Assessment Report on Biodiversity and Ecosystem Services identified invasive alien species/invaders as the primary cause of biodiversity loss (Stevance et al. 2020). Thus, invasive species have considerable impact on socioeconomic and ecological dimensions on the Earth, and appropriate management approaches are the urgent need of the hour.

11.5 Suggestions for Managing Invasive Species in Mountain Ecosystems

The literature on invasive species distribution and impacts emphasises on the significance of a variety of anthropogenic disturbances, such as human settlements, road construction, intensive agricultural practices, and tourism. These anthropogenic activities act as drivers of the invasive alien plant invasions in the mountain ecosystems all over the world. Therefore, it is crucial to create environmental education programmes about the ecological risks and various effects of non-native invasive species, that inform visitors, park rangers, farmers, and inhabitants of these sensitive habitats. Moreover, it is also required to develop management and control protocols/measures that have regulation on the permit of non-native invasive species growth and dispersal in non-native and/or sensitive ecosystems. There is a need to develop programs that highlight the preservation of unaltered native vegetation in the periphery of national reserves and along their roadsides, in order to prevent the invasive species from further encroaching upon these protected ecosystems. In addition, it is crucial to regulate the agricultural and forestry activity in the national parks due to their significant impacts on propagule dissemination and creation of microhabitat conditions for the invasive species spread and establishment (Pauchard and Alaback 2004; Pauchard et al. 2016). Generally, the native flora that has not been altered can act as a biotic barrier to keep invasive species out. Overall, the establishment of invasive alien species and their subsequent dispersal towards comparatively less invaded high-elevation ranges could, therefore, be reduced by developing initiatives for conserving the undisturbed areas with native flora in the low elevation ranges and by avoiding the further land use changes.

11.6 Conclusion and Future Prospects

The spread and establishment of various non-native species to different elevation ranges are species-specific and mainly governed by the anthropogenic activities and climatic conditions. Compared to the abiotic factors, anthropogenic activities such as distance from human settlements and roads play major role in species diversity and distribution patterns at local scales. In addition to the anthropogenic activities, several abiotic factors such as pH and N content also play major role in explaining the distribution of non-native species at regional scales. It reflects that the anthropogenic activities and inherent abiotic factors are playing major role in comparison to the climatic variability in regulating the spread of invasive alien species along the altitudinal gradients in the mountain regions. However, the relative contribution of anthropogenic, biotic, and even abiotic (e.g. pH, N, etc.) underlying elements to different elevation ranges which may help in explaining the invasive species distribution patterns in mountain ecosystems is needed to be explored further. Overall, limiting the spread of human influences (tourism, infrastructure, and presence of animals), either by implementing direct control measures for the tourism and agricultural operations or through biosafety rules for regulating the alien species infestations is the need of the hour. We would not be able to regulate the spread of invasive species in the mountainous ecosystems until such steps are implemented.

Further, climatic barriers such as temperature and other harsh conditions act as natural regulators of plant invasion at the high-elevation ranges in the mountain ecosystems. Recently, a number of invasive species from the lowlands are reported to intrude the mid- to high-elevation ranges in different mountainous regions. Changes in temperature and precipitation patterns in addition to the wider phenotypic plasticity of the invasive plants provide an advantage to the invaders in shifting their ecological niches to the higher elevation ranges as compared to the native species. Therefore, there is need of coordinated exploration of anthropogenic measures along with the changing climatic scenarios on plant invasion dynamics in the mountain ecosystems. Several studies are being carried out in finding species distribution patterns under different representative concentration projection scenarios in various habitat conditions. However, there is a dire need to further emphasise on the global distribution patterns of some of the major invasive species with respect to changing climatic conditions.

Acknowledgements M.A. Dar is thankful to the Council of Scientific and Industrial Research (CSIR), New Delhi, India, and R. Chuskit thankfully acknowledge the University Grants Commission (UGC), New Delhi, India for providing financial assistance as Senior Research Fellowship. R. Singh gratefully acknowledges the financial support from DST-SERB, New Delhi as National Post-Doctoral Fellowship (Grant number: PDF/2020/001607).

References

Adhikari P, Jeon JY, Kim HW, Shin MS, Adhikari P, Seo C (2019) Potential impact of climate change on plant invasion in the Republic of Korea. J Ecol Environ 43:1–12

Adhikari S, Burke IC, Eigenbrode SD (2020) Mayweed chamomile (*Anthemis cotula* L.) biology and management—a review of an emerging global invader. Weed Res 60:313–322

Ahmad R, Khuroo AA, Hamid M, Rashid I (2019a) Plant invasion alters the physico-chemical dynamics of soil system: insights from invasive Leucanthemum vulgare in the Indian Himalaya. Environ Monit Assess 191:792

Ahmad R, Khuroo AA, Charles B, Hamid M, Rashid I, Aravind NA (2019b) Global distribution modelling, invasion risk assessment and niche dynamics of *Leucanthemum vulgare* (Ox-eye Daisy) under climate change. Sci Rep 9:11395

Alexander JM, Lembrechts JJ, Cavieres LA, Daehler C, Haider S, Kueffer C, Liu G, McDougall K, Milbau A, Pauchard A, Rew LJ (2016) Plant invasions into mountains and alpine ecosystems: current status and future challenges. Alp Bot 126:89–103

Allaie RR, Reshi Z, Rashid I, Wafai BA (2006) Effect of aqueous leaf leachate of *Anthemis cotula*—an alien invasive species on germination behaviour of some field crops. J Agron Crop Sci 192:186–191

Allaie RR, Reshi Z, Wafai BA (2005) Demographic studies on alien invasive *Anthemis cotula* L. Kashmir Himalaya Trends Life Sci 20:49–56

Bajwa AA, Farooq M, Nawaz A, Yadav L, Chauhan BS, Adkins S (2019) Impact of invasive plant species on the livelihoods of farming households: evidence from *Parthenium hysterophorus* invasion in rural Punjab, Pakistan. Biol Invasions 21:3285–3304

Banerjee AK, Khuroo AA, Dehnen-Schmutz K, Pant V, Patwardhan C, Bhowmick AR, Mukherjee A (2021) An integrated policy framework and plan of action to prevent and control plant invasions in India. Environ Sci Pol 124:64–72

Bartz R, Kowarik I (2019) Assessing the environmental impacts of invasive alien plants: a review of assessment approaches. NeoBiota 43:69–99

Barua IC, Deka J, Devi M, Deka RL, Moran J (2017) Weeds as emerging threat to biodiversity: a consequence of spread of Ludwigia peruviana in Dhansiri and Kopili catchment areas of Assam, North East India. Curr Sci 112:1904–1914

Bellard C, Bernery C, Leclerc C (2021) Looming extinctions due to invasive species: irreversible loss of ecological strategy and evolutionary history. Glob Change Biol 27:4967–4979

Bogale GA, Tolossa TT (2021) Climate change intensification impacts and challenges of invasive species and adaptation measures in Eastern Ethiopia. Sustain Environ 7:1875555

Bughani I, Rajwar GS (2005) Primary productivity and the impacts of the exotic weed *Eupatorium glandulosum* in a montane grassland of Garhwal Himalaya. Environ Syst Decis 25:31

Carvalho F, Pascoal C, Cássio F, Teixeira A, Sousa R (2022) Combined per-capita and abundance effects of an invasive species on native invertebrate diversity and a key ecosystem process. Freshw Biol 67:828–841

Catford JA, Bode M, Tilman D (2018) Introduced species that overcome life history tradeoffs can cause native extinctions. Nat Commun 9:1–7

Chamier J, Schachtschneider K, Le Maitre DC, Ashton PJ, Van Wilgen BW (2012) Impacts of invasive alien plants on water quality, with particular emphasis on South Africa. Water SA 38: 345–356

Changjun G, Yanli T, Linshan L, Bo W, Yili Z, Haibin Y, Xilong W, Zhuoga Y, Binghua Z, Bohao C (2021) Predicting the potential global distribution of *Ageratina adenophora* under current and future climate change scenarios. Ecol Evol 11:12092–120113

Charles H, Dukes JS (2007) Impacts of invasive species on ecosystem services. In: Nentwig W (ed) Biol invasions. Springer, Berlin, Heidelberg, pp 217–237

Chaudhary A, Sarkar MS, Adhikari BS, Rawat GS (2021) *Ageratina adenophora* and *Lantana camara* in Kailash Sacred Landscape, India: current distribution and future climatic scenarios through modelling. PLoS One 16:e0239690

Chhogyel N, Kumar L, Bajgai Y (2021) Invasion status and impacts of parthenium weed (*Parthenium hysterophorus*) in West-Central region of Bhutan. Biol Invasions 23:2763–2779

Clements DR, Ditommaso A (2011) Climate change and weed adaptation: can evolution of invasive plants lead to greater range expansion than forecasted? Weed Res 51:227–240

Datta A, Kühn I, Ahmad M, Michalski S, Auge H (2017) Processes affecting altitudinal distribution of invasive *Ageratina adenophora* in western Himalaya: the role of local adaptation and the importance of different life-cycle stages. PLoS One 12:e0187708

Dar MA, Ahmad M, Singh R, Kohli RK, Singh HP, Batish DR (2023) Invasive plants alter soil properties and nutrient dynamics: A case study of Anthemis cotula invasion in Kashmir Himalaya. Catena 226:107069

Debnath A, Debnath B (2017) Diversity, invasion status and usages of alien plant species in northeastern hilly state of Tripura: a confluence of Indo-Barman hotspot. Am J Plant Sci 8: 212–235

Debnath B, Debnath A, Paul C (2015a) Diversity of invasive alien weeds in the major roadside areas of Tripura and their effect and uses. J Chem Biol Phys Sci 5:3091–3102

Debnath B, Debnath A, Paul C (2015b) Diversity of invasive plant species in Trishna Wildlife Sanctuary, Tripura, North East India. Life Sci Leafl 70:9–21

Diagne C, Leroy B, Gozlan RE, Vaissière AC, Assailly C, Nuninger L, Roiz D, Jourdain F, Jarić I, Courchamp F (2020) InvaCost, a public database of the economic costs of biological invasions worldwide. Sci Data 7:277

Diez JM, D'Antonio CM, Dukes JS, Grosholz ED, Olden JD, Sorte CJ, Blumenthal DM, Bradley BA, Early R, Ibáñez I, Jones SJ (2012) Will extreme climatic events facilitate biological invasions? Front Ecol Environ 10:249–257

Dobhal PK, Kohli RK, Batish DR (2011) Impact of *Lantana camara* L. invasion on riparian vegetation of Nayar region in Garhwal Himalayas (Uttarakhand, India). J Ecol Nat Environ 3: 11–22

Dogra KS, Kohli RK, Sood SK, Dobhal PK (2009) Impact of *Ageratum conyzoides* L. on the diversity and composition of vegetation in the Shivalik hills of Himachal Pradesh (Northwestern Himalaya). India Int J Biodivers Conserv 1:135–145

Esler KJ, Prozesky H, Sharma GP, McGeoch M (2010) How wide is the "knowing-doing" gap in invasion biology? Biol Invasions 12:4065–4075

Evans HC (1997) *Parthenium hysterophorus*. a review of its weed status and the possibilities for biological control. Biocontrol News Inform 18:89–98

Fantle-Lepczyk JE, Haubrock PJ, Kramer AM, Cuthbert RN, Turbelin AJ, Crystal-Ornelas R, Diagne C, Courchamp F (2022) Economic costs of biological invasions in the United States. Sci Total Environ 806:151318

Foster JG, Gervan CA, Coghill MG, Fraser LH (2021) Are arthropod communities in grassland ecosystems affected by the abundance of an invasive plant? Oecologia 196:1–2

Fuentes-Lillo E, Lembrechts JJ, Cavieres LA, Jiménez A, Haider S, Barros A, Pauchard A (2021) Anthropogenic factors overrule local abiotic variables in determining non-native plant invasions in mountains. Biol Invasions 23:3671–3686

Gazoulis I, Antonopoulos N, Kanatas P, Karavas N, Bertoncelj I, Travlos I (2022) Invasive alien plant species—raising awareness of a threat to biodiversity and Ecological Connectivity (EC) in the Adriatic-Ionian Region. Diversity 14:387

Gnanavel I, Natarajan SK (2013) *Parthenium hysterophorus* L.: a major threat to natural and agro eco-systems in India. Int J Agric Environ Biotech 6:261–269

Goyal N, Shah K, Sharma GP (2018) Does intrinsic light heterogeneity in *Ricinus communis* L. monospecific thickets drive species' population dynamics? Environ Monitor Assess 190:1–2

Gritti ES, Smith B, Sykes MT (2006) Vulnerability of Mediterranean Basin ecosystems to climate change and invasion by exotic plant species. J Biogeogr 33:145–157

Haider S, Alexander J, Dietz H, Trepl L, Edwards PJ, Kueffer C (2010) The role of bioclimatic origin, residence time and habitat context in shaping non-native plant distributions along an altitudinal gradient. Biol Invasions 12:4003–4018

Hakim N, Ahmad M, Rathee S, Sharma P, Kaur S, Batish DR, Singh HP (2023) Invasive Cirsium arvense displays different resource-use strategies along local habitat heterogeneity in the trans-Himalayan region of Ladakh. Environ Monit Assess 195:730

Hamad AA, Kashaigili JJ, Eckert S, Eschen R, Schaffner U, Mbwambo JR (2022) Impact of invasive *Lantana camara* on maize and cassava growth in East Usambara, Tanzania. Plant-Environ Interact 3:193–202

Heckscher CM, Taylor SM (2014) Veery (*Catharus fuscescens*) nest architecture and the use of alien plant parts. Am Midl Nat 171:157–164

Hui C, Richardson DM, Landi P, Minoarivelo HO, Garnas J, Roy HE (2016) Defining invasiveness and invasibility in ecological networks. Biol Invasions 18:971–983

Hulme PE (2006) Beyond control: wider implications for the management of biological invasions. J Appl Ecol 43:835–847

Hulme PE (2017) Climate change and biological invasions: evidence, expectations, and response options. Biol Rev 92:1297–1313

Hulme PE (2021) Unwelcome exchange: international trade as a direct and indirect driver of biological invasions worldwide. One Earth 4:666–679

IPCC (2014). Fifth Assessment Report. Chapter 1: Introduction Chapter, pp. 1–65

Jaryan V, Chopra S, Uniyal SK, Singh RD (2007) Spreading fast yet unnoticed: are we in for another invasion? Curr Sci 93:1483

Jaryan V, Uniyal SK, Gupta RC, Singh RD (2013) Alien flora of Indian Himalayan state of Himachal Pradesh. Environ Monit Assess 185:6129–6153

Joshi S, Shrestha BB, Shrestha L, Rashid I, Adkins S (2022) Plant invasions in mountains. In: Clements DR, Upadhyaya MK, Joshi S, Shrestha A (eds) Global plant invasions. Springer, Cham, pp 279–300

Chandra Sekar K (2012) Invasive alien plants of Indian Himalayan region—diversity and implication. Am J Plant Sci 3:177–184

Kariyawasam CS, Kumar L, Ratnayake SS (2021) Potential distribution of aquatic invasive alien plants, *Eichhornia crassipes* and *Salvinia molesta* under climate change in Sri Lanka. Wetl Ecol Manag 29:531–545

Kato-Noguchi H, Kurniadie D (2021) Allelopathy of *Lantana camara* as an invasive plant. Plan Theory 10:1028

Kaur A, Kaur S, Singh HP, Batish DR, Kohli RK (2019) Phenotypic variations alter the ecological impact of invasive alien species: lessons from *Parthenium hysterophorus*. J Environ Manag 241:187–197

Kay QON (1971) Biological flora of the British Isles *Anthemis arvensis* L. J Ecol 59:637–648

Khanduri AR, Biswas SA, Vasistha HB, Rathod D, Jha SK (2017) A status of invasive alien species plant diversity in Tehri district forest ecosystem of Garhwal Himalayan Region. Curr World Environ 12:377–388

Khuroo AA, Malik AH, Reshi ZA, Dar GH (2010) From ornamental to detrimental: plant invasion of *Leucanthemum vulgare* Lam. (Ox-eye Daisy) in Kashmir valley. India Curr Sci 98:600–602

Khuroo AA, Rashid I, Reshi Z, Dar GH, Wafai BA (2007) The alien flora of Kashmir Himalaya. Biol Invasions 9:269–292

Khuroo AA, Reshi ZA, Malik AH, Weber E, Rashid I, Dar GH (2012a) Alien flora of India: taxonomic composition, invasion status and biogeographic affiliations. Biol Invasions 14:99–113

Khuroo AA, Weber E, Malik AH, Reshi ZA, Dar GH (2011) Altitudinal distribution patterns of the native and alien woody flora in Kashmir Himalaya, India. Environ Res 111:967–977

Khuroo AA, Reshi ZA, Dar GH, Hamal IA (2012b) Plant invasions in Jammu and Kashmir state, India. In: Invasive alien plants: an ecological appraisal for the Indian subcontinent. CABI, Wallingford, UK, pp 216–226

Khuroo AA, Reshi Z, Rashid I, Dar GH, Malik AH (2009) Plant invasions in montane ecosystems. Front Ecol Environ 7:408–408

Kohli RK, Dogra KS, Batish DR, Singh HP (2004) Impact of invasive plants on the structure and composition of natural vegetation of northwestern Indian Himalayas. Weed Technol 18:1296–1300

Kong CH (2010) Ecological pest management and control by using allelopathic weeds (*Ageratum conyzoides*, *Ambrosia trifida*, and *Lantana camara*) and their allelochemicals in China. Weed Biol Manag 10:73–80

Kosaka Y, Saikia B, Mingki T, Tag H, Riba T, Ando K (2010) Roadside distribution patterns of invasive alien plants along an altitudinal gradient in Arunachal Himalaya, India. Mt Res Dev 30: 252–258

Kueffer C, McDougall K, Alexander J, Daehler C, Edwards P, Haider S, Milbau A, Parks C, Pauchard A, Reshi ZA, Rew LJ (2013) Plant invasions into mountain protected areas: assessment, prevention and control at multiple spatial scales. In: Foxcroft L, Pyšek P, Richardson D, Genovesi P (eds) Invasion ecology: plant invasions in protected areas. Invading Nature. Springer, pp 89–113

Kumar M, Garkoti SC (2021) Functional traits, growth patterns, and litter dynamics of invasive alien and co-occurring native shrub species of chir pine forest in the central Himalaya, India. Plant Ecol 222:723–735

Lamsal P, Kumar L, Aryal A, Atreya K (2018) Invasive alien plant species dynamics in the Himalayan region under climate change. Ambio 47:697–710

Lázaro-Lobo A, Ervin GN (2021) Wetland invasion: a multi-faceted challenge during a time of rapid global change. Wetlands 41:64

Lugo AE (2009) Novel tropical forests: the natural outcome of climate and land cover changes. International Institute of Tropical Forestry, Río Piedras, San Juan, Puerto Rico

Lustig A, Worner SP, Pitt JP, Doscher C, Stouffer DB, Senay SD (2017) A modeling framework for the establishment and spread of invasive species in heterogeneous environments. Ecol Evol 7: 8338–8348

Mandal G, Joshi SP (2015) Eco-physiology and habitat invasibility of an invasive, tropical shrub (*Lantana camara*) in western Himalayan forests of India. Forest Sci Technol 11:182–196

Manish K (2021) Species richness, phylogenetic diversity and phylogenetic structure patterns of exotic and native plants along an elevational gradient in the Himalaya. Ecol Process 10:1–3

Marini L, Gaston KJ, Prosser F, Hulme PE (2009) Contrasting response of native and alien plant species richness to environmental energy and human impact along alpine elevation gradients. Glob Ecol Biogeogr 18:652–661

McDougall K, Wright G, Peach E (2018) Coming to terms with Ox-eye Daisy (*Leucanthemum vulgare*) in Kosciuszko National Park, New South Wales. Ecol Manage Restor 19:4–13

McDougall KL, Khuroo AA, Loope LL, Parks CG, Pauchard A, Reshi ZA, Rushworth I, Kueffer C (2011) Plant invasions in mountains: global lessons for better management. Mt Res Dev 31: 380–387

McGeoch M, Jetz W (2019) Measure and reduce the harm caused by biological invasions. One Earth 1:171–174

Mittermeier RA, Turner WR, Larsen FW, Brooks TM, Gascon C (2011) Global biodiversity conservation: the critical role of hotspots. In: Zachos F, Habel J (eds) Biodiversity hotspots. Springer, pp 3–22

Moktan S, Das AP (2013) Diversity and distribution of invasive alien plants along the altitudinal gradient in Darjiling Himalaya, India. Pleione 7:305–313

Naithani HB (1987) *Parthenium hysterophorus* a pernicious weed in Arunachal Pradesh and Nagaland. Indian For 113:709–710

Nandi SK, Rawat S (2019) In vitro propagation of selected Himalayan medicinal herbs: conservation and sustainable utilization. In: Tripathi SK, Upadhyaya K, Hegde N (eds) Medicinal plants of India: conservation and sustainable use. Today & Tomorrow's Printers and Publishers, New Delhi, pp 395–407

Negi VS, Tiwari DC, Singh L, Thakur S, Bhatt ID (2022) Review and synthesis of climate change studies in the Himalayan region. Environ Dev Sustain 24:10471–10502

Nguyen T, Bajwa AA, Navie S, O'donnell C, Adkins S (2017) Parthenium weed (*Parthenium hysterophorus* L.) and climate change: the effect of CO2 concentration, temperature, and water deficit on growth and reproduction of two biotypes. Environ Sci Pollut Res 24:10727–10739

Niphadkar M, Nagendra H (2016) Remote sensing of invasive plants: incorporating functional traits into the picture. Int J Remote Sens 37:3074–3085

NITI Aayog (2023) Sustainable development in the Indian Himalayan region. https://niti.gov.in/sustainable-development-indian-himalayan-region. Accessed on 30.04.2023

Palni LM, Rawal RS (2010) Conservation of Himalayan bioresources: an ecological, economical and evolutionary perspective. In: Sharma VP (ed) Nature at work: ongoing saga of evolution. Springer, New Delhi, pp 369–402

Pandey DK, Palni LM, Joshi SC (2003) Growth, reproduction, and photosynthesis of ragweed parthenium (*Parthenium hysterophorus*). Weed Sci 51:191–201

Pathak HN, Bhuju DR, Shrestha BB, Ranjitkar S (2021) Impacts of invasive alien plants on ecosystem services of Ramsar lake cluster in middle mountain Nepal. Glob Ecol Conserv 27: e01597

Pathak R, Negi VS, Rawal RS, Bhatt ID (2019) Alien plant invasion in the Indian Himalayan Region: state of knowledge and research priorities. Biodivers Conserv 28:3073–3102

Pauchard A, Alaback PB (2004) Influence of elevation, land use, and landscape context on patterns of alien plant invasions along roadsides in protected areas of South-Central Chile. Conserv Biol 18:238–248

Pauchard A, Kueffer C, Dietz H, Daehler CC, Alexander J, Edwards PJ, Arévalo JR, Cavieres LA, Guisan A, Haider S, Jakobs G (2009) Ain't no mountain high enough: plant invasions reaching new elevations. Front Ecol Environ 7:479–486

Pauchard A, Milbau A, Albihn A, Alexander J, Burgess T, Daehler C, Englund G, Essl F, Evengård B, Greenwood GB, Haider S (2016) Non-native and native organisms moving into high elevation and high latitude ecosystems in an era of climate change: new challenges for ecology and conservation. Biol Invasions 18:345–353

Petitpierre B, McDougall K, Seipel T, Broennimann O, Guisan A, Kueffer C (2016) Will climate change increase the risk of plant invasions into mountains? Ecol Appl 26:530–544

Pradhan SP, Siddique T (2020) Stability assessment of landslide-prone road cut rock slopes in Himalayan terrain: a finite element method based approach. J Rock Mech Geotech Eng 12:59–73

Pyšek P, Hulme PE, Simberloff D, Bacher S, Blackburn TM, Carlton JT, Dawson W, Essl F, Foxcroft LC, Genovesi P, Jeschke JM (2020) Scientists' warning on invasive alien species. Biol Rev 95:1511–1534

Pyšek P, Richardson DM (2010) Invasive species, environmental change and management, and health. Annu Rev Environ Res 35:25–55

Rai PK (2015) Plant invasion ecology of an indo-Burma hot spot region along the disturbance gradient: a case study. Int J Environ Sci 4:108–114

Rai PK, Singh JS (2020) Invasive alien plant species: their impact on environment, ecosystem services and human health. Ecol Indic 111:106020

Rai PK, Singh JS (2021) Plant invasion in protected areas, the Indian Himalayan region, and the North East India: progress and prospects. Proc Indian Natl Sci Acad 87:19–35

Rai PK, Singh MM (2015) Lantana camara invasion in urban forests of an Indo–Burma hotspot region and its ecosustainable management implication through biomonitoring of particulate matter. J Asian Pac Biodivers 8(4):375–381

Rashid I, Reshi Z, Allaie RR, Wafai BA (2007) Germination ecology of invasive alien Anthemis cotula helps it synchronise its successful recruitment with favourable habitat conditions. Ann Appl Biol 150:361–369

Rathee S, Ahmad M, Sharma P, Singh HP, Batish DR, Kaur S, Kohli RK (2021) Biomass allocation and phenotypic plasticity are key elements of successful invasion of *Parthenium hysterophorus* at high elevation. Environ Exp Bot 184:104392

Rawal RS, Bhatt ID, Sekar KC, Nandi SK (2013) The Himalayan biodiversity: richness, representativeness, uniqueness and life-support values. GB Pant Institute of Himalayan Environment and Development (GBPIHED), Kosi-Katarmal, Almora, Uttarakhand, India, p 84

Reisinger A, Clark H (2018) How much do direct livestock emissions actually contribute to global warming? Glob Chang Biol 24:1749–1761

Reynolds C, Venter N, Cowie BW, Marlin D, Mayonde S, Tocco C, Byrne MJ (2020) Mapping the socio-ecological impacts of invasive plants in South Africa: are poorer households with high ecosystem service use most at risk? Ecosyst Serv 42:101075

Sekar KC, Aseesh P, Srivastava S, Giri L (2015) Invasive alien plants of Himachal Pradesh, India. Indian For 141:520–527

Shabbir A, Bajwa R (2006) Distribution of parthenium weed (*Parthenium hysterophorus* L.), an alien invasive weed species threatening the biodiversity of Islamabad. Weed Biol Manag 6:89–95

Shah K, Sharma GP (2022) A missing cog in the wheel: an Allee effects perspective in biological invasion paradigm. Environ Monit Asses 194:7

Shah MA, Reshi Z (2007) Invasion by alien *Anthemis cotula* L. in a biodiversity hotspot: release from native foes or relief from alien friends? Curr Sci 92:21–22

Shankar U, Yadav AS, Rai JP, Tripathi RS (2012) Status of alien plant invasions in the North-eastern region of India. In: Bhatt JR, Singh JS, Singh SP, Tripathi RS, Kohli RK (eds) Invasive alien plants: an ecological appraisal for the Indian subcontinent. CABI, pp 174–188

Sharma GP, Raghubanshi AS, Singh JS (2005) Lantana invasion: an overview. Weed Biol Manag 5:157–165

Sheppard CS, Stanley MC (2014) Does elevated temperature and doubled CO2 increase growth of three potentially invasive plants? Invasive Plant Sci Manag 7:237–246

Shrestha BB, Pokhrel K, Paudel N, Poudel S, Shabbir A, Adkins SW (2019) Distribution of *Parthenium hysterophorus* and one of its biological control agents (Coleoptera: *Zygogramma bicolorata)* in Nepal. Weed Res 59:467–478

Shrestha UB, Shrestha BB (2019) Climate change amplifies plant invasion hotspots in Nepal. Divers Distrib 25:1599–1612

Shrestha BB, Shrestha KK (2021) Invasions of alien plant species in Nepal: patterns and process. In: Pullaiah T, Lelmini MR (eds) Invasive alien species: observations and issues from around the world. Wiley, pp 168–183

Sikkim Biodiversity Action Plan (2011) Sikkim Biodiversity Conservation and Forest Management Project (SBFP) Forest, Environment and Wildlife Management Department Government of Sikkim, p. 44.

Singh A, Dangwal LR (2014) Noxious weeds of district Rajouri, Jammu and Kasmir, India. World J Pharm Pharm Sci 3:1442–1451

Singh B, Phukan SJ, Sinha BK, Singh VN, Borthakur SK (2011a) Conservation strategies for nepenthes Khasiana in the Nokrek biosphere reserve of Garo Hills, Northeast, India. Int J Conserv Sci 2:55–64
Singh C, Chauhan N, Upadhyay SK, Singh R, Rani A (2021) The Himalayan natural resources: challenges and conservation for sustainable development. J Pharmacogn Phytochem 10:1643–1648
Singh HP, Batish DR, Dogra KS, Kaur S, Kohli RK, Negi A (2014) Negative effect of litter of invasive weed Lantana camara on structure and composition of vegetation in the lower Siwalik Hills, northern India. Environ Monit Assess 186:3379–3389
Singh SP, Bassignana-Khadka I, Singh Karky B, Sharma E (2011b) Climate change in the Hindu Kush-Himalayas: the state of current knowledge. International Centre for Integrated Mountain Development (ICIMOD)
Singh TB, Das AK, Singh PK, Singh TB, Das AK, Singh PK (2015) Study of alien and invasive flora of Valley District of Manipur and their control. Int J Innov Res Sci Tech 1:616–626
Sintayehu DW (2018) Impact of climate change on biodiversity and associated key ecosystem services in Africa: a systematic review. Ecosyst Health Sust 4:225–239
Srivastava P, Raghubanshi AS (2021) Impact of Parthenium hysterophorus L. invasion on soil nitrogen dynamics of grassland vegetation of Indo-Gangetic plains, India. Environ Monit Assess 193:286
State Action Plan on Climate Change: Mizoram (2012–2017) Directorate of Science and Technology, Govenrment of Mizoram
Stevance AS, Bridgewater P, Louafi S, King N, Beard TD Jr, Van Jaarsveld AS, Ofir Z, Kohsaka R, Jenderedijan K, Rosales Benites M, Mulongoy KJ (2020) The 2019 review of IPBES and future priorities: reaching beyond assessment to enhance policy impact. Ecosyst People 16:70–77
Stewart PS, Hill RA, Stephens PA, Whittingham MJ, Dawson W (2021) Impacts of invasive plants on animal behaviour. Ecol Lett 24:891–907
Stewart RR, Ali SI, Nasir E (1972) An annotated catalogue of the vascular plants of West Pakistan and Kashmir. Printed at Fakhri Print
Stutz S, Mraz P, Hinz HL, Müller-Schärer H, Schaffner U (2018) Biological invasion of oxeye daisy (Leucanthemum vulgare) in North America: pre-adaptation, post-introduction evolution, or both? PLoS One 13:e0190705
Taylor S, Kumar L, Reid N, Kriticos DJ (2012) Climate change and the potential distribution of an invasive shrub, Lantana camara L. PLoS One 7:e35565
Tito R, Vasconcelos HL, Feeley KJ (2020) Mountain ecosystems as natural laboratories for climate change experiments. Front Forests Glob Change 3:1–8
Tripathi AK, Pandey PC, Sharma JK, Triantakonstantis D, Srivastava PK (2022) Climate change and its impact on forest of Indian Himalayan Region: a review. In: Rani S, Kumar R (eds) Climate change: impacts, responses and sustainability in the Indian Himalaya. Springer, pp 207–222
Tripathi RS, Kushwaha SP, Yadav AS (2006) Ecology of three invasive species of Eupatorium: a review. Int J Ecol Environ Sci 32:301–326
Tripathi RS, Yadav AS (1982) Population regulation of *Eupatorium adenophorum* Spreng. and E. riparium Regel: effect of population density, soil nitrogen and light intensity. Plant Soil 65: 35–49
Tripathi RS (2013) Alien plant invasion and its ecological implications: an Indian perspective with particular reference to biodiversity-rich regions. In: Invasive plant ecology 2013 Jan 9. CRC Press, pp 152–161
Van Kleunen M, Dawson W, Essl F, Pergl J, Winter M, Weber E, Kreft H, Weigelt P, Kartesz J, Nishino M, Antonova LA (2015) Global exchange and accumulation of non-native plants. Nature 525:100–103
Vilà M, Corbin JD, Dukes JS, Pino J, Smith SD (2007) Linking plant invasions to global environmental change. In: Canadell JG, Pataki DE, Pitelka LF (eds) Global change—the IGBP series: terrestrial ecosystems in a changing world. Springer, pp 93–102

Wu D, Zhao X, Liang S, Zhou T, Huang K, Tang B, Zhao W (2015) Time-lag effects of global vegetation responses to climate change. Glob Chang Biol 21:3520–3531

Wu ZY, Milne RI, Liu J, Nathan R, Corlett RT, Li DZ (2022) The establishment of plants following long-distance dispersal. Trends Ecol Evol 38:289–300

Ziada MA, El-Sherbeny GA, Askar MR (2014) Ecology and phytochemistry of *Anthemis cotula* L. in Egypt. Asian J Plant Sci 13:156–163

Ziska LH, Blumenthal DM, Franks SJ (2019) Understanding the nexus of rising CO_2, climate change, and evolution in weed biology. Invasive Plant Sci Manag 12:79–88

Ziska LH (2011) Global climate change and carbon dioxide: assessing weed biology and management. In: Handbook on climate change and agroecosystems: impacts, adaptation, and mitigation. Imperial College Press, London, pp 191–208

Part III

Plant Invasion: Assessment, Mapping and Forecasting

The Role of Epigenetics on Plant Invasions Under Climate Change Scenario

12

Mehmet Arslan, İlhan Üremiş, and Ahmet Uludağ

Abstract

Epigenetic mechanisms such as DNA methylation, histone modifications, and changes in the expression of non-coding RNAs are sensitive to the environmental variations which permit exotic species to adapt and invade new environments or vice versa. Different mechanisms of invasiveness such as phenotypic plasticity, enemy release, empty niches, propagule pressure, adaptive mutations, genetic variations, and epigenetic changes enable the introduced organisms to become invasive in their new environments. Among the diverse mechanisms that govern invasion, epigenetics can assist invasion by regulating gene expression without altering the DNA sequence. Plants have the ability to adapt to their new environments by modifying gene expression patterns by epigenetic modifications that affect plant growth and development. Epigenetic modifications are inherited through mitotic cell divisions, and they can be transmitted to the next generation. The role of epigenetic mechanisms in the adaptation of invasive plant species is one of the most exciting areas in weed science. Recent advances in molecular genetics have highlighted the role of epigenetic modifications on invasiveness. Environmental exposures can affect genes' function without changing the DNA sequence. Epigenetic mechanisms are considered essential for stress memories and adaptation in plants under stressed environments, which will increase under climate change in many areas of the world. Epigenetic mechanisms have been

M. Arslan
Erciyes University, Talas/Kayseri, Türkiye
e-mail: mehmetarslan@erciyes.eud.tr

İ. Üremiş
Hatay Mustafa Kemal University, Antakya, Hatay, Türkiye

A. Uludağ (✉)
Çanakkale Onsekiz Mart University, Çanakkale, Türkiye

S. Tripathi et al. (eds.), *Plant Invasions and Global Climate Change*,
https://doi.org/10.1007/978-981-99-5910-5_12

reported in most invasive plant species. To predict and prevent future invasions and effectively manage existing invaders, it is crucial to understand the relative contributions of the epigenetic basis of phenotypic variations occurring in the course of adaption to a new environment. To understand invasions, we present the role of epigenetic mechanisms that would allow the alien species to become invasive in the newly introduced environments.

Keywords

Adaptation · Epigenetics · Invasive alien plant · Invasion · Molecular ecology

12.1 Introduction

Invasive plant species are non-indigenous species that have been introduced, accidentally or deliberately into new natural environments where they proliferate, persist, and threaten human and animal health, forests, agriculture, and native biodiversity worldwide (Thomas et al. 2004; Mooney et al. 2005; Pysek et al. 2012; Early et al. 2016). The number of invasive plants in each environment further increased with the increased rate of trade and transport globally. Currently, the number of estimated invasive plant species is about 6500 in the terrestrial and aquatic ecosystems (Early et al. 2016). It was estimated that the number of introduced alien plant species in each continent may raise on average by 18% from 2005 to 2050 (Seebens et al. 2020). The properties that are responsible for the invasiveness were attributed to rapid adaptation, high growth with a high reproduction rate, higher phenotypic plasticity, rapid evolutionary change, high tolerance to biotic and abiotic stresses, and great dispersal ability in the new environments (Hellmann et al. 2008; Wolkovich and Cleland 2014). Invasive plants have significant effects on ecosystems since they can noticeably alter the species composition, structure, and function of the ecosystems (Vilà et al. 2011; Pyšek et al. 2012, 2020; Linders et al. 2019).

Climate changes can fluctuate significantly with geography and time. Global climate changes have altered precipitation patterns: arid and semiarid regions of the earth are becoming drier, while other regions, especially mid-to-high latitudes, are becoming wetter. In the regions in which the precipitation increased, there was an unequal increase in the occurrence of the heaviest precipitation events (Alexander et al. 2006; Brunner et al. 2021). The small changes in global climate can cause noticeable effects on a wide variety of natural ecosystems (Canturk and Kulaç 2021; Pörtner et al. 2022).

Climate change is an irrefutable, pervasive, and treacherous global crisis that somewhat affects every life form living on the earth, despite efforts to reduce the risks. The climate of the earth has changed throughout earth's history due to volcanic activities, the energy output of the sun, changes in the Earth's orbit, the terrestrial movement of the earth masses (Houghton et al. 1996; Paillard 1998, 2001). This type of long-term climate change is called natural climate change. Since the final decades of the twentieth century, the earth has experienced an unprecedented climate change due to increased emissions of greenhouse gases such as carbon dioxide, methane, nitrous oxides, sulfur dioxide, ozone, and water vapor due to human activities called anthropogenic climate change (Karl and Trenberth 2003; Huntley 2007). Continued greenhouse gas emissions at the current rates or above will cause further global climate change. Development of national and international policies to reduce greenhouse gases emission will not solve the problem in the near future since the previously accumulated greenhouse gases in the atmosphere have been continuing to affect the earth's climate (Gregory and Huybrechts 2006).

According to the reliable weather records that started to be kept in the middle of the nineteenth century, the temperature of the earth has been significantly increasing. In the past hundred years, the global average temperature has risen about 1 °C that caused noticeable effects on ecosystems. Recent researches on global climate change allowed us to understand how changing climatic factors increased the distribution range, growth, and survival of invasive species. Global climate change, caused by either natural or anthropogenic, alters the patterns of precipitation, wind, temperature, humidity, melting of snow and ice, and cloudiness that significantly affect the flora and fauna of the earth (Soberón and Townsend Peterson 2005). Compared to natural climate change, anthropogenic climate change is getting the biggest threat to alter patterns of precipitation, humidity, and temperature that make arid and semiarid regions drier and hotter, while mid-to-high latitudes are wetter and warmer. Human-induced climate alterations may increase the invasiveness of some native and non-native plant species since more favorable climate conditions allow some species to disperse and establish into new habitats (Rogers and McCarty 2000; Moore 2004; IPCC 2007; Thuiller et al. 2007). Changes in both precipitation and temperature have long been known to highly affect the presence, absence, phenology, genetic composition, distribution, and reproductive success of invasive and native plant species (Walther et al. 2002; Root et al. 2003; Huntley 2007; Thuiller et al. 2008). Altered redistribution of invasive plant species toward higher latitudes and elevations is the clear evidence of their response to climate change.

Climate change-caused environmental extremes influence both the direction and severity of evolution. Invasive plant species have a large genetic diversity to cope with environmental extremes compared to non-invasive native species. Large genetic diversity alone is too slow to cope with environmental extremes. Epigenetic variations for evolving a plant species into an invasive one are higher than gene mutations, gene drifts, and selections (Zhang et al. 2010; Medrano et al. 2014; Lele et al. 2018). Therefore, a great attention has been given to epigenetic modifications in plant ecological genetics.

The term epigenetics was coined by Waddington (1900–1975) as the alteration of gene expression in a cell during development (Loison 2022). Epigenetic modifications are heritable and reversible changes in gene expression without changing DNA sequences (Kumari et al. 2022). DNA methylation, histone modifications, chromatin configuration, and actions of non-coding RNA species are the major epigenetic modifications (Kim 2021). DNA methylation is the most known and well-studied mechanism that controls gene expression, gene imprinting,

DNA stability, DNA conformation, and transposon silencing and occurs through the addition of a methyl group to the fifth carbon of the pyrimidine ring of cytosine nucleotides (Mattei et al. 2022).

Epigenetic modifications may cause remarkable phenotypic variations within and among the natural plant populations (Zhang et al. 2013; Boquete et al. 2021; Husby 2022). Environmental extremes such as increased temperature, CO_2, and altered precipitation regimes highly influence epigenetic modifications. Epigenetic alteration of gene expression can be associated with climate change. The potential of environmental extremes on epigenetic modifications was reported or reviewed by several researchers (Sani et al. 2013; Ni et al. 2018; Kim 2021; Uludag et al. 2022).

This chapter is divided into four sections followed by an introduction: the first covers climate change-triggered epigenetic variations, the second deals with the impacts of temperature, the third summarizes elevated CO_2 impacts, and the fourth section deals with the impacts of precipitation alteration on invasive plants as well as concluding remarks. In this chapter, we explain how a better understanding of how global climate change and epigenetic modifications affect the spreading, growth, reproduction, and competitive ability of invasive plant species help to develop management strategies for their prevention, eradication, and control.

12.2 Climate Change Altered Epigenetic Variations on Adaptation and Dispersal Potential of Invasive Plant Species

The plant species can be accidentally or deliberately introduced into a distant location due to increased human activities. The introduced species may adapt, survive, and proliferate easily in the recipient ecosystems that likely differ from the native environment (Sakai et al. 2001). Genetic variations play a significant role in the process of successful adaptation (Muller-Scharer et al. 2004). However, genetic variations due to mutation, genetic drift are too slow to cope with fast adaptations (Norouzitallab et al. 2019). Therefore, epigenetics may compensate for reduced genetic diversity in invasive species in the invaded areas (Table 12.1). Epigenetic mechanisms provide new and critical ways in which the plant genome responds to the environmental extremes (Bossdorf et al. 2008; Richards et al. 2012).

Epigenetics is a study of heritable and reversible changes in organisms caused by modification of gene expression rather than the absence of changes in DNA sequences (Russo et al. 1996). Epigenetic regulation of fundamental genes involved in adaptation is a hallmark of plant invasion, particularly under adverse environmental conditions (Wang et al. 2012). It is now well known that epigenetic alterations can be caused by the environmentally triggered epimutations.

The following three epigenetic mechanisms (Fig. 12.2) affect gene expression and are associated with adaptation and invasion of invasive plants: (a) DNA methylation, the addition of a methyl group to cytosine at the fifth carbon atom, (b) histone modification, and (c) non-coding RNAs, including small RNAs (Turner 2000; Rice and Allis 2001; Lim and Maher 2010; Pauli et al. 2011; Feil and Fraga

Table 12.1 Examples of the epigenetic modifications of invasive plants caused by environmental stresses

Species	Methodology	Stress type	Summary of the major findings	Reference
Ageratina adenophora	Bisulfite sequencing	Cold stress	Demethylation was responsible for evolution	Xie et al. (2015)
Alternanthera philoxeroides	Methylation-Sensitive Amplified Polymorphism (MSAP)	Salinity stress	Epigenetic diversity in response to environmental stress compensate for genetic disadvantage and contribute to the evolution in clonal species	Shi et al. (2019)
Alternanthera philoxeroides	MSAP	Drought stress	DNA methylation alterations due to different water availability caused epigenetic reprogramming and reversible phenotypic response	Gao et al. (2010)
Arabidopsis thaliana	MethylC-seq and RNA-seq	Long day conditions	Epigenetic variation in DNA methylation generates new epi-alleles that provide phenotypic diversity	Schmitz et al. (2011)
Arabidopsis thaliana	ChIP-seq analysis of four histone modifications	Salinity stress	Epigenome alters physiological responses to salt stress and causes significant changes in genome-wide profiles of four histone modifications	Sani et al. (2013)
Dactylorhiza majalis	MSAP	Changing environments	The stable epigenetic variance was mainly responsible for persistent ecological differences	Paun et al. (2010)
Helleborus foetidus	Amplified Fragment Length Polymorphism (AFLP) and methylation-sensitive AFLP markers	Changing environments	Epigenetic similarity between individuals was much greater than genetic similarity at shortest distances	Herrera et al. (2016)
Fragaria vesca	Whole genome bisulfite sequencing	Drought stress	Population history, rather than short-term environmental stress, played a major role in	De Kort et al. (2020)

(continued)

Table 12.1 (continued)

Species	Methodology	Stress type	Summary of the major findings	Reference
			shaping epigenetic signatures	
Laguncularia racemosa	MSAP	Salinity and nutrient variations	Individuals from salt marsh and riverside presented abundant DNA methylation that help individuals to cope with different environments	Lira-Medeiros et al. (2010)
Plantago lanceolata	MSAP	Changing environments	Epigenetic diversity exists in natural population and it is related to environmental variation	Gáspár et al. (2019)
Reynoutria japonica	AFLP and methylation sensitive-AFLP (MS-AFLP)	Changing environments	Epigenetic differentiations were present among locations, and epigenetic loci responded to local microhabitat conditions.	Richards et al. (2012)
Spartina anglica	MSAP	Salinity stress	High level of epigenetic regulation was responsible for the morphological plasticity of it and its larger ecological amplitude	Salmon et al. (2005)
Taraxacum officinale (Fig. 12.1)	AFLP and epigenetic MS-AFLP markers	Changing environments	Heritable DNA methylation contributed to population differentiation along ecological gradients	Preite et al. (2015)
Viola cazorlensis	MSAP	Changing environments	Epigenetic differentiation in the populations was correlated with adaptive genetic divergence	Herrera and Bazaga (2010)

2012). These epigenetic mechanisms can lead to altered gene expression, consequently resulting in the evolution of plant species and the development of epigenetic adaptation (Massicotte and Angers 2012; Verhoeven and Preite 2014). Among the epigenetic mechanisms, DNA methylation is the best studied mechanism and has a

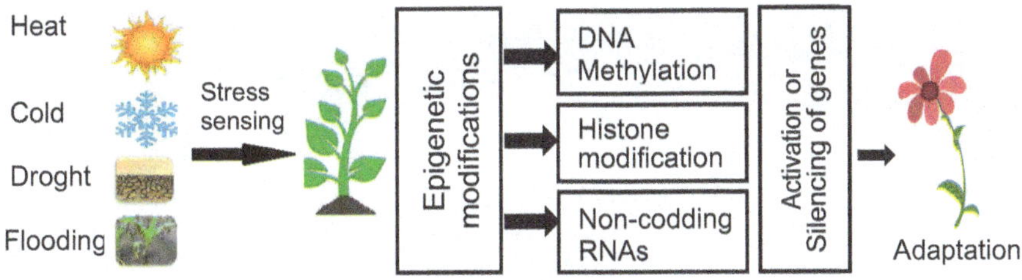

Fig. 12.2 Environmental factors causing epigenetic modification on invasive plants and their epigenetic mechanisms

Fig. 12.1 Common dandelion (*Taraxacum officinale*)

significant role in the control of gene regulations. Flowering symmetry in yellow toadflax (*Linaria vulgaris*) is given as a good example of phenotypic plasticity. Flowering symmetry of yellow toadflax can be hereditarily bilateral or radial. The shift from one to the other flowering pattern was caused by DNA methylation at a gene encoding a transcription factor, without known genetic mutations (Cubas et al. 1999). However, histone methylation patterns and non-coding RNAs, including small RNAs also can contribute to the regulation of the epigenome (Mirouze and Paszkowski 2011; Pauli et al. 2011; Zhou et al. 2011).

Recent studies have revealed the effects of epigenetics on stress response and adaptation to climate extremes that have long-lasting effects on gene regulation and chromatin alterations. The high adaptive potential of Spanish violet (*Viola cazorlensis*) was associated with methylation-based epigenetic modifications (Herrera and Bazaga 2010). Phenotypic variation among geographically and ecologically diverse orchid taxa (*Dactylorhiza majalis, D. traunsteineri*, and *D. majalis* ssp. *ebudensis*) was attributed to the epigenetic factors in response to an environmental stimulus (Paun et al. 2010). Similarly, two populations of *Laguncularia racemosa* grown in saline and non-saline environments exhibited different phenotypic appearances due to epigenetic modifications (Lira-Medeiros et al. 2010).

Epigenetic mutations and phenotypic plasticity are the major sources for phenotypic variations in invasive plant species to adapt to climatic extremes. Clonally

Fig. 12.3 Water hyacinth (*Eichhornia crassipes*)

reproducing plant species have narrow genetic diversity that limits their adaptive potential in altering environments. Despite low levels of genetic diversity of clonally reproducing plant species, successful invasions were recorded for some clonal plant species, such as Japanese knotweed (*Reynoutria japonica*), alligator weed (*Alternanthera philoxeroides*), common cordgrass (*Spartina anglica*), and water hyacinth (*Eichhornia crassipes,* Fig. 12.3) (Ren et al. 2005; Gao et al. 2010; Shi et al. 2012). The adaptive and invasion potential of these species under novel and fluctuating environments was attributed to accumulated epigenetic variations. It was stated that genetic diversity was not always essential for the adaptation and invasion of clonally reproducing species. Epigenetic mutations in clonally propagated species could compensate for narrow genetic variations to persist for millions of years in highly fluctuated environments (Schlichting 1986; Sultan 2004).

12.3 Impact of Increased Temperature on Invasive Plant Species

Studies have shown that, over the past century, the world has experienced a global warming (IPCC 2007; Arndt et al. 2010). The temperature increase was 0.78 °C, and the temperature is expected to increase by 2.6–4.8 °C by 2100 (IPCC 2014). Most of this warming is attributed to the result of increased concentrations of greenhouse gases in the atmosphere. Global warming enhances the reproduction of invasive species in temperate climates and expands the dispersion range of some invasive species. It was estimated that up to 10 °C temperature increase would be seen in high latitudes by the year 2100, while less temperature increases (3–4 °C) would be seen in the tropics (Ciais et al. 2013). The availability of novel niches, lack of natural enemies, and the potential of invaders to adapt to new habitats can increase the capacity of invaders to respond positively to temperature increases in temperate climates (Jarnevich et al. 2014). Therefore, the potential for the expansion and colonization of invasive plant species due to temperature alteration in new habitats is assumed to be high. Global warming may extend the growing season by approximately 4–6 weeks in temperate regions. Extended growing seasons and shorter milder winters could stimulate biomass and seed production, resulting in increased population sizes of the invasive species (Walther et al. 2007). Consequently, elevated temperatures decrease the compotation potential of native species by stressing them, but not the invaders (Zerebecki and Sorte 2011).

Generally, climate change negatively affects native species since they have no experience with the new environmental extremes (Byers et al. 2002). Climate change may generate more suitable conditions for the dispersal and establishment of invasive plant species. Increased temperatures, reduced snowfall, altered frequency of freeze-thaw cycles, and earlier ice cover melts in the northeastern United States stimulated overwintering and survival of some aquatic invasive species such as hydrilla (*Hydrilla verticillata*) and water hyacinth (*Eichhornia crassipes*) (Hayhoe et al. 2007; U.S. EPA 2008). Overwintering and survival of both aquatic invaders can be attributable to the milder winter conditions that extend the growing seasons and increase the invasion of these species. Similarly, increased winter temperatures extended the expansion range of buffelgrass (*Pennisetum ciliare*) in the upslopes of the Southwestern United States (Archer and Predick 2008). Similarly, extended range expansion of buffelgrass due to climate change was reported in Australia (Martin et al. 2015). Another study conducted in Europe revealed that the distribution of Japanese knotweed (*Fallopia japonica*) was significantly altered by increased average minimum cold temperature threshold limits and length of the growing season, however, the distribution of Himalayan balsam (*Impatiens glandulifera)* was altered only from the length of the growing season (Beerling 1993). Likewise, decreased winter frost and fluctuating water levels increased the invasion of water hyacinth (*E. crassipes*) in the Netherlands (PlantLife 2005). Summer climate alterations from rainy to warm and dry periods likely increased the invasion of both the water net (*Hydrodictyon reticulatum*) and blanket weed (*Cladophora glomerata*) in the wetlands of the United Kingdom (PlantLife 2005).

Global warming is a particular concern in temperate regions since many invasive species have minimum cold temperature threshold limits to survive (Ayres and Lombardero 2000; Owens et al. 2004). The global temperature increase may shift the daily temperature regimes from cold to warm that lowers the vulnerability of low temperatures while rises the vulnerability of high temperatures in cold and hot environments, respectively (DeGaetano et al. 2002). Studies in the warm extremes are higher than those of the cold extremes. Kukla and Karl (1993) stated that temperature increases were greater in winter than in summer. At higher elevations, cold temperatures are considered to be the dominant factors preventing the growth and establishment of some plant species. Survival and establishment of yellow star thistle (*Centaurea solstitialis,* Fig. 12.4) at higher elevations are prevented by cold temperatures, however, its presence was reported at 2590 m elevation due to elevated winter temperature (D'Antonio et al. 2004). The spreading and invasion of common ragweed (*Ambrosia artemisiifolia*) in Europe, New Zealand, Hawaii, Madagascar, and Mauritius was another good evidence of global warming (Brandes and Nitzsche 2007; Richter et al. 2013; Storkey et al. 2014).

In a study on two invasive species, garlic mustard (*Alliaria petiolata*) and Japanese barberry (*Berberis thunbergii)* in New England, United States, it was predicted that climate change noticeably reduced the establishment of garlic mustard in warmer climates while increasing the invasion of *Berberis thunbergii* due to

Fig. 12.4 Yellow star thistle (*Centaurea solstitialis*)

higher growth and germination that could increase its establishment potential (Merow et al. 2017). With moderate warming, biomass production, reproduction and establishment, and survival of invasive plants are expected to increase in some colder regions, but the overall effects of global warming on invasive species in the hotter region are expected to be negative (Burgiel and Muir 2010; Masters and Norgrove 2010). The effect of increased temperature on epigenetic modifications has been gained great attention due to increasing global temperatures. Hu et al. (2015) stated that increased acetylation in the promoter regions of heat stress response genes in *Arabidopsis thaliana* promoted seed germination and plant survival under heat stress.

12.4 Impact of Increased CO_2 Concentration on Invasive Plant Species

The atmospheric CO_2 concentration has been increasing due to human activities with an annual rate of 1.8 ppm year^{-1} over the last 4 decades. According to the IPCC, the CO_2 concentration in the atmosphere will increase to about 600 μmol mol^{-1} from the current 417 μmol mol^{-1} at the end of the next century (Betts 2021). Currently, ambient CO_2 levels are 30% higher than the pre-industrial level. Besides the greenhouse gas effect, increasing concentration of CO_2 in the atmosphere can significantly stimulate the growth and development of plants by increasing net photosynthesis, and water-use efficiency and decreasing transpiration (Wand et al. 1999; Ainsworth and Long 2005; Leakey et al. 2009).

The impacts of increased CO_2 concentration on plants have been intensively studied (Prior et al. 2011; Sundar 2015; Shanker et al. 2022). Some of the invasive plant species may have high genetic potential and plasticity to respond to rising atmospheric CO_2 more rapidly than native species (Liu et al. 2017). Generally, plants in the C3 photosynthetic pathway benefit from increasing CO_2 concentrations better than plants in C4 and CAM pathways (Ziska and Bunce 1997; Dukes and Mooney 1999; Dukes 2000) that enhance competition, dispersal, and establishment of C3 invasive species only in certain environments since the increased temperature in those environments may inhibit the stimulating effects of elevated CO_2 on the photosynthesis of C3 plants (Batts et al. 1997; Morison and Lawlor 1999). However, both C3 and C4 invasive plant species may benefit from the raised atmospheric carbon dioxide (CO_2) more than native plant species (Willis et al. 2010; Liu et al. 2017). Therefore, the effects of increased CO_2 on invasive plant species are important to consider. When the raised CO_2 increases the availability of plant resources, invasive plant species may have some advantages from these new conditions. Increased atmospheric CO_2 concentration favored invasive herb species yellow star thistle (*C. solstitialis*), and more than six times growth was observed under CO_2 elevated conditions (Dukes et al. 2011). The higher biomass production of Japanese honeysuckle (*Lonicera japonica*), an alien invasive species, growing in elevated CO_2 concentrations had higher advantages for fast growth and development compared with its native relative coral honeysuckle (*L. sempervirens*) (Sasek and Strain 1991).

12.5 Impact of Altered Precipitation on Invasive Plant Species

Precipitation is not distributed evenly on the earth, and its amount, intensity, frequency, and distribution are controlled primarily by atmospheric air circulation patterns. Climate change, increased temperature, altered precipitation patterns, and increased frequency and severity of storms can affect the invasive plant species. Altered precipitation patterns may disturb ecosystems that provide notable opportunities for the growth, survival, and dispersal of invasive plant species. For instance, drought can promote invasions, by weakening native species and increasing the success of colonizing some alien plant species. This can create opportunities for alien plant invasions (Baruch and Fernandez 1993; Nernberg and Dale 1997; Diez et al. 2012).

The global temperature increase is expected to decrease precipitation rates in Central America, North Africa, southern Europe, and parts of southern Asia. However, global temperature increase is expected to increase precipitation rates in some areas, mostly the higher latitudes such as Alaska, northern and central Asia, along with eastern parts of North and South America (Finch et al. 2021). Extensive increases in heavy precipitation events have been observed, even in places where total precipitation amounts have decreased. It was estimated that precipitation will support plant invasion by creating new habitats and niches for expansion. Water is one of the most determining factors in plant growth and development in semiarid ecosystems. Changes in the amount and timing of precipitation determine plant distribution (Archer and Predick 2008; Bradley et al. 2009). Small changes in precipitation could significantly change species composition (Knapp and Smith 2001; Huxman et al. 2004; Byrne et al. 2013). Kharivha et al. (2022) reported that high rainfall did not enhance black wattle (*Acacia mearnsii*) invasion through altering germination and growth, but reduced rainfall decreased its germination and invasiveness. Invasive species, such as cheatgrass (*Bromus tectorum*, Fig. 12.5), may be able to get advantages in changing precipitation patterns that native species cannot. Decreased population growth of cheatgrass was attributed to

Fig. 12.5 Cheatgrass (*Bromus tectorum*)

Fig. 12.6 Tree of heaven (*Ailanthus altissima*)

winter drought that reduced invasion success (Prevéy and Seastedt 2015). These findings indicated that both winter and spring precipitation played a significant role in the success of *B. tectorum* (Bradley 2009).

Drought stress is one of the most important limiting factors for seed germination, plant growth, reproduction, and dispersal. Drought stress also limits nutrient uptake and transportation, affecting the process of photosynthesis negatively. Some invasive plant species are naturally more tolerant to drought stress than native species due to higher phenological plasticity, adapting the reproduction time in drier seasons (Rice et al. 1992), higher water use, (Cavaleri et al. 2010), and higher water use efficiency (Heberling and Fridley 2013). For example, tree of heaven (*Ailanthus altissima*, Fig. 12.6) is sensitive to drought stress in its native ranges, but it is highly drought tolerant in invaded ranges (Albright et al. 2010). Higher phenotypic plasticity in kudzu (*Pueraria lobata*) (Pereira-Netto et al. 1999), and rapid evolution of Japanese stiltgrass (*Microstegium vimineum*) (Droste et al. 2010; Ziska et al. 2015) were attributed to their adaptive response to drought stress.

A number of studies on the influence of altered environment such as exposure to drought, high temperature or cold have provided strong evidence of environmental stimuli for epigenetic alterations to extend adaptability during exposure to abiotic stress factors (Steward et al. 2002; Castonguay and Angers 2012; Dubin et al. 2015; Akhter et al. 2021). It was stated that abrupt climatic changes cause selection pressure on plants that affect the direction and degree of adaptation (Eveno et al.

2008). Environmental factors may trigger specific loci that alter gene expression through epigenetic mechanisms and these epigenetic modifications increase the adaptation ability of invasive plant species as rapid response to environmental extremes (Dowen et al. 2012).

12.6 Concluding Remarks

The climate of the earth, which has been changing for thousands of years, has been remarkably changing in recent years due to increased anthropogenic activities. Especially after the industrial revolution, raised CO_2 and temperature and altered precipitation patterns greatly affected phenology, genetic composition, reproductive success, distribution, and colonization of both native and invasive plant species. It is widely believed that invasive plant species benefit from global warming, atmospheric CO_2 enrichment, and precipitation pattern alterations more than native plants. The extreme climate alterations that damage many native plants provide suitable habitats for the establishment and colonization of invasive species due to their higher tolerance ranges. Environmental extremes highly influence epigenetic modifications. Epigenetic alteration of gene expression can be associated with climate change. By better understanding the adaptation and survival mechanisms of invasive plants under altered climatic conditions, the invasiveness potential of plant species might be assessed to predict and track future invasions. In addition, it can provide invaluable insights to identify how to enable native species to compete more effectively with invasive rivals. We believe that understanding the effects of all global climate changes on invasive plants can, therefore, help in creating more sustainable management because we probably will need more if we will not be able to slow down climate change. Future epigenetic studies further elucidate the molecular mechanisms of rapid adaptation and successful invasion of invasive species under stressful conditions caused by global climate change.

References

Ainsworth EA, Long SP (2005) What have we learned from 15 years of free air CO_2 enrichment (FACE)? A meta-analytic review of the responses of photosynthesis, canopy properties and plant production to rising CO_2. New Phytol 165:351–372

Akhter Z, Bi Z, Ali K, Sun C, Fiaz S, Haider FU, Bai J (2021) In response to abiotic stress, DNA methylation confers epigenetic changes in plants. Plan Theory 10(6):1096

Albright TP, Chen H, Chen L, Guo Q (2010) The ecological niche and reciprocal prediction of the disjunct distribution of an invasive species: the example of *Ailanthus altissima*. Biol Invasions 12(8):2413–2427

Alexander LV, Zhang X, Peterson TC, Caesar J, Gleason B, Klein Tank AMG et al (2006) Global observed changes in daily climate extremes of temperature and precipitation. J Geophys Res Atmos 111(D5). https://doi.org/10.1029/2005JD006290

Archer SR, Predick KI (2008) Climate change and ecosystems of the southwestern United States. Rangelands 30:23–28

Arndt DS, Baringer MO, Johnson MR (2010) State of the climate in 2009. Bull Am Meteorol Soc 91:S1–S224

Ayres MP, Lombardero MJ (2000) Assessing the consequences of global change for forest disturbance from herbivores and pathogens. Sci Total Environ 262:263–286

Baruch Z, Fernandez DS (1993) Water relations of native and introduced C4 grasses in a neotropical savanna. Oecologia 96:179–185

Batts GR, Morison JIL, Ellis RH, Hadley P, Wheeler TR (1997) Effects of CO_2 and temperature on growth and yield of crops of winter wheat over four seasons. Eur J Agron 7(1–3):43–52

Beerling DJ (1993) The Impact of temperature on the northern distribution limits of the introduced species *Fallopia japonica* and *Impatiens glandulifer*a in North-West Europe. J Biogeogr 20(1): 45–53

Betts R (2021) Met Office: atmospheric CO2 now hitting 50% higher than pre-industrial levels. Carbon Brief. https://www.carbonbrief.org/met-office-atmospheric-CO_2-now-hitting-50-higher-than-pre-industrial-levels

Boquete MT, Muyle A, Alonso C (2021) Plant epigenetics: phenotypic and functional diversity beyond the DNA sequence. Am J Bot 108(4):553–558

Bossdorf O, Richards CL, Pigliucci M (2008) Epigenetics for ecologists. Ecol Lett 11(2):106–115. https://doi.org/10.1111/j.1461-0248.2007.01130.x

Bradley BA (2009) Regional analysis of the impacts of climate change on cheatgrass invasion shows potential risk and opportunity. Glob Chang Biol 15:196–208. https://doi.org/10.1111/j.1365-2486.2008.01709.x

Bradley BA, Oppenheimer M, Wilcove DS (2009) Climate change and plant invasions: restoration opportunities ahead? Glob Chang Biol 15:1511–1521

Brandes D, Nitzsche J (2007) Biology, introduction, dispersal, and distribution of common ragweed (*Ambrosia artemisiifolia* L.) with special regard to Germany. Nachrichtenblatt des Deutschen Pflanzenschutzdienstes 58:286–291

Brunner MI, Swain DL, Wood RR, Willkofer F, Done JM, Gilleland E, Ludwig R (2021) An extremeness threshold determines the regional response of floods to changes in rainfall extremes. Commun Earth Environ 2(1):1–11

Burgiel SW, Muir AA (2010) Invasive species, climate change and ecosystem- based adaptation: addressing multiple drivers of global change. Global Invasive Species Programme (GISP), Washington, DC, US, and Nairobi, Kenya

Byers JE, Reichard S, Randall JM et al (2002) Directing research to reduce the impacts of nonindigenous species. Conserv Biol 16:630–640

Byrne KM, Lauenroth WK, Adler PB (2013) Contrasting effects of precipitation manipulations on production in two sites within the central grass lands region, USA. Ecosystems 16(69): 1039–1051

Canturk U, Kulaç Ş (2021) The effects of climate change scenarios on Tilia ssp. in Turkey. Environ Monit Assess 193(12):1–15

Castonguay E, Angers B (2012) The key role of epigenetics in the persistence of asexual lineages. Genet Res Int 2012:534289. https://doi.org/10.1155/2012/534289

Cavaleri MA, Oberbauer SF, Clark DB, Clark DA, Ryan MG (2010) Height is more important than light in determining leaf morphology in a tropical forest. Ecology 91:1730–1739. https://doi.org/10.1890/09-1326.1

Ciais P, Sabine C, Bala G, Bopp L, Brovkin V, Canadell J, Chhabra A, DeFries R, Galloway J, Heimann M, Jones C, Le Quéré C, Myneni RB, Piao S, Thornton P (2013) Carbon and other biogeochemical cycles. In: Stocker TF, Qin D, Plattner G-K, Tignor M, Allen SK, Boschung J, Nauels A, Xia Y, Bex V, Midgley PM (eds) Climate change 2013: the physical science basis. Contribution of working group I to the fifth assessment report of the intergovernmental panel on climate change. Cambridge University Press, Cambridge, United Kingdom and New York, NY, USA

Cubas P, Lauter N, Doebley J, Coen E (1999) The TCP domain: a motif found in proteins regulating plant growth and development. Plant J 18:215–222

D'Antonio CM Berlow EL, Haubensak KL (2004) Invasive exotic plant species in Sierra Nevada ecosystems. In: Proceedings from the Sierra Science Conference of 2002: science for management and conservation. USDA Forest Service Gen. Tech. Rep. PSW-GTR-103

De Kort H, Panis B, Deforce D, Van Nieuwerburgh F, Honnay O (2020) Ecological divergence of wild strawberry DNA methylation patterns at distinct spatial scales. Mol Ecol 29:4871–4881

DeGaetano AT, Allen RJ, Gallo KP (2002) A homogenized historical temperature extreme dataset for the United States. J Atmos Oceanic Technol 19:1267–1284

Diez JM, D'Antonio CM, Dukes JS, Grosholz ED, Olden JD, Sorte CJ et al (2012) Will extreme climatic events facilitate biological invasions? Front Ecol Environ 10:249–257. https://doi.org/10.1890/110137

Dowen RH, Pelizzola M, Schmitz RJ, Lister R, Dowen JM, Nery JR, Dixon JE, Ecker JR (2012) Widespread dynamic DNA methylation in response to biotic stress. Proc Natl Acad Sci U S A 109:E2183–E2191. https://doi.org/10.1073/pnas.1209329109

Droste T, Flory SL, Clay K (2010) Variation for phenotypic plasticity among populations of an invasive exotic grass. Plant Ecol 207(2):297–306

Dubin MJ, Zhang P, Meng D, Remigereau MS, Osborne EJ, Casale P, Drewe P, Kahles A, Jean G, Vilhjálmsson B, Jagoda J, Irez S, Voronin V, Song Q, Long Q, Ratsch G, Stegle O, Clark RM, Nordborg M (2015) DNA methylation in *Arabidopsis* has a genetic basis and shows evidence of local adaptation. eLife 4:e05255. https://doi.org/10.7554/eLife.05255

Dukes JS (2000) Will the increasing atmospheric CO_2 concentration affect the success of invasive species? In: Mooney HA, Hobbs RJ (eds) Invasive species in a changing world. Island Press, Washington, DC, Covelo, CA, pp 95–113

Dukes JS, Mooney HA (1999) Does global change increase the success of biological invaders ? Trends Ecol Evol 14(4):135–139

Dukes JS, Chiariello NR, Loarie SR, Field CB (2011) Strong response of an invasive plant species (*Centaurea solstitialis* L.) to global environmental changes. Ecol Appl 21:1887–1894

Early R, Bradley BA, Dukes JS, Lawler JJ, Olden JD, Blumenthal DM, Gonzalez P, Grosholz ED, Ibañez I, Miller LP, Sorte CJB, Tatem AJ (2016) Global threats from invasive alien species in the twenty-first century and national response capacities. Nat Commun 7:12485. https://doi.org/10.1038/ncomms12485

Eveno E, Collada C, Guevara MA, Léger V, Soto A, Díaz L, Léger P, González Martínez SC, Cervera MT, Plomion C, Garnier-Géré PH (2008) Contrasting patterns of selection at *Pinus pinaster* Ait. drought stress candidate genes as revealed by genetic differentiation analyses. Mol Biol Evol 25:417–437

Feil R, Fraga MF (2012) Epigenetics and the environment: emerging patterns and implications. Nat Rev Genet 13(2):97–109. https://doi.org/10.1038/nrg3142

Finch DM, Butler JL, Runyon JB, Fettig CJ, Kilkenny FF, Jose S, Frankel SJ, Cushman SA, Cobb RC, Dukes JS et al (2021) Effects of climate change on invasive species. In: Poland TM, Patel-Weynand T, Finch DM, Ford MC, Hayes DC, Lopez VM (eds) Invasive species in forests and rangelands of the United States: a comprehensive science synthesis for the United States forest sector. Springer International Publishing, Heidelberg, Germany, pp 57–84

Gao L, Geng Y, Li B, Chen J, Yang J (2010) Genome-wide DNA methylation alterations of *Alternanthera philoxeroides* in natural and manipulated habitats: implications for epigenetic regulation of rapid responses to environmental fluctuation and phenotypic variation. Plant Cell Environ 33:1820–1827

Gáspár B, Bossdorf O, Durka W (2019) Structure, stability and ecological significance of natural epigenetic variation: a large-scale survey in *Plantago lanceolata*. New Phytol 221:1585–1596

Gregory JM, Huybrechts P (2006) Ice-sheet contributions to future sea-level change. Philos Trans Royal Soc A: Math Phys Eng Sci 36:1709–1732

Hayhoe, K., Wake, C., Huntington, T., Luo, L., Schwartz, M., Sheffield, J., Wood, E., Anderson, B., Bradbury, J., DeGaetano, A., Troy, T., . and Wolfe, D. 2007. Past and future changes in climate and hydrological indicators in the U.S. Northeast Clim Dyn, 28: 381–407

Heberling JM, Fridley JD (2013) Resource-use strategies of native and invasive plants in Eastern North American forests. New Phytol 200:523–533

Hellmann JJ, Byers JE, Bierwagen BG, Dukes JS (2008) Five potential consequences of climate change for invasive species. Conserv Biol 22:534–543

Herrera CM, Bazaga P (2010) Epigenetic differentiation and relationship to adaptive genetic divergence in discrete populations of the violet *Viola cazorlensis*. New Phytol, 876 187:867

Herrera CM, Medrano M, Bazaga P (2016) Comparative spatial genetics and epigenetics of plant populations: heuristic value and a proof of concept. Mol Ecol 25:1653–1664

Houghton JT, Filho LGM, Callander BA, Harris N, Kattenberg A, Maskell K (1996) Climate change 1995: the science of climate change, intergovernmental panel on climate change. Cambridge University Press, Cambridge, pp 572

Hu Z, Song N, Zheng M, Liu X, Liu Z, Xing J, Ma J, Guo W, Yao Y, Peng H, Xin M, Zhou DX, Ni Z, Sun Q (2015) Histone acetyltransferase GCN5 is essential for heat stress-responsive gene activation and thermotolerance in *Arabidopsis*. Plant J 84:1178–1191

Huntley B (2007) Climatic change and the conservation of European biodiversity: towards the development of adaptation strategies. T-PVS/Inf.(2007)3. In: Document prepared for the 27th meeting of The Standing Committee of the Convention on the Conservation of European Wildlife and Natural Habitats. Strasbourg, 26–19 November 2007. 59 pp.

Husby A (2022) Wild epigenetics: insights from epigenetic studies on natural populations. Proc R Soc B 289(1968):20211633

Huxman TE, Snyder KA, Tissue D, Leffler AJ, Ogle K, Pockman WT, Sandquist R, Potts DL, Schwinning S (2004) Precipitation pulses and carbon fluxes in semiarid and arid ecosystems. Oecologia 141:254–268

IPCC (2007) Summary for Policymakers. In: Parry ML, Canziani OF, Palutikof JP, van der Linden PJ, Hanson CE (eds) Climate change 2007 impacts, adaptation and vulnerability. Contribution of working group II to the fourth assessment report of the intergovernmental panel on climate change. Cambridge University Press, Cambridge, UK, pp 7–22

IPCC (2014) 2013 Supplement to the 2006 IPCC Guidelines for national greenhouse gas inventories. In: Hiraishi T, Krug T, Tanabe K, Srivastava N, Baasansuren J, Fukuda M, Troxler TG (eds) Wetlands. IPCC, Switzerland. 354 pp., ISBN: 978-92-9169-139-5

Jarnevich CS, Holcombe TR, Bella EM et al (2014) Cross-scale assessment of potential habitat shifts in a rapidly changing climate. Invas Plant Sci Manage 7:491–502

Karl TR, Trenberth E (2003) Modern global climate change. Science 302(5651):1719–1723

Kharivha T, Ruwanza S, Thondhlana G (2022) Effects of elevated temperature and high and low rainfall on the germination and growth of the invasive alien plant *Acacia mearnsii*. Plan Theory 11:2633. https://doi.org/10.3390/plants11192633

Kim JH (2021) Multifaceted chromatin structure and transcription changes in plant stress response. Int J Mol Sci 22(4):2013

Knapp AK, Smith MD (2001) Variation among biomes in temporal dynamics of aboveground primary production. Science 291:481–484

Kukla G, Karl TR (1993) Nighttime warming and the greenhouse effect. Environ Sci Technol 27: 1468–1474

Kumari P, Khan S, Wani IA, Gupta R, Verma S, Alam P, Alaklabi A (2022) Unravelling the role of epigenetic modifications in development and reproduction of angiosperms: a critical appraisal. Front Genet 13:819941. https://doi.org/10.3389/fgene.2022.819941

Leakey AD, Ainsworth EA, Bernacchi CJ, Rogers A, Long SP, Ort DR (2009) Elevated CO_2 effects on plant carbon, nitrogen, and water relations: six important lessons from FACE. J Exp Bot 60: 2859–2876

Lele L, Ning D, Cuiping P, Xiao G, Weihua G (2018) Genetic and epigenetic variations associated with adaptation to heterogeneous habitat conditions in a deciduous shrub. Ecol Evol 8:2594–2606

Lim DCH, Maher ER (2010) DNA methylation: a form of epigenetic control of gene expression. Obstetr Gynaecol 12(1):37–42. https://doi.org/10.1576/toag.12.1.037.27556

Linders TEW, Schaffner U, Eschen R, Abebe A, Choge SK, Nigatu L, Mbaabu PR, Shiferaw H, Allan E (2019) Direct and indirect effects of invasive species: biodiversity loss is a major mechanism by which an invasive tree affects ecosystem functioning. J Ecol 107:2660–2672. https://doi.org/10.1111/1365-2745.13268

Lira-Medeiros CF, Parisod C, Fernandes RA, Mata CS, Cardoso MA, Ferreira PCG (2010) Epigenetic variation in mangrove plants occurring in contrasting natural environment. PLoS One 5:e10326

Liu YJ, Oduor AMO, Zhang Z, Manea A, Tooth IM, Leishman MR, Xu XL, van Kleunen M (2017) Do invasive alien plants benefit more from global environmental change than native plants? Glob Chang Biol 23:3363–3370. https://doi.org/10.1111/gcb.13579

Loison L (2022) The environment: an ambiguous concept in Waddington's biology. Stud Hist Phil Sci 91:181–190

Martin TG, Murphy H, Liedloff A et al (2015) Buffel grass and climate change: a framework for projecting invasive species distributions when data are scarce. Biol Invasions 17:3197–3210

Massicotte R, Angers B (2012) General-purpose genotype or how epigenetics extend the flexibility of a genotype. Genet Res Int 2012:1–7. 317175

Masters G, Norgrove L (2010) Climate change and invasive alien species, vol 1. CABI Working Paper, UK, p 30

Mattei AL, Bailly N, Meissner A (2022) DNA methylation: a historical perspective. Trends Genet 38(7):676–707. https://doi.org/10.1016/j.tig.2022.03.010

Medrano M, Herrera CM, Bazaga P (2014) Epigenetic variation predicts regional and local intraspecific functional diversity in a perennial herb. Mol Ecol 23:4926–4938

Merow C, Bois ST, Allen JM, Xie Y, Silander JA Jr (2017) Climate change both facilitates and inhibits invasive plant ranges in New England. Proc Natl Acad Sci 114:E3276–E3284

Mirouze M, Paszkowski J (2011) Epigenetic contribution to stress adaptation in plants. Curr Opin Plant Biol 14:267–274

Mooney HA, Mack RN, McNeely JA, Neville LE, Schei PJ, Waage JK (2005) Invasive alien species: a new synthesis. Island Press, London, UK

Moore AB (2004) Alien invasive species: impacts on forest and forestry. Forest Health and Biosecurity Working Papers FAO., 66 p, Rome, Italia

Morison JIL, Lawlor DW (1999) Interactions between increasing CO_2 concentration and temperature on plant growth. Plant Cell Environ 22(6):659–682

Müller-Schärer H, Schaffner U, Steinger T (2004) Evolution in invasive plants: implications for biological control. Trends Ecol Evol 19:417–422

Nernberg D, Dale MRT (1997) Competition of five native prairie grasses with *Bromus inermis* under three moisture regimes. Can J Bot 75:2140–2145. https://doi.org/10.1139/b97-925

Ni P, Li SG, Lin YP, Xiong W, Huang XN, Zhan AB (2018) Methylation divergence of invasive *Ciona ascidians*: significant population structure and local environmental influence. Ecol Evol 8:10272–10287

Norouzitallab P, Baruah K, Vanrompay D, Bossier P (2019) Can epigenetics translate environmental cues into phenotypes? Sci Total Environ 647:1281–1293

Owens CS, Smart RM, Stewart RM (2004) Low temperature limits of giant salvinia. J Aquat Plant Manag 42:91–94

Paillard D (1998) The timing of *Pleistocene glaciations* from a simple multiple state climate model. Nature 391:378–381

Paillard D (2001) Glacial cycles: toward a new paradigm. Rev Geophys 39(3):325–346

Pauli A, Rinn JL, Schier AF (2011) Non-coding RNAs as regulators of embryogenesis. Nat Rev Genet 12:136–149

Paun O, Bateman RM, Fay MF, Hedren M, Civeyrel L, Chase MW (2010) Stable epigenetic effects impact adaptation in allopolyploid orchids (*Dactylorhiza*: Orchidaceae) research article. Mol Biol Evol 27:2465–2473

Pereira-Netto ABD, Gabriele AC, Pinto HS (1999) Aspects of leaf anatomy of kudzu (*Pueraria lobata*, Leguminosae Faboideae) related to water and energy balance. Pesquisa Agropecuaria Bras 34(8):1361–1365

PlantLife (2005) The wild plant conservation charity. 2005. Under pressure. Climate change and the UK' s wild plants. PlantLife—The wild plant conservation charity (Ed.), Salisbury, UK. 11pp

Pörtner HO, Roberts DC, Adams H, Adler C, Aldunce P, Ali E, Birkmann J, Begum RA et al (2022) Climate change 2022: impacts, adaptation and vulnerability. In: Contribution of Working Group II to the Sixth Assessment Report of the intergovernmental panel on climate change. IPCC Sixth Assessment Report, Intergovernmental Panel on Climate Change (IPCC), Geneva. 36 p

Preite V, Snoek LB, Oplaat C, Biere A, van der Putten WH, Verhoeven KJF (2015) The epigenetic footprint of poleward range-expanding plants in apomictic dandelions. Mol Ecol 24:4406–4418

Prevéy JS, Seastedt TR (2015) Effects of precipitation change and neighboring plants on population dynamics of *Bromus tectorum*. Oecologia 179:1–15. https://doi.org/10.1007/s00442-015-3398-z

Prior SA, Runion GB, Marble SC, Rogers HH, Gilliam CH, Torbert HA (2011) A review of elevated atmospheric CO2 effects on plant growth and water relations: implications for horticulture. HortScience 46:158–162

Pyšek P, Jarošík V, Hulme PE, Pergl J, Hejda M, Schaffner U, Vilà M (2012) A global assessment of invasive plant impacts on resident species, communities and ecosystems: the interaction of impact measures, invading species' traits and environment. Glob Chang Biol 18:1725–1737. https://doi.org/10.1111/j.1365-2486.2011.02636.x

Pyšek P, Hulme PE, Simberloff D, Bacher S, Blackburn TM, Carlton JT, Dawson W, Essl F, Foxcroft LC, Genovesi P et al (2020) Scientists' warning on invasive alien species. Biol Rev 95: 1511–1534. https://doi.org/10.1111/brv.12627

Ren MX, Zhang QG, Zhang DY (2005) Random amplified polymorphic DNA markers reveal low genetic variation and a single dominant genotype in *Eichhornia crassipes* populations throughout China. Weed Res 45:236–244. https://doi.org/10.1111/j.1365-3180.2005.00445

Rice JC, Allis CD (2001) Histone methylation versus histone acetylation: new insights into epigenetic regulation. Curr Opin Cell Biol 13:263–273

Rice KJ, Black RA, Radamaker G, Evans RD (1992) Photosynthesis, growth, and biomass allocation in habitat ecotypes of cheatgrass (*Bromus tectorum*). Funct Ecol 6(1):32–40

Richards CL, Schrey AW, Pigliucci M (2012) Invasion of diverse habitats by few Japanese knotweed genotypes is correlated with epigenetic differentiation. Ecol Lett 15:1016–1025

Richter R, Berger UE, Dullinger S, Essl F, Leitner M, Smith M, Vogl G (2013) Spread of invasive ragweed: climate change, management and how to reduce allergy costs. J Appl Ecol 50:1422–1430

Rogers CE, McCarty JP (2000) Climate change and ecosystems of the mid-Atlantic region. Clim Res 14:235–244

Root TL, Price JT, Hall KR, Schneider SH, Rosenzweig C, Pounds JA (2003) Fingerprints of global warming on animals and plants. Nature 421:57–60

Russo VEA, Martienssen RA, Riggs AD (1996) Epigenetic mechanisms of gene regulation. Cold Spring Harbor Laboratory Press, New York

Sakai T, Kagawa T, Kasahara M, Swartz TE, Christie JM, Briggs WR et al (2001) *Arabidopsis* nph1 and npl1: blue light receptors that mediate both phototropism and chloroplast relocation. Proc Natl Acad Sci U S A 98:6969–6974

Salmon A, Ainouche ML, Wendel JF (2005) Genetic and epigenetic consequences of recent hybridization and polyploidy in Spartina (Poaceae). Mol Ecol 14:1163–1175

Sani E, Herzyk P, Perrella G, Colot V, Amtmann A (2013) Hyperosmotic priming of *Arabidopsis* seedlings establishes a long-term somatic memory accompanied by specific changes of the epigenome. Genome Biol 2013(14):R59. https://doi.org/10.1186/gb-2013-14-6-r59

Sasek TW, Strain BR (1991) Effects of CO_2 enrichment on the growth and morphology of a native and an introduced honeysuckle vine. Am J Bot 78:69–75

Schlichting CD (1986) The evolution of phenotypic plasticity in plants. Annu Rev Ecol Evol Syst 17:667–693

Schmitz RJ, Schultz MD, Lewsey MG, O'Malley RC, Urich MA, Libiger O, Schork NJ, Ecker JR (2011) Transgenerational epigenetic instability is a source of novel methylation variants. Science 334:369–373

Seebens H, Bacher S, Blackburn TM, Capinha C, Dawson W, Dullinger S, Genovesi P, Hulme PE, van Kleunen M, Kuhn I et al (2020) Projecting the continental accumulation of alien species through to 2050. Glob Chang Biol 26:1–13. https://doi.org/10.1111/gcb.15333

Shanker AK, Gunnapaneni D, Bhanu D, Vanaja M, Lakshmi NJ, Yadav SK, Prabhakar M, Singh VK (2022) Elevated CO_2 and water stress in combination in plants: brothers in arms or partners in crime? Biology 11:1330. https://doi.org/10.3390/biology11091330

Shi L, Li S, Shen F, Li H, Qian S, Lee DHS, Wu JZ, Yang W (2012) Characterization of nucleosome positioning in hepadnaviral covalently closed circular DNA minichromosomes. J Virol 86(18):10059

Shi W, Hu X, Chen X, Ou X, Yang J, Geng Y (2019) Increased population epigenetic diversity of the clonal invasive species *Alternanthera philoxeroides* in response to salinity stress. Genes Genet Syst 93:259–269

Soberón J, Townsend Peterson A (2005) Interpretation of models of fundamental ecological niches and species' distributional areas. Biodivers Inform 2:1–10

Steward N, Ito M, Yamaguchi Y, Koizumi N, Sano H (2002) Periodic DNA methylation in maize nucleosomes and demethylation by environmental stress. J Biol Chem 277:37741–37746

Storkey J, Stratonovitch P, Chapman DS, Vidotto F, Semenov MA (2014) A process-based approach to predicting the effect of climate change on the distribution of an invasive allergenic plant in Europe. PLoS One 9:e88156

Sultan SE (2004) Promising directions in plant phenotypic plasticity. Perspect Plant Ecol Evol Syst 6:227–233

Sundar S (2015) Effect of elevated carbon dioxide concentration on plant growth: a mathematical model. Am J Appl Math Stat 3:59–67

Thomas CD, Cameron A, Green RE, Bakkenes M, Beaumont LJ, Collingham YC, Erasmus BFN, de Siqueira MF, Grainger A, Hannah L, Hughes L, Huntley B, van Jaarsveld AS, Midgley GF, Miles L, Ortega-Huert AMA, Peterson AT, Phillips OL, Williams SE (2004) Extinction risk from climate change. Nature 427:145–148

Thuiller W, Richardson DM, Midgley GF (2007) Will climate change promote alien plant invasions? In: Nentwig W (ed) Ecological studies. Springer, Berlin, Heidelberg, pp 197–211

Thuiller W, Albert C, Araújo MB, Berry PM, Cabeza M, Guisan A, Hickler T, Midgley GF, Paterson J, Schurr FM, Sykes MT, Zimmermann NE (2008) Predictingglobal change impacts on plant species' distributions: future challenges. Perspect Plant Ecol Evol Syst 9:137–152

Turner BM (2000) Histone acetylation and an epigenetic code. BioEssays 22(9):836–845

U.S. EPA (Environmental Protection Agency) (2008) Effects of climate change on aquatic invasive species and implications for management and research. EPA, Office of Research and Development, Washington, DC

Uludag A, Arslan M, Üremiş İ, Aksoy N (2022) Advances in understanding invasive characteristics in weed species. In: Kudsk P (ed) Advances in integrated weed management. Burleigh Dodds Science Publishing Limited, Cambridge, UK, pp 121–140

Verhoeven KJF, Preite V (2014) Epigenetic variation in asexually reproducing organisms. Evolution 68:644–655. https://doi.org/10.1111/evo.12320

Vilà M, Espinar JL, Hejda M, Hulme PE, Jarošík V, Maron JL, Pergl J, Schaffner U, Sun Y, Pyšek P (2011) Ecological impacts of invasive alien plants: a meta analysis of their effects on species, communities and ecosystems. Ecol Lett 14:702–708. https://doi.org/10.1111/j.1461-0248.2011.01628.x

Walther GR, Post E, Convey P, Menzel A, Parmesan C, Beebee TJC, Fromentin JM, Hoegh-Guldberg O, Bairlein F (2002) Ecological responses to recent climate change. Nature 416:389–395

Walther GR, Gritti ES, Berger S, Hickler T, Tang Z, Sykes MT (2007) Palms tracking climate change. Glob Ecol Biogeogr 16:801–809

Wand SJE, Midgley GF, Jones MH, Curtis PS (1999) Responses of wild C_4 and C_3 grass (Poaceae) species to elevated atmospheric CO_2 concentration: a meta-analytic test of current theories and perceptions. Glob Chang Biol 5:723–741

Wang H, Maurano MT, Qu H et al (2012) Widespread plasticity in CTCF occupancy linked to DNA methylation. Genome Res 22:1680–1688

Willis CG, Ruhfel BR, Primack RB, Miller-Rushing AJ, Losos JB, Davis CC (2010) Favorable climate change response explains non-native species' success in Thoreau's woods. PLoS One 5(1):e8878

Wolkovich EM, Cleland EE (2014) Phenological niches and the future of invaded ecosystems with climate change. AoB Plants 6:plu013

Xie HJ, Li AH, Liu AD, Dai WM, He JY, Lin S et al (2015) ICE1 demethylation drives the range expansion of a plant invader through cold tolerance divergence. Mol Ecol 24:835–850

Zerebecki RA, Sorte CJB (2011) Temperature tolerance and stress proteins as mechanisms of invasive species success. PLoS One 6(4):e14806. https://doi.org/10.1371/journal.pone.0014806

Zhang YY, Zhang DY, Barrett SCH (2010) Genetic uniformity characterizes the invasive spread of water hyacinth (*Eichhornia crassipes*), a clonal aquatic plant. Mol Ecol 19:1774–1786

Zhang YY, Fischer M, Colot V, Bossdorf O (2013) Epigenetic variation creates potential for evolution of plant phenotypic plasticity. New Phytol 197:314–322

Zhou VW, Goren A, Bernstein BE (2011) Charting histone modifications and the functional organization of mammalian genomes. Nat Rev Genet 12:7–18

Ziska LH, Bunce JA (1997) Influence of increasing carbon dioxide concentration on the photosynthetic and growth stimulation of selected C4 crops and weeds. Photosynthesis Res 54(3): 199–208

Ziska LH, Tomecek MB, Valerio M, Thompson JP (2015) Evidence for recent evolution in an invasive species, *Microstegium vimineum*, Japanese stiltgrass. Weed Res 55(3):260–267

Comparative Assessment of Machine Learning Algorithms for Habitat Suitability of *Tribulus terrestris* (Linn): An Economically Important Weed

13

Manish Mathur and Preet Mathur

Abstract

Weed species have the potential to alter the structure and functions of the ecosystem and besides their antagonistic ecological relationships with main crops, simultaneously they are also valued for their secondary metabolites of pharmaceutical and nutraceutical values. Climate and community-associated changes may alter the presence of such species as well as the concentration and quality of their active chemical constituents. In the present study, we carried out a comparative study to assess the proportional performance of different algorithms (both regression and machine learning based) for the assessment of habitat suitability of *Tribulus terrestris* within Indian arid and semi-arid areas. Furthermore, the impact of niche modeling on the Extent of Occurrence (EOO) and Area of Occupancy (AOO) of this species with three bioclimatic timeframe projections was also quantified. We hypothesized that these objectives will enable us to identify the major bioclimatic and community predictors that determine the habitat suitability of *T. terrestris* and also give projected area cover with this species under different suitability classes. For the above objectives, we implemented the ensemble techniques in which different algorithms (General linear model; GLM), (Generalized additive model; GAM), (Classification tree analysis; CTA), (Artificial neural network; ANN), (Support vector machine; SVM), (Multivariate adaptive spline; MARS), (Random forest; RF), and (Maximum entropy; MAXENT) were utilized and their prediction performance was assessed by using Kappa statistic, Area Under the receiver operating characteristic Curve (AUC), sensitivity, specificity, and True Skill Statistic (TSS). Niche

M. Mathur
ICAR-Central Arid Zone Research Institute, Jodhpur, India

P. Mathur (✉)
Jodhpur Institute of Engineering and Technology, Jodhpur, India

S. Tripathi et al. (eds.), *Plant Invasions and Global Climate Change*,
https://doi.org/10.1007/978-981-99-5910-5_13

overlap was carried out to visualize the amount of area retained by this species under different predictions. Comparative evaluation of different approaches revealed the best performance of random forest among all other algorithms that produced excellent model qualities for all three studied bioclimatic variables while good model quality for Habitat Heterogeneity Indices (HHI). Our results also revealed that HHI are less dynamic for species distribution modeling (SDM) of this species as compared to bioclimatic variables. Precipitation of Coldest Quarter (BC-19), Precipitation Seasonality (BC-15), and Annual Precipitation (BC-12) were the most significant variables that affect the SDM of this species. With current climatic conditions, we observed that optimum areas are located in the northern region of the arid and semi-arid areas of India covering 92,400 km^2 areas. While during 2050 projection area under this class increases up to 100,800 km^2 which suggests a 9.09% increase. While during 2070, this class covers 91,900 km^2 which showed −8.83% area decreases with respect to the previously projected timeframe and only 0.54% decrease compared to the current BC. With HHI variables, we found the disintegration of different classes in small patches as compared to bioclimatic variables. Overall, 111.25 km centroid shifting will be anticipated from the current to 2070-time era. In this analysis, we also find a significant negative pattern between EOO and AOO ($R^2 = 0.87$). Our results can be used to enhance ecologically (regarded as weed species) as well as economic (regarded as medicinally most important species) management in order to curb this or for harvesting the higher biomass (standing state) for its important secondary metabolites.

Keywords

Bioclimatic variables · Ensemble ecological niche modeling · IUCN · Niche overlap · Raster similarity random forest

13.1 Introduction

Historically, weeds are defined as "*any plant that is objectionable or interferes with the activities or welfare of man*" (Radosevich et al. 2007). Biological traits of "weediness" include (1) Larger seed production (R selection) and seed bank with longer survival that are resilient against various biotic and abiotic constraints, (2) quick germination and rapidly completes their growth cycle, (3) having various mechanical (e.g., spines), chemical (e.g., poison) defense mechanism against herbivory, (4) high niche breadth, (5) low dormancy period and rapid phenological transitions, (6) seed dispersal through various agencies like livestock, wind, etc.

Weeds are also valuable sources of feed and food, medicine (Table 13.1), improving soil quality and fertility, serve as cover crops, help to reduce soil erosion, slowdown nutrient loss, increase soil organic matter, improve nitrogen content, and conserve soil water, take part in phytoremediation, and mycoremediation (bio-herbicide; Mathur and Gehlot 2018). Like other crops, weed species are also

Table 13.1 A partial list of plant-derived pharmaceutical drugs and their clinical uses (based on Authors' original works, Mathur (2012); Mathur and Sundarmoorthy (2017)Vyas et al. (2017))

Plant species	Drug	Action/clinical use
Tribulus terrestris (L) Zygophyllaceae	Roncuvita	Cure male infertility (aphrodisiac)
Digitalis lanata Ehrh Plantaginaceae	Acetyldigoxin	Cardiotonic
Brassica nigra (L) Brassicaceae	Allyl isothiocyanate	Rubefacient
Artemisia annua (L) Asteraceae	Artemisinin	Antimalarial
Datura stramonium (L) Solanaceae	Atropine	Anticholinergic
Berberis vulgaris (L) Berberidaceae	Berberine	Bacillary dysentery
Papaver somniferum (L) Papaveraceae	Codeine	Analgesic
Mucuna pruriens (L) Fabaceae	L-Dopa	Anti-Parkinson
Ephedra sinica Stapf Ephedraceae	Ephedrine	Antihistamine
Lycoris squamigera Maxim Amaryllidaceae	Galanthamine	Cholinesterase inhibitor
Piper methysticum G.Forst Piperaceae	Kawain	Tranquilizer
Tabebuia avellanedae A.P. de Candolle Bignoniaceae	Lapachol	Anticancer
Strophanthus gratus (Wall. And Hook.) Baill. Apocynaceae	Ouabain	Cardiotonic
Salix alba (L) Salicaceae	Salicin	Analgesic
Podophyllum peltatum (L) Berberidaceae	Taxol	Antitumor
Vinca minor (L) Apocynaceae	Vasicine	Cerebral stimulant

vulnerable to climate change that can significantly alter the weed community composition, their phenological cycle and their primary and secondary metabolite. With predictive climate change scenarios, some weed species may go to vanish, while some become more invasive (Anwar et al. 2021). With reference to their enormous ecosystem services (provisional, cultural, and regulatory) many weed species, as well as their active ingredients, were screened with a climatic change perspective. For example, an increase in the anti-depressants' hypericin and pseudo-hypericin was noted in St. Johnswort at a CO_2 concentration of 1000 ppm relative to CO_2 conditions (Zobayed and Saxena 2004).

The study of Ziska et al. (2005) demonstrated that CO_2 and temperature distinctly and synchronously had significant effects on the concentration and yield of all alkaloids from *Nicotiana tabacum* (L) and *Datura stramonium* (L). Ziska et al. (2008) also quantified changes in the growth and alkaloid content of *Papaver setigerum*, with recent and projected increases in atmospheric CO_2. Their results revealed that increasing CO_2 from 300 to 600 ppm increased the number of capsules, capsule weight, and latex production by 3.6, 3.0, and 3.7 times, respectively. Mathur (2013) assessed spatial and modular variability in phytosterol composition in *Tribulus terrestris* (L) and reported the synergistic effect of soil organic carbon and clay content on this metabolite. Similarly, Mathur and Sundarmoorthy (2017) studied the effects of solar energy variables on secondary metabolites of a cosmopolitan plant species *Corchorus depressus* (L). Their results revealed that outgoing net longwave radiation, extra-terrestrial radiation, actual vapor pressure, and incoming solar radiation significantly controlled total carbohydrate and steroidal sapogenin in every plant module of this species.

In comparison to native plant species, parameters of global change, such as increased temperature and CO_2 enrichment, enhance the performance of weed species (Liu et al. 2017). Therefore, predicting the distribution of weeds under climate change scenarios and identifying the areas of their habitat are vital to exploiting the service served by such underutilized economic species (Poudel et al. 2020).

Across a landscape, ecological niche models (ENMs) are the initial step to predict suitable ecological niches for a species. The ENM is a statistical approximation regarding the distribution of a species as well as it links their location data to environment variables by using statistical techniques in order to describe, understand, and/or predict the distribution of species (Sillero and Barbosa 2021). The mathematical output of ENMs can either be an equation that correlates the expected distribution of the species to a set of environmental predictors, or a response curve that describes how the predictors regulate species distribution. The mathematical model can be specialized into a cartographic model, i.e. a map showing habitat suitability, probability of species occurrence, or the favourability for species occurrence. Therefore, ENMs are forecasted in the environmental space and projected into the geographical space. A niche consists of sets of biotic and abiotic conditions of the environment that define the limits of a species' ability to survive. Alternatively, the niche is a set of resources occupied by an organism (Putman and Wratten 1984). The niche was initially visualized as a species habitat requirement and as a trait of the biotic community. Later niche formalization and hypervolume introduce the concept of niche variables and axes. Each dimension of the hypervolume is a variable, and variables are interconnected to match species to environmental gradients. Niche space contains many distinct regions that are summarized in multiple ways, including habitat, trophic, and multidimensional (Dash 2007). The basic idea is that there are several important axes to be considered, those which correlate well with species survival. For example, soil moisture and nitrate ions concentration, as effective predictors of the presence of arbuscular mycorrhizae.

Ecological niche modeling (ENM) analysis for weed species have been carried out across the globe that includes study of *Senna obtusifolia* L. (Dunlop et al. 2006; Australia), *Lantana camara* (Taylor and Kumar 2013; Australia), *Ambrosia artemisiifolia* L. and *Ambrosia trifida* L. (Qin et al. 2014), *Chromolaena odorata* (L.) R. M. King & H. Rob. (Suarez-Mota et al. 2016; South Africa), *Amaranthus retroflexus* L., *A. spinosus* L., *A. viridis* L., *Bidens pilosa* L., *Conyza bonariensis* L., *C. canadensis* L., *Galinsoga parviflora* Cav., and *Physalis angulata* L. (Wan et al. 2017; China), *Cassia tora* L. and *L. camara* L. (Panda et al. 2017; India), *Parthenium hysterophorus* L. (Ahmad et al. 2019a, b; India), *Ageratina adenophora* (Spreng.) King & H.Rob. (Poudel et al. 2020; Nepal), *Cardaria draba* L., *Centaurea maculosa Hayek*, *Cirsium arvense* L., *Convolvulus arvensis* L, *Cynoglossum officinale* L., *Euphorbia esula* L., *Hypericum perforatum* L., *Leucanthemum vulgare* Lam., *Linaria dalmatica* Mill., *Potentilla recta* L., and *Tanacetum vulgare*

L. (Adhikari et al. 2020; USA), *Erigeron canadensis* (L.) Cronquist (Yan et al. 2020; China), *Apium leptophyllum* (Pers.) F.Muell, *Astragalus sinicus* L., *Bromus unioloides* Hack., *Chenopodium ambrosioides* L., *Coronopus didymus* L., *Gnaphalium calviceps* (Fernald) Cabrera, *Lolium multiflorum* (Braun) Schinz & Keller, *Modiola caroliniana* (L.) G.Don, *Oenothera laciniata* Hill., *Paspalum dilatatum* (Poiret) Coste, *Silene gallica* L., *Sisymbrium officinale (Linnaeus) Scopoli., Sisyrinchium angustifolium* Mill., *Spergularia rubra* Merino and *Malva parviflora* L. (Hong et al. 2021; South Korea), *A. adenophora* Spreng, *Alternanthera philoxeroides* (Mart.) Griseb, *Ambrosia artemisiifolia* L., and *Mikania micrantha* Kunth (Tu et al. 2021; China).

In general, species distribution modeling (SDM) involves two different sub-groups using presence-absence data, i.e. regression-based and machine learning. Regression-based techniques include Generalized Linear Models (GLM), Generalized Additive Models (GAM), and Multivariate Adaptive Regression Splines (MARS). Machine learning algorithms include Artificial Neural Network (ANN), Classification Trees (CART), Maximum Entropy (MaxEnt), Genetic Algorithm (GARP), and Random Forest (RF). Details of these techniques can be found in Pecchi et al. (2019). Briefly, these approaches are different from each other in terms of species records (absence/presence or presence-only) as well as the factors used to make predictions (mechanistic-physiological constrain or empirical-climatic approach). Each model is associated with drawbacks that limit the accuracy of predictions (Elith et al. 2006). Consequently, the most reliably modeled potential distribution of a species could be identified through comparing predictions obtained from more than one algorithm (Zhang et al. 2020).

One of the inconveniences that arise when applying SDM is that there are a great number of available alternatives, which, in some cases, provide different results; this complicates the choice of the best option for each case (Thuiller et al. 2009). According to these authors, this kind of situation happens when the priority is to predict the distribution of a species as a function of different scenarios of climate change. Another disadvantage may appear when many predictive environmental variables are used, producing an over-adjustment (Breiner et al. 2015). Over-adjustments frequently reduce the applicability of the models to a new set of data (Merow et al. 2014). One way to overcome this problem is by using ensemble methods which provide greater precision than the individual counterparts. Ensemble modeling is advised as both a technique to produce more robust model predictions and to provide a measure for the degree of similarity among different model results.

Tribulus terrestris belongs to the family Zygophyllaceae and is known as Puncture-vine—(English), "Gokhru" and "Chota Gokhru" (Hindi) and "Goksharu" – (Sanskrit). It is a cosmopolitan species distributed throughout the warm regions from its centre of origin in the Mediterranean region (https://www.cabi.org/isc/datasheet/54447#tosummaryOfInvasiveness Mathur 2014a and Mathur 2020). It is found throughout India ascending to 3385 metres. It is a prostrate, spreading herb particularly adapted to dry regions, which produces fruit with sharp, rigid spines. The entire plant, particularly the fruit is used in traditional medicine (regarded as herbal Viagra). A review of the literature suggested that several pharmacological and phytochemical (particularly dealing with steroidal saponin and phytosterol) studies have been carried out in various parts of the world so as to identify the chemical constitutes (Mathur 2013, 2014b, 2017). The majority of the research on this species were aimed toward its phytochemical and pharmacological and aphrodisiac properties (Mathur and Sundarmoorthy 2013a). Our review of the literature revealed only one habitat suitability model for this species conducted from Alberta area of the Canada (Chai et al. 2016) and no such effort have been carried out in the Indian region.

According to the existing literature, the ensemble forecasting model from different SDM techniques is recognized as the most powerful, stable, and well-referenced method to analyse the potential impact of climate change on plant species (Pecchi et al. 2019). An ensemble (or sometimes consensus) modeling is based on the idea that each different modeling output represents a possible state of the real distribution. With this technique, single-model projections are combined into a final surface where the predictions are averaged. Our primary aim was to assess the predictive performance of different modeling techniques by evaluating the Kappa and AUC and TSS values. Our specific objectives were (a) to assess the comparative performance of different algorithms (both regression and machine learning based) for the assessment of habitat suitability of *T. terrestris* within Indian arid and semi-arid areas. This objective was carried out with ensemble techniques, and the final area assessment was carried out with the most perfect model, (b) to assess the impact of niche modeling on the extent of occurrence and area of occupancy of this species with three timeframe projections. We hypothesized that these objectives will enable us to identify the major bioclimatic and community predictors that determine the habitat suitability of *T. terrestris* and also give projected area cover by this species under different suitability classes. For the above objectives, we implemented the ensemble techniques in which different algorithms (General linear model; GLM), (Generalized additive model; GAM), (Classification tree analysis; CTA), (Artificial neural network; ANN), (Support vector machine; SVM), (Multivariate adaptive spline; MARS), (Random forest; RF), and (Maximum entropy; MAXENT) were utilized and were evaluated based on values of certain indices. Further, the algorithm that provides the highest accuracy was processed for the estimation of different types of habitats and the area covered by them.

13.2 Material and Methods

13.2.1 Data Collections

Distributional records for this species were obtained from data repositories like the Global Biodiversity Information Facility (www.gbif.org/), Indian Biodiversity Portal (https://indiabiodiversity.org/species/show/33318), and published literature (Mathur and Sundarmoorthy 2013a; Kaur et al. 2016). The coordinates of these points were marked on a WGS84 coordinate system using high-resolution Google Earth satellite image data and GIS ArcMap (Coban et al. 2020) software. Further, where occurrence records lacked exact geo-coordinates, we used Google Earth (http://ditu.google.cn/) to determine their latitude and longitude values (Xu et al. 2021). Using the above sources, the distributional localities were compiled into a database in CSV format (.csv). Duplicate records were filtered spatially and deleted to keep only one occurrence by using the Spatial Thin window of R based Graphical User Interface Wallace Software (Kass et al. 2018).

13.2.2 Bioclimatic (BC) and Non-bioclimatic Variables (NBC)

The sixth Intergovernmental Panel on Climate Change (IPCC) assessment report publishes four climate change scenarios, namely SSP126 scenario, SSP245 scenario, SSP370 scenario, and SSP585 scenario. We selected the SSP245 scenario, where greenhouse gas emissions are about the same as the current condition (1970–2000) and the global average temperature tends to reduce with human intervention. Current (Mishra et al. 2021) and 2 future scenarios (2050-time frame represent the mean values from 2041 to 2060, while 2070 represents the mean values from 2061 to 2080) (Coban et al. 2020; Ye et al. 2020). The 19 bioclimatic variables (Hijmans et al. 2005) are one of the outputs of WorldClim Version 1.4, which was downloaded (accessed on 22nd March 2022) and clipped from world data for India at a spatial resolution of 30 arc sec (~1 × 1 km resolution) and converted to ASCII (or ESRI ASCII) in DIVA-GIS version 7.5 (Hijmans et al. 2001; Zhang et al. 2021). Details of each bioclimatic parameter, their units and mathematical expressions are provided in Table 13.2.

13.2.3 Habitat Heterogeneity Index (HHI)

Niche theory predicts a positive heterogeneity–diversity relationship, because a more heterogeneous area may provide more niche space and allow more species to co-exist through niche partitioning. Habitat heterogeneity has long been recognized as a key landscape characteristic with strong relevance for biodiversity and its functions. Tuanmu and Jetz (2015) developed 14 metrics based on the textural features of Enhanced Vegetation Index (EVI, i.e. the frequency distribution of pixel values) imagery from Moderate Resolution Imaging Spectroradiometer (MODIS) to characterize global habitat heterogeneity at 1-km resolution.

Six first-order and eight second-order texture measures are available (http://www.earthenv.org/texture) at 30 arc second (~1 km at the equator), 2.5 arc minute (~5 km) and 12.5 arc-minute (~25 km) resolutions. The first-order texture measures are statistics describing the frequency distribution of Enhanced Vegetation Index (EVI) values and measuring compositional variability in EVI within an area. The second-order texture measures are statistics of the occurrence probabilities of different EVI combinations among pixel pairs within an area, and thus, also reflect spatial arrangement and dependency of the EVI values. In this study, we used 30 arc second data set related to first-order texture measures (Coefficient of variation = Normalized

Table 13.2 Predictive variables (current and future) bioclimatic data variables. Calculation criterion of each variable can be found at https://pubs.usgs.gov/ds/691/ds691.pdf

Code	Environmental variables	Scaling factor	Unit
BC-1	Annual mean temperature	10	°C
BC-2	Mean diurnal range (mean of monthly (max temp – min temp))	10	°C
BC-3	Isothermality (BC2/BC7) (×100)	100	–
BC-4	Temperature seasonality (standard deviation ×100)	100	–
BC-5	Max temperature of warmest month	10	°C
BC-6	Min temperature of coldest month	10	°C
BC-7	Temperature annual range (BC 5–BC 6)	10	°C
BC-8	Mean temperature of wettest quarter	10	°C
BC-9	Mean temperature of driest quarter	10	°C
BC-10	Mean temperature of warmest quarter	10	°C
BC-11	Mean temperature of coldest quarter	10	°C
BC-12	Annual precipitation	1	mm
BC-13	Precipitation of wettest month	1	mm
BC-14	Precipitation of driest month	100	mm
BC-15	Precipitation seasonality (coefficient of variation)	1	Percent
BC-16	Precipitation of wettest quarter	1	mm
BC-17	Precipitation of driest quarter	1	mm
BC-18	Precipitation of warmest quarter	1	mm
BC-19	Precipitation of coldest quarter	1	mm

dispersion of EVI; Evenness = Evenness of EVI; Range = Range of EVI; Shannon and Simpson Indices = Diversity of EVI; Standard deviation = Dispersion of EVI) and to second-order texture measures (Uniformity = Orderliness of EVI; Maximum = Dominance of EVI combinations).

13.2.4 Data Processing

13.2.4.1 Issue of Multicollinearity

In this study, a multicollinearity test was conducted to minimize the risk of overfitting by using Pearson's Correlation Coefficient (r) to examine the cross-correlation. Further, the variables with cross-correlation coefficient value of > ±0.85 were stepwise excluded (Pradhan 2016). This analysis was carried out by using Niche Tool Box (Osorio-Olvera et al. 2020, https://github.com/luismurao/ntbox).

Multicollinearity among predictor variables was reduced following the methods suggested by Kumar et al. (2006). Among two highly cross-correlated variables, one was selected which is biologically relevant to the species and offers ease in the interpretation of the model (Kumar and Stohlgren 2009; Padalia et al. 2014). For instance, if variables annual precipitation and precipitation of wettest month were found highly correlated, we retained precipitation of wettest month since it captures seasonal variability in precipitation. Only one variable from each set of highly cross-correlated variables ($r^2 > 0.85$) was kept for further analysis (Ma and Sun 2018). In this research, 70% and 30% of data were assigned for model training and validation, respectively (Obiakara and Fourcade 2018).

13.2.5 Species Distribution Modeling

Modeling was conducted using the following algorithms available within the Dismo 1.1–4 package (Hijmans et al. 2017): Generalized Linear Models using Gaussian distribution "GLM", Generalized Additive Model "GAM" (Wood 2019), Support Vector Machines "SVM" (Vapnik 1998), Random Forest "RF" (Breiman 2001; Cutler and Wiener 2018), Multivariate Adaptive Spline "MARS" (Friedman 2001), and Maximum Entropy (Maxent v. 3.4.1; Phillips et al. 2017). Artificial Neural Network "ANN" (Ripley and Venables 2020), and Classification Tree Analysis "CTA" (Therneau et al. 2019), the last two being executed through the biomod2 3.4.6 package (Thuiller et al. 2020). We used the default for all studied algorithms. We ran the SDM analysis in two different steps: (a) to identify the best-performed model/algorithm that was judged on some pre-decided evaluation criteria (described below) and (b) SDM modeling with the best-performed algorithm using three climatic timeframes and HHI.

Models were evaluated using K-fold cross-validation with 10 folds and 10 replications for each algorithm; for each replicates the data are divided randomly into 10 folds, one of which is used to evaluate the model calibrated using the other 9 folds, so as to give more precise projections (Elith et al. 2011). Model prediction performance assessed by Kappa statistic, area under the receiver operating characteristic curve (AUC; this is a threshold-independent statistic), sensitivity, specificity, and True Skill Statistic (TSS which is threshold-dependent statistic, Monserud and Leemans 1992; Duan et al. 2014). These statistics are considered to be the best evaluation standard, and they were widely used in SDMs (Hanley and McNeil 1983; Fielding and Bell 1997; Li et al. 2012).

Further, models making presence-absence predictions are typically appraised by comparing the predictions with a set of validation locations and making a confusion matrix that records the number of true positives (a), false positive (b), false negative (c), and true negative (d) cases forecast by the model (Table 13.3). One simple measure of correctness that can be derived from the confusion matrix is the quantity of correctly predicted sites (overall accuracy; Table 13.3). However, this measure was disapproved for assigning high precisions for rare species. Two alternative measures that are regularly derived from the confusion matrix are sensitivity and specificity. Sensitivity is the number of observed presences that are predicted as such, and therefore, quantifies omission errors. Specificity is the proportion of observed absences that are predicted as such, and therefore, quantifies commission errors (Table 13.3). Sensitivity and specificity are independent of each other when

Table 13.3 An error milieu is used to appraise the analytical correctness of presence-absence models. (a) Number of cells for which presence was properly forecasted by the model; (b) number of cells for which the species was not found but the model predicted presence; (c) number of cells for which the species was found but the model predicted absence; and (d) number of cells for which absence was correctly predicted by the model

Model		**Validation data set**	
		Presence	**Absence**
	Presence	a	b
	Absence	c	d

Table 13.4 Procedures of predictive accuracy are intended from a 2 × 2 error matrix (Table 13.2). Overall correctness is the rate of properly classified cells. Specificity is the probability that the model will correctly classify an absence. Sensitivity is the probability that the model will correctly classify a presence. The kappa statistic and true skill statistic (TSS) normalize the overall accuracy by the accuracy that might have occurred by chance alone. In all formulae $n = a + b + c + d$

Measure	Formula
Overall accuracy	$\frac{a+d}{n}$
Sensitivity	$\frac{a}{a+c}$
Specificity	$\frac{d}{b+d}$
Kappa statistic	$\frac{\left[\frac{a+d}{n}\right]-\frac{(a+b)(a+c)+(c+d)(d+b)}{n^2}}{1-\frac{(a+b)(a+c)+(c+d)(d+b)}{n^2}}$
TSS	Sensitivity + specificity-1

compared across models and are also independent of prevalence ($(a + c)/n$), the proportion of sites in which the species was recorded as the present; Table 13.4).

Both TSS and kappa are threshold-dependent metrics of model evaluation and range from −1 to +1. Generally, the values of TSS and kappa below 0.40 indicate poor model performance, values ranging from 0.40 to 0.75 specify good model performance, and values above 0.75 indicate excellent model performance (Beaumont et al. 2016; Ahmad et al. 2019a, b). Evaluation criteria for the AUC statistic are as follows: excellent (0.90–1.00), very good (0.8–0.9), good (0.7–0.8), fair (0.6–0.7), and poor (0.5–0.6).

The contribution of each bioclimatic variable (19 three-time projections) and HHI (six variables) to ecological niche modeling of *T. terrestris* was identified by their respective variable importance ranking (Irving et al. 2019).

13.2.6 Post Ensemble Analysis

13.2.6.1 Habitat Suitability

Habitat suitability of *T. terrestris* was quantified (square kilometre) under four predefined classes that were optimum, moderate, marginal, and low. This was carried out by transferring the raster output of the best model to ArcMap and the area under each class was quantified by using a raster calculator. This exercise was also carried out between the two-time frames, i.e. transitional area changes under different classes by overlaying the ArcMap of future timeframe upon the ArcMap of the previous time frame (i.e. 2050-current; 2070–2050 and 2070-current). Centroid shifting between different bioclimatic timeframes was carried out using SDM toolbox (Brown 2014). This tool calculates the distributional changes between two binary SDMs (i.e. current vs. future SDMs). Further, raster output was overlaid on google earth to identify the GPS locations of each centroid. To calculate the impact of the climate scenarios on predicted habitat suitability, we measured the percent changes in mean habitat suitability by using the following formula (Mathur and Sundarmoorthy 2013b; Mathur 2014b; Wright et al. 2016; Kaky et al. 2020)

$$\left[\left(\frac{(\text{Future} - \text{Current})}{\text{Current}}\right) \times 100\right]$$

13.2.6.2 Raster Similarity Analysis

A fuzzy kappa map comparison was carried out to quantify the grades of similarity between pairs of cells in two maps. Fuzzy kappa is based on a cell-by-cell map comparison, which takes the neighbourhood of a cell in account to express the similarity of that cell in a value between 0 (fully distinct) and 1 (fully identical). This was carried out using Map Comparison Kit (MCK http://mck.riks.nl/downloads) for among individual BC pertains to different time frame.

13.2.6.3 Niche Overlap

Niche overlap compared the inferred and true distributions of suitability scores over geographic space in the present day. The output of each bioclimatic higher performance model in ASCII format was utilized to quantify niche overlap between two studied parameters (related with bioclimatic parameters of different timeframes). This was done by using ENM Tools (Warren and Seifert 2011). This analysis was carried out to visualize the amount of area retained by this species under different predictions. Schoener' *D* (*D* measures consistency of niche overlap per pair) and Hellinger's-based *I* (that measures overlap degree of the geographical distribution) values were employed to represent the ecological niche overlap. *D* and *I* values ranged from 0 to 1.

13.2.6.4 Automated Conservation Assessments (AA): I and II

In the present study, we first explored the current status of the *T. terrestris* by using our spatially thinned geographical locations. We quantified Extent of Occurrence (EOO Sq km), Area of Occupancy (AOO Sq km), number of unique occurrences, number of subpopulations, number of locations, IUCN (2014) threat category according to Criterion B and finally, IUCN annotation (Category Code) using R programme "*ConR*" (Dubey et al. 2017) with using IUCN defined cell width, i.e. 2 × 2 km (Kass et al. 2021). Additionally, we also carried out a similar exercise to evaluate the impact of niche modeling on EOO and AOO with the algorithm that showed maximum values of various SDM evaluation criteria. For this, we transform the binary map of the most suitable model into "XYZ" file with help of QGIS. This file further transfers to the text file and lastly, it was opened with Microsoft Excel and with help of these coordinates with higher probability value, i.e. 1 was selected for estimation of EOO and AOO using ConR software. Such approaches were advocated by Marcer et al. (2013), Adhikari et al. (2018), and Marco et al. (2018).

13.3 Results and Discussion

With our different data sources, we have collected two hundred seven coordinates of *T. terrestris* (Fig. 13.1). To address the spatial autocorrelation issue, we treated our data set with SpatialThin menu with 5-km area. This test provided spatially thin data set with 151 coordinates, and these data points were utilized for further analysis.

Results of correlation analysis among different variables of bioclimatic and non-bioclimatic variables are presented in Fig. 13.2a–d. To address the issue of multicollinearity in species distribution modeling, we follow the procedures described by Kumar et al. (2006) and Pradhan (2016). Within current bioclimatic variables, we found significant correlation (> 0.85) between temperature seasonality (BC-4) and minimum temperature of coldest month (B-6 $R^2 = -0.89$) and between BC-4 and temperature annual range (BC-7 $R^2 = 0.94$). Similarly, precipitation of wettest quarter (BC-16) significantly correlated with BC-12 (annual precipitation $R^2 = 0.95$) and with BC-13 (precipitation of wettest month $R^2 = 0.95$). The maximum temperature of the warmest month and mean temperature of the warmest quarter were also significantly related to each other ($R^2 = 0.91$ Fig. 13.2a).

In stepwise exclusion, we removed BC – 4, 5, 6 and 16 from SDM analysis. Similarly, highly correlated variables related to BC-2050 were also removed (temperature seasonality BC-4; the maximum temperature of warmest month BC-5; a minimum temperature of coldest month BC-6; temperature annual range BC-7, and precipitation of wettest quarter BC-16 (Fig. 13.2b)). Among bioclimatic variables of

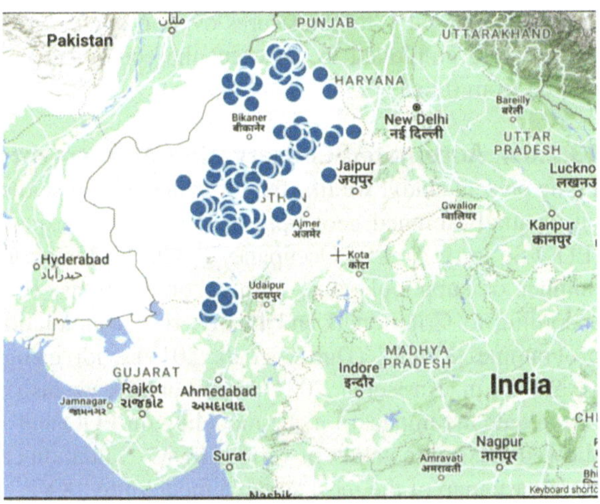

Fig. 13.1 Existing field locations of *Tribulus terrestris* (L) in India developed with ArcMap-GIS with uploading the latitude and longitude of existing locations of this species on the shape file

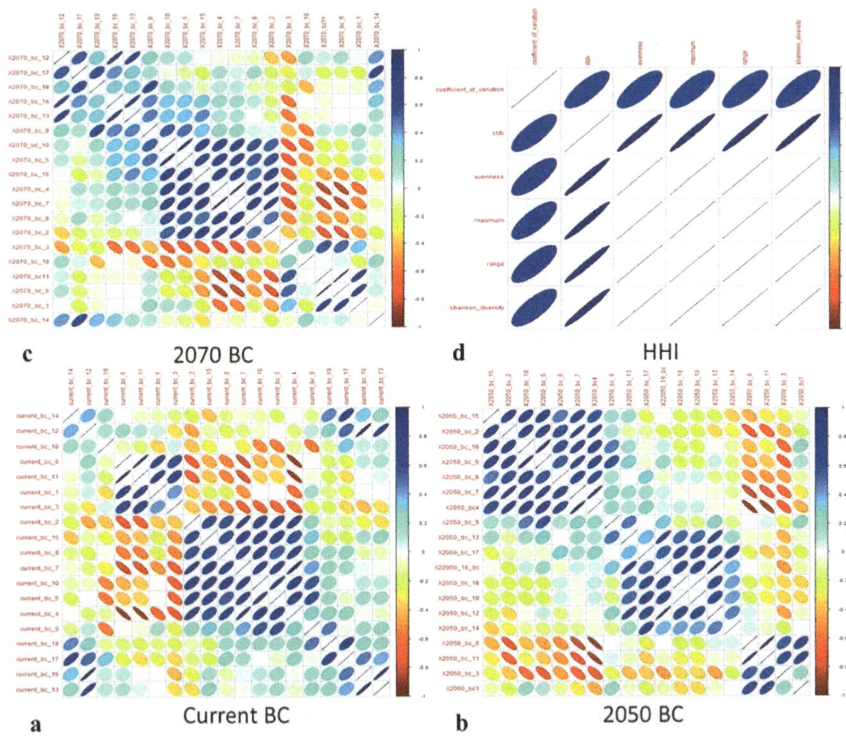

Fig. 13.2 (**a** and **b**) Correlation analysis among variables related to current (**a**) and future (2050, **b**) bioclimatic variables. (**c** and **d**) Correlation analysis among variables related to future (2070 **c**) and habitat heterogeneity indices (**d**)

2070, BC-5, BC-7 and BC-11 (mean temperature of the coldest quarter) and BC-17 (precipitation of driest quarter) were excluded (Fig. 13.2c). No such relationships were observed among variables related to habitat heterogeneity indices (Fig. 13.2d).

Results of Ensemble Species Distribution Evaluation (ESDEVL) are presented in Table 13.5. AUC value of each bioclimatic timeframe revealed the excellent model quality, while HHI ESDEVL AUC suggested the ensemble model of very good quality. Similar to AUC, TSS values of current, 2050 and 2070 bioclimatic also revealed excellent model qualities ($>$ 0.75) and HII values (0.628) suggested good model performance. However, kappa values for all timeframe/parameters were recorded $<$0.75 that suggests good model qualities.

The above evaluation criteria for individual algorithm quantified under the current timeframe is presented in Table 13.6. Results suggested that random forest performed best among all the algorithms with higher AUC (0.95), kappa (0.90), and TSS (0.90) values. While lowest model quality was assessed with GLM that had have lowest AUC (0.89), kappa (0.61), and TSS (0.79). Further, among these algorithms, the lowest kappa values (0.29) were recorded with MaxEnt while its

Table 13.5 Ensemble species distribution model evaluation criterions quantified with different bioclimatic timeframe and with habitat heterogeneity indices (HHI)

Timeframe/parameter	AUC	Sensitivity	Specificity	Kappa	TSS
Current ensemble ESDEVL	0.931	0.950	0.911	0.731	0.861
2050 ensemble ESDEVL	0.933	0.948	0.917	0.740	0.865
2070 ensemble ESDEVL	0.927	0.949	0.906	0.722	0.855
HHI ensemble ESDEVL	0.814	0.814	0.814	0.508	0.628

Table 13.6 Model evolution parameters of individual algorithm calculated through ensemble modelling with the current bioclimatic timeframe

Algorithms	AUC	Sensitivity	Specificity	Kappa	TSS
GLM	0.89	0.91	0.88	0.61	0.79
GAM	0.93	0.92	0.93	0.75	0.86
MARS	0.94	0.96	0.92	0.74	0.88
CTA	0.92	0.95	0.89	0.85	0.85
RF	0.95	0.95	0.94	0.90	0.90
MAXENT	0.94	0.96	0.93	0.29	0.89
ANN	0.92	0.98	0.87	0.85	0.85
SVM	0.91	0.93	0.89	0.83	0.83

TSS values (0.89) were recorded as the second highest after random forest. Thus, within ensemble SDM analysis such types of comparative analysis will always be beneficial to correctly identify the habitat suitability with using perfect algorithm.

Similar to the current timeframe, model evaluation results for the individual algorithm with a 2050-time frame are presented in Table 13.7. Similar to current bioclimatic variables, we got similar model outputs with the future 2050 projection. Random forest was identified as the most suitable algorithm with their higher AUC (0.95), kappa (0.90), and TSS (0.90) values. Again, MaxEnt have the similar high (0.95) values with the lowest (0.35) kappa value. For this projection, the lowest TSS (0.80) was recorded GLM. ANN and SVM also performed more or less equal to RF, however, their kappa values are not at par, therefore, similar to current BC, we re-run RF with this projection for further analysis.

For the 2070 future projection, we recorded more or less similar AUC and TSS values with RF and MaxEnt. However, the kappa value of the latter one was the lowest (0.33) among all algorithms (Table 13.8). Thus, similar to the above, here we also decided to use RF for further analysis. ANN and SVM performed in a more or less similar manner but their model evaluation values are lower than RF.

With habitat heterogeneity indices (HHI), we recorded lower values for all the model quality parameters with eight studied algorithms, and this suggested that this parameter may not be vital or suitable for the prediction of distribution modeling of *T. terrestris* as compared to bioclimatic variables. However, similar to the above all analysis we got higher AUC (0.85), kappa and TSS (0.70) with RF (Table 13.9). Again, MaxEnt's kappa value (0.06) was the lowest among all other algorithms.

Table 13.7 Model evolution parameters of individual algorithm calculated through ensemble modelling with 2050 bioclimatic timeframe

Algorithms	AUC	Sensitivity	Specificity	Kappa	TSS
GLM	0.90	0.92	0.88	0.62	0.80
GAM	0.93	0.95	0.92	0.73	0.87
MARS	0.95	0.96	0.94	0.78	0.90
CTA	0.91	0.92	0.90	0.82	0.82
RF	0.95	0.94	0.95	0.90	0.90
MAXENT	0.95	0.96	0.95	0.35	0.89
ANN	0.92	0.96	0.88	0.84	0.84
SVM	0.92	0.94	0.90	0.84	0.84

Table 13.8 Model evolution parameters of individual algorithm calculated through ensemble modeling with 2070 bioclimatic timeframe

Algorithms	AUC	Sensitivity	Specificity	Kappa	TSS
GLM	0.87	0.89	0.85	0.54	0.74
GAM	0.93	0.94	0.91	0.71	0.86
MARS	0.93	0.95	0.91	0.72	0.87
CTA	0.93	0.96	0.90	0.86	0.86
RF	0.95	0.96	0.95	0.91	0.91
MAXENT	0.95	0.96	0.94	0.33	0.91
ANN	0.91	0.98	0.85	0.83	0.83
SVM	0.91	0.92	0.90	0.83	0.83

Table 13.9 Model evolution parameters of individual algorithm calculated through ensemble modelling with habitat heterogeneity indices (HHI)

Algorithms	AUC	Sensitivity	Specificity	Kappa	TSS
GAM	0.78	0.79	0.78	0.38	0.57
MARS	0.81	0.81	0.81	0.44	0.63
CTA	0.83	0.87	0.80	0.67	0.67
RF	0.85	0.81	0.88	0.70	0.70
MAXENT	0.76	0.75	0.78	0.06	0.53
ANN	0.82	0.830	0.81	0.64	0.64
SVM	0.81	0.820	0.81	0.63	0.63

Such ranking of modeling techniques for plant species were conducted by Bio et al. (1998; *Deschampsia cespitosa* GAM > GLM); Franklin (1998; *Arctostaphylos glandulosa* CART > GAM > GLM); Elith and Burgman (2002; *Grevillea barklyana, Oxalis magellanica, Tetratheca stenocarpa, Wittsteinia vacciniacea, Helichrysum scorpioides, Leptospermum grandifolium, Nothofagus cunninghamii,* and *Phebalium bilobum* GARP > GAM > GLM > ANUCLIM); Farber and Kadmon (2003; many species; Mahalanobis distance > BIOCLIM); Thuiller (2003; *Quercus petraea, Castanea sativa, Pinus halepensis* NN > GAM >

Table 13.10 Evolution parameters of random forest algorithm with three bioclimatic timeframes and habitat heterogeneity indices (HHI)

Variables	AUC	Sensitivity	Specificity	Kappa	TSS
Current BC	0.963	0.964	0.962	0.927	0.927
2050-BC	0.956	0.967	0.944	0.911	0.911
2070-BC	0.968	0.980	0.956	0.936	0.936
HHI	0.867	0.867	0.867	0.733	0.733

GLM > CART); Thuiller et al. (2003; GAM > GLM > CART); Robertson et al. (2004; *Lantana camara*, *Ricinus communis* and *Solanum mauritianum* FEM > BIOCLIM); Randin et al. (2006; *Fagus sylvatica*; GAM ≈ GLM); Schussman et al. (2006; *Eragrostis lehmanniana* GLM > GARP); and Broennimann et al. (2007; *Centanurea maculosa*; Random Forests > BoostedRT > GAM ≈ MARS > GLM > MixureDA > NN > CART). Attorre et al. (2011) compared RF, GAM, and CART to evaluate the potential effects of climate change on the abundance of 27 species on the Italian peninsula. In Garzón et al. (2006), RF, ANN, and CART are used to study the potential distribution area of *Pinus sylvestris*. RF demonstrated the best predictive performance. RF is also used in Koo et al. (2017) where six other algorithms were combined to model the geographical distribution of *Machilus thunbergii* Siebold & Zucc, a typical evergreen broadleaved tree in Korea Peninsula.

Thus, with reference to results depicted in Tables 13.5, 13.6, 13.7, 13.8, and 13.9, we decided to re-run over SDM analysis with the individual best-performed algorithm that is random forest. Results of this individual RF are presented in Table 13.10. With the interpretation criterion of AUC, kappa, and TSS, we got excellent model qualities with RF for all three studied timeframes while good model quality for HHI (Table 13.10).

13.3.1 Variable Importance Ranking (VIR)

Results of VIR for different bioclimatic timeframes and HHI for habitat suitability of *T. terrestris* are presented in Figs. 13.3, 13.4, 13.5, and 13.6. With current bioclimatic timeframes, variables like Precipitation of Coldest Quarter (BC-19 with VIR = 22.55), Precipitation Seasonality (i.e. Coefficient of Variation BC-15 with VIR = 13.78), and Annual Precipitation (BC-12 with VIR = 13.71) were the most significant variables that affect the SDM or habitat suitability of this species. While other variables had less than 10 VIR (Fig. 13.3). With 2050 and 2070 bioclimatic projections, we found that Precipitation Seasonality (BC-15 with VIR = 27.44) and Precipitation of Coldest Quarter (BC-19 with VIR = 27.47) were the most influencing variables, respectively (Figs. 13.4 and 13.5). While other variables had less than 10 contributions for its SDM.

Among the habitat heterogeneity indices, coefficient of variation (i.e., normalized dispersion of enhanced vegetation index with VIR = 73.95) and standard deviation (i.e., dispersion of EVI with VIR = 15) were identified as the most influencing

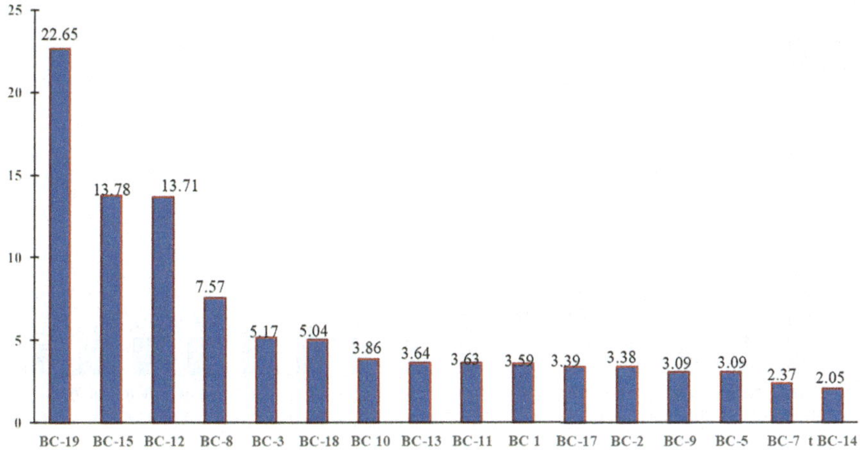

Fig. 13.3 Variable importance ranking of bioclimatic parameters of current time frame for habitat suitability of *Tribulus terrestris*

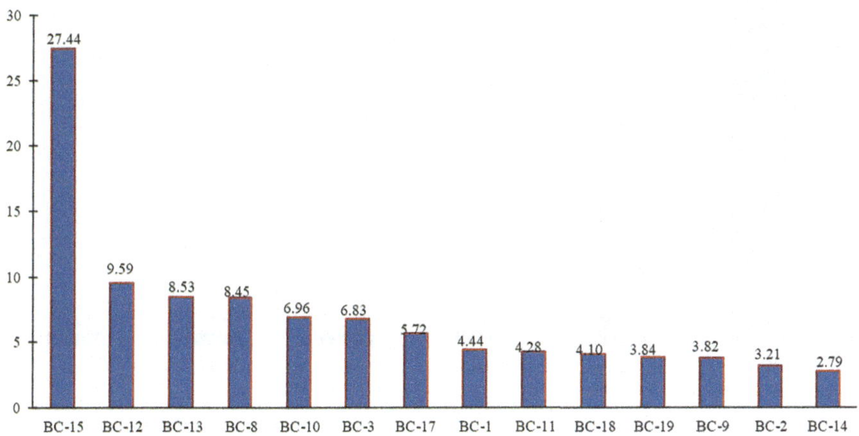

Fig. 13.4 Variable importance ranking of bioclimatic parameters of 2050 time frame for habitat suitability of *Tribulus terrestris*

factors (Fig. 13.6). While other factors, i.e. maximum, evenness, Shannon diversity, and range were less than 5 VIR.

Among the climatic parameters, temperature, and precipitation, which greatly vary over space and time, particularly in hot arid and semi-arid areas, are well-known main factors influencing the dynamics of plant communities. Our modeling approach indicates that precipitation in the coldest quarter (BC = 19; total precipitation dwindling in the coldest quarter (13-week period) of the year), precipitation seasonality (BC = 15; the variation in weekly precipitation totals over a year based on the standard deviation of weekly total precipitation), and annual precipitation

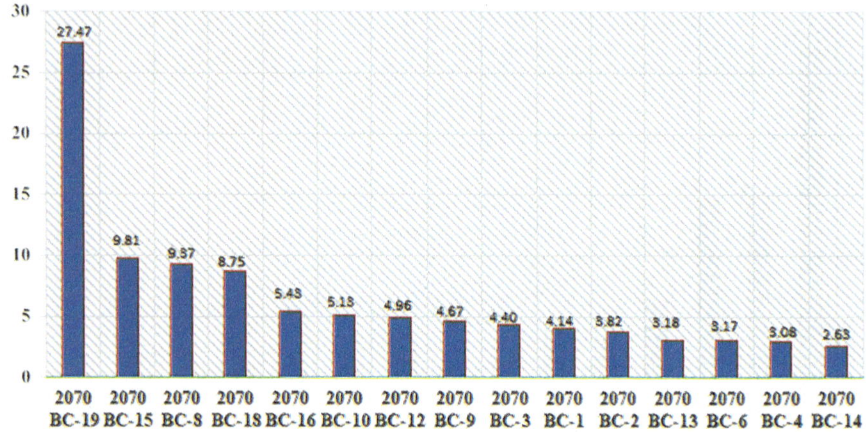

Fig. 13.5 Variable importance ranking of bioclimatic parameters of 2070 time frame for habitat suitability of *Tribulus terrestris*

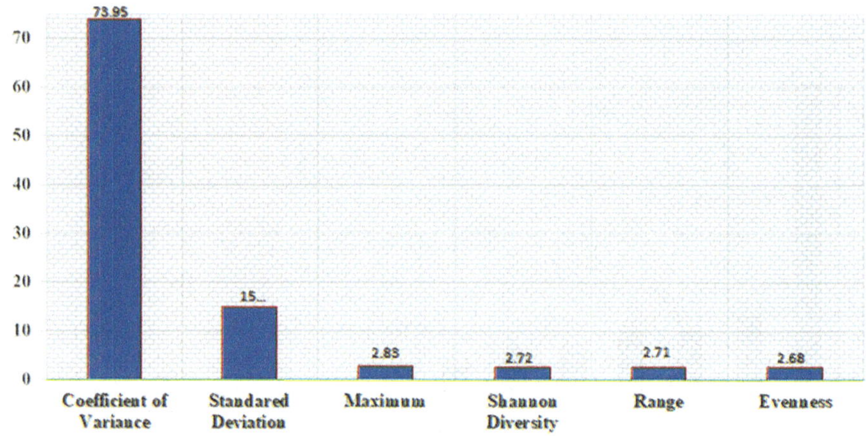

Fig. 13.6 Variable importance ranking of bioclimatic parameters of HHI for habitat suitability of *Tribulus terrestris*

(BC = 12; the sum of all weekly precipitation values over a year) together influenced the habitat suitability of this species compared to all other climatic variables during current conditions. Such a result revealed that currently, moisture availability throughout the years is essential for the distribution of this species. While during climate change timeframes, precipitation seasonality (BC = 15) and precipitation of the coldest quarter (BC = 19) during 2050 and 2070, respectively, revealed that with the 2050 timeframe, the monsoonal moisture availability will be crucial for this species, however, during 2070, winter moisture availability will control its distribution pattern significantly.

Such climate-based distribution patterns on plant weed species were conducted for *Amaranthus retroflexus, Arundo reynaudiana, Conyza canadensis, Elaeagnus angustifolia, Euonymus alata, Euphorbia esula, Hibiscus trionum, Lactuca serriola, Linaria vulgaris, Lythrum salicaria, Myriophyllum heterophyllum, Oenothera biennis, Phragmites australis, Populus alba, Salix babylonica, Solanum rostratum, Solidago canadensis,* and *Sonchus oleraceus* (Xu 2015, from Northwest China); *Ambrosia artemisiifolia, A. trifida, Ageratina altissima, Paspalum distichum, Sicyos angulatus, Hypochaeris radicata, Solidago altissima,* and *Lactuca serriola* (Adhikari et al. 2019); *Aegilops tauschii* (Wang and Chen 2019); *Parthenium hysterophorus* (Ruheili et al. 2021); *Cenchrus spinifex* (CAO et al. 2021).

Bioclimatic conditions during the coldest and/or driest month (or quarter, which is a period of 3 months) may significantly affect potential distributions for this species. Across the globe, the different quantitative analysis suggested that the number of weed species is likely to increase considerably with the increase in precipitation during the driest month, particularly with a variation of 3 mm in precipitation during the driest month. More alien species may invade regions where the precipitation during the driest quarter under future conditions is 10 mm higher than the current conditions. The effect of precipitation during the coldest quarter on species invasion is similar to that of precipitation during the driest quarter: the higher the precipitation during the driest quarter under future conditions (more than 9 mm than that under current conditions), the higher will be the rate of alien species invasion in the study area (Xu 2015).

13.3.2 Habitat Suitability

The result of habitat suitability with random forest algorithm processed with ArcMap are depicted in Fig. 13.7a (current), Fig. 13.7b (2050), Fig. 13.8a (2070), and Fig. 13.8b (HHI) and in Table 13.10. We classified habitat suitability into four classes, namely optimum, moderate, marginal, and low which have specific values and colours. With current bioclimatic conditions (Fig. 13.7a; Table 13.10), the optimum areas were located in more northern regions of the arid and semi-arid areas of India covering 92,400 km^2 area. While during 2050 projection area under this class increases up to 100,800 km^2 (Fig. 13.7b) which suggests 9.09% increase (Fig. 13.9). While during 2070, this class covers 91,900 km^2 (Fig. 13.8a) that showed −8.83% area decrease with respect to last project and only 0.54% decrease compared to current BC.

In this study, moderate area during all the three BC projections is located to adjoining of optimum areas and no separate patches of this class were observed at study areas. Area under this class during three-time projections was 56,500 km^2, 67,300 km^2, and 77,500 km^2, respectively (Table 13.11). Which revealed 19.12%, 15.16%, and 37.17% increase between 2050-current bioclimatic, 2070–2050 and 2070-current bioclimatic, respectively (Fig. 13.9). Marginal areas were more or less similar during all three projections; however, a 3.15% area decrease was recorded between 2050 to current. The highest lower suitable class was recorded in 2070

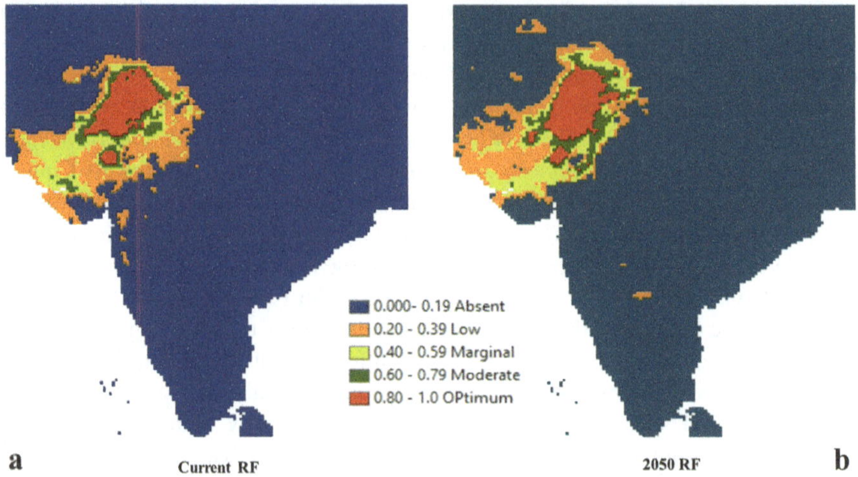

Fig. 13.7 (**a** and **b**) Raster output of ArcMap exhibits the areas under different classes with current (**a**) and 2050 (**b**) projections

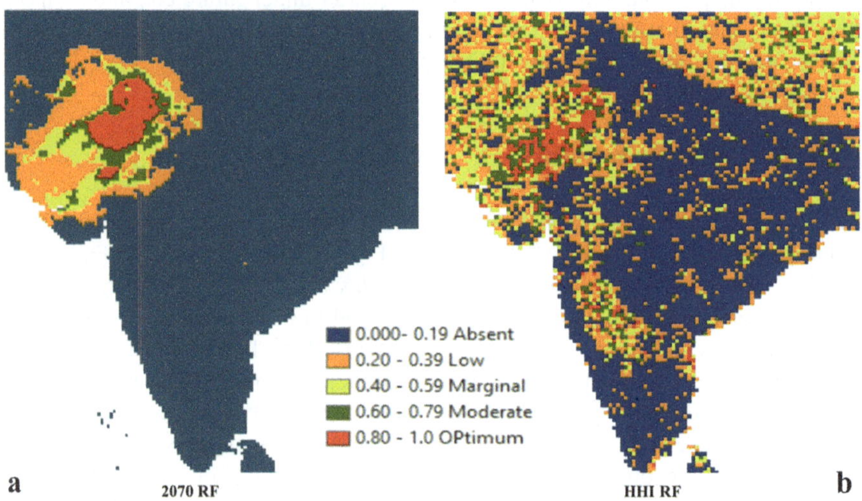

Fig. 13.8 (**a** and **b**) Raster output of ArcMap exhibits the areas under different classes with 2070 projection (**a**) and HHI (**b**)

(274,700 km^2). 4.79% reduction was recorded between 2050-current, while the highest percent (35.86) area increase for this class was recorded between 2070 and 2050. Overall, for this species, holistically, 3.89% area reduction was recorded for all classes between 2050-current. While 16.03 and 11.52% increases in overall areas were recorded between 2070–2050 and 2070-current, respectively (Fig. 13.9).

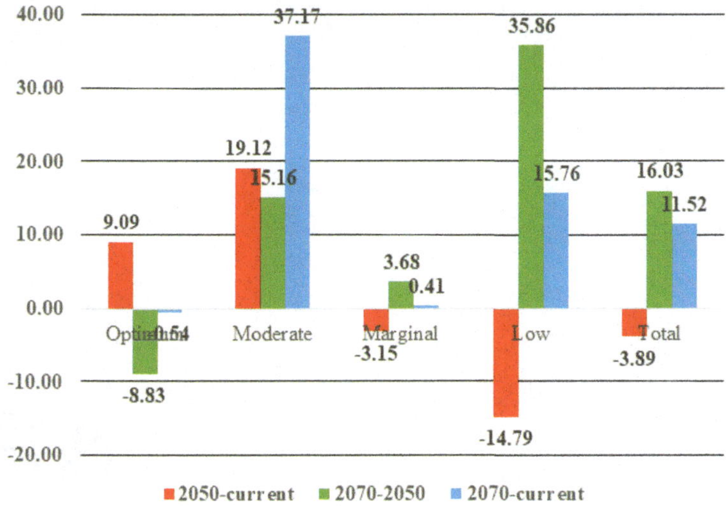

Fig. 13.9 Percent changes in mean habitat suitability with respect to two timeframe projections

Table 13.11 Area (km^2) under each habitat suitability classes pertain to *Tribulus terrestris* with three bioclimatic timeframes and with Habitat Heterogeneity Indices (HHI)

Classes	Current RF	2050 RF	2070 RF	HHI
Optimum	92,400	100,800	91,900	108,800
Moderate	56,500	67,300	77,500	220,400
Marginal	120,600	116,800	121,100	484,800
Low	237,300	202,200	274,700	804,000

With HHI variables, we found the disintegration of different classes in small patches as compared to bioclimatic variables (Fig. 13.8b). Interestingly with such a plant community-based variable, we noticed the expansion of this species outside the arid and semi-arid areas also reflected in the area covered under different classes (Table 13.11).

With our GIS analysis, we were able to identify the centroid of *T. terrestris* specifically under optimum class for all three-time projections (Fig. 13.10). Under current scenarios, its optimum centroid is located at 28°22′17.07″N, 74°14′16.24″E Ghadsisar, Bikaner, Rajasthan will shift to 104.95 km away at 27°44′42.53"N, 73° 21′21.12″E Rasisar, Bikaner during 2050 and subsequently, 69.56 km distance from 2050 during 2070 which will be located at 27°22′10.9″N, 73°56′44.14″E Gadriya, Nagor. Overall, 111.25 km centroid shifting will be anticipated from current to 2070 (Fig. 13.10).

We also quantified the area under the different classes of habitat suitability during three bioclimatic transitional periods. Results are summarized in Table 13.12 and graphically displayed in Fig. 13.11 (2050-current), Fig. 13.12 (2070-current), and Fig. 13.13 (2070–2050).

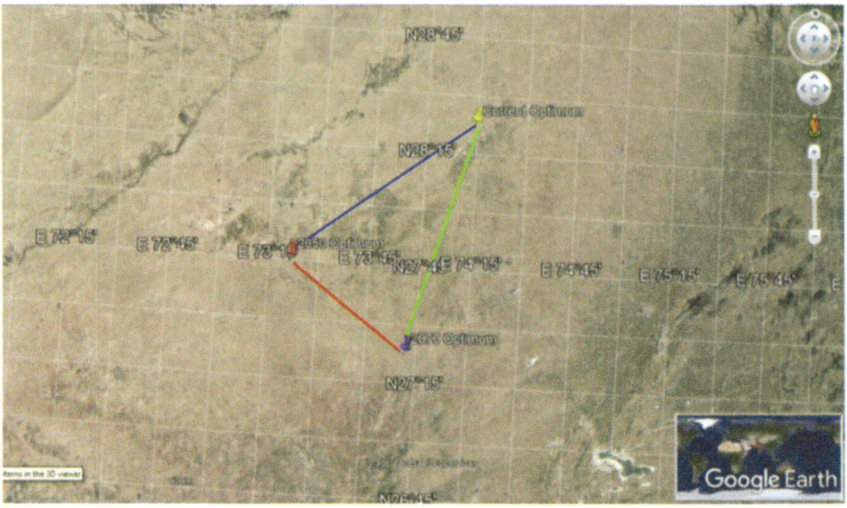

Fig. 13.10 Locations of centroid's (optimum class) during three timeframe projections

Table 13.12 Area (km^2) under each habitat suitability classes pertain to *Tribulus terrestris* with three bioclimatic transitional periods

	Classes				
Transitional periods	Optimum	Moderate	Marginal	Low	Total
2050-Current	75,500	220,300	128,500	119,000	543,300
2070-Current	52,700	317,400	92,900	354,500	817,500
2070–2050	68,000	448,000	79,800	429,500	1,025,300

During the current period, the optimum area covers 75,000 km^2 while the other three classes cover 220,300, 128,500, and 119,000 km^2 area, respectively. Compared to individual projection, transitional projection study revealed the patchiness of different classes extending in Gujarat, Rajasthan, Haryana, and Delhi states (Fig. 13.11).

Between, 2070-current, optimum and marginal area showed shrinkage covering 52,700 and 92,900, respectively. While moderate and low areas showed a larger area increase compared to 2050-current, covering 317,400 km^2 and 354,500 km^2, respectively (Fig. 13.12).

Between 2070–2050, optimum areas disintegrate into many small patches covering 68,000 km^2 area and a similar trend was exhibited with the marginal class that has 79,800 km^2 (Fig. 13.13). With such a transitional period, the highest area was recorded with moderate class covering 448,000 km^2 followed by low class with 429,500 km^2.

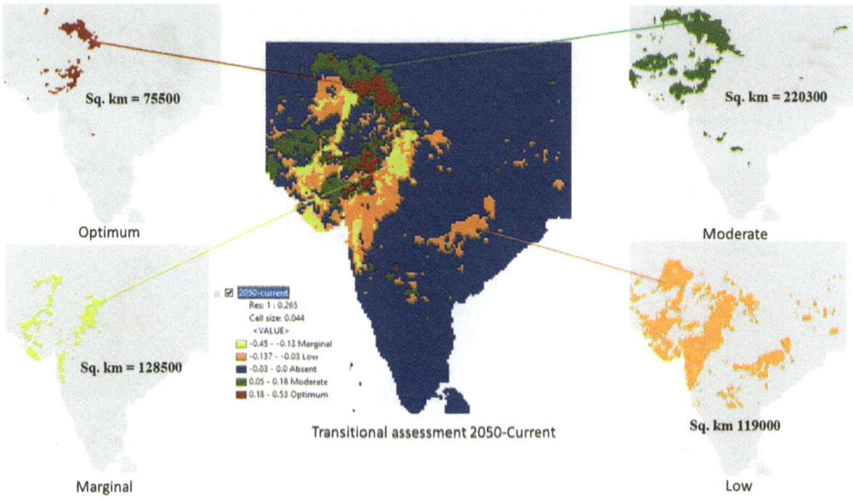

Fig. 13.11 Transitional assessment 2050-current depicted individual class along with area cover by them

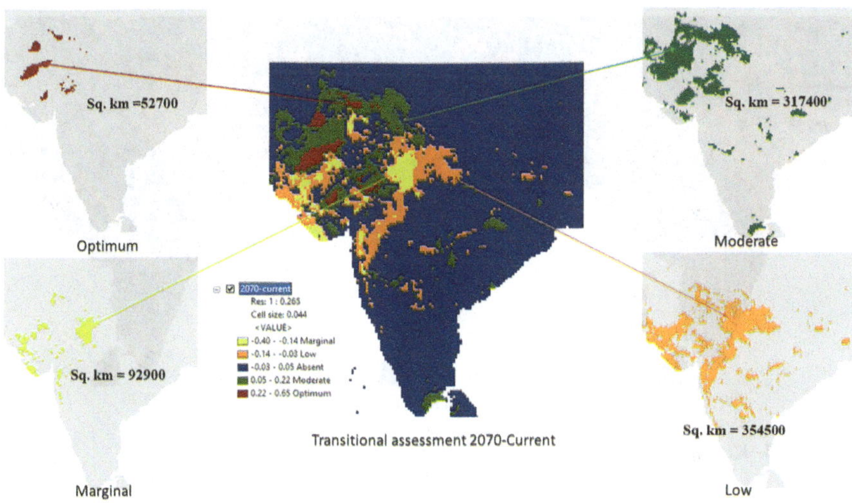

Fig. 13.12 Transitional assessment 2070-current depicted individual class along with area cover by them

13.3.3 Raster Similarity Analysis

Finally, raster similarity analysis between these bioclimatic projections suggested the area of unequal and equal with values ranges from 1 to 0. The results of these analyses are presented in Fig. 13.14 and Table 13.13. Highest similarity (0.507) was

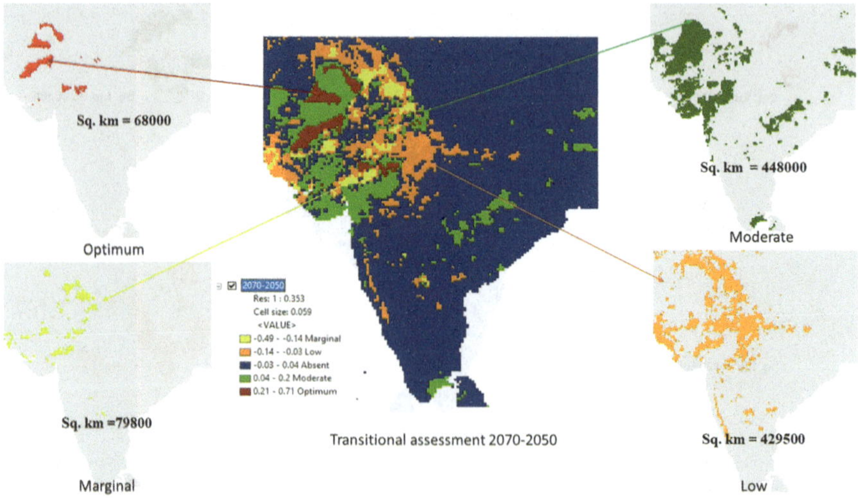

Fig. 13.13 Transitional assessment 2070–2050 depicted individual class along with area cover by them

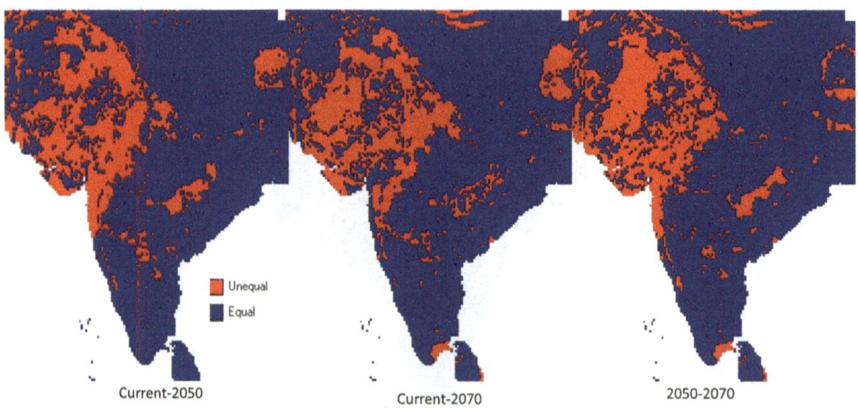

Fig. 13.14 Kappa simulation between two climatic timeframes

found between current-2070, followed with 2050–2070 (0.491) and current-2050 (0.486).

Values of Schoener' *D* and Hellinger's *I* indices are presented in Tables 13.14 and 13.15, respectively. Highest *D* (0.826) and *I* (0.964) values were recorded between current and 2050 BC, respectively. While the lowest values of both these indices (0.41 and 0.69) were recorded between 2050 and HHI variables.

Table 13.13 Kappa simulation similarity values among different bioclimatic timeframes

Projections	Current	2050
2050	0.486	–
2070	0.507	0.491

Table 13.14 Niche overlap *D* values between various climatic and non-climatic variables (random forest)

Variables	2050	2070	Current
HHI	0.414	0.447	0.427
2050	–	0.816	0.826
2070	–	–	0.823

Table 13.15 Niche overlap *I* value between various climatic and non-climatic (random forest)

Variables	2050	2070	Current
HHI	0.692	0.707	0.699
2050	–	0.960	0.964
2070	–	–	0.962

13.3.4 Automated Conservation Assessments (AA): I and II

The Red List Categories and its associated five criteria developed by the International Union for Conservation of Nature (IUCN) provide an authoritative and comprehensive methodology to assess the conservation status of species. Red List criterion B, which principally uses distribution data, is the most widely used to assess conservation status, particularly of plant species. The IUCN has five complementary criteria (A, B, C, D, and E) under which a species can be evaluated, and, when not already extinct, assessments assign species to three threatened categories (Critically Endangered (CR); Endangered (E); VU (Vulnerable), or otherwise to LC (Least Concerned), NT (Near Threatened), or DD (Data Deficient, when insufficient data are available). Criterion B is suitable for estimating conservation status even when the distribution of a taxon is only known from georeferenced herbarium or museum collections and with limited information on local threats and potential continuing decline, and it plays a prominent role in describing global trends in extinction risk. Criterion B involves two sub-criteria (B1 and B2), which reflect two different kinds of geographic range size estimates [sub-criterion B1 is based on extent of occurrence (EOO) while B2 is based on area of occupancy (AOO)].

The extent of occurrence (EOO) is defined as "*the area contained within the shortest continuous imaginary boundary that can be drawn to encompass all the known, inferred or projected sites of present occurrence of a taxon, excluding cases of vagrancy*". EOO is generally measured by a minimum convex polygon, or convex hull, defined as "*the smallest polygon in which no internal angle exceeds 180° and which contains all the sites of occurrence*". AOO differs from EOO as it reflects the fact that a taxon will not usually occur all over its EOO, that is, there will be areas where the taxon is absent, including unsuitable areas. The AOO will be a function of the scale or grid cell size at which it is measured, and which should reflect relevant biological aspects of the taxon (i.e., AOO = number of occupied cells x area of an individual cell). The intent of EOO is to "measure the degree to which risks from threatening factors are spread spatially across the taxon's geographic distribution", while the primary intent of AOO is as a measure of the "insurance effect", whereby taxa that occur within many patches or large patches across a landscape or seascape are "insured" against risks from spatially explicit threats. AOO is defined as the area within extent of occurrence that is engaged by a taxon. We consider that habitat loss will unswervingly reduce the AOO by reducing available suitable patches within the landscape. Otherwise, AOO has been used in similar studies that use SDM as descriptors of species ranges for conservation prioritization (Guillera-Arroita et al. 2015).

The current status of the *T. terrestris* as well as impact of niche modeling on EOO and AOO with the output of random forest algorithm is depicted in Table 13.16. Results suggested higher EOO (km^2) with current spatially thinned data (318189), however, the niche modeled random forest output revealed shrinkage in EOO during the same time period and measured as 1,147,372 (km^2). The percent changes in EOO with niche modeled data of three-time projection showed 64.06, 55.27, and 48.53% reductions with respect to current spatially thinned data set (Fig. 13.15). Similar trends were also observed for number of subpopulations that showed percent reduction of 27.84, 2.06, and 6.19 during three BC projections, respectively (Table 13.16 and Fig. 13.15).

However, area of occupancy, number of unique occurrences, and number of locations showed increasing trends with niche modeled data as compared to current spatially thin data (Table 13.15).

Further, 221.9% increase in AOO was recorded with current data set modeled with random forest, while almost similar percent gain (316.8 and 317.3) for AOO

Table 13.16 Current status of the *Tribulus terrestris* as well as impact of niche modeling on different IUCN categories

IUCN criterions	Current spatially thinned data	Current RF projected	2050 RF projected	2070 RF projected
Extent of occurrence (EOO km^2)	318,189	114,372	142,327	163,762
Area of occupancy (AOO km^2)	784	2524	3268	3272
Number of unique occurrences	206	631	817	818
Number of subpopulations	97	70	95	91
Number of locations	142	369	469	481
IUCN threat category according to criterion B	LC or NT			
IUCN annotation (category code)	B1a + B2a			

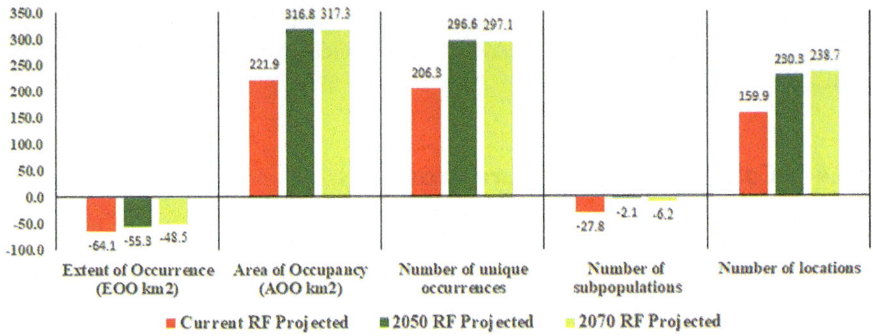

Fig. 13.15 Percent changes in different evaluation criterion of IUCN with respect to current spatially thinned data set

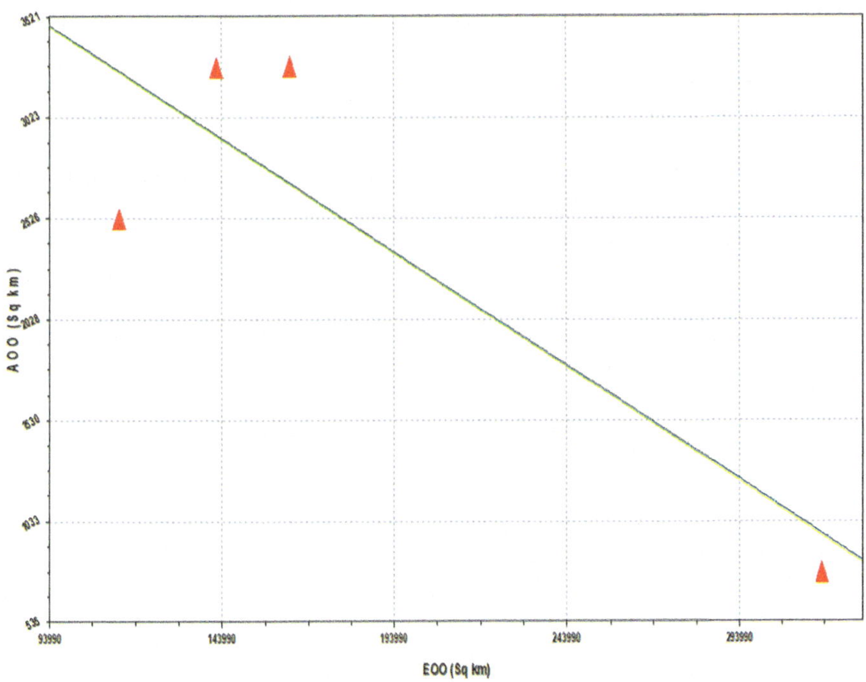

Fig. 13.16 Relationships between EOO and AOO of *Tribulus terrestris* calculated with current thinned data set and with random forest output with three studied timeframes

was observed with 2050 and 2070 RF projected, respectively. Similar trends were recorded for the number of unique occurrences and locations (Fig. 13.15). In this analysis, we also find a significant negative pattern between EOO and AOO ($R^2 = 0.87$, Fig. 13.16). Such an observed pattern between these two variables suggests the tendency of species that describes "smaller EOO to show a larger variation in the proportion of AOO" and which was advocated by Marco et al. (2018).

13.4 Conclusion and Future Prospects

In this study, we successfully predicted the habitat suitability of *T. terrestris* for the current and future climate change scenarios. Our results indicated that the areas of its habitat suitable classes were changed with different bioclimatic projections. Overall, 111.25 km^2 centroid changes were recorded from the current climatic condition to 2070. The key environmental variables influencing the distribution of *T. terrestris* were precipitation of the coldest quarter, precipitation seasonality and annual precipitation. While among the habitat heterogeneity indices, normalized dispersion of enhanced vegetation index, and dispersion of EVI were identified as the most influencing factors. Our results can be used to enhance ecological (regarded as weed species) as well as economic (regarded as medicinally most important species) management in order to curb this or for harvesting the higher biomass (standing state) for its important secondary metabolites.

- For SDM of this species, bioclimatic parameters are more fluid than habitat heterogeneity indices (HHIs).
- Among the individual algorithms, random forest was identified as most efficient tool for ecological niche modeling of this species using the climatic and non-climatic predictors.
- The SDM for this species is influenced by the coldest quarter precipitation (BC-19), the seasonality of precipitation (BC-15), and the annual precipitation (BC-12).
- Our findings can be applied to better control this or exploit the increased biomass (standing state) for its significant secondary metabolites.
- One possible direction for future research is to examine how this species interacts with other members of its community. This may entail analysing the types (fundamental and realized niche) and range extensions of various ecological niches. This type of study helps shed light on the dynamics of succession patterns in various regions.

Authors Contributions Dr. Manish Mathur conceptualized the chapter theme and interpretation of output of various machine learning techniques. Ms. Preet Mathur prepared various types of language codes in python, Java, and in R scripts and convert the various file format from ASCII to KML, Raster, dbf, CSV, etc. for software's like QGIS 3.10.0; Wallace; DIVA-GIS version 7.5; OpenModeller; MaxEnt 3.4.1 software; SDM toolbox; Map Comparison Kit; ENMTools and Ntbox; randomForest R packages.

References

Adhikari D, Reshi Z, Datta BK, Samant SS, Chettri A, Upadhaya K, Shah MA, Singh PP, Tiwary R, Majumdar K, Pradahan A, Thakur ML, Salam Z, Zahoor Z, Mir SH, Kaloo ZA, Barik SK (2018) Inventory and characterization of new population through ecological niche modeling improve threat assessment. Curr Sci 114(3):519–531

Adhikari P, Jeon JY, Kim HW, Shin MS, Adhikari P, Seo C (2019) Potential impact of climate change on plant invasion in the Republic of Korea. J Ecol Environ 43:1–2. https://doi.org/10.1186/s41610-019-0134-3

Adhikari A, Rew LJ, Mainali KP, Adhikari S, Maxwell BD (2020) Future distribution of invasive weed species across the major road network in the state of Montana, USA. Reg Environ Chang 20:60. https://doi.org/10.1007/s10113-020-01647-0

Ahmad R, Khuroo AA, Charles B, Hamid M, Rashid I, Aravind NA (2019a) Global distribution modeling, invasion risk assessment and niche dynamics of Leucanthemum vulgare (ox-eye-daisy) under climate change. Sci Rep 9:11395. https://doi.org/10.1038/s41598-019-47859-1

Ahmad R, Khuroo AA, Hamid M, Charles B, Rashid I (2019b) Predicting invasion potential and niche dynamics of Parthenium hysterophorus (congress grass) in India under projected climate change. Biodivers Conserv 28:2319–2344. https://doi.org/10.1007/s10531-019-01775-y

Anwar MP, Islam AKMM, Yeasmin S, Rashid MH, Juraimi AS, Ahmed S, Shrestha A (2021) Weeds and their responses to management efforts in a changing climate. Agronomy 11:1921. https://doi.org/10.3390/agronomy11101921

Attorre F, Alfo M, De Sanctis M, Francesconi F, Valenti R, Vitale M, Bruno F (2011) Evaluating the effects of climate change on tree species abundance and distribution in the Italian peninsula. Appl Veg Sci 14:242–255. https://doi.org/10.1111/j.1654-109X.2010.01114.x

Beaumont LJ, Graham E, Duursma DE, Wilson PD, Cabrelli A, Baumgartner JNB, Hallgren W, Esperon-Rodriguez M, Nipperess DA, Warren DL, Laffan SW, VanDerWal J (2016) Which species distribution models are more (or less) likely to project broad-scale, climate-induced shifts in species ranges? Ecol Model 342:135–114

Bio AMF, Alkemande R, Barendregt A (1998) Determining alternative models for vegetation response analysis – a non-parametric approach. J Veg Sci 9:5–16

Breiman L (2001) Random forests. Mach Learn 45:5–32

Breiner F, Guisan A, Bergamini A, Nobis M (2015) Overcoming limitations of modeling rare species by using ensembles of small models. Methods Ecol Evol 6:1210–1218

Broennimann O, Treier UA, Muller-Scharer H, Thuiller W, Peterson AT, Guisan A (2007) Evidence of climatic niche shift during biological invasion. Ecol Lett 10:701–709

Brown JL (2014) SDMtoolbox: a python-based GIS toolkit for landscape genetic, biogeographic, and species distribution model analyses. Methods Ecol Evol 5(7):694–700. https://doi.org/10.1111/2041-210X.12200

Cao J, Xu J, Pan X, Monaco TA, Zhao K, Wang D, Rong Y (2021) Potential impact of climate change on the global geographical distribution of the invasive species, Cenchrus spinifex (field sandbur, Gramineae). Ecol Indic 131:108204. https://doi.org/10.1016/j.ecolind.2021.108204

Chai SL, Zhang J, Nixon A, Nielsen S (2016) Using risk assessment and habitat suitability models to prioritise invasive species for management in a changing climate. PLoS One 11(10): e0165292. https://doi.org/10.1371/journal.pone.0165292

Coban HO, Orucu OK, Arslan ES (2020) MaxEnt modeling for predicting the current and future potential geographical distribution of Quercus libani Olivier. Sustainability 12(7):2671. https://doi.org/10.3390/su12072671

Cutler F, Wiener R (2018) randomForest: Breiman and Cutler's random forests for classification and regression

Dash MC (2007) Fundamentals of ecology. Tate McGraw Hill, New Delhi

Duan RY, Kong XQ, Huang MY, Fan WY, Wang ZG (2014) The predictive performance and stability of six species distribution models. PLoS One 9(11):e112764. https://doi.org/10.1371/journal.pone.0112764

Dubey PN, Saxena SN, Mishra BK, Solanki RK, Vishal MK, Singh B, Sharma LK, John S, Agarwal D, Yogi A (2017) Preponderance of cumin (Cumnum cyminum L.) essential oil constituents across cumin growing agro-ecological sub regions, India. Ind Crop Prod 95:50–59. https://doi.org/10.1016/j.indcrop.2016.10.011

Dunlop EA, Wilson JC, Mackey AP (2006) The potential geographic distribution of invasive weed Senna obtusifolia in Australia. Weed Res 46:404–413

Elith J, Burgman M (2002) Predictions and their validation: rare plants in the central highlands, Victoria, Australia. In: Scott JM et al (eds) Predicting species occurrences: issues of accuracy and scale. Island Press, Washington, DC, pp 303–313

Elith J, Graham CH, Anderson RP, Dudik M, Ferrier S, Guisan A, Hijmans R, Huettmann F, Leathwick JR, Lehmann A (2006) Novel methods improve prediction of species' distributions from occurrence data. Ecography 29:129–151

Elith J, Phillips SJ, Hastie T, Dudík M, Chee YE, Yates CJ (2011) A statistical explanation of MaxEnt for ecologists. Divers Distribut 17:43–57. https://doi.org/10.1111/j.1472-4642.2010.00725.x

Farber O, Kadmon R (2003) Assessment of alternative approaches for bioclimatic modeling with special emphasis on the Mahalanobis distance. Ecol Model 160:115–130

Fielding AH, Bell JF (1997) A review of methods for the assessment of prediction errors in conservation presence/absence models. Environ Conserv 24:38–49

Franklin J (1998) Predicting the distribution of shrub species in southern California from climate and terrain-derived variables. J Veg Sci 9:733–748

Friedman JH (2001) Greedy function approximation: a gradient boosting machine. Ann Stat 29:1189–1232

Garzón MB, Blazek R, Neteler M, Dios RS, de Ollero HS, Furlanello C (2006) Predicting habitat suitability with machine learning models: the potential area of Pinus sylvestris L. in the Iberian Peninsula. Ecol Model 197:383–393. https://doi.org/10.1016/j.ecolmodel.2006.03.015

Guillera-Arroita G, Lahoz-Monfort JJ, Elith J, Gordon A, Kujala H, Lentini PE, McCarthy MA, Tingley R, Wintle BA (2015) Is my species distribution model fit for purpose? Matching data and models to applications. Glob Ecol Biogeogr 24:276–292. https://doi.org/10.1111/geb.12268

Hanley JA, McNeil BJ (1983) A method of comparing the areas under receiver operating characteristic curves derived from the same cases. Radiology 148:839–843

Hijmans RJ, Guarino L, Cruz M, Rojas E (2001) Computer tools for spatial analysis of plant genetic resources data: 1. DIVA-GIS. Plant Gen Resour Newslett 127:15–19

Hijmans RJ, Cameron SE, Parra JL, Jones PG, Jarvis A (2005) Very high-resolution in-terpolated climate surfaces for global land area. Int J Climatol 25(15):1965–1968

Hijmans RJ, Phillips S, Leathwick J, Elith J (2017) Dismo: species distribution modeling. https://cran.r-project.org/web/packages/dismo/dismo.pdf

Hong SH, Lee YH, Lee G, Lee DH, Adhikari P (2021) Predicting impacts of climate change on northward range expansion of invasive weeds in South Korea. Plan Theory 10:1604. https://doi.org/10.3390/plants10081604

Irving K, Jahnig SC, Kuemmerlen M (2019) Identifying and applying an optimum set of environmental variable in species distribution models. Inland Waters 10:11–28. https://doi.org/10.1080/20442041.2019.1653111

IUCN Standards and Petitions Subcommittee (2014) Guidelines for using the IUCN red list categories and criteria THE IUCN RED LIST OF THREATENED SPECIES™

Kaky E, Nolan V, Alatawi A, Gilbert F (2020) A comparison between ensemble and MaxEnt distribution modeling approaches for conservation: a case study with Egyptian medicinal plants. Eco Inform 60:101150. https://doi.org/10.1016/j.ecoinf.2020.101150

Kass JM, Vilela B, Aiello-Lammens ME, Muscarella R, Merow C, Anderson RP (2018) Wallace: a flexible platform for reproducible modeling of species niches and distributions built for community expansion. Methods Ecol Evol 9:1151–1156. https://doi.org/10.1111/2041-210X.12945

Kass JM, Meenan SI, Tinoco N, Burneo SF, Anderson RP (2021) Improving area of occupancy estimates for parapatric species using distribution models and support vector machines. Ecol Appl 31(1):1–15

Kaur K, Sharma V, Singh V, Wani MS, Gupta RC (2016) Development of novel SSR markers for evaluation of genetic diversity and population structure in Tribulus terrestris L. (Zygophyllaceae). Biotechnology 6:156. https://doi.org/10.1007/s13205-016-0469-8

Koo KA, Park SU, Kong WS, Hong S, Jang I, Seo C (2017) Potential climate change effects on tree distributions in the Korean peninsula: understanding model & amp; climate uncertainties. Ecol Model 353:17–27. https://doi.org/10.1016/j.ecolmodel.2016.10.007

Kumar S, Stohlgren TJ (2009) MaxEnt modeling for predicting suitable habitat for threatened and endangered tree Canacomyrica monticola in New Caledonia. J Ecol Nat Environ 1:94–98

Kumar S, Stohlgren TJ, Chong GW (2006) Spatial heterogeneity influences native and non-native plant species richness. Ecology 87:3186–3199

Li G, Liu C, Liu Y, Yang J, Zhang X, Guo K (2012) Effects of climate, disturbance and soil factors on the potential distribution of Liaotung oak (Quercus wutaishanica Mayr) in China. Ecol Res 27:427–436

Liu Y, Oduor AM, Zhang Z, Manea A, Tooth IA, Leishman MR, van Kleunen M (2017) Do invasive alien plants benefit more from global environmental change than native plants? Glob Chang Biol 23:3363–3370

Ma B, Sun J (2018) Predicting the distribution of Stipa purpurea across the Tibetan Plateau via MaxEnt model. BMC Ecol 18:10. https://doi.org/10.1186/s12898-018-0165-0

Marcer A, Sáez L, Molowny-Horas R, Pons X, Pino J (2013) Using species distribution modeling to disentangle realised versus potential distributions for rare species conservation. Biol Conserv 166:221–230. https://doi.org/10.1016/j.biocon.2013.07.001

Marco P, Villen S, Mendes P, Noberga C, Cortes L, Castro T, Souza R (2018) Vulnerability of cerrado threatened mammals: an integrative landscape and climate modeling approach. Biodivers Conserv 29:1637–1658. https://doi.org/10.1007/s10531-018-1615-x

Mathur M (2012) Complementary and alternative medicine use for gastro-intestinal disorders. Int J Pharma Res Rev 1(7):16–36

Mathur M (2013) Spatial and modular variability in phytosterol composition in Tribulus terrestris. Trends Biosci 6(1):59–62

Mathur M (2014a) Attributes of plant spatial analysis. Today and Tomorrow Printer and Publisher, New Delhi. ISBN 817019489X

Mathur M (2014b) Spatio-temporal Variability's in distribution patterns of Tribulus terrestris: linking patterns and processes. J Agric Sci Technol 16:1187–1201

Mathur M (2017) Influences of soil, community composition and species traits on steroidal Sapogenin, zinc and iron concentration in fruits of Tribulus terrestris. Studies Ethno Med 11(1):82–90

Mathur M (2020) Comportments of arid grazing land plant diversity: a temporal assessment with bottom-up and top-down factors. Range Manag Agroforest 41(2):200–208

Mathur M, Gehlot P (2018) Recruit the plant pathogen for weed management: bioherbicides-a sustainable strategy. In: Gehlot P, Singh J (eds) Fungi and their role in sustainable development: current perspectives. Springer, Singapore, pp 159–181. https://doi.org/10.1007/978-981-13-0393-7_10. ISBN: 978-981-13-0392-0 Print/ISBN 978-981-13-0393-7

Mathur M, Sundarmoorthy S (2013a) Mineral concentration and their deviation from optimum percentage in Tribulus terrestris. Indian. J Ecol 40(1):19–23. NAAS Ranking 5.10

Mathur M, Sundarmoorthy S (2013b) Inter-specific association of herbaceous vegetation in semi-arid thar desert. India Range Manag Agroforest 34(1):26–32

Mathur M, Sundarmoorthy S (2017) Effects of solar energy variables on secondary metabolites: a comparative modular assessment for Corchorus depressus. Int J Environ Sci 8(1):27–40

Merow C, Smith M, Guisan A, Mcmahon S, Normand S, Thuiller W, Rafael W, Zimmermann N, Elith J (2014) What do we gain from simplicity versus complexity in species distribution models? Ecography 37:1267–1281

Mishra SN, Kumar D, Kumar B, Tiwari S (2021) Assessing impact of varying climatic conditions on distribution of Buchanania cochinchinensis in Jharkhand using species distribution modeling approach. Curr Res Environ Sustain. https://doi.org/10.1016/j.crsust.2021.100025

Monserud RA, Leemans R (1992) Comparing global vegetation maps with the kappa statistic. Ecol Model 62:275–293

Obiakara MC, Fourcade Y (2018) Climatic niche and potential distribution of Tithonia diversifolia (Hemsl.) A. gray in Africa. PLoS One 13(9):e0202421. https://doi.org/10.1371/journal.pone.020242

Osorio-Olvera L, Lira-Noriega A, Soberón J, Townsend Peterson A, Falconi M, Contreras Díaz RG, Martínez-Meyer E, Barve V, Barve N (2020) Ntbox: an R package with graphical user interface for modeling and evaluating multidimensional ecological niches. Methods Ecol Evol 11:1199–1206. https://doi.org/10.1111/2041-210X.13452. https://github.com/luismurao/ntbox

Padalia H, Srivastava V, Kushwaha SPS (2014) Modeling potential invasion range of alien invasion species, Hyptis suaveolens (L) in India: comparison of MaxEnt and GARP. Eco Inform 22:36–43

Panda RM, Behera MD, Roy PS (2017) Assessing distributions of two invasive species of contrasting habits in future climate. J Environ Manag 213:478–488. https://doi.org/10.1016/j.jenvman.2017.12.053

Pecchi M, Marchi M, Burton V, Giannetti F, Moriondo M, Bernetti I, Bindi M, Chirici G (2019) Species distribution modeling to support forest management. A literature reviews. Ecol Model 411:108817

Phillips SJ, Anderson RP, Dudík M, Schapire RE, Blair ME (2017) Opening the black box: an open-source release of Maxent. Ecography 40:887–893

Poudel AS, Shrestha BB, Joshi MD, Muniappan R, Adiga A, Venkatramanan S, Jha PK (2020) Predicting the current and future distribution of the invasive weed Ageratina adenophora in the Chitwan-Annapurna landscape, Nepal. Mount Res Dev 40(2):61–71

Pradhan P (2016) Strengthening Maxent modeling through screening of redundant explanatory bioclimatic variables with variance inflation factor analysis. Researcher 8(5):29–34

Putman R, Wratten SD (1984) Principles of ecology. University of Chicago Press, Chicago, IL

Qin Z, Ditommaso A, Wu RS, Huang HY (2014) Potential distribution of two ambrosia species in China under projected climate change. Weed Res 54:520–531

Radosevich SR, Holt JS, Ghersa CM (2007) Weeds and invasive plants. In: Steven RR, Jodie SH, Claudio MG (eds) Ecology of weeds and invasive plants: relationship to agriculture and natural resource management, 3rd edn. John Wiley & Sons, Inc., Hoboken, NJ, p 474. ISBN 978-0-471-76779-4

Randin CF, Dirbnock T, Dullinger S, Zimmermann NE, Zappa M, Giusan A (2006) Are niche-based species distribution models transferable in space? J Biogeogr 33:1689–1703

Ripley B, Venables W (2020) nnet: Feed-Forward Neural Networks and Multinomial Log-Linear Models. https://cran.r-project.org/web/packages/nnet/nnet.pdf

Robertson MP, Villet MH, Palmer AR (2004) A fuzzy classification technique for predicting species' distributions: applications using invasive alien plants and indigenous insects. Divers Distribut 10:461–474

Ruheili AM, Sariri T, Subbhi M (2021) Predicting the potential habitat distribution of parthenium weed (Parthenium hysterophorus) globally and in Oman under projected climate change. J Saudi Soc Agric Sci 21:469–478. https://doi.org/10.1016/j.jssas.2021.12.004

Schussman H, Geiger E, Mau-Crimmins T, Ward J (2006) Spread and current potential distribution of an alien grass, Eragrostis lehmanniana Nees, in the southwestern USA: comparing historical data and ecological niche models. Divers Distribut 12:81–89

Sillero N, Barbosa AM (2021) Common mistakes in ecological niche models. Int J Geogr Inf Sci 35(2):213–226. https://doi.org/10.1080/13658816.2020.1798968

Suarez-Mota ME, Ortiz E, Villasenor JL, Espinosa-Garcia FJ (2016) Ecological niche modeling of invasive plant species according to invasion status and management needs: the case of Chromalaena odorata (Asteraceae) in South Africa. Pol J Ecol 64:369–383
Taylor S, Kumar L (2013) Potential distribution of an invasive species under climate change scenarios using CLIMEX and soil drainage: a case study of Lantana camara L. in Queensland. Aust J Environ Manag 114:414–422
Therneau T, Atkinson B, Port BR (2019) rpart: Recursive Partitioning and Regression Trees. https://cran.r-project.org/web/packages/rpart/rpart.pdf
Thuiller W, Araujo MB, Lavorel S (2003) Generalized models vs. classification tree analysis: predict ng spatial distributions of plant species at different scales. J Veg Sci 14:669–680
Thuiller W, Lafourcade B, Engler R, Araujo M (2009) BIOMOD—A platform for ensemble forecasting of species distributions. Ecography 32:369–373
Thuiller W, Georges D, Engler R (2020) Biomod2: ensemble platform for species distribution modeling. R Development Core Team, Vienna, Austria
Tu W, Xiong Q, Qiu X, Zhang Y (2021) Dynamics of invasive alien plant species in China under climate change scenarios. Ecol Indic 129:107919. https://doi.org/10.1016/j.ecolind.2021.107919
Tuanmu MN, Jetz W (2015) A global, remote sensing-based characterization of terrestrial habitat heterogeneity for biodiversity and ecosystem modeling. Glob Ecol Biogeogr 24:1329. https://doi.org/10.1111/geb.12365
Vapnik VN (1998) Statistical learning theory. Wiley, Hoboken, NJ
Vyas G, Mathur M, Patel NA, Patel RP (2017) Aphrodisiac efficacy of Blepharis sindica seeds: a comparative assessment using different solvent types. Indian J Biochem Biophys 54:223–230
Wan J-Z, Wang CJ, Tan JF, Yu FH (2017) Climatic niche divergence and habitat suitability of eight alien invasive weeds in China under climate change. Ecol Evol 7:1541–1552
Wang N, Chen H (2019) Increased nitrogen deposition increased the competitive effects of the invasive plant Aegilops tauschii on wheat. Acta Physiol Plant 41(176). https://doi.org/10.1007/s11738-019-2968-9
Warren DL, Seifert SN (2011) Ecological niche modeling in Maxent: the importance of model complexity and the performance of model selection criteria. Ecol Appl 21:335–342
Wood S (2019) Mgcv: mixed GAM computation vehicle with automatic smoothness estimation. https://stat.ethz.ch/R-manual/R-devel/library/mgcv/html/00Index.html
Wright AN, Schwartz MW, Hijmans RJ, Shaffer HB (2016) Advances in climate models from CMIP3 to CMIP5 do not change predictions of future habitat suitability for California reptiles and amphibians. Climate Change 134:579–591. https://doi.org/10.1007/s10584-015-1552-6
Xu Z (2015) Potential distribution of invasive alien species in the upper Ili river basin: determination and mechanism of bioclimatic variables under climate change. Environ Earth Sci 73:779–786
Xu W, Jin J, Cheng J (2021) Predicting the potential geographic distribution and habitat suitability of two economic Forest trees on the loess plateau, China. Forests 12:747. https://doi.org/10.3390/f12060747
Yan H, Feng L, Zhao Y, Feng L, Zhu C, Qu Y, Wang H (2020) Predicting the potential distribution of an invasive species, Erigeron canadensis L., in China with a maximum entropy model. Global Ecol Conserv 21:e00822. https://doi.org/10.1016/j.gecco.2019.e00822
Ye XZ, Zhao GH, Zhang MZ, Vui XY, Fan HH, Liu B (2020) Distribution pattern of endangered plants Semiliquidambar catayensis (Hamamelidaceae) in response to climate change after the last interglacial period. Forest 11:434. https://doi.org/10.3390/f11040434
Zhang K, Sun L, Tao J (2020) Impact of climate change on the distribution of Euscaphis japonica (Staphyleaceae) trees. Forests 11:525. https://doi.org/10.3390/f11050525
Zhang Y, Tang J, Ren G, Zhao K, Wang X (2021) Global potential distribution prediction of Xanthium italicum based on Maxent model. Sci Rep 11:16545. https://doi.org/10.1038/s41598-021-96041-z

Ziska LH, Emche SD, Johnson EL, George K, Reed DR, Sicher RC (2005) Alterations in the production and concentration of selected alkaloids as a function of rising atmospheric carbon dioxide and air temperature: implications for ethno-pharmacology. Glob Chang Biol 11:1798–1807

Ziska LH, Panicker S, Wojno HL (2008) Recent and projected increases in atmospheric carbon dioxide and the potential impacts on growth and alkaloid production in wild poppy (Papaver setigerum DC.). Climate Change 91:395–403

Zobayed SMA, Saxena PK (2004) Production of St. John's wort plants under controlled environment for maximizing biomass and secondary metabolites. In Vitro Cell Dev Biol Plant 40:108–114

Part IV

Plant invasion: Management and Control

The Role of Halophytic Plant Invasions for the Conservation and Restoration of Degraded Agricultural Lands

14

Rida Zainab, Maria Hasnain, Zainul Abideen, and Hammad Afzal Kayani

Abstract

Worldwide ecological systems associated with plants have changed because of recent abrupt and fast climatic changes. Physiological, phenological, species distributional, interspecific, and disturbance regime alterations have all been connected to global warming. Future climate change projections will probably cause even more pronounced changes in the conditions of many ecosystems. Loss in gross crop productivity is a growing concern in agriculture due to growing soil salinity, but promotes invasion of salt-tolerant plant species in these degraded lands. Invasive halophytes due to emerging potential industrial uses can be helpful in restoration of such soils. Managing natural resources and planning for conservation is particularly difficult considering these changes. In the face of a changing climate, new methods are needed to manage natural resources and ecosystems. This chapter discusses a variety of methods to highlight the invasive ecosystem dynamics and enlist the various adaptation of halophytic plants restoration under changing climate. The invasion of wild salt resistance plant can survive at suboptimum conditions can be a useful remedy for the restoration of poor lands and to reduce their invasion to cultivated lands. The practical adaptation of invasive wild stress-resistant plants could be a win-win strategy to control

R. Zainab · M. Hasnain
Department of Biotechnology, Lahore College for Women University, Lahore, Pakistan

Z. Abideen (✉)
Dr. Muhammad Ajmal Khan Institute of Sustainable Halophyte Utilization, University of Karachi, Karachi, Pakistan
e-mail: Zuabideen@uok.edu.pk

H. A. Kayani
Department of Biosciences, The Shaheed Zulfikar Ali Bhutto Institute of Science and Technology, Karachi, Pakistan
e-mail: hammad.afzal@szabist.pk

S. Tripathi et al. (eds.), *Plant Invasions and Global Climate Change*,
https://doi.org/10.1007/978-981-99-5910-5_14

plant invasion and management for the conservation and restoration of degraded agricultural lands, especially in developing countries.

Keywords

Adaption approaches · Biochemical · Climate change · Conservation · Ecosystem · Natural resources

Abbreviations

CAI	Cotton/alfalfa intercropping
CIMMYT	International Maize and Wheat Improvement Center
CSSI	Cotton/*S. salsa* intercropping
GHG	Greenhouse gases
IWP	Irrigation water productivity
MC	Monoculture cotton
PEPC	Phosphoenolpyruvate carboxylase
PGPRs	Plant growth promoting rhizobacteria

14.1 Introduction

The greatest ecological and societal issue of the twenty-first century due to increasing CO_2 levels and world temperatures is climate change. Climate change affects weather patterns, increases atmospheric CO_2 levels, raises temperatures, increases the frequency of droughts and the quantity of precipitation and/or flooding, as well as causes heat waves, rapid sea level rise, contaminates soils with salt, and increases the number of fires (Jansson and Hofmockel 2020). The atmosphere and oceans are hugely affected by these climatic changes, that have a profound effect on the ecosystem of the world, flora and fauna. Since the nineteenth century, anthropogenic activities have caused an increase in the average temperature of 0.9 °C, mostly because of atmospheric emissions of greenhouse gases (GHGs). It will increase by 1.5 °C or more by 2050 given the rate of deforestation, the growth in GHG emissions, and the pollution of the land, water, and air (Arora 2019). Ocean acidification and climate change by anthropogenic GHG emissions have put the resilience and survival of natural ecosystems in danger, affecting human societies (Malhi et al. 2020). Climate change also has a substantial influence on nitrogen cycling due to the production of CO_2, N_2O, and CH_4, which modifies N flow to rivers by affecting surface runoff and other N transit channels. Climate change causes nutrients to move from the land to the watershed, which eventually affects the retention time of nitrogen in the river (Xia et al. 2018).

Sustainable alternatives against climate change are difficult to achieve due to the declining freshwater resources for irrigation, continuous urbanization, degradation of arable land, and the rising global population. Agriculture-related enterprises, food and crop production, as well as ecological balance have all been impacted negatively by climate change. The time duration (e.g., inundation duration), size (e.g., inundation depth), and flood frequency all have an impact on vegetation production in wetlands. The average inundation depth was from 3.9 to 4.0 m, and the inundation duration ranged from 39 to 41%, making these the ideal flooding conditions for the wetland's plants to produce biomass (Dai et al. 2020). The future of agriculture and food security is in problem due to poor agricultural practice and global warming. Therefore, making suitable changes to present agricultural techniques and using new salt-resistant plant species that can withstand various biotic and abiotic environmental pressures might improve the climate change in favor of plant biodiversity (Bhadouria et al. 2019). Droughts cause the soil in agricultural and forestry areas to deteriorate in both direct and indirect ways forcing rural residents, while also reduce the food yield per hectare. Droughts can result in increased heat stress on plants, altered contents of soil moisture, increased erosion due to wind and rain, soil nutrient depletion and salinization, as well as a reduce biomass and vegetation cover and plant productivity (Hermans and McLeman 2021). Future droughts are predicted to reduce rice and wheat harvests as the quantitative assessment of the impact of drought on agronomic characteristics such as plant height, biomass, yield, and yield components showed a decrease by 27.5% and 25.4%, respectively, in wheat and rice yields (Zhang et al. 2018). Additionally, during droughts, the fractions of stem, leaf, and reproductive mass decline while the fraction of root mass grows considerably. The roots of herbaceous plants are more susceptible to drought, which reduces their potential for reproduction than the roots of woody plants (Eziz et al. 2017). In general, the effects of warming and high precipitation are cumulative. The combined effects of heat and drought on above- and below-ground biomass were more damaging in plant mixes than in monocultures, and less destructive in systems of woody plants than in those of herbaceous plants (Wilschut et al. 2022).

Several crops, food consumption, and waste practices have been modified to counteract the growing food crisis. Research in genomics and agronomy has aided in minimizing a few negative effects of climate change on agricultural production (Anderson et al. 2020). One of the main strategies for facilitating agricultural adaptation to climate change is breeding. However, a major barrier that is lowering the efficacy of breeding is phenotyping due to the associated cost as well as the scarcity of appropriate procedures (Araus and Kefauver 2018). Better crop varieties based on genotypic diversity may mitigate the harmful effects of impending climate change. With elevated CO_2 and high temperature, the genotype of bread wheat from the International Maize and Wheat Improvement Center (CIMMYT) has outperformed Gazul in terms of grain yield and biomass, that demonstrated the effectiveness of breeding efforts in warmer climates (Marcos-Barbero et al. 2021). Expanding agriculture to new regions is impossible in many underdeveloped nations, particularly those with numerous inhabitants or varied habitats that must be preserved. In addition, farmers' activities that contribute to land degradation, poor soil quality, and the depletion of natural resources require attention to stop and reverse these trends considering the growing demand for agricultural products across the world (Hossain et al. 2020). Hence, the efficiency of adaptation choices is decreased by land degradation, which makes sustainable farming systems more susceptible to climate change. More than 25% (37.25 million km^2) of the land surface is affected by land deterioration, which includes deteriorating biological and economic productivity as well as a decline in soil quality owing to physical and chemical changes and erosion. These alterations are taking place in all agricultural areas, including croplands, agro-forestry systems, and dry and semi-arid rangelands and pasturelands (Webb et al. 2017).

There are currently 17 land degradation processes that are active at different geographical scales worldwide, including salinization, coastal erosion, aridity, waterlogging, land subsidence, land erosion by water and wind, land pollution, landslides, vegetation degradation, biological invasions, permafrost thawing, soil acidification, soil organic carbon loss, soil biodiversity loss, soil compaction, and soil sealing (Prăvălie 2021). With the increased need for food, clean water, and fuel throughout the world, finding alternative species that are not only tolerant of environmental salinity and drought but also able to flourish in a range of environments is urgently needed to reduce demands placed on cropping systems and restore salt-degraded lands (Liu and Wang 2021). Approximately 7% of the earth's geographical surface, or one billion hectares, is now impacted by salt. While most of it is caused by natural geochemical processes, secondary human-induced salinization is estimated to be blamed for 30% of the irrigated land globally being damaged by salt (Hopmans et al. 2021). The detrimental effects of salinity on agricultural output is a contributing factor to migration in Bangladesh (Chen and Mueller 2018).

Soil salinization presents a major environmental issue that has reduced crop cultivation and loss of land suitable for agricultural production, and hence salt-tolerant plants are gaining more importance. Furthermore, the existence of salinized areas in hot and dry climates necessitates not just salt tolerance but also resistance to the stresses of heat and drought. Numerous salt-tolerant varieties of cereals, including rice and wheat, have been created and are produced in salty environments. Sorghum, the fifth-largest cereal crop in the world, does well in hot, arid climates and was previously thought to be only moderately salt tolerant. Future research in the sorghum genome's sequencing is expected to speed up salt tolerance breeding for increased sorghum biomass to satisfy renewable energy, livestock feed, human food, and fiber needs (Yamazaki et al. 2020). Additionally, *Arabidopsis* and rice with overexpressed OsGATA8 in the *Saltol* QTL area showed better yield and tolerance to salt and drought (Pareek et al. 2020). However, although there has been a significant advancement in the development of crop varieties by the introduction of salt tolerance-associated characteristics, most crop varieties grown in salty soils still show a reduction in yield, creating a need for alternatives. In this chapter, a brief insight has been given on management of saline soils using invasive halophytes, followed by detailed characteristics of halophytes and their economic importance and suitability for restoring the salt-affected lands, whereas factors affecting the efficiency of halophytes in sustainable remediation have been elaborated in the later sections.

14.2 Management of Saline Lands by Using Invasive Plants

Soil salinity has a negative impact on 800 million hectares of agricultural land globally (Li et al. 2022b). There are approximately 1125 million hectares of salty soil worldwide (Agarwal et al. 2022). Additionally, high salinity affects 20% of total cultivated lands and 33% of irrigated agricultural areas (Kapadia et al. 2022). 8,517,000 and 7,238,000 hectares of salty soil, respectively, are found in the United States and Canada. There are also seven million hectares of saline-alkaline terrain in India (Liu and Wang 2021). The disruption of soil by increased anthropogenic activity severely impacts the ecosystem function that helps competitive invasive plant species to outgrow native plant species (see Fig. 14.1). The arid and hostile environment present in polluted areas impedes the establishment of natural vegetation, yet invasive plants are successful in growing in such environments because they are opportunistic and incursive (Syed et al. 2021). Invasive plants already inhabit around 100 million acres in the United States, and they are expanding by 14 million acres annually. The European Union made the cultivation, production, transportation, sale, exchange, or unintentional release of any of 14 non-native invasive plants illegal in January 2016 in response to the threat presented by invasive species (Prabakaran et al. 2019). Salt-resistant energy plants, notably halophytes, were previously regarded to be unsuitable for agricultural production. However, presently the use of halophytes, agricultural lands and freshwater conservation is more prominent for better food and fodder production (Ali et al. 2021).

All populations of high-salinity species showed higher tolerance for salt compared to low-salinity populations when invasive and native plants from areas with low and high levels of salt were cultivated under controlled and stressed conditions. The performance of the improved salt tolerance in native species, which lowered

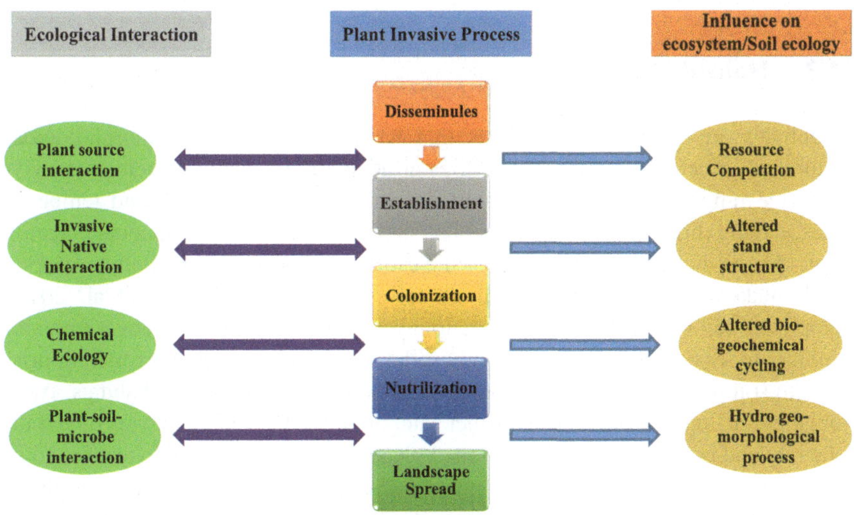

Fig. 14.1 Plant invasive process with ecological interaction and their influence on ecosystem

total biomass under control conditions, related to the morphology of leaf and rate of carbon absorption, whereas the ability of native species to withstand stress was connected to the rate of leaf formation and the amount of resources allocated to roots (Liu et al. 2019). Due to their strong plasticity and capacity for adaptability, invasive species frequently have a high potential for colonization. The ecophysiological fitness of *Spartina patens* (Aiton) Muhl, an invasive species found in various Mediterranean nations, including Spain and France, and *Cyperus longus* L., an aquatic species found in the Mediterranean, showed that *S. patens* can colonize saline habitats because it is physiologically well-adapted to them, but *C. longus*, an opportunistic invader, only enters the marsh when salinities are low, as during wet seasons (Duarte et al. 2015).

The invasive plant species *Phragmites karka* (Retz.) Trin. ex Steud. is widely distributed in tropical and subtropical environments, freshwater areas, and brackish marshland areas, including riverbanks and lake margins. The plant can adapt and endure highly salinized environments. The discovery of numerous genes in root and leaf tissue that had distinct regulation patterns in stress was identified in *P. karka* under salt stress with several important metabolic pathways over-represented. Additionally, several distinct transcription factor families such as CCCH, WRKY, NAC, MYB, etc. were distinctly expressed (Nayak et al. 2020). *Dittrichia viscosa* (L.) Greuter is a native species that is indigenous to the Mediterranean area that may be invasive because of its capacity to outcompete other species. Its resistance to stress was compared to that of a taxonomically similar species, *Limbarda crithmoides* (L.) Dumort; both species were formerly classified in the genus *Inula*. *Dittrichia viscosa* is known to only pose a danger to halophyte flora at lower salinities, such as those found along salt marsh borders, as it can only withstand mild salt stress, that will be useful for managing salt marshes since they provide details on how invasive *D. viscosa* may be in these crucial environments (Al Hassan et al. 2021).

14.3 Halophytes

Halophytes are a category of plants that have evolved genetic, morphological, anatomical, biochemical, and physiological adaptations enabling them to survive in a wide range of habitats, including deserts, wetlands, tropical, and temperate zones having high salts, heavy metals, and other toxic anthropogenic agents. Angiosperms make up the majority of the estimated 2000–3000 halophyte species, which make up fewer than 2% of all terrestrial plants (Sharma et al. 2016). Halophytes can be divided into xero-halophytes (suited to very arid environments), psammophytes (usually present in sandy soils), and hydro-halophytes (often flourish in damp soil or in aquatic environments) depending on their various habitats. Apart from these, halophytes have been further categorized as weedy halophytes (generally penetrate and occupy extremely disturbed places or areas), chasmophytes (found on the top of cliffs, both rocky and sandy beaches, and saltmarshes), phreatophytes (plants with deep roots that draw their water from a potentially saline deep underground source), and xero-halophytes (suited to saline environments and inland salt deserts) making a total of seven groups (Bueno and Cordovilla 2020; Rahman et al. 2021).

Halophytes have evolved several tactics to conclude their life cycle in challenging conditions, including the process of succulence formation, compartmentalization of harmful ions, synthesis of osmolytes, increased antioxidant activity, and synthesis of compatible solutes (Grigore and Toma 2020; Barros et al. 2021; Singh et al. 2023). Halophytes serve a variety of agricultural and non-agricultural purposes as well as help to preserve ecological balance, clean up the environment and provide grains, vegetables, fruits, animal feed, and coastal protection. It may be grown on saline land with saltwater irrigation for food, fuel, fodder/forage, medicinal crops and produce considerable amounts of bioactive metabolites (Hasanuzzaman et al. 2019; Nikalje et al. 2019). *Suaeda fruticosa* (L.) Delile, a halophytic species, is potentially employed as a model system for studies on salt tolerance since it can live and reproduce normally in soil having a pH of 10.5, a salinity of 65 dS m^{-1}, and little to no water (Pareek et al. 2020).

Halophytes are salt-tolerant organisms that can survive under stressful conditions; as a result, they show preserved and differential metabolic reactions from those of conventional plants. When present at the right concentrations, salt promotes halophyte vegetative growth while preventing non-halophyte development. Inland deserts, dunes, saline depressions, and coastal salt marshes are just a few of the diverse environments where halophytes may be found (Caparrós et al. 2022). They possess many strategies to survive under stressful conditions such as growth and growth regulators modulation, phenotypic plasticity, salt excretion, somatic and CO_2 resistance, saline dilution, Na^+ compartmentalization in vacuoles, water-use efficiency, transpiration control, activation of antioxidant systems, high K^+/Na^+ selectivity, osmolyte synthesis to favor osmotic adjustment, the different C3-C4-CAM pathway depending on circumstances of the environment, and the expression of specific genes (Bueno and Cordovilla 2020).

14.3.1 Characteristics of Halophytes for Proper Invasion of Saline Lands

These extraordinary plants have evolved a variety of defense mechanisms to withstand salinity and flourish in situations with high levels of salt. They have efficient mechanisms for tolerating salt, which allows them to generate high-quality seeds, sustain the early stages of development in salty conditions, and finish their life cycles at high salinity when there is more than 200 mM of NaCl. Halophyte reproductive development is made more effective by salt through later and more frequent flowering increased pollen vitality and increased seed production (Yuan et al. 2019a). The coastal halophyte *Plantago coronopus* L. has a high capacity for seed production; one plant was reported to have produced more than 1200 seeds. Despite having three times as many seeds as *C. danica*, *P. coronopus* had a far poorer tolerance for salt (Fekete et al. 2021). With seven genera (*Atriplex, Suaeda, Beta, Kali, Halimione, Salicornia, Oxybasis*) and 15 species blooming on the French

Flanders coast, the largest represented family of halophytes, Amaranthaceae, is located near the North Sea (Lefèvre and Rivière 2020).

14.3.1.1 Growth

Dune spinach (*Tetragonia decumbens* Mill.) showed a notable rise in overall yield, branch development, and the antioxidant potential to reduce ferric ions in nutrient solution combined with 50 mM NaCl to study the salinity effect on the composition of plant minerals, growth, and antioxidant activity. However, the salinity rise (200 mM) led to a reduction in the quantity of chlorophyll but also raised phenolic together with the amounts of sodium, phosphorus, and nitrogen (Sogoni et al. 2021). It has been claimed that halophytic plants, notably hydro-halophytes and certain phreatophytes, use ultrafiltration processes caused by their root systems to keep out excess salts. The resistance to bypass flow is increased by installing apoplastic barriers at the roots which effectively excludes salts from the roots and, therefore, lowers the buildup of harmful ions to go through the transpiration stream in the aboveground shoots (Rahman et al. 2021). The addition of two halophytes, Seashore mallow (*Kosteletzkya virginica* (L.) C.Presl ex A.Gray.) and Sesbania (*Sesbania cannabina* (Retz.) Poir.), showed the promotion of plant growth (biomass, root development, and germination) when biochar and inorganic fertilizer were applied separately or together to a coastal soil. This was mostly ascribed to the increased nutrient availability as a result of better soil health brought about by biochar (Zheng et al. 2018). Application of plant growth promoting rhizobacteria (PGPRs) minimizing salt stress in agricultural production is gaining popularity. It was found that crop PGPR strains of the same species have comparatively well-preserved genomes in comparison to same types of bacteria that have been isolated from local plants (i.e., aromatic plants or halophytes) (Leontidou et al. 2020). Halophilic PGPRs may be employed as efficient bioinoculants to encourage the development of non-halophytic species in salty soils, which is a useful tactic for the sustained improvement of non-halophytic crop growth (Etesami and Beattie 2018).

14.3.1.2 Water Relation

Many obligatory halophytes that live in salt marshes and sabkhas absorb the water they require together with the salt they need. Most halophytes have diverse mechanisms that may remove excess salts. These plants need to absorb a lot of salt in addition to water, thus they have salt glands, salt bladders, and potentially additional structures that allow them to extrude more salt. These plants also have fleshy leaves that can exude additional salts, like those of *Limonium* and *Atriplex*. An excessive amount of water is stored to make up for high internal NaCl levels, which results in a high water to dry weight ratio. These plants become more succulent when soil salinity rises due to increased water and salt absorption (Yasseen and Al-Thani 2022). To keep their cells turgor, salt-accumulating euhalophytes such as *S. salsa* and *Kalidium foliatum* (Pall.) Moq. compartmentalize enormous quantities of ions in vacuoles. When the possibility for water in the soil is minimal, some plants also produce leaf or stem succulence (Yuan et al. 2019b). Maintaining osmotic balance is very important for resistance against salt stress as it was discovered that *S. patens*, an invasive species, has salt stress resistance mostly because of greater proline levels in its leaves, which enable it to keep its osmotic balance stable and shield its photochemical systems from damage. As opposed to that, the freshwater species *C. longus* was severely harmed by high salt concentrations, mostly because it lacked osmotic balance and was unable to counterbalance the high ionic strength of the surrounding medium when osmocompatible solutes are present (Duarte et al. 2015).

14.3.1.3 Mineral Nutrition

Three halophyte species—*Suaeda maritima* (L.) Dumort, *Mesembryanthemum nodiflorum* L., and *Sarcocornia fruticosa* (L.) A.J.Scott—gathered in Portuguese and Spanish salt marshes were evaluated for their nutritional value and antioxidant capacity in comparison to cultivated plants. In comparison to wild plants, cultivated *S. fruticosa* and *S. maritima* displayed greater moisture content values, while *M. nodiflorum* exhibited no variations. Every species tested is an excellent provider of minerals, fiber, and proteins. Additionally, they have high levels of vitamins A, C, and B_6, carotenoids, and notably *S. maritima* has significant antioxidant potential. The most important phenolic chemicals found in the studied halophytes were ferulic and caffeic acids (Castañeda-Loaiza et al. 2020).

14.3.1.4 Photosynthesis

Spartina patens, a salt-excreting grass among the most widely spread genera of halophytes worldwide, has demonstrated in the past that it can deal with salt stress quite effectively, sustaining very high rates of photosynthesis despite obvious indicators of stress, primarily because of methods for internal osmoregulation based on proline. Being an aggressive salt marsh colonizer, these species have a considerable competitive advantage due to their C_4-type photosynthesis, which is dependent on phosphoenolpyruvate carboxylase (PEPC) activity that enables them to concentrate CO_2 more quickly than what a C_3 organism would do (Duarte et al. 2018). Additionally, halophytes develop processes of succulence for ion homeostasis and osmoprotectants should be accumulated in saline conditions to maintain cell turgor pressure. It has been discovered that certain halophytes such as *S. fruticosa, Achras sapota* L., and *Salsola drummondii* Ulbr., have effective succulent mechanisms to provide salt tolerance by maintaining increased photosynthetic efficiency while storing more salt in their leaves and stems (Rahman et al. 2021).

14.3.2 Economic Use of Halophytes

14.3.2.1 Bioactive and Phenolic Compounds

Halophytes have significant concentrations of phenolic and bioactive compounds. Phenolic compounds are utilized for use as raw ingredients in the cosmetic, pharmaceutical, and agro-food sectors due to their strong associations with important biological processes like antioxidant and antimicrobial properties (Lopes et al. 2023). It is well known that *Atriplex halimus* L. is rich in phenolic chemicals and exhibits a range of biological properties, such as antimicrobial, antioxidant, and immunomodulatory activities. *Beta vulgaris* L. has significant commercial value because of the sugar produced from its fleshy edible roots (Lefèvre and Rivière 2020).

14.3.2.2 Desalinization and Effect as Heavy Metal Phytoremediators

Halophytic species from several plant families, with the most notable species being categorized as phytoextractors or phytostabilizers, may remove heavy metals (Table 14.1) from the environment (Singh et al. 2023). The well-known physiological and morphological characteristics, including the root system's restriction of the introduction of heavy metals, osmolyte production and preservation like proline, and the complexation, chelation, and division of metal ions into compartments inside the cells, all play a significant role (Caparrós et al. 2022). The Mediterranean salt marshes are home to an especially large population of *Halimione portulacoides* (L.) Aellen, as it has potential in phytoremediation and the bioindication of metal pollution. This kind of plant is a halophyte that accumulates mercury (Lefèvre and Rivière 2020). In Qatar, halophytes such as *Tetraena qatarensis* (Hadidi) Beier &

Table 14.1 Halophyte species, types of metal accumulations by them and their accumulation potential

Halophyte	Metal	Accumulation (μg/g)	Reference
Avicennia officinalis	Copper	15	Caparrós et al. (2022)
	Zinc	107	
	Lead	23	
Rhizophora apiculata	Copper	10	Afifudin et al. (2022)
	Zinc	16	
	Lead	12	
Rhizophora mucronata	Copper	19	Nikalje and Suprasanna (2018)
	Zinc	40	
	Lead	12	
Excoecaria agallocha	Copper	8	El Shaer (2021)
	Zinc	76	
	Lead	27	
Bruguiera cylindrica	Copper	17	Munir et al. (2022)
	Zinc	116	
	Lead	17	
Ceriops decandra	Copper	95	Aziz and Mujeeb (2022)
	Zinc	9	
	Lead	11	
Aegiceras corniculatum	Copper	13	Yao et al. (2022)
	Zinc	12	
	Lead	12	
Acanthus ilicifolius	Copper	13	Sarath et al. (2022)
	Zinc	67	
	Lead	16	

Thulin, *Salsola soda* L., *Salicornia europaea,* and *Halopeplis perfoliata* are important in the phytoremediation of contaminated soils and streams. These halophytes' associated microorganisms, such as endophytic bacteria, may help these plants improve salinized and contaminated soils. Several of these bacteria, including *Pseudomonas* spp. and *Bacillus* spp., play crucial roles in numerous aspects of life (Yasseen and Al-Thani 2022).

14.3.2.3 Bioenergy

Bioenergy has the potential to significantly contribute to the long-term control of climate change by reducing the cost of achieving climate goals (Daioglou et al. 2020). Low-cost biofuel production allows for the growth and maintenance of bioenergy crops, which also have a positive impact on the environment by reducing soil erosion, GHG emissions, and CO_2 levels (Yadav et al. 2019). The lignocellulosic biomass of halophytes and the oil generated from their seeds may both be used to make biofuel. As a kind of adaptation, certain halophytes—referred to as salt accumulators—store salt in their organs, while others have found mechanisms to exclude it. The non-combustible element of the biomass generated by accumulators may cause fouling issues, hence salt excluders are often a preferable option for biofuels. Bioenergy halophyte species, such as *Suaeda aralocaspica, Crithmum maritimum, Salicornia bigelovii, Descurainia sophia, Ricinus communis, Kosteletzkya virginica,* and *Euphorbia tirucalli,* may store significant concentrations of oils that make up a percentage of the dry seed weight greater than 20% (Sharma et al. 2016).

14.3.2.4 Salinity Tolerance: Physiological Mechanisms and Genetic Basis

Ecologists divide halophytes into three major categories: euhalophytes, recretohalophytes, and pseudo-halophytes. Euhalophytes exhibit a high level of salt tolerance and are capable of dilution of salt in their stems or leaves. Recretohalophytes, which can release salt from their leaves, are widely distributed around the world on salty soils and in saltwater. To safeguard metabolic tissues, pseudo-halophytes maintain ions in the roots in addition to other functions; they also make an effort to confine their movement to aerial areas (Badri and Ludidi 2020). There have been reports of many salt-responsive genes and promoters in halophytes such as *Aeluropus, Thellungiella, Mesembryanthemum, Suaeda, Atriplex, Salicornia,* and *Cakile*. Several well-known genes, including those that code for antioxidants (*BADH, CAT, APX, GST, SOD*), ion channels (aquaporins, Ca^{2+}, Cl^-), antiporters (*VTPase, SOS, NHX, HKT*), and additional new genes including *USP, SRP, SDR1,* etc., from halophytes were isolated and tested for improving crop plant (glycophytes) resistance to stress. Stress activates salt sensors, which then cause the up-or down-regulation of a number of genes, the separate or combined actions of genes sensitive to stress, and the activation of stress tolerance mechanisms (Mishra and Tanna 2017).

14.3.2.5 Use as Animal Feed

In Africa, halophytes, especially perennials and shrubs, make up a sizable portion of the native vegetation used as animal feed. Forage halophytes can be a significant protein supplement. The digestion of highly energizing substances such as hay, straw, or dry grass can be improved by consuming greens as little as 20 g of leaves each day (Badri and Ludidi 2020).

14.3.2.6 Intercropping

Growing two or more crops either concurrently or in succession is known as intercropping. Halophytes are prospective plants for resilient agriculture, alongside glycophytes intercropping, to improve their production in salty soils because of their adaptability to extremely saline environments. This method has historically been employed by small farmers all over the world to supply food demands and lower the possibility of a single crop failure. When watermelon (*Citrullus lanatus* (Thunb.) Matsum. & Nakai) was intercropped with six halophytic species— garden orache, four-wing saltbush, barley, purslane, wheat, and saltwort—the garden orache (*Atriplex hortensis* L.) showed the fastest rates of growth, and purslane (*Portulaca oleracea* L.), under saline irrigation, a lot of salt accumulates in the plant tissues. Saline irrigation was found to have no effect on watermelon fruit quality, stem water potential, and yields; nevertheless, the intercropping watermelon/orache combination produced noticeably greater yields (Simpson et al. 2018). *Salicornia europaea* L. intercropped with tomato plants was examined for their nutritional profile and the presence of various health-promoting chemicals. It was discovered that, except for flavonoids, nutrient concentration was unaffected by the farming method and bioactive compounds. *Salicornia* showed strong anti-inflammatory effects and antibacterial activity against *Bacillus subtilis* without any detectable cytotoxic effects, suggesting the significance for both animal and human health of this halophyte (Castagna et al. 2022). When intercropped with *S. soda* in hydroponic culture, saline-grown *Lactuca sativa* L. plants showed decreased growth than *L. sativa* cultivated alone. The favorable effects of intercropping on total phenolic compounds, calcium, iron, and phosphorus imply that several qualitative features of *L. sativa* may be enhanced. Halophyte species like *S. soda* need special consideration since they could offer a significant way to integrate into the human diet while also protecting the environment (Atzori et al. 2022).

The classic monoculture cotton (MC) is improved by a novel cotton/halophyte intercropping system, such as the cotton/*S. salsa* intercropping (CSSI) and cotton/alfalfa intercropping (CAI) systems. In comparison to the MC system, CSSI and CAI systems might reduce soil $EC_{1:5}$, salt buildup, bulk density, Na^+ concentration, and pH in strips without mulch while increasing salt removal, soil porosity, and organic carbon content. The CSSI system enhanced root mass density in comparison to the MC system. In 2014, there were no appreciable variations in seed cotton yield and irrigation water productivity (IWP) between the MC, CSSI, and CAI systems. However, in 2015 and 2016, irrigation water productivity, cumulative aboveground biomass, and seed cotton production were all significantly higher in the CSSI and CAI systems compared to the MC system. According to this study, the intercropping system of cotton and halophytes removed more salt from the soil and produced more crops than the MC system. As a result, systems using CSSI and CAI as a long-term agronomic remedy for enhancing cotton production and soil salinization in the arid northwest Chinese area are being promoted (Liang and Shi 2021).

In another study, the growth, fruit, and biochemical characteristics of strawberry plants either grown with a halophytic companion plant or without one (*Portulaca oleracea* L.) were examined. The growth, physiological (stomatal conductance and electrolyte leakage), and biochemical parameters of strawberry plants were adversely impacted by salt stress. The presence of *P. oleracea* improved physiological and biochemical parameters while also increasing the dry weight, fresh weight, average fruit weight, and the total number of strawberries produced. This study demonstrated that the production and quality of strawberry fruit may be successfully boosted by co-cultivating strawberry plants under salt stress with *P. oleracea.* It is recommended that *P. oleracea* be used in situations where salinity is a major problem as it may be an environmentally beneficial strategy (Karakas et al. 2021).

14.3.3 Controlling Halophytes Plant Invasions: Eradication, Biocontrol, and Biochemical Approaches

One of the biggest dangers to biodiversity protection is the invasion of exotic species. Once an invasive species establishes itself, it can be challenging to control it, and eradication is typically not achievable. The effects on ecosystem functions and natural communities can also be quite detrimental. Therefore, it is crucial to create future work that enables the early identification of invasions. A major danger to biodiversity that is posed by an invading species, *Spartina alterniflora*, to biodiversity in China has been successfully managed and controlled in several provinces, such as Zhejiang, Shanghai, and Jiangsu, as shown by the sharp decrease in *S. alterniflora* area from 2015 to 2020. However, in areas where invasion is more serious, like Shandong, Guangdong, Fujian and provinces, stronger control is required. *Spartina alterniflora* management in China has been successfully achieved through the implementation of pertinent laws, rules, and ecological restoration initiatives by local or national governments (Li et al. 2022a).

An annual halophytic plant, *S. soda* (Amaranthaceae), native to the Old World, has biological traits that pose a severe danger. Its capacity for mass seed production, easy water and wind transfer, and ability to establish itself in typical marine coastal climates all indicate its strong potential for growth and invasion in similar conditions. Therefore, the control and eradication of *S. soda* in invaded regions may depend on the plant's yearly life cycle and the short seed survival period (Marbán and Zalba 2019). Major riparian invaders on land in South Africa are *Tamarix* species. *Diorhabda carinulata* (Coleoptera: Chrysomelidae), a biocontrol agent that has shown effective against the invasive *Tamarix* plant in North America, was studied in South Africa as the initial potential agent. The viability of *Tamarix* in terms of serving as a host for *D. carinulata* and the amount of its invasive potential may be influenced by its capacity to thrive across a wide soil salinity gradient. The preference of *D. carinulata* for its hosts and the suitability of the *Tamarix* species as hosts is thus unaffected by increased soil salt. A factor in the invasiveness of the foreign *Tamarix* taxa in South Africa may be that they are less vulnerable than *T. usneoides* to stress brought on by salt (Drude et al. 2020).

14.4 Conclusion and Future Prospects

Soil salinization, a serious environmental problem caused by the disturbance of soil due to climate change and growing anthropogenic activities, has restricted crop production and led to the loss of agriculturally productive lands. One of the biggest threats to the maintenance of biodiversity is the introduction of exotic species in saline environments. Salt-tolerant competitive invasive plant species have replaced native plant species, having a detrimental impact on the ecological function. Halophytes are salt-tolerant plants that have undergone genetic, morphological, biochemical, anatomical, and physiological modifications allowing them to thrive in a variety of environments. It can be challenging to control an invasive species after it has established itself. Therefore, eradication, biocontrol, and biochemical approaches are all used to manage halophyte plant infestations. Halophytes are useful for creating a variety of foods and animal feed, as well as functioning as phytoremediators to remove heavy metals and causing soil desalinization. They also supply bioactive substances, phenolic compounds, and biofuel. This provides more insight into the economic potential of halophytes to be used as salt-tolerant plants in degraded agricultural lands.

References

Afifudin AFM, Irawanto R, Purwitasari N (2022) Selection of potential plants as phytoremediation for heavy metals in estuarine ecosystem: a systematic review. In: 4th international conference on life sciences and biotechnology (ICOLIB 2021). Atlantis Press, pp. 420–434

Agarwal P, Mishra K, Singh PC (2022) Integrated application of Trichoderma mixture and NPK enhances the rice productivity in sodic soil. Int J Plant Environ 8:128–132

Al Hassan M, Boscaiu M, Mayoral O (2021) Competition between halophytes and invasive species: Dittrichia viscosa and Limbarda crithmoides: a study case from the Valencian salt marshes. Handbook of halophytes: from molecules to ecosystems towards biosaline agriculture, pp. 599–621

Ali M, Mustafa A, Abideen Z, Gul B (2021) Bioenergy production from halophytes crops for sustainable development. In: Energy and environmental security in developing countries. Springer, Cham, pp 571–586

Anderson R, Bayer PE, Edwards D (2020) Climate change and the need for agricultural adaptation. Curr Opin Plant Biol 56:197–202

Araus JL, Kefauver SC (2018) Breeding to adapt agriculture to climate change: affordable phenotyping solutions. Curr Opin Plant Biol 45:237–247

Arora NK (2019) Impact of climate change on agriculture production and its sustainable solutions. Environ Sustain 2:95–96

Atzori G, Guidi Nissim W, Mancuso S, Palm E (2022) Intercropping salt-sensitive Lactuca sativa L. and salt-tolerant Salsola soda L. in a saline hydroponic medium: an agronomic and physiological assessment. Plants 11:2924
Aziz I, Mujeeb A (2022) Halophytes for phytoremediation of hazardous metal (loid) s: a terse review on metal tolerance, bio-indication and hyperaccumulation. J Hazard Mater 424:127309
Badri M, Ludidi N (2020) Halophytes as a resource for livestock in Africa: present status and prospects: present status and prospects. Handbook of halophytes: from molecules to ecosystems towards biosaline agriculture, pp. 1–17
Barros NLF, Marques DN, Tadaiesky LBA, de Souza CRB (2021) Halophytes and other molecular strategies for the generation of salt-tolerant crops. Plant Physiol Biochem 162:581–591
Bhadouria R, Singh R, Singh VK et al (2019) Agriculture in the era of climate change: consequences and effects. In: Climate change and agricultural ecosystems. Elsevier, pp 1–23
Bueno M, Cordovilla MP (2020) Ecophysiology and uses of halophytes in diverse habitats. Handbook of halophytes: from molecules to ecosystems towards biosaline agriculture, pp. 1–25
Caparrós PG, Ozturk M, Gul A et al (2022) Halophytes have potential as heavy metal phytoremediators: a comprehensive review. Environ Exp Bot 193:104666
Castagna A, Mariottini G, Gabriele M et al (2022) Nutritional composition and bioactivity of Salicornia europaea L. plants grown in monoculture or intercropped with tomato plants in salt-affected soils. Horticulturae 8:828
Castañeda-Loaiza V, Oliveira M, Santos T et al (2020) Wild vs cultivated halophytes: nutritional and functional differences. Food Chem 333:127536
Chen J, Mueller V (2018) Coastal climate change, soil salinity and human migration in Bangladesh. Nat Clim Chang 8:981–985
Dai X, Yu Z, Yang G, Wan R (2020) Role of flooding patterns in the biomass production of vegetation in a typical herbaceous wetland, Poyang Lake wetland, China. Front Plant Sci 11: 521358
Daioglou V, Rose SK, Bauer N et al (2020) Bioenergy technologies in long-run climate change mitigation: results from the EMF-33 study. Clim Chang 163:1603–1620
Drude L, Marlin D, Venter N, Byrne MJ (2020) Increased soil salinity has no effect on the host preference and suitability of the Tamarix biological control agent, Diorhabda carinulata. Biocontrol Sci Tech 30:1340–1350
Duarte B, Santos D, Marques JC, Caçador I (2015) Ecophysiological constraints of two invasive plant species under a saline gradient: halophytes versus glycophytes. Estuar Coast Shelf Sci 167: 154–165
Duarte B, Matos AR, Marques JC, Caçador I (2018) Leaf fatty acid remodeling in the salt-excreting halophytic grass Spartina patens along a salinity gradient. Plant Physiol Biochem 124:112–116
El Shaer HM (2021) Potential use of halophytes and salt-tolerant forages as animal feed in the Arab region: an overview. Handbook of halophytes: from molecules to ecosystems towards biosaline agriculture, pp. 2479–2498
Etesami H, Beattie GA (2018) Mining halophytes for plant growth-promoting halotolerant bacteria to enhance the salinity tolerance of non-halophytic crops. Front Microbiol 9:148
Eziz A, Yan Z, Tian D et al (2017) Drought effect on plant biomass allocation: a meta-analysis. Ecol Evol 7:11002–11010
Fekete R, Haszonits G, Schmidt D et al (2021) Rapid continental spread of a salt-tolerant plant along the European road network. Biol Invasions 23:2661–2674
Grigore M-N, Toma C (2020) Integrative anatomy of halophytes from Mediterranean climate BT - handbook of halophytes: from molecules to ecosystems towards biosaline agriculture. In: Grigore M-N (ed) . Springer International Publishing, Cham, pp 1–35
Hasanuzzaman M, Nahar K, Öztürk M (2019) Ecophysiology, abiotic stress responses and utilization of halophytes. Springer, Singapore
Hermans K, McLeman R (2021) Climate change, drought, land degradation and migration: exploring the linkages. Curr Opin Environ Sustain 50:236–244

Hopmans JW, Qureshi AS, Kisekka I et al (2021) Critical knowledge gaps and research priorities in global soil salinity. Adv Agron 169:1–191
Hossain A, Krupnik TJ, Timsina J et al (2020) Agricultural land degradation: processes and problems undermining future food security. In: Environment, climate, plant and vegetation growth. Springer, pp 17–61
Jansson JK, Hofmockel KS (2020) Soil microbiomes and climate change. Nat Rev Microbiol 18: 35–46
Kapadia C, Patel N, Rana A et al (2022) Evaluation of plant growth-promoting and salinity ameliorating potential of halophilic bacteria isolated from saline soil. Front Plant Sci 13:946217
Karakas S, Bolat I, Dikilitas M (2021) The use of halophytic companion plant (Portulaca oleracea L.) on some growth, fruit, and biochemical parameters of strawberry plants under salt stress. Horticulturae 7:63
Lefèvre G, Rivière C (2020) Amaranthaceae halophytes from the French Flanders coast of the North Sea: a review of their phytochemistry and biological activities. Phytochem Rev 19:1263–1302
Leontidou K, Genitsaris S, Papadopoulou A et al (2020) Plant growth promoting rhizobacteria isolated from halophytes and drought-tolerant plants: genomic characterisation and exploration of phyto-beneficial traits. Sci Rep 10:1–15
Li H, Mao D, Wang Z et al (2022a) Invasion of Spartina alterniflora in the coastal zone of mainland China: control achievements from 2015 to 2020 towards the sustainable development goals. J Environ Manag 323:116242
Li S, Chang L, Sun R et al (2022b) Combined transcriptomic and metabolomic analysis reveals a role for adenosine triphosphate-binding cassette transporters and cell wall remodeling in response to salt stress in strawberry. Front Plant Sci 13:996765
Liang J, Shi W (2021) Cotton/halophytes intercropping decreases salt accumulation and improves soil physicochemical properties and crop productivity in saline-alkali soils under mulched drip irrigation: a three-year field experiment. Field Crop Res 262:108027
Liu L, Wang B (2021) Protection of halophytes and their uses for cultivation of saline-alkali soil in China. Biology (Basel) 10:353
Liu M, Liao H, Peng S (2019) Salt-tolerant native plants have greater responses to other environments when compared to salt-tolerant invasive plants. Ecol Evol 9:7808–7818
Lopes M, Sanches-Silva A, Castilho M et al (2023) Halophytes as source of bioactive phenolic compounds and their potential applications. Crit Rev Food Sci Nutr 63(8):1078–1101
Malhi Y, Franklin J, Seddon N et al (2020) Climate change and ecosystems: threats, opportunities and solutions. Philos Trans R Soc B 375:20190104
Marbán LM, Zalba SM (2019) When the seeds go floating in: a salt marsh invasion. Estuar Coast Shelf Sci 231:106442
Marcos-Barbero EL, Pérez P, Martínez-Carrasco R et al (2021) Screening for higher grain yield and biomass among sixty bread wheat genotypes grown under elevated CO2 and high-temperature conditions. Plan Theory 10:1596
Mishra A, Tanna B (2017) Halophytes: potential resources for salt stress tolerance genes and promoters. Front Plant Sci 8:829
Munir N, Hasnain M, Roessner U, Abideen Z (2022) Strategies in improving plant salinity resistance and use of salinity resistant plants for economic sustainability. Crit Rev Environ Sci Technol 52:2150–2196
Nayak SS, Pradhan S, Sahoo D, Parida A (2020) De novo transcriptome assembly and analysis of Phragmites karka, an invasive halophyte, to study the mechanism of salinity stress tolerance. Sci Rep 10:1–12
Nikalje GC, Suprasanna P (2018) Coping with metal toxicity–cues from halophytes. Front Plant Sci 9:777
Nikalje GC, Bhaskar SD, Yadav K, Penna S (2019) Halophytes: prospective plants for future. In: Ecophysiology, abiotic stress responses and utilization of halophytes. Springer, pp 221–234

Pareek A, Dhankher OP, Foyer CH (2020) Mitigating the impact of climate change on plant productivity and ecosystem sustainability. J Exp Bot 71:451–456
Prabakaran K, Li J, Anandkumar A et al (2019) Managing environmental contamination through phytoremediation by invasive plants: a review. Ecol Eng 138:28–37
Prăvălie R (2021) Exploring the multiple land degradation pathways across the planet. Earth Sci Rev 220:103689
Rahman MM, Mostofa MG, Keya SS et al (2021) Adaptive mechanisms of halophytes and their potential in improving salinity tolerance in plants. Int J Mol Sci 22:10733
Sarath NG, Shackira AM, El-Serehy HA et al (2022) Phytostabilization of arsenic and associated physio-anatomical changes in Acanthus ilicifolius L. Environ Pollut 298:118828
Sharma R, Wungrampha S, Singh V et al (2016) Halophytes as bioenergy crops. Front Plant Sci 7: 1372
Simpson CR, Franco JG, King SR, Volder A (2018) Intercropping halophytes to mitigate salinity stress in watermelon. Sustainability 10:681
Singh VK, Singh R, Rajput VD, Singh VK (2023) Halophytes for the sustainable remediation of heavy metal-contaminated sites: recent developments and future perspectives. Chemosphere 313:137524
Sogoni A, Jimoh MO, Kambizi L, Laubscher CP (2021) The impact of salt stress on plant growth, mineral composition, and antioxidant activity in tetragonia decumbens mill.: an underutilized edible halophyte in South Africa. Horticulturae 7:140
Syed S, Xu M, Wang Z et al (2021) Invasive Spartina alterniflora in controlled cultivation: environmental implications of converging future technologies. Ecol Indic 130:108027
Turner KG, Hufbauer RA, Rieseberg LH (2014) Rapid evolution of an invasive weed. New Phytol 202:309–321
Webb NP, Marshall NA, Stringer LC et al (2017) Land degradation and climate change: building climate resilience in agriculture. Front Ecol Environ 15:450–459
Wilschut RA, De Long JR, Geisen S et al (2022) Combined effects of warming and drought on plant biomass depend on plant woodiness and community type: a meta-analysis. Proc R Soc B 289: 20221178
Xia X, Zhang S, Li S et al (2018) The cycle of nitrogen in river systems: sources, transformation, and flux. Environ Sci Process Impacts 20:863–891
Yadav P, Priyanka P, Kumar D et al (2019) Bioenergy crops: recent advances and future outlook. Prospect Renew Bioprocess Futur Energy Syst:315–335
Yamazaki K, Ishimori M, Kajiya-Kanegae H et al (2020) Effect of salt tolerance on biomass production in a large population of sorghum accessions. Breed Sci 70:167–175
Yao L, Wang J, Yang K et al (2022) Proteomic analysis reveals molecular mechanism of Cd2+ tolerance in the leaves of halophyte Halogeton glomeratus. J Proteome 104703
Yasseen BT, Al-Thani RF (2022) Endophytes and halophytes to remediate industrial wastewater and saline soils: perspectives from Qatar. Plan Theory 11:1497
Yuan F, Guo J, Shabala S, Wang B (2019a) Reproductive physiology of halophytes: current standing. Front Plant Sci 9:1954
Yuan F, Xu Y, Leng B, Wang B (2019b) Beneficial effects of salt on halophyte growth: morphology, cells, and genes. Open Life Sci 14:191–200
Zhang J, Zhang S, Cheng M et al (2018) Effect of drought on agronomic traits of rice and wheat: a meta-analysis. Int J Environ Res Public Health 15:839
Zheng H, Wang X, Chen L et al (2018) Enhanced growth of halophyte plants in biochar-amended coastal soil: roles of nutrient availability and rhizosphere microbial modulation. Plant Cell Environ 41:517–532

Plant Invasion and Climate Change: An Overview on History, Impacts, and Management Practices

15

Ritturaj Singh, Ashutosh Kumar Singh, Prakash Rajak, and Hema Singh

Abstract

Species that implement new home ranges and then multiply, proliferate, and endure there at the cost of their surroundings are said to be alien species. Anthropogenic activities are one of the most important reasons for the unheard changes to the dispersion of the world biota. Invasion of plants (as well as animals) has been significantly expanded in the last few decades as a consequence of the quickly developing international trade and transportation. Among the primary causes of species extinction, invasive alien plant species (IAPS) are thought to affect social and economic conditions and ecological services through a variety of processes. The health of humans is also significantly impacted by alien species, both positively and negatively, but mostly negatively. A comprehensive understanding of the dynamic mechanisms associated in the invasion procedure must be developed in order to establish an effective management strategy for invasive species. Changes in temperature and precipitation regimes and related processes are further accelerating the invasion success of several species and causing severe threat to the native ecosystems and their species composition. For avoiding severe cumulative effects of plant invasion and climate change, it is essential to judicially handle unwanted alien species in native and foreign habitats. Direct monitoring of invasive alien species usually requires an integrated strategy comprising of the coordinated application of a number of techniques. In this chapter, emphasis has been given on understanding the impacts of invasive species on different ecological and socio-economic aspects, followed by outlining some prudent measures for their management in light of

R. Singh · P. Rajak · H. Singh (✉)
Ecosystem Analysis Laboratory, Department of Botany, Banaras Hindu University, Varanasi, India

A. K. Singh
Centre of Science and Society, University of Allahabad, Prayagraj, India

S. Tripathi et al. (eds.), *Plant Invasions and Global Climate Change*,
https://doi.org/10.1007/978-981-99-5910-5_15

changing climate scenario. Several techniques that are accessible are often divided into three categories, viz. mechanical, chemical, and biocontrol techniques. Here, we conclude that using alien species in various value-added processes may be crucial in limiting their spread.

Keywords

Alien species · Biodiversity · Characteristics · Ecosystems services · Human health · Invasibility · Invasive traits

15.1 Introduction: History of Invasion Biology

Specifically, maintaining a strong historical perspective is very important as a researcher because it reduces the likelihood of recycling ideas and losing true scientific momentum (Graham and Dayton 2002; Pianka and Horn 2005). Moreover, knowing what has come before us helps us to identify key questions and to determine what important knowledge and understanding is missing or, to put it bluntly, ignorance of history means ignorance of current scientific developments (Cuddington and Beisner 2005). Chew (2006) provided a detailed and comprehensive review of interest in extra-terrestrial species in the 200 years before Elton. Alexander Humboldt, an explorer and naturalist whose life spanned the latter three decades between the 18th and 19th, was aware of the redistribution of flora and animals over the globe. He highlighted, for instance, how, by the middle of the 1800s, the American *Opuntia* (cactus) had spread over the different continent (Humboldt et al. 1850). Swiss botanist Albert Thellung (1911–1912), who pioneered the integration of the ideas and theories of invasion ecology in the eighteenth century, highlighted the history of invasion ecology (Kowarik and Pyšek 2012). Natural invasions have also been researched for generations, but the topic of invasion biology was founded by a book "*The Ecology of Invasions by Animals and Plants*" written by Charles S. Elton in 1958 (Rejmánek et al. 2005). The curiosity in biological invasion has increased dramatically over the last 20 years. The SCOPE (Scientific Committee on Problems of the Environment) initiative laid the groundwork in the 1980s by reconsidering some of the fundamental premises and generalisations made by Elton (1958), evaluating the recent invasion situation in several regions around the globe, and asking some of the world's top conservationists to implement.

For ecologists, conservationists, geneticists, public health experts, biogeographers, and environmental historians, the study of exotic or invasive species (plants, animals, and micro-organisms) is a recent but difficult and contentious field. Yet recently, invasive species have caused a serious threat in the region's flora, natural systems, ecological sustainability, and population health (Pejchar and Mooney 2009; Jones and McDermott 2018; Bartz and Kowarik 2019). According to United Nation's Intergovernmental Panel on Biodiversity and Ecosystem Services (IPBES), biotic intruders threaten around one-fifth of the surface of the planet, especially worldwide ecological regions (IPBES 2019) through competition, predation, hybridisation, and through several indirect effects, introduced species have a main effect on community structure, genetic diversity, and global biological diversity (McGeoch et al. 2010; Pyšek and Richardson 2010). Nonetheless, invasion ecologists generally agree that anthropogenic disruptions are hastening the worldwide invasive alien plant species (IAPS) issue (Young and Larson 2011). Biological invasions have appeared extremely quickly as a result of the combination of variables that jeopardise diversity in almost all ecosystems and at different scales from micro to macro level (Carlton and Geller 1993; Vitousek et al. 1996, 1997; Rejmánek et al. 2005; Sax et al. 2005).

Introduced, exotic, and alien species are other terms used to describe invasive species. Without careful clarification, both phrases have been used interchangeably frequently. In cases where the invasive status of immigrant species is unclear, we will refer to them as "nonindigenous" as a general phrase. But these phrases have different conceptual meanings. There are some organisms that have been found beyond their typical environment and range known as alien species. Non-native plants known as accidental or causal aliens can survive and even sporadically replicate in the wild, but they got extinct because they cannot establish self-replacing populations and must be repeatedly planted in order to survive. Naturalised plants are alien species that naturally produce replacement communities by at least a decade without significant anthropogenic presence (or without human intervention), primarily from spores or ramets capable of spontaneous growth (Pyšek et al. 2004). A portion of the naturalised plants known as invasive plants can spread across a vast area because they generate fertile offspring far from their parents, frequently in great numbers.

Invasion ecology examines the movement of species facilitated by people, particularly to regions that are obviously outside of their possible range, which is determined by their conventional sources of dissemination and phylogeographic obstacles. The study of entry of species, their capacity to naturalise and dominate the region of interest, their interactions with other living things, and increasingly, the risks and advantages of their abundance and existence in connection to people, are all included in this field value structures (Pyšek and Richardson 2010). Because of their inherent threat to replace and damage ecosystems, non-native species create a big issue for humans (Mack et al. 2000; Pimentel et al. 2005). "*The Ecology of Invasions by Animals and Plants*" by Charles S. Elton (Elton 1958) reignited interest in invasions, but it was not until the 1980s partially due to the SCOPE International Program about Biological Encroachment (Drake et al. 1989) that invaders started having a big impact on conventional biology. The invasion ecology is also now deeply embedded in sustainable development for comprehending the impacts of invading species, and several studies have been conducted to create an effective method to slow the spread of introduced species.

Invasion biology has developed as a consequence of two causes: (1) establishing the science behind invasion ecology and (2) the situation of introduced species is urgent (Reichard and White 2003). Invasion ecology has expanded to include and take ideas, techniques, and strategies from a variety of fields, including distribution and abundance, population dynamics, epidemics, conservation science, and many others (McGeoch et al. 2010). When a species is introduced to an area with less rivalry and higher stability (from grazing animals or predation) than its wild places, it may become invasive (Abbasi and Nipaney 1986; Ganesh et al. 2005; Walter 2011; Martin et al. 2019). Nonetheless, it is known that just 2% of the organisms brought to these alien settings have flourished (Hierro et al. 2005). Invaders have the power to alter basic ecological characteristics such as the dominant species within a neighbourhood, the environment's physical characteristics, the cycling of nutrients, and the yield of plants (Bertness 1984; Vitousek 1990; Funk et al. 2020). Broad impacts of invasions make it difficult to maintain agriculturally productive systems, retain healthy ecosystems, ensure human safety, and protected biological diversity (Vitousek et al. 1996; Walker and Steffen 1997). 32 of the 100 species listed by the Advisory Group on Invasive Species as most harmful unwanted species on the Earth are plant species (Lowe et al. 2000). The efficiency of management measures as well as the dispersion, activity, and damage made by introduced species can all be affected by the ongoing climatic change. By significantly modifying their behaviour with the other species, and the instruments used to control them may also be affected by climate change, which makes invasion control even more challenging (Runyon et al. 2012).

In this chapter, after providing the brief history of invasion ecology we have emphasised on the key characteristics of invasive species (in brief); invasive plants and climate change; impacts of invasive species on ecological, social and economic components; major management approaches applied and the management perspectives.

15.2 Characteristics of Invasive Species

Every organism has to be capable of surviving in a specific surrounding in order to avoid dying. The essential question is how "prepared" or what attributes an organism needs to have in order to sustain in a particular surrounding. It is essential to discover whether any other organisms are fundamentally more suited to spread quickly when people transfer them into new places in the current anthropogenic biodiversity catastrophe, where non-native species play a significant role. Early studies on invasions mostly focused on identifying characters responsible for invasiveness (Booth et al. 2003). Several characteristics of invasive organisms (Fig. 15.1) are responsible for their spread (Willis and Blossey 1999; Patnaik 2017). For instance, they can withstand a variety of environmental factors like moisture, heat, soil, and water quality. By generating a significant number of seeds with a significant level of germination, quick growth, and outperforming indigenous species in terms of vigour, biomass, length, and endurance, invasive species dominate in the invaded areas. Several studies indicate that allelopathy determine an important role in some invasive organisms' capacity to dominate invading plant communities (Osvald 1948; Kanchan 1979; El-Ghareeb 1991; Vaughn and Berhow 1999; Ridenour and Callaway 2001; Wang et al. 2022). The new weapon hypothesis of wild plant growth

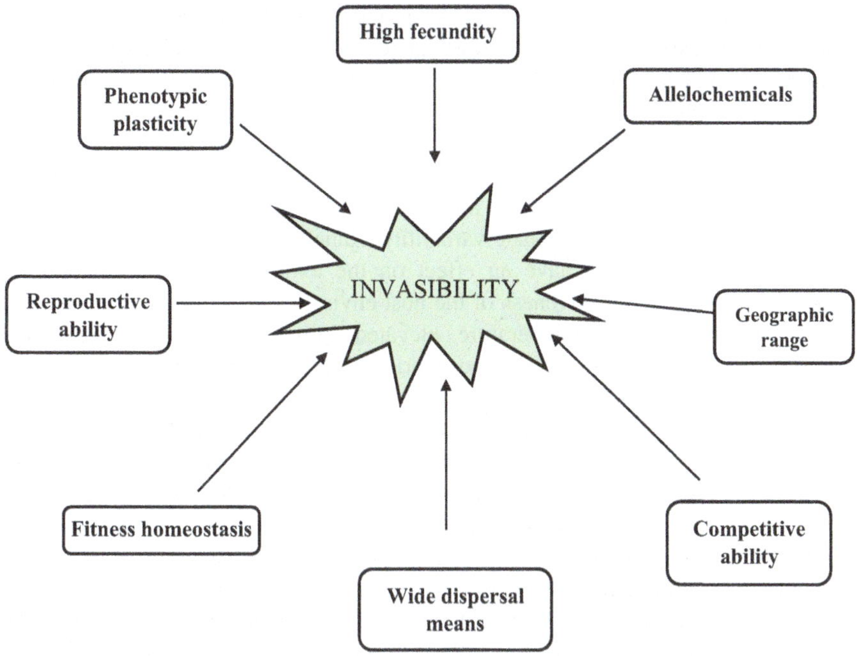

Fig. 15.1 Characteristics of unwanted species responsible for invasion in non-native areas

(Callaway and Aschehoug 2000) of invasive plants producing allelochemicals that prevent the growth of native plants is one technique to do this.

15.3 Invasive Species and Climate Change

Given that introduced species and climate variability are among the most significant impacts on biodiversity, their interaction may be predicted to result in catastrophic consequences. The likelihood of greatly enhanced susceptibility to climate change is expected to result from cumulative combinations. Introduced species will be significantly impacted by higher CO_2 levels, rising temperatures, altered rainfall pattern, and a spike in the occurrence of catastrophic events like fire and floods (Masters and Norgrove 2010). Water is probably a crucial factor that will facilitate incursion the most, especially by lowering the resilience of farming and natural ecosystems. Weather shifts will additionally influence invasive species, and their actions will have a large effect on the economy and surrounding. Since invading species are maintained significantly differently than the majority of indigenous species, they raise nearly opposing issues under the changing climate scenario. Few scientists have discovered precise impacts that climate variation is having on exotic plants, even if several earlier articles (e.g. Dukes and Mooney 1999; Thuiller 2007) imply that certain introduced species are probably going to benefit from it. The notion of introduced species being also put to the test by climate variability, since many indigenous species will alter their host range and move into regions where they had been previously inaccessible. Other groups that were formerly invasive may see a reduction in their effects. Several theories (Williamson 1993; Williamson and Fitter 1996; Theoharides and Dukes 2007) demonstrate that organisms must successfully navigate a variety of ecological filters before becoming intrusive. Each of the above steps depends heavily on a unique array of processes, a number that are probable to be impacted by climate variability (Rahel and Olden 2008).

Climate alteration will have an effect on the exotic plants, its potential for invasiveness, and the invasiveness of the host environment, whether it be native or derived, via rising mean temperature, elevated variance of annual precipitation (occurrence; strength), enhanced atmospheric CO_2 levels, elevated magnitude and frequency of hurricanes, and increasing sea levels (Rajaka et al. 2021). Variations in the degree of severity and occurrence of severe weather events that disrupt ecosystems and make them prone to intrusions may have the largest effects on introduced species. These changes also present unique chances for exotic species to spread and develop. Wildland fire threat is heightened in several Mediterranean areas by reduced precipitation, more frequent droughts, and hotter days. These adjustments will strengthen some invasive species and benefit others (Dukes and Mooney 1999). Climate changes might, however, make it easier for invasive species to endure and proliferate. Typically, invasions occur in two stages: a quiescent or lag phase, within which boundaries only slightly change, and a proactive or expansion phase, throughout which accelerated development is started. The interval here between stages can be anywhere around a few decades and a century. By disrupting the ecology and allowing inactive invasive alien species (IAS) to take advantage of such perturbations, climate variability may probably add adequate triggers. IAS may quickly take advantage of the niches created by climatic disruption and its ecological effects on the environment (Masters and Norgrove 2010).

There are three ways in which climate change might result in the emergence of new introduced species. First, organisms that are reportedly unable to sustain and colonise a region due to climatic restrictions may become best equipped to do so. For example, a species which has not yet established itself in Antarctica due to the region's extreme temperatures may do so in the future as a result of warming (Lee and Chown 2007). IAS will have possibilities to colonise new glacier areas at the latitudes as a response of melting of glaciers brought on by rising temperatures. From the late 1990s, vegetation and small flowering plants have also been observed populating Antarctica. Second, new species that could also adapt to the environment may have a good opportunity of surpassing biotic development restrictions and establishing long-lasting populations as a result of climate change. Competitive resilience from indigenous species may decrease since it is anticipated that climate change will cause native organisms to become less adapted to their current environments (Byers 2002). Third, if climate variability improves the competitiveness or pace of growth of pre-existing non-native lifeforms, they may turn invasive. It will be essential to organise reactions at broad spatial scales, conduct latest research, and conduct more thorough monitoring due to the possibility that fluctuations in temperature, rainfall, and sea level will have an effect on the treatment of exotic plants. In the next section, a detailed insight has been given on the impacts of invasive species on different ecological and socio-economic dimensions.

15.4 Impacts of Invasive Species

Continual and rising redistribution of species by humans for the support of farming, forest habitat, marine, recreation, and horticulture, as well as the unintentional spread of species, result in invasive alien species (Montagnani et al. 2022). They include pests, weeds used in agriculture, and disease-causing microbes. It is commonly known that these species damage natural resources, endanger ecosystem integrity, lower economic output, and create a human health problem. Because of anthropogenic perturbations, global climate change, biogeochemical cycles, and enhanced dispersal, ecosystems are increasingly endangered by alien species invasions and the severity and geographic breadth of the problem grow as the global trade and travel are expanding (Le Maitre et al. 2004; Fernández-Palacios et al. 2021). Impact of invasive species on different aspects is presented in the following sub-section (Table 15.1).

15.4.1 Impact on Environment and Ecosystem Services

The introduction of unwanted species, that have frequently multiplied to the point where they affect the functioning and organisation of ecosystems, is a significant catalyst for environmental alteration on a global level (Ogle et al. 2003; Meffin et al. 2010). Although species invasions impact almost all environments on this Planet, the range of the impact varies substantially between various habitats and locales (Foxcroft et al. 2010). Impact of non-native organisms' incursions on natural systems and public health is also possible (Charles et al. 2007). Ecosystem services can be divided into two categories: those that maintain, regulate, and supply basic human requirements, and those that advance human health (cultural assistance). While regulatory facilities manage the distribution of benefits and the management of waste, ailments and infestations, supporting facilities support the fundamental critical processes required to maintain all ecosystems. Ecological services improve people's wellness and their standard of living, whereas supply services provide things for people to use. Some of the major impacts of plant invasion on key ecological components are briefly described below.

15.4.1.1 Biodiversity

Similar to habitat loss, invasive alien plants constitute a serious danger to biodiversity. They cause a decline in biological variety that can possibly could cause the destruction of some creatures. Also, they contributed a 42% lessening of the proportion of threatened and endangered lifeforms in the US (Wilcove et al. 1998). In the latest years, the impact of hazardous alien lifeforms on biodiversity

Table 15.1 List of a few alien plant species and their impact different ecological components

S. N	Species	Impact	Reference
1	*Acacia mangium*	Contaminate the water condition in Brazil to have a negative economic impact.	de Souza Machado et al. (2018)
2	*Acacia mangium*	Changed rural water quality, which in turn altered rural livelihood	Pathak et al. (2021)
3	*Ageratum houstonianum*	Have severe detrimental effects on agriculture product	Shrestha et al. (2018)
4	*Ambrosia artemisiifolia*	Allergic reaction in human caused by this invasive species	Chen et al. (2018)
5	*Anthemis cotula*	Change the physicochemical properties of soil by altering soil pH, porosity, electrical conductivity, and water saturation level in Himalayan region	Dar et al. (2023)
6	*Arundo donax*	Modifying fire regimes and indigenous population effects on ecosystem processes	Plaza et al. (2018)
7	*Eichhornia crassipes*	Reduce the amount of paddy crop production	Kariyawasam et al. (2021)
8	*Eucalyptus tereticornis*	Physicochemical properties of the soil will be altered	Qu et al. (2021)
9	*Euphorbia esula*	Have an effect on the grassland's soil health and cause destruction of the environment.	Gibbons et al. (2017)
10	*Fallopia japonica*	Produce the secondary metabolite that interfere with the food web in the ecosystem	Abgrall (2018)
11	*Gutenbergia cordifolia*	Influence the rumen chemistry and bacterial population of African cattle	Ngondya et al. (2019)
12	*Impatiens glandulifera*	Higher soil erosion in the region that was invaded	Greenwood et al. (2018)
13	*Lantana camara*	Physicochemical properties of the soil increase in the Himalayan region due to this invasive species	Kumar et al. (2021)
14	*Opuntia stricta*	Cause the deterioration of animal health and forage in the African continent	Shackleton et al. (2017)
15	*Parthenium hysterophorus*	Aid in the spread of malaria by luring the parasites as the host	Stone et al. (2018)
16	*Prosopis juliflora*	Decrease the exchangeable form of calcium, sodium, and magnesium of soil	Shiferaw et al. (2021)
17	*Rhododendron ponticum*	Cardiac issue brought on by tainted honey that included poisons generated and transmitted by this invasive species	Pyšek and Richardson (2010)
18	*Rhus typhina*	*Tagetes erecta* growth would be affected by this invasive species	Qu et al. (2021)
19	*Senecio inaequidens*	An alkaloid called retrorsine has harmful effects on human health produced by this invasive species.	Eller and Chizzola (2016)
20	*Solidago* sp.	It causes human skin contact illness	Denisow-Pietrzyk et al. (2019)

in various ecosystems has varied. While in some locations, the influence is minor and has little bearing on biodiversity, in others the effect is considerable and there have been substantial changes in the natural biodiversity. According to United Nations Environment Programme (UNEP 2005), the Millennium Assessment identified invasive alien species as among the primary causes of decrease of species diversity during the last 50–100 years.

15.4.1.2 Soil

Plant invasion can change soil assemblages and processes by interfering with mineralisation, energy flow, temperature, moisture, soil enzymatic activity regarding the make-up of the microbial population in the soil (Sinsabaugh et al. 2003; McKinley 2019). Because soil qualities take a while to develop, invading plants can create changes in soil processes that last long after the invader has been exterminated.

15.4.1.3 Fire Regimes

According to several experts, among the planet's most important species for changing systems is the fire system-altering intruder (D'Antonio and Vitousek 1992; Vitousek 1990). In western North America, the alien yearly herb *Bromus tectorum* is an example of a pervasive immigrant that has dramatically changed fire regimes as well as other environmental traits. Due to its meddling in this vast landscape, there are now so many fires that the original scrub steppes cannot recover (Whisenant 1990). The animals that depend on this type of habitat for food and shelter are negatively impacted as a result.

15.4.1.4 Impact on Aquatic Ecosystems

Due to their new physiological traits (large biomass, deep roots, and high evaporation capacity), IAPS can enter water bodies and block water flow, rendering the water unusable for irrigation and drinking (van Wilgen et al. 1998; Pejchar and Mooney 2009). Alien plant species have an important role to make floods more frequent by making smaller stream channels and altering the characteristics of the soil (lower capacity to store water and accelerated surface runoff). This subsequently adversely affects indigenous plant communities along riparian zones in addition to public health.

15.4.2 Impacts on Human Health

Human health is intricately tied to biodiversity and its changes in both positive and negative ways (Daszak et al. 2000; Young et al. 2017; Aerts et al. 2018; Stone et al. 2018). Among their positive effects, alien species implementations in vector-borne prevention and ethnomedicinal usage are reported (Rai 2018; Rai and Lalramnghinglova 2011; Tourapi and Tsioutis 2022). For example, *Lantana camara* is used to make a herbal insect disinfectant (Mng'ong'o et al. 2011). Variety of exotic ornamental plants, like giant thistle (*Heracleum mantegazzianum*) and goldenrod (*Laburnum anagyroides*), can be harmful to people's health.

15.4.3 Economic Impact of Invasive Species

Invasive plants have drawn a lot of attention because they can damage biodiversity (Daehler and Strong 1994; Wilcove et al. 1998; Hughes 2017; Reid et al. 2019), change the environment and ecological services, have a large impact on the human health, and also alter the soil enzymatic activity (Kourtev et al. 2002; Allison et al. 2006; Chapuis-Lardy et al. 2006; Li et al. 2006; Raizada et al. 2008). Due to all these damages, they have a high economic cost impact (Pimentel 2002; Malhi et al. 2021). For the period 1960–2020, invasive species has charged the Economy of India approximately US$ 127.3 billion to 182.6 billion and these expenditures have risen over period (Bang et al. 2022).

15.5 Management of Invasive Species

A significant concern facing society in the twenty-first century is how to control the expanding negative effects of unwanted exotic lifeforms on the eco-sociological surrounding. Both the Convention on Biological Diversity (CBD 2010) and the Sustainable Development Goals (UN 2015) focus on this problem and call on signatories to take steps to halt IAS migration, significantly minimise its consequences, and manage or remove priority species. Many tasks are involved in management at various phase of penetration processes (Wilson et al. 2017). The seamless integration of IAPS management through all aspects of managing woods and protected regions is a top goal, according to the IUCN's Vth World Parks Congress (2003). As during 2012 IUCN World Conservation Congress, and the 2014 IUCN World Parks Congress, this issue was discussed in relation to conservation areas. Due to the IAPS's ecosystem services, both positive and negative, it needs to be distinctly characterised in order to assist decision-makers and users, to particularly explain their expense (Zengeya 2017; Everard et al. 2018).

"*Any lethal or non-lethal activity intended at the elimination, population reduction, or quarantine of a population of an unwanted alien species*" is what management is considered as. In the USA, introduced species management is explained legally as "*eliminating, repressing, lowering, or managing exotic species populations, restricting the spread of introduced species from regions where they are visible, and taking measures, such as restoration of native habitats and species, in order to minimize the impact of introduced species and to inhibit further invasions*" (Robertson et al. 2020). Effective management can, therefore, stop a possible alien species from spreading or entering an entirely unfamiliar region, eradicate it if it is introduced, and lessen the effect of a current alien lifeforms by lowering its distribution and availability. Environmental restoration following species eradication or effect adaption without species involvement are other examples of management (Da Rosa et al. 2017; Robertson et al. 2020).

Conflict might arise from the variety of terminologies currently being used to define management. For instance, the term "containment" can be used to describe either carefully monitoring an IAS in captivity (Scott Schneider et al. 2004; Dobson et al. 2013) or limiting the community expansion in the environment (Grice et al. 2011). The term "eradication" is frequently used to refer to the full elimination of a species (Bomford and O'Brien 1995). According to certain theoretical analyses, eliminating noxious weeds is the most economical control strategy (Olson and Santanu 2002; Shay 2022). If fresh introductions are detected in on-time, eliminating introduced species is an aim that is reachable. When colonies surpass a certain size, elimination could not be possible. Nevertheless, this description excludes circumstances in which a community has been eliminated from a region but continuous management of latent life phases, such as spores, remains necessary (Panetta 2015), or the continuing entry of dispersed people from nearby regions (Robertson et al. 2020).

Despite the projected billions of dollars in damages brought on by invasive species worldwide, there are not any real, well-coordinated efforts being made to address the issue. Our present challenge related to invasive species has several fundamental causes, including inadequate policy, consistent study and management financing, an absence of social organisation to combat these incursions, and ongoing gaping holes in science (Simberloff et al. 2005; Miller and Schelhas 2009). The autecology of exotic lifeforms and their ecological impact is frequently unknown to natural resource planners and the general public. Although there are control methods for some organism, they can sometimes be insufficient to get rid of the foreign invading species.

In spite of the fact that prevention and control should be using a holistic strategy and all support available, including biological, mechanical, chemical, and cultural control techniques, that must be guided by the life process features of mainly exotic lifeforms and current best technology to ascertain that stop strategy or group of approaches that is the best efficient and cost-effective for the lifeforms. Introduced species control necessitates a comprehensive method including sustainable management pattern conducted through the collaborations all over terrains for the control of unwanted species and regeneration of affected bars and ecosystem functions, but there is an absence of general understanding and ongoing extensive sales and revegetation of weed species in several regions around the globe (Miller and Schelhas 2009). Different approaches for the management of invasive species have been presented in the following sub-sections:

15.5.1 Mechanical Methods

Weeds present in the grasslands are controlled mechanically using a variety of methods, such as hand-pulling, hoeing, tilling, mowing, grubbing, chaining, and bulldozing. Small vegetation in soft soils with shallow roots is especially suitable for manual removal. As a result, hand eradication costs money and takes time, yet it may be an important part of alien plant management. When there are only just few species left, they are frequently used in a programme for follow-up maintenance (Sheley et al. 1998). When intruders are first discovered, hand plucking or grubbing is frequently the easiest and most rapid method to stop them, which makes it a very effective instrument for workers. Nonetheless, roots that separate upon removal will occasionally produce new plants. Hand extraction can also disrupt the soil unnecessarily, creating an environment that is conducive to the reinvasion of invasive plants. Achievement with manual operation frequently requires repeated visits over the period of a few years. Moreover, mowing is frequently employed as a strategy to reduce hazardous range annuals as well as some woody tree (Tyser and Key 1988; Benefield et al. 1999; Entsminger et al. 2019; Klimešová and Martínková 2022).

It may stop the formation of seeds, cut back on glycogen stores, and benefit attractive perennial weeds. The plants' base branching arrangement and time are two factors that frequently affect how well a cut goes. Mowing can encourage a noxious weed issue if done incorrectly or by selecting the wrong organisms. Whenever manual control is necessary, mattocks are indeed the preferred tool. When managing invasive species of plants, a mattock cutting implemented with a hatchet with one side and a digging tool on the opposite side is recommended. The extensive root system should have been cut out for flora that easily resprout from that of the roots. Nevertheless, for lifeforms with root crowns, only the crown and any supported vine nodules need to be removed. Mechanical techniques can be used to manage bushes or trees, such as wood-cutting, bulldozing, chaining, roller chopping, shredding, and power grubbing (Cross and Wiedemann 1985). A somewhat flat landscape is needed for the majority of these manual procedures. Shredding is limited to pruning tiny shrubs, whereas fuel-wood cutting, bulldozing, and chaining are typically only successful on big trees or bushes that do not quickly resprout mostly from roots (Cross and Wiedemann 1985). Tree cuttings that are having potential of sprouting new growth can be removed with bulldozers (Dawkins and Esiobu 2016; Adams et al. 2019).

15.5.2 Biological Control Methods

The biocontrol of weed lifeforms has been considered a rather secure and efficient method (Ehler 1998; McFadyen 1998; Pemberton 2000). Using species-specific bugs, other invertebrates, and illnesses from the foreign plant's native region constitutes biological control. The majority of unwanted alien plants exhibit no weedy tendency in its natural habitats, and a variety of coevolved creatures regulate their growth and prevent them from producing vast quantities of seeds. As well, it is asserted that biocontrol is an economical method for regionally eradicating noxious plants (Hill and Greathead 2000; Nordblom et al. 2002). This one is especially true when weeds gain chemical tolerance. A biocontrol program's objective is to create enough environmental pressures to lessen the target weed's predominance in the vegetation cover rather than to completely eradicate it (Wilson 1999). Insect predators can accomplish this by drilling into the roots, branches, and stems of plants, defoliating leaves, consuming seeds, or drawing out plant fluids. Each of these consequences can lessen the plant's capacity to compete with the plants around it. The great majority of agents that have been released for weed control on land are intended to combat invasive weeds in rangelands (Julien 1992). The majority of efforts at biological management of rangeland vegetation have failed, despite several tries. Just 29% of the 23 pest species under biocontrol agents have shown full or considerable control rates over large areas (DeLoach 1991). Nonetheless, when it's effective, biological control may serve as a long-lasting and autonomous alternative for management (Blossey et al. 1994). Control agents that are biological in nature do have limitations, though. For example, weevils have unpredictable and erratic impacts on vegetation system (Giga and Mazarura 1991; Derera 2000).

15.5.3 Chemical Methods

Herbicides can be used to destroy saplings after fire or falling in order to avoid sprouting from cut stump. Pesticides, for instance, can attack grasses or life forms with broad leaves while sparing other vegetation. Dicamba, triclopyr, Picloram, 2,4-D, and clopyralid are some of these substances. The scheduling of herbicide sprays can affect the treatment's efficacy. While using a herbicide makes most plants and bushes simpler to control. April is the best time to apply it for *Euphorbia esula* regulation (Lym and Messersmith 1994). Based upon that herbicide, duration could also change. Although hazardous range weeds are effectively controlled by herbicides, they rarely offer lengthy weed management when used exclusively (Bussan et al. 1999). A hazardous weed could be substituted by some other equally unwanted species that is resistant to the chemical treatment if there is not a healthy plant population made up of favourable species. Moreover, persistent application of a particular herbicide may lead to resistance development inside the intended weed species. With repeated application of a particular herbicide, changing demographics may also lessen species diversity and result in nutritional changes that lessen the range's overall vigour. The use of chemicals, however, raises valid worries about possible environmental effects. Even if more recent herbicides are now more targeted, less toxic, and have shorter retention durations, there are still worries about their negative effects on the ecosystem. Each of these issues can limit the application of chemical control on a broad scale: regulation frequently governs the use of herbicides, and the successful and safe utilisation of herbicides necessitates a significant amount of training. Chemical weed management is frequently more economical and much more certain than natural weed management at the grazing level, although it can still be costlier and have unfavourable side effects. As a result, it is must to control the use of herbicides to lessen weeds (Chikowo et al. 2009).

15.5.4 Cultural Control Methods

Fire, browsing, or revegetation operations are the most frequent cultural management methods used in rangelands. In order to control invasive species and improve suitable vegetation, each of these tactics calls for modifying disturbance patterns. Cultural measures also include the execution by owners and practitioner of local quarantine laws as well as other legal directions. In grasslands, effective management of grazing may reduce noxious weed distribution. Olson (1999) outlines three grazing management techniques for weed species, viz. (1) medium grazing stages to minimise the physiological effect on indigenous vegetation and to decrease soil disruption, (2) intense feeding has equal effects on all fodder species, including weeds, in order to negate cattle's natural nutritional preferences, and (3) interspecies grazing, which allocates the effect of cattle grazing quite evenly among both wanted and unwanted species. Multi-species grazing makes use of the natural grazing inclinations of many cattle types (Walker and Steffen 1997). In every scenario, it is crucial to choose the best grazer for such circumstances at hand. Rangeland ecosystems have benefited greatly from the development and ongoing operation of fire. Like any disruption, fire's regularity, severity, annual timing, and connections with some other perturbations all have an impact on how it affects ecosystems and how invasive plants are managed. Prescribed fire generally encourages perennials to resprout mostly from base and is particularly effective in controlling late-season herbaceous plants. The most long-term, sensible solution to limit or prevent plant incursions while supplying plants with increased fodder value and improved wildlife habitat is revegetation or re-establishment of preferred and competing species of plants. In order to restore the site's production capabilities when grasslands degradation is extreme and there are few acceptable species present, revegetation may be required. Revegetation is usually very expensive.

15.5.5 Future Prospects for Management

Good management must be cost-efficient, specific, and maximise the use of scarce resources. This necessitates that it be included into a broader context of improving operations, such as public schooling, risk perception, identification, tracking, and risk evaluation, emergency preparedness, economic evaluation, and risk assessment, which together assist and educate active management. There are also ways to make use of the intruder stands which already exist such that regular harvesting can place a restraint on its growth. The choice of alternatives is used which will be relying on what is most advantageous in a particular situation. Wakie et al. (2016), for instance, have demonstrated in an economic feasibility that is particular to Ethiopia that it might be economically feasible to convert *Prosopis juliflora* infested regions into irrigated cotton plantations there. In addition to this, producing charcoal from *P. juliflora* stands that are currently present, can also be a profitable approach. Now, invasion biologists understand that not every IAPS represent environmental concerns (Young and Larson 2011). It is widely accepted that about 99% of a chosen alien species are grown for agricultural product on a worldwide scale (Pejchar and Mooney 2009). Perhaps some invasive species (*L. camara* and *Ageratum conyzoides*) get some ethnomedicinal uses in healthcare system, according to some reports (Rai and Lalramnghinglova 2011).

Both IAPS and hyperaccumulators share the stress resistance towards contaminants, resilience to pathogens/herbivores, and allelopathy (Rai 2018; Prabakaran et al. 2019). Consequently, the majority of significant risk of IAPS could be utilised as tools for environmental remediation. Many IAPS, including *L. camara, Pistia stratiotes,* and *Eichhornia crassipes* (water hyacinth), are productive ecosystems for bioremediation (Prabakaran et al. 2019; Rai and Kim 2020). It is fascinating to consider that these IAPS are included among the most popular damaging 100 alien species worldwide (Lowe et al. 2000). Consequently, by converting nuisance trash into a product that reduces pollution, our strategy should concentrate on the efficient and environmentally feasible possibilities of IAPS. For the bioremediation of a health-damaging heavy metal cadmium (Cd), three IAPS (*Chromolaena odorata, Praxelis clematidea,* and *Bidens pilosa*) are also known as hyperaccumulators (Wei et al. 2018). It has been shown that the IAPS, *Pistia stratiotes* may gather silver (Ag) nanoparticles from the surroundings (Hanks et al. 2015). In addition, preparation of biochar from the massive biomass of invasive species is considered as an effective technique for managing the plant invasion and improving the soil quality (Ghosh and Maity 2021). The effective implementation of introduced species management strategies depends on public involvement. Also, it is necessary to meet statutory compliance requirements in addition to an ethical standpoint (Boudjelas 2009). A global hotspot has to be shielded from extra-terrestrial life. In order to effectively control introduced species, the tactics used by the nations sharing borders must be unified. Biological intrusion is a trans-boundary concern. There must be an integration of methods because no one strategy can stop the proliferation of invasive species. The appropriate methods vary depending on the kind of invasive species. Protecting the world's hotspots against alien creatures is an urgent need of the hour.

15.6 Conclusions and Future Prospects

Although invasion is not an entirely new issue, the risk posed by unwanted species that are invasive has grown significantly along with the accelerated pace of globalisation. Extreme climate events (such as a severe temperature, storms, flooding, or droughts) may encourage biological infestations, but anthropogenic perturbations (planned invasion) are a key factor in the process. It represents one of the utmost significant effects that people have ever had on the biosphere of the planet. The pace at which biotic invasions are changing the environmental make-up of the earth's natural populations is astonishing. We are at threat of depleting and homogenising the very ecological systems that we depend on to survive our agricultural production, forest products, fishing, and other assets as well as to provide us with priceless natural facilities if we do not put effective tactics in place to stop the most harmful effects of invaders. They seriously endanger the richness of local species, which could potentially result in the loss of species that are threatened or rare. Significant financial and environmental loss are the overall effects. Realising that generating strategic decisions and using scientific decision-making require a deeper knowledge of the biological traits of exotic species and the environmental principles that underlie the invasion processes.

Although there are techniques for controlling invasive species, management plans that involve integrated management programmes will be the only way to achieve satisfactory long-term containment. This comprises using many control methods simultaneously or in succession to handle the invasive weed while increasing the desired vegetation. In the last 25 years, a lot has been accomplished in gathering data on the dangers posed by invasive species. The present chapter shows that there is a dearth of scientific and financial study on how invasive species and climate change interact, and the effects this has on ecological systems and the advantages they provide. The effects of invasive species on ecosystem services, economics (livelihood), and public health should, thus, be addressed in future studies using a variety of environmental stresses. The IAPS management's strategic path must also take societal approval and financial factors into account. It is quite expensive to manage the IAPS through their removal. Cost-benefit evaluation is crucial in additional research since IAPS influence on ecosystems is very diverse in relation to their social and economic factors. To generate a favourable atmosphere in which study on species incursion can be encouraged and sponsored, public knowledge about ecological change and deterioration, prevalent issues for the establishment of environmentally friendly land use systems, and knowledge about the impact of introduced species on the system must be merged.

Acknowledgements We are thankful to Joint-CSIR University Grant Commission (UGC-Delhi), DST-SERB, and Institute of Eminence for providing Fellowship, Support and facilities.

References

Abbasi SA, Nipaney PC (1986) Infestation by aquatic weeds of the fern genus Salvinia: its status and control. Environ Conserv 3:235–241

Abgrall R (2018) A general framework to construct schemes satisfying additional conservation relations: application to entropy conservative and entropy dissipative schemes. J Comput Phys 372:640–666

Adams MB, Sanderson T, Sena K, Barton CD, Agouridis CT, Angel PN, Zipper CE (2019) Managing invasive exotic plant species on legacy mine lands

Aerts R, Honnay O, Van Nieuwenhuyse A (2018) Biodiversity and human health: mechanisms and evidence of the positive health effects of diversity in nature and green spaces. Br Med Bull 1:5–22

Allison SD, Nielsen C, Hughes RF (2006) Elevated enzyme activities in soils under the invasive nitrogen-fixing tree Falcataria moluccana. Soil Biol Biochem 7:1537–1544

Bang A, Cuthbert RN, Haubrock PJ, Fernandez RD, Moodley D, Diagne C, Turbelin AJ, Renault D, Dalu T, Courchamp F (2022) Massive economic costs of biological invasions despite widespread knowledge gaps: a dual setback for India. Biol Invasions 7:2017–2039

Bartz R, Kowarik I (2019) Assessing the environmental impacts of invasive alien plants: a review of assessment approaches. NeoBiota 43:69–99

Benefield C, DiTomaso J, Kyser G, Orloff S, Churches K, Marcum D, Nader G (1999) Success of mowing to control yellow starthistle depends on timing and plant's branching form. Calif Agric 2:17–21

Bertness MD (1984) Habitat and community modification by an introduced herbivorous snail. Ecology 2:370–381

Blossey B, Schroeder D, Hight SD, Maleck RA (1994) Host specificity and environmental impact of two leaf beetles (Galerucella calmariensis and G. pusilla) for biological control of purple loosestrife (Lythrum salicaria). Weed Sci 42:1134–1140

Bomford M, O'Brien P (1995) Eradication or control for vertebrate pests. Wildl Soc Bull:249–255

Booth BD, Murphy SD, Swanton CJ (2003) Weed ecology in natural and agricultural systems. CABI, Wallingford

Boudjelas S (2009) Public participation in invasive species management. Invasive species management: A handbook of principles and techniques

Bussan AJ, Dyer WE, Sheley RL, Petroff JK (1999) Herbicides and rangeland. In: Biology and Management of Noxious Rangeland Weeds. Oregon State University, Corvallis, OR, pp 116–132

Byers JE (2002) Impact of non-indigenous species on natives enhanced by anthropogenic alteration of selection regimes. Oikos 3:449–458

Callaway RM, Aschehoug ET (2000) Invasive plants versus their new and old neighbors: a mechanism for exotic invasion. Science 5491:521–523

Carlton JT, Geller JB (1993) Ecological roulette: the global transport of nonindigenous marine organisms. Science 5117:78–82

CBD (2010) Strategic plan for biodiversity 2011–2020 and the Aichi biodiversity targets. (Decision X/2). Secretariat of the convention on biological diversity (CBD), Canada. https://www.cbd.int/decision/cop/?id=12268. Accessed 31 May 2018

Chapuis-Lardy L, Vanderhoeven SONIA, Dassonville N, Koutika LS, Meerts P (2006) Effect of the exotic invasive plant Solidago gigantea on soil phosphorus status. Biol Fertil Soils 6:481–489

Charles H, Dukes JS, Nentwig W (2007) Biological invasions: impacts of invasive species on ecosystem services: Germany. Springer-Verlag, Berlin, Heidelberg, pp 217–237

Chen KW, Marusciac L, Tamas PT, Valenta R, Panaitescu C (2018) Ragweed pollen allergy: burden, characteristics, and management of an imported allergen source in Europe. Int Arch Allergy Immunol 176(3–4):163–180

Chew MK (2006) Ending with Elton preludes to invasion biology (Doctoral dissertation, Arizona State University)

Chikowo R, Faloya V, Petit S, Munier-Jolain NM (2009) Integrated Weed Management systems allow reduced reliance on herbicides and long-term weed control. Agric Ecosyst Environ 132: 237–242

Cross BT, Wiedemann HT (1985) Grubbing for control of blackbrush acacia (Acacia rigidula) invading rootplowed rangeland. Weed Sci 33:263–266

Cuddington K, Beisner BE (2005) Kuhnian paradigms lost: embracing the pluralism of ecological theory. Ecological paradigms lost: routes of theory change 419

D'Antonio CM, Vitousek PM (1992) Biological invasions by exotic grasses, the grass/fire cycle and global change. Ann Rev Ecol Syst 23:63–87

Da Rosa CA, de Almeida Curi NH, Puertas F, Passamani M (2017) Alien terrestrial mammals in Brazil: current status and management. Biol Invasions 19:2101–2123

Daehler CC, Strong DR (1994) Variable reproductive output among clones of Spartina alterniflora (Poaceae) invading San Francisco Bay, California: the influence of herbivory, pollination, and establishment site. Am J Bot 3:307–313

Dar MA, Ahmad M, Singh R, Kohli RK, Singh HP, Batish DR (2023) Invasive plants alter soil properties and nutrient dynamics: a case study of Anthemis cotula invasion in Kashmir Himalaya. Catena 226:107069

Daszak P, Cunningham AA, Hyatt AD (2000) Emerging infectious diseases of wildlife—threats to biodiversity and human health. Science 5452:443–449

Dawkins K, Esiobu N (2016) Emerging insights on Brazilian pepper tree (Schinus terebinthifolius) invasion: the potential role of soil microorganisms. Front Plant Sci 7:712

de Souza Machado AA, Kloas W, Zarfl C, Hempel S, Rillig MC (2018) Microplastics as an emerging threat to terrestrial ecosystems. Glob Chang Biol 4:1405–1416

DeLoach CJ (1991) Past successes and current prospects in biological control of weeds in the United States and Canada. Nat Areas J 3:129–142

Denisow-Pietrzyk M, Pietrzyk Ł, Denisow B (2019) Asteraceae species as potential environmental factors of allergy. Environ Sci Pollut Res 26:6290–6300

Derera NF (2000) Condiment Paprika: Breeding, Harvesting & Commercialisation: a Report for the Rural Industries Research and Development Corporation RIRDC

Dobson A, Barker K, Taylor SL (2013) Biosecurity: the socio-politics of invasive species and infectious diseases. Routledge, New York

Drake JA, Mooney HA, di Castri F, Groves RH, Kruger FJ, Rejmánek M, Williamson M (1989) Biological invasions: a global perspective. John Wiley & Sons, Chichester

Dukes JS, Mooney HA (1999) Does global change increase the success of biological invaders? Trends Ecol Evol 14(4):135–139

Ehler L (1998) Conservation biological control: past, present, and future. In Conservation biological control. Academic Press, San Diego, CA, pp 1–8

El-Ghareeb RM (1991) Suppression of annuals by Tribulus terrestris in an abandoned field in the sandy desert of Kuwait. J Veg Sci 2:147–154

Eller A, Chizzola R (2016) Seasonal variability in pyrrolizidine alkaloids in Senecio inaequidens from the Val Venosta (northern Italy). Plant Biosyst 6:1306–1312

Elton CS (1958) The reasons for conservation. In: The ecology of invasions by animals and plants. Springer, Boston, MA, pp 143–153

Entsminger ED, Jones JC, Guyton JW, Leopold BD, Strickland BK (2019) Mowing effects on woody stem density and woody and herbaceous vegetation heights along Mississippi highway right-of-ways. J Fish Wildl Manag 10:19–37

Everard M, Sharma OP, Vishwakarma VK, Khandal D, Sahu YK, Bhatnagar R, Singh JK, Kumar R, Nawab A, Kumar A, Kumar V (2018) Assessing the feasibility of integrating ecosystem-based with engineered water resource governance and management for water security in semi-arid landscapes: a case study in the Banas catchment, Rajasthan, India. Sci Total Environ 612:1249–1265

Fernández-Palacios JM, Kreft H, Irl SD, Norder S, Ah-Peng C, Borges PA, Burns KC, de Nascimento L, Meyer JY, Montes E, Drake DR (2021) Scientists' warning–the outstanding biodiversity of islands is in peril. Global Ecol Conserv 31:01847

Foxcroft LC, Richardson DM, Rejmánek M, Pyšek P (2010) Alien plant invasions in tropical and sub-tropical savannas: patterns, processes and prospects. Biol Invasions 12:3913–3933

Funk JL, Parker IM, Matzek V, Flory SL, Aschehoug ET, D'Antonio CM, Dawson W, Thomson DM, Valliere J (2020) Keys to enhancing the value of invasion ecology research for management. Biol Invasions 22:2431–2445

Ganesh PS, Ramasamy EV, Gajalakshmi S, Abbasi SA (2005) Extraction of volatile fatty acids (VFAs) from water hyacinth using inexpensive contraptions, and the use of the VFAs as feed supplement in conventional biogas digesters with concomitant final disposal of water hyacinth as vermicompost. Biochem Eng J 1:17–23

Ghosh D, Maity SK (2021) Effect of invasive weed biochar amendment on soil enzymatic activity and respiration of coal mine spoil: a laboratory experiment study. Biochar 3(4):519–533

Gibbons SM, Lekberg Y, Mummey DL, Sangwan N, Ramsey PW, Gilbert JA (2017) Invasive plants rapidly reshape soil properties in a grassland ecosystem. MSystems 2:178–116

Giga DP, Mazarura UW (1991) Levels of resistance to the maize weevil, Sitophilus zeamais (Motsch.) in exotic, local open-pollinated and hybrid maize germplasm. Int J Trop Insect Sci 12:159–169

Graham MH, Dayton PK (2002) On the evolution of ecological ideas: paradigms and scientific progress. Ecology 6:1481–1489

Greenwood P, Baumann P, Pulley S, Kuhn NJ (2018) The invasive alien plant, Impatiens glandulifera (Himalayan Balsam), and increased soil erosion: causation or association? Case studies from a river system in Switzerland and the UK. J Soils Sediments 18:3463–3477

Grice AC, Clarkson JR, Calvert M (2011) Geographic differentiation of management objectives for invasive species: a case study of Hymenachne amplexicaulis in Australia. Environ Sci Pol 8: 986–997

Hanks NA, Caruso JA, Zhang P (2015) Assessing Pistia stratiotes for phytoremediation of silver nanoparticles and Ag (I) contaminated waters. J Environ Manag 164:41–45

Hierro JL, Maron JL, Callaway RM (2005) A biogeographical approach to plant invasions: the importance of studying exotics in their introduced and native range. J Ecol 1:5–15

Hill G, Greathead D (2000) Economic evaluation in classical biological control. The economics of biological invasions. Edward Elgar Publishing Inc., Northampton, MA, pp 208–226

Hughes AC (2017) Understanding the drivers of Southeast Asian biodiversity loss. Ecosphere 2017(8):01624

Humboldt AV, Baxter G, Bohn HG, Otté EC (1850) Views of nature, or, Contemplations on the sublime phenomena of creation: with scientific illustrations

IPBES (2019) Summary for policymakers of the global assessment report on biodiversity and ecosystem services of the Intergovernmental science-policy platform on biodiversity and ecosystem services. In: Diaz S, Settele J, Brondizio ES, Ngo TH, Gueze M, Agard J, Arneth A, Balvanera P, Brauman AK, Butchart SHM et al (eds) IPBES secretariat, Bonn, Germany. Accessed 27 Sep 2019

Jones BA, McDermott SM (2018) Health impacts of invasive species through an altered natural environment: assessing air pollution sinks as a causal pathway. Environ Resour Econ 1:23–43

Julien MH (1992) Biological control of weeds: a world catalogue of agents and their target weeds

Kanchan SD (1979) Allelopathic effects of Parthenium hysterophorus L. Plant Soil 1:37–47

Kariyawasam JC, Jayarajah U, Riza R, Abeysuriya V, Seneviratne SL (2021) Gastrointestinal manifestations in COVID-19. Trans R Soc Trop Med Hyg 12:1362–1388

Klimešová J, Martínková J (2022) Clonal growth, resprouting, and vegetative propagation of weeds. Persist Strat Weeds 4:200–218

Kourtev PS, Ehrenfeld JG, Häggblom M (2002) Exotic plant species alter the microbial community structure and function in the soil. Ecology 11:3152–3166

Kowarik I, Pyšek P (2012) The first steps towards unifying concepts in invasion ecology were made one hundred years ago: revisiting the work of the S wiss botanist A lbert T hellung. Divers Distrib 12:1243–1252

Kumar M, Kumar S, Verma AK, Joshi RK, Garkoti SC (2021) Invasion of Lantana camara and Ageratina adenophora alters the soil physico-chemical characteristics and microbial biomass of chir pine forests in the central Himalaya, India. Catena 207:105–624

Le Maitre DC, Richardson DM, Chapman RA (2004) Alien plant invasions in South Africa: driving forces and the human dimension: working for water. S Afr J Sci 100:103–112

Lee JE, Chown SL (2007) Mytilus on the move: transport of an invasive bivalve to the Antarctic. Mar Ecol Progress Ser 339:307–310

Li J, Zhang G, Qi S, Li X, Peng X (2006) Concentrations, enantiomeric compositions, and sources of HCH, DDT and chlordane in soils from the Pearl River Delta, South China. Sci Total Environ 1:215–224

Lowe S, Browne M, Boudjelas S, De Poorter M (2000) 100 of the world's worst invasive alien species: a selection from the global invasive species database. Invasive Species Specialist Group, Auckland

Lym RG, Messersmith CG (1994) Leafy spurge (Euphorbia esula) control, forage production, and economic return with fall-applied herbicides. Weed Technol 4:824–829

Mack RN, Simberloff D, Mark Lonsdale W, Evans H, Clout M, Bazzaz FA (2000) Biotic invasions: causes, epidemiology, global consequences, and control. Ecol Appl 3:689–710

Malhi GS, Kaur M, Kaushik P (2021) Impact of climate change on agriculture and its mitigation strategies: a review. Sustainability 3:1318
Martin P, Choy DL, LeGal E (2019) Discussion paper: effective citizen action on invasive species
Masters G, Norgrove L (2010) Climate change and invasive alien species, vol 1. CABI Working Paper, UK, p 30
McFadyen REC (1998) Biological control of weeds. Annu Rev Entomol 43:369–393
McGeoch MA, Butchart SH, Spear D, Marais E, Kleynhans EJ, Symes A, Chanson J, Hoffmann M (2010) Global indicators of biological invasion: species numbers, biodiversity impact and policy responses. Divers Distrib 16:95–108
McKinley VL (2019) Effects of land use and restoration on soil microbial communities. Understanding Terrestrial Microbial Communities. 173–242
Meffin R, Miller AL, Hulme PE, Duncan RP (2010) Experimental introduction of the alien. Divers Distrib 16:804–815
Miller JH, Schelhas J (2009) Adaptive collaborative restoration: a key concept in invasive plant management. Invasive plants and forest ecosystems/edited by Ravinder Kumar Kohli [et al.]
Mng'ong'o FC, Sambali JJ, Sabas E, Rubanga J, Magoma J, Ntamatungiro AJ, Turner EL, Nyogea D, Ensink JH, Moore SJ (2011) Repellent plants provide affordable natural screening to prevent mosquito house entry in tropical rural settings—results from a pilot efficacy study. PLoS One 10:25927
Montagnani C, Gentili R, Brundu G, Caronni S, Citterio S (2022) Accidental introduction and spread of top invasive alien plants in the European Union through human-mediated agricultural pathways: what should we expect. Agronomy 12:423
Ngondya IB, Treydte AC, Ndakidemi PA, Munishi LK (2019) Can Cynodon dactylon suppress the growth and development of the invasive weeds Tagetes minuta and Gutenbergia cordifolia. Plan Theory 12:576
Nordblom TL, Smyth MJ, Swirepik A, Sheppard AW, Briese DT (2002) Cooperative Research Centre for Australian Weed Management (Weeds CRC): Spatial economics of biological control: investing in new releases of insects for earlier limitation of Paterson's curse in Australia. Agric Econ 3:403–424
Ogle SM, Reiners WA, Gerow KG (2003) Impacts of exotic annual brome grasses (Bromus spp.) on ecosystem properties of northern mixed grass prairie. Am Midland Natur 149:46–58
Olson JS (1999) Historical dictionary of the 1960s. Greenwood Publishing Group, New York
Olson LJ, Roy S (2002) The economics of controlling a stochastic biological invasion. Am J Agric Econ 5:1311–1316
Osvald H (1948) Toxic exudates from the roots of Agropyron repens. J Ecol 38:192–193
Panetta FD (2015) Weed eradication feasibility: lessons of the 21st century. Weed Res 55:226–238
Pathak HN, Bhuju DR, Shrestha BB, Ranjitkar S (2021) Impacts of invasive alien plants on ecosystem services of Ramsar lake cluster in middle mountain Nepal. Global Ecol Conserv 27:e01597
Patnaik P (2017) Handbook of environmental analysis: chemical pollutants in air, water, soil, and solid wastes. CRC Press, Boca Raton, FL
Pejchar L, Mooney HA (2009) Invasive species, ecosystem services and human Well-being. Trends Ecol Evol 9:497–504
Pemberton RW (2000) Predictable risk to native plants in weed biological control. Oecologia 125: 489–494
Pianka ER, Horn HS (2005) Ecology's legacy from Robert MacArthur. Ecological paradigms lost: routes of theory change 213
Pimentel D (2002) Non-native invasive species of Arthropods and plant pathogens in the British Isles. Biological invasions. CRC Press, Boca Raton, FL, pp 151–158
Pimentel D, Zuniga R, Morrison D (2005) Update on the environmental and economic costs associated with alien-invasive species in the United States. Ecol Econ 3:273–288

Plaza C, Zaccone C, Sawicka K, Méndez AM, Tarquis A, Gascó G, Heuvelink GB, Schuur EA, Maestre FT (2018) Soil resources and element stocks in drylands to face global issues. Sci Rep 8:13788
Prabakaran K, Li J, Anandkumar A, Leng Z, Zou CB, Du D (2019) Managing environmental contamination through phytoremediation by invasive plants: a review. Ecol Eng 138:28–37
Pyšek P, Richardson DM (2010) Invasive species, environmental change and management, and health
Pyšek P, Richardson DM, Rejmánek M, Webster GL, Williamson M, Kirschner J (2004) Alien plants in checklists and floras: towards better communication between taxonomists and ecologists. Taxon 1:131–143
Qu T, Du X, Peng Y, Guo W, Zhao C, Losapio G (2021) Invasive species allelopathy decreases plant growth and soil microbial activity. PLoS One 16:e0246685
Rahel FJ, Olden JD (2008) Assessing the effects of climate change on aquatic invasive species. Conserv Biol 3:521–533
Rai PK (2018) Phytoremediation of emerging contaminants in wetlands. CRC Press
Rai PK, Kim KH (2020) Invasive alien plants and environmental remediation: a new paradigm for sustainable restoration ecology. Restor Ecol 28:3–7
Rai PK, Lalramnghinglova H (2011) Ethnomedicinal plants of India with special reference to an indo-Burma hotspot region: an overview. Ethnobot Res Appl 9:379–420
Raizada P, Raghubanshi AS, Singh JS (2008) Impact of invasive alien plant species on soil processes: a review. Proc Natl Acad Sci India Sect B–Biol Sci 78:288–298
Rajaka P, Afreen T, Singha H (2021) Assessing the impact of invasive alien plant species on environment, ecosystem services and human health. Plant Arch 21:2155–2163
Reichard SH, White PS (2003) Invasion biology: an emerging field of study. Ann Mo Bot Gard:64–66
Reid AJ, Carlson AK, Creed IF, Eliason EJ, Gell PA, Johnson PT, Kidd KA, MacCormack TJ, Olden JD, Ormerod SJ, Smol JP (2019) Emerging threats and persistent conservation challenges for freshwater biodiversity. Biol Rev 3:849–873
Rejmánek M, Richardson DM, Higgins SI, Pitcairn MJ, Grotkopp E (2005) Ecology of invasive plants: state of the art. Scope scientific committee on problems of the environment international council of scientific unions 63:104
Ridenour WM, Callaway RM (2001) The relative importance of allelopathy in interference: the effects of an invasive weed on a native bunchgrass. Oecologia 126:444–450
Robertson PA, Mill A, Novoa A, Jeschke JM, Essl F, Gallardo B, Geist J, Jarić I, Lambin X, Musseau C, Pergl J (2020) A proposed unified framework to describe the management of biological invasions. Biol Invasions 9:2633–2645
Runyon JB, Butler JL, Friggens MM, Meyer SE, Sing SE (2012) Invasive species and climate change. Climate change in grasslands, shrublands, and deserts of the interior American west: A review and needs assessment 97
Sax DF, Stachowicz JJ, Gaines SD (2005) Species invasions: insights into ecology, evolution and biogeography. Sinauer, Sunderland
Scott Schneider S, DeGrandi-Hoffman G, Smith DR (2004) The African honey bee: factors contributing to a successful biological invasion. Ann Rev Entomol 49:351–376
Shackleton RT, Witt AB, Aool W, Pratt CF (2017) Distribution of the invasive alien weed, Lantana camara, and its ecological and livelihood impacts in eastern Africa. Afr J Range Forage Sci. 1–11
Shay NJ (2022) Novel Pest management strategies in southern forage systems (Doctoral dissertation, University of Georgia)
Sheley RL, Jacobs JS, Carpinelli MF (1998) Distribution, biology, and management of diffuse knapweed (Centaurea diffusa) and spotted knapweed (Centaurea maculosa). Weed Technol 12: 353–362
Shiferaw K, Mengiste B, Gobena T, Dheresa M (2021) The effect of antenatal care on perinatal outcomes in Ethiopia: a systematic review and meta-analysis. PLoS One 16:0245003

Shrestha UB, Sharma KP, Devkota A, Siwakoti M, Shrestha BB (2018) Potential impact of climate change on the distribution of six invasive alien plants in Nepal. Ecol Indic 95:99–107
Simberloff D, Parker IM, Windle PN (2005) Introduced species policy, management, and future research needs. Front Ecol Environ 3:12–20
Sinsabaugh RL, Saiya-Cork K, Long T, Osgood MP, Neher DA, Zak DR, Norby RJ (2003) Soil microbial activity in a Liquidambar plantation unresponsive to CO2-driven increases in primary production. Appl Soil Ecol 24(3):263–271
Stone CM, Witt AB, Walsh GC, Foster WA, Murphy ST (2018) Would the control of invasive alien plants reduce malaria transmission? A review. Parasit Vectors 11:1–18
Thellung A (1911–1912) La flore adventice de Montpellier. Me'moires de la Socie'te´ Nationale des Sciences Naturelles et Mathemaiques de Cherbourg 38:57–728
Theoharides KA, Dukes JS (2007) Plant invasion across space and time: factors affecting nonindigenous species success during four stages of invasion. New Phytol 176:256–273
Thuiller W (2007) Climate change and the ecologist. Nature 448:550–552
Tourapi C, Tsioutis C (2022) Circular policy: a new approach to vector and vector-borne diseases' management in line with the global vector control response (2017–2030). Trop Med Infect Dis 7:125
Tyser RW, Key CH (1988) Spotted knapweed in natural area fescue grasslands: an ecological assessment. Northwest Sci 62:4
UNEP (2005) Implications of the findings of the Millenium Ecosystem Assessment for the future work of the convention-Addendum Summary for decision makers of the biodiversity synthesis report. UNEP/CBD/SBSTTA/11/7/Add 1 31 August 2005
UN (2015) Transforming our world: the 2030 agenda for sustainable development. In: Resolution adopted by the general assembly on 25 September 2015, A/RES/70/1.United Nations General Assembly
Van Wilgen BW, Biggs HC, Potgieter A (1998) Fire management and research in the Kruger National Park, with suggestions on the detection of thresholds of potential concern. Koedoe 41: 69–87
Vaughn SF, Berhow MA (1999) Allelochemicals isolated from tissues of the invasive weed garlic mustard (Alliaria petiolata). J Chem Ecol 11:2495–2504
Vitousek PM (1990) Biological invasions and ecosystem processes: towards an integration of population biology and ecosystem studies. In Ecosystem management. Springer, New York, NY, pp 183–191
Vitousek PM, D'Antonio CM, Loope LL, Westbrooks R (1996) Biological invasions as global environmental change
Vitousek PM, D'antonio CM, Loope LL, Rejmanek M, Westbrooks R (1997) Introduced species: a significant component of human-caused global change. N Z J Ecol:1–16
Wakie TT, Hoag D, Evangelista PH, Luizza M, Laituri M (2016) Is control through utilization a cost effective Prosopis juliflora management strategy. J Environ Manag 168:74–86
Walker B, Steffen W (1997) An overview of the implications of global change for natural and managed terrestrial ecosystems. Conserv Ecol 1(2)
Walter K (2011) Prosopis, an alien among the sacred trees of South India. University of Helsinki Tropical Forestry Reports. Consultado
Wang A, Melton AE, Soltis DE, Soltis PS (2022) Potential distributional shifts in North America of allelopathic invasive plant species under climate change models. Plant Diversity 44:11–19
Wei H, Huang M, Quan G, Zhang J, Liu Z, Ma R (2018) Turn bane into a boon: application of invasive plant species to remedy soil cadmium contamination. Chemosphere 210:1013–1020
Whisenant SG (1990) Changing fire frequencies on Idaho's snake river plains: ecological and management. In Proceedings—Symposium on Cheatgrass Invasion, Shrub Die-off and Other Aspects of Shrub Biology and Management: Las Vegas, NV. Intermountain Research Station, Forest Service, US Department of Agriculture 276: 4
Wilcove DS, Rothstein D, Dubow J, Phillips A, Losos E (1998) Quantifying threats to imperiled species in the United States. Bioscience 48:607–615

Williamson M (1993) Invaders, weeds and the risk from genetically manipulated organisms. Experientia 49:219–224

Williamson M, Fitter A (1996) The varying success of invaders. Ecology 6:1661–1666

Willis AJ, Blossey B (1999) Benign environments do not explain the increased vigour of non-indigenous plants: a cross-continental transplant experiment. Biocontrol Sci Tech 4:567–577

Wilson LM (1999) A prelude to biological control of weeds: studies on meadow hawkweed (Hieracium pratense Tausch.) compensation. University of Idaho, Moscow, ID

Wilson JR, Klein H, Manyama P, Zimmermann HG, Henderson L, Kaplan H, Richardson DM, Novoa A, Ivey P (2017) A proposed national strategic framework for the management of Cactaceae in South Africa. Bothalia-Afr Biodivers Conserv 47:1–12

World Parks Congress Vth (2003). http://www.iucn.org/themes/wcpa/wpc2003/pdfs/outputs/wpc/recommendationspdf. Accessed 17 Dec 2003

Young AM, Larson BM (2011) Clarifying debates in invasion biology: a survey of invasion biologists. Environ Res 7:893–898

Young HS, Parker IM, Gilbert GS, Guerra AS, Nunn CL (2017) Introduced species, disease ecology, and biodiversity–disease relationships. Trends Ecol Evol 32(1):41–54

Zengeya T (2017) Managing conflict-generating invasive species in South Africa: challenges and trade-offs. Bothalia-Afr Biodivers Conserv 47:1–11

Biochar: A Tool for Combatting Both Invasive Species and Climate Change

16

Leeladarshini Sujeeun and Sean C. Thomas

Abstract

Allelopathy, or chemically mediated interference between co-occurring species, is present in more than half of invasive plant species globally and plays an important role in invasive species dominance in native plant communities. Allelopathy commonly increases the competitive advantage of invasive plants and their ability to displace native species. In extreme cases, invasive plants can cause native species to go extinct and this effect is particularly pronounced in small island ecosystems or isolated and fragmented ecosystems. Extirpation of native species from local communities greatly reduces biodiversity and ecosystem stability and can potentially reduce system productivity and thus C sequestration. Invasive allelopathic plants can also have wide-ranging effects on plant communities and ecosystem processes such as herbivory, decomposition, and nutrient mineralization. Invasive plants are notoriously difficult to control, and management strategies can be expensive, labor-intensive, and often marginally effective. Biochar, or charcoal used as a soil amendment, is primarily known for its potential to enhance productivity and carbon sequestration, but it also has a high capacity to sorb toxic organic compounds, including allelochemicals. Biochar is a form of pyrogenic carbon (PyC) that shares properties with naturally occurring forms of PyC that are widespread in forest soils, particularly in systems with natural fire regimes. Sorption of allelopathic compounds by natural PyC can widely influence overall productivity and species composition in plant communities. Studies to date indicate that biochar can greatly reduce the allelopathic effects of invasive allelopathic plants, including *Psidium cattleianum*, *Acer platanoides*, and *Alliaria petiolata*. Biochar has also been shown to promote

L. Sujeeun (✉) · S. C. Thomas
Institute of Forestry and Conservation, University of Toronto, Toronto, ON, Canada
e-mail: lutchmee.sujeeun@mail.utoronto.ca; sc.thomas@utoronto.ca

S. Tripathi et al. (eds.), *Plant Invasions and Global Climate Change*,
https://doi.org/10.1007/978-981-99-5910-5_16

native tree growth in invaded plant communities and can also suppress the regeneration of invasive plants. In a seemingly hopeless battle against invasive species, biochar may be a critical tool for successfully combatting invasive allelopathic plants and climate change while simultaneously promoting native biodiversity and carbon sequestration. This chapter reviews the ecological impacts of invasive plants, the facilitation of plant invasions via allelopathy, and the potential of biochar to mitigate the effects of allelopathic invasive plants and climate change.

Keywords

Allelopathy · Biochar · Climate change · Ecosystem restoration · Invasive species · Native species diversity

16.1 Introduction

Allelochemicals or growth-inhibiting secondary metabolites are a critical weapon used by plants engaging in "chemical warfare" to increase their competitiveness (Rice 1984). Allelopathic species reduce the germination, growth, and metabolism of neighboring plants by releasing organic compounds into the environment through root exudation, foliar leaching, litter decomposition, and/or volatilization (Muller 1966; Inderjit and Keating 1999). The ability of exotic invasive plants to successfully establish outside their native range is often attributed to the lack of natural predators that allow them to fully exploit their potential for resource competition in the introduced habitat and displace native species (Callaway and Aschehoug 2000). However, recent research supports an alternative theory that many non-native species use "novel chemistry" via allelopathy to suppress native plants and spread aggressively in the new habitat (Callaway and Aschehoug 2000; Cappuccino and Arnason 2006). Allelopathy confers many invasive species a major advantage over neighboring plants that is thought to contribute to their dominance in their introduced habitat (Lodhi 1978; van Kleunen et al. 2015) and can increase their ability to alter native plant community composition and diversity. The "novel chemistry" hypothesis suggests that allelopathy may generally be less important in plant communities in which plant species have coevolved than in communities with both native and invasive species (Reigosa et al. 1999, 2002; Callaway and Hierro 2006). Native species may be severely affected if they are exposed to allelochemicals to which they have not evolved mechanisms for tolerance or resistance. In addition, soil microorganisms that would break down allelochemicals from an invasive plant in its native habitat may be absent from the new environment, which can increase the allelopathic potential of the invasive species (Sinkkonen 2006).

Invasive plant species have a high potential to expand within biodiversity hotspots and reduce plant diversity globally (Wan and Wang 2018). Allelopathy occurs throughout all major plant phylogenetic lineages and is estimated to be present in approximately half of invasive plants globally (Kalisz et al. 2021) (Table 16.1). Some of the most notorious invasive plants globally are potently allelopathic (Callaway and Aschehoug 2000), including *Imperata cylindrica*, *Lantana camara*, and *Chromolaena odorata* in the tropics, and *Acer platanoides*, *Rhamnus cathartica*, and *Alliaria petiolata* in North America. These species generally have strong inhibitory effects on native plant species (Hagan et al. 2013; Hu and Zhang 2013; Kato-Noguchi and Kurniadie 2021). In some cases, non-native allelopathic species can invade even undisturbed forested areas (Huenneke and Vitousek 1990).

Invasive plants commonly reduce local plant species diversity (Vilà et al. 2006; Gaertner et al. 2009; Hejda et al. 2009; Powell et al. 2011) and alter rates of nutrient cycling (Liao et al. 2008; Ehrenfeld 2010) and disturbance regimes (Brooks et al. 2004). Invasive plants, including allelopathic species, may also dramatically alter ecosystem productivity and impact other ecosystem services (Pejchar and Mooney 2009) (Table 16.2). In addition to these ecological effects, invasive allelopathic plants have large negative impacts on the economy through crop losses, reduced forage yields, toxicity to livestock, and reduced land values (Di Tomaso 2000) (Fig. 16.1). Further economic losses are associated with invasive plant control through herbicides and other measures; in the 1990s in the US, these costs were estimated at US $3 billion/year (Pimentel et al. 2000). The total negative economic impacts of non-native invasive plants to the US economy alone have more recently been estimated at US $120 billion/year (Duenas et al. 2018). Moreover, these costs do not take into account non-monetary losses such as reduced quality of life resulting from the lack of undisturbed native wildlands accessible to humans and the loss of unique cultural landscapes (Henderson et al. 2006).

Several management approaches have been widely used to control the spread of invasive plants, including chemical, biological, and mechanical control (Annighöfer et al. 2012; Seastedt 2015; Weidlich et al. 2020). However, invasive plant management can be costly, ineffective, and harmful to the environment and native species (Smith et al. 2006; Stricker et al. 2015). Chemical treatment is commonly expensive and can have negative effects on non-target species (Cronk and Fuller 1995). Widespread use of herbicides for control of invasive plants can also result in a "pesticide treadmill" effect in which efficacy declines due to natural selection for herbicide resistance, requiring higher dosages or changes in herbicide types (Foster and Magdoff 1998). Mechanical removal of invasive plants is generally labor-intensive and increases canopy openings that can promote the establishment of other invasive species; it can also increase the spread of new propagules of clonal invasive species (Webb et al. 2001; Nunez-Gonzalez et al. 2021). Biological control of invasive plants can be highly effective (Clewley et al. 2012), but in some cases, biocontrol agents can adversely impact native flora and the risks are substantial when the invasive species occur in areas dominated by closely related native species (Webb et al. 2001). In addition, allelochemical production can increase in response to mechanical control, herbicides, or biocontrol agents (Siemens et al. 2002; Thelen et al. 2005). Therefore, establishing management strategies that are effective against allelopathic invasive species in the long term remains a substantial challenge.

Table 16.1 List of twenty widespread non-native invasive plants for which allelopathy has been demonstrated or is strongly suspected

Species	Native range	Introduced range	Reference
Alliaria petiolata	Eurasia	North America	Cipollini (2016)
Fallopia japonica	East Asia	Europe; North America	Šoln et al. (2022)
Lantana camara	Central and South America	Tropical and subtropical areas	Kato-Noguchi and Kurniadie (2021)
Chromolaena odorata L.	North, Central, and South America	Asia, Oceania, and Africa	Karim (2017)
Psidium cattleianum	Brazil	Tropical and subtropical areas	Sujeeun and Thomas (2017)
Cymbopogon flexuosus	Southern India, Sri Lanka	Mexico, St. Lucia	Sujeeun and Thomas (2017)
Imperata cylindrica	Southeast Asia; East Africa	Southern USA; Mediterranean region; Northern Africa; Middle East; tropical and subtropical Asia; Australia; Pacific Islands	Kato-Noguchi (2022)
Ulex europaeus	British Isles and Western Europe	Washington, Oregon, California, Hawaii, and British Columbia	Pardo-Muras et al. (2018)
Acacia mearnsii	South-eastern Australia and Tasmania	North America, South America, Asia, Europe, Pacific and Indian Ocean islands, Africa, and New Zealand	Fatunbi et al. (2009)
Pueraria montana var. *lobata*	East Asia	Eastern U.S., Ukraine, Caucasus, central Asia, southern Africa, Hawaii, Hispaniola, and Panama	Rashid et al. (2010)
Schinus terebinthifolius	Central and eastern South America	North America, Africa, and Australasia	Morgan and Overholt (2005)
Arundo donax	Eastern Asia	Warm temperate, subtropical and tropical regions	Abu-Romman and Ammari (2015)
Leucaena leucocephala	Mexico	Africa, Asia and Oceania	Ahmed et al. (2008)
Sphagneticola trilobata	Mexico, Central America, and the Caribbean islands	Asia, Neotropics	Hernández-Aro et al. (2016)
Euphorbia esula	Eurasia	North America	Steenhagen and Zimdahl (1979)
Mikania micrantha	Central and South America	The Pacific region	Wu et al. (2009)
Mimosa pigra	Tropical America	Africa, South East Asia and Australia	Koodkaew et al. (2018)

(continued)

Table 16.1 (continued)

Species	Native range	Introduced range	Reference
Ligustrum robustum	Sri Lanka	Mascarene islands	Lavergne et al. (1999)
Melaleuca quinquenervia	Australia	USA, Puerto Rico and the Bahamas	DiStefano (1982)
Amaranthus retroflexus	Central and eastern North America	Temperate regions of the northern and southern hemispheres	Qasem (1995)

Table 16.2 Summary of meta-analyses and other integrated analyses examining effects of invasive species on community and ecosystem processes and services, and evidence for allelopathy in driving these effects

Process/ service	Summary of effect	Role of allelopathy	References
Native plant abundance	Reduction	Suppression of root elongation and shoot growth Reduction of ectomycorrhizal fungi and thus, reduction in nutrient acquisition	Ridenour and Callaway (2001)
Native plant species diversity	Reduction	Reduction of the competitive ability of native plants Increased competitive ability of invasive species	Chen et al. (2017); Koocheki et al. (2013)
Native animal abundance	Reduction	Reduction of native plant abundance	Nissanka et al. (2005)
Native animal diversity	Reduction	Reduction in plant diversity and available niches Inhibition of plant mycorrhizal associations	Nissanka et al. (2005)
Soil microbial abundance	Reduction	Reduction in growth of mycorrhizal fungi	Grove et al. (2012)
Soil microbial diversity	Reduction	Elevated soil N levels, causing some plants to suppress mycorrhizal associations	Broz et al. (2007)
Net primary productivity	Mixed	Reduced productivity through reduced photosynthetic rates in native species or increased productivity through higher growth rate of invasive species than native species	Liao et al. (2008)
Carbon sequestration	Mixed	Reduced C sequestration through suppression of native tree growth or increased C sequestration through higher growth rate of invasive species than native species	Liao et al. (2008)
Soil GHG flux	Mixed	Positive or negative effects on microbial organisms and respiration	Bezabih Beyene et al. (2022)

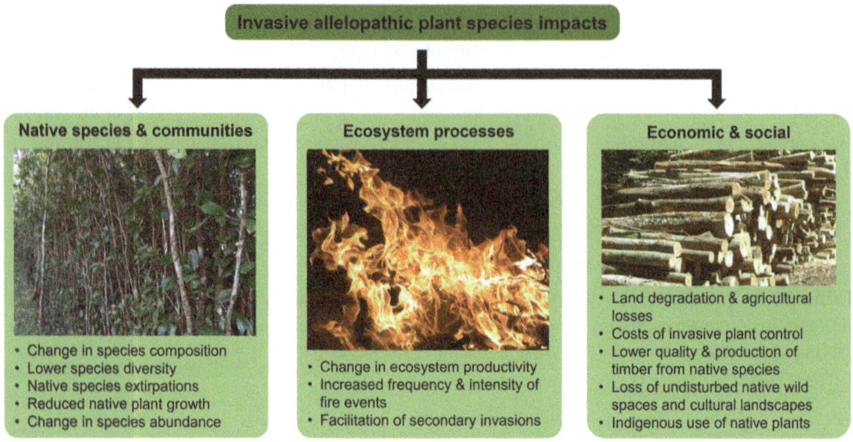

Fig. 16.1 Ecological, economic, and social impacts of invasive allelopathic plant species

A promising new mitigation strategy to control the spread of some invasive species and promote native plant growth is the addition of "biochar," which can be produced from a wide range of biomass sources at low cost and can be highly effective without the risks of biocontrol or herbicide use (Sujeeun and Thomas 2022). Biochar—or charcoal used as a soil amendment—is primarily known for its potential to enhance productivity and sequester carbon (Lehmann and Joseph 2015). However, due to its high versatility and physicochemical properties (Qambrani et al. 2017), biochar is increasingly being used for the mitigation of soil contaminants, such as heavy metals (Beesley et al. 2015) and toxic organic compounds (Hale et al. 2015), and has promise for mitigating the effects of allelochemicals from invasive plants (Gámiz et al. 2021; Bieser et al. 2022; Sujeeun and Thomas 2022). Biochar has been found to reduce the inhibitory effects of several allelopathic species, including highly invasive plants, through the sorption of allelopathic compounds in leaf extracts and crop residues (Mahall and Callaway 1992; Rogovska et al. 2012; Sujeeun and Thomas 2017).

The present chapter addresses the ecological impacts of allelopathic invasive plants, the contribution of allelopathy to invasiveness in species globally, and biochar as a tool for mitigating both allelopathy and climate change. We specifically address evidence that allelopathic invasive plants displace native species and overall plant diversity, their effects on ecosystem productivity, their impacts on fire regimes, and the issue of long-term effects mediated by secondary invasions. In addition, we review recent studies that examine biochar sorption of allelochemicals from invasive species, biochar effects on microbial breakdown of allelochemicals, and the potential for widespread use of biochar on invasive species to mitigate climate change through increased carbon sequestration. Finally, we examine the use of biochar for the control of allelopathic invasive species, highlighting our experimental work on the control of strawberry guava (*Psidium cattleianum*) in the small island nation of Mauritius.

16.2 Ecological Impacts of Allelopathic Invasive Plants

16.2.1 Native Species Suppression

Anthropogenic introductions (both intentional and accidental) of plants outside their native range have occurred for centuries and have favored the establishment of non-native invasive species globally (Pyšek et al. 2017). Exotic plant invasions and their impacts are worsening due to increased global trade that facilitates the movement of species across geographical barriers, and favorable conditions for the establishment and spread of non-native plants promoted by climate change and human disturbance (Henderson et al. 2006). Non-native plant species are estimated to be responsible for the population declines of 431 native species within the USA alone; over 90% of the species threatened by invaders are plants, which are affected by direct competition from invasive plants or indirect changes to the introduced habitat (Gurevitch and Padilla 2004). For example, alien plants can benefit from induced changes to soil microbial community, pH, nutrients, and secondary metabolites (Batten et al. 2006; Reinhart and Callaway 2006; Weidenhamer and Callaway 2010). These biogeochemical effects can facilitate invasions by non-native plants, inhibit the re-establishment of native plants, and change ecosystem structure and function. Small island ecosystems are particularly vulnerable to extirpations or extinctions of native species caused by plant invasions. For example, one of the world's most invasive plants, *Psidium cattleianum*, is only established in logged and disturbed rainforests in Madagascar (Brown and Gurevitch 2004), but on the smaller islands of Hawaii and Mauritius, this species can invade fragments of intact forests and potentially cause local extinctions of native species (Huenneke and Vitousek 1990; Baider et al. 2010).

Most studies on the community-level impacts of invasive plant species show that the survival, abundance, richness, and diversity of both native plant and animal species are significantly reduced in invaded communities (Pyšek et al. 2012). A pan-European analysis of plant invasion impacts indicated that invaded areas have greatly reduced abundance (by 43.5%), diversity (by 50.7%), and fitness (by 41.7%) of native plant species and reduced fitness (by 16.5%) and abundance (by 17.5%) of native animal species (Vilà et al. 2010). Invasive plants can reduce overall diversity at a much lower level of relative abundance (less than 35%) than native plants (over 60% abundance), and some invasive species can reduce plant diversity at 10% relative abundance (Qi et al. 2014). Invasive species that have the ability to exceed the cover and height of dominant native species generally have the most severe impacts on plant diversity and community composition, with potential cascading effects on diversity at the landscape level (Hejda et al. 2009). Plant invasions can threaten the generation of ecosystem services associated with high plant diversity, including provisioning (e.g., food, fiber, biofuel, timber, and firewood), regulating (e.g., soil fertility, resistance to invasions, and water availability), and cultural services (e.g., indigenous use of native plants, recreation and tourism, cultural heritage, and land stewardship) (Quijas et al. 2012).

16.2.2 Ecosystem Productivity

The effects of invasive plants on community productivity can potentially be positive, negative, or neutral, but in most cases, plant invasions significantly decrease the primary productivity of invaded communities (Pyšek et al. 2012). An invasive plant species reduces the productivity of a community when it grows at a slower rate than the native species it replaced (Walker and Smith 1997; Vilà and Weiner 2004). Invading plants can also suppress biomass accumulation in native species through allelopathy (Rice 1984) and salt deposition (Vivrette and Muller 1977), by affecting resources (e.g., total water or light availability), or by degrading the overall environment (e.g., enhancing soil erosion) (Crooks 2002), or altering the disturbance regime (e.g., increasing fire frequency or intensity) (D'Antonio and Vitousek 1992; Mack et al. 2000). In the long term, invasive plants may also reduce productivity through changes in community structure, such as by reducing abundance and diversity of native plants and animals (Crooks 2002), or by increasing the prevalence of other invasive species (Mack et al. 2000). Giant hogweed, an allelopathic invasive plant in several European countries, greatly reduces the diversity and productivity of native species in invaded communities, resulting in declines in productivity that persist for decades following species removal (Dostál et al. 2013). Litter from invasive species commonly decomposes more rapidly than that of native species and results in greater loss of organic matter and carbon from ecosystems, compared to uninvaded systems (Peltzer et al. 2010). Therefore, plant invasions can reduce C sequestration and the potential of native communities to mitigate climate change.

16.2.3 Fire Regime

Invasive plants, in particular grasses, can affect the fire regime of an ecosystem by changing the structure and composition of fuel loads (Brooks et al. 2004). Increased frequency and/or intensity of fire events often promote the expansion of invasive plants and threaten the survival of native species (Crooks 2002). Examples of non-native invasive grasses that have long-term ecosystem-level changes can be found globally, some of which can increase fire frequency to the point that native tree species cannot recover (D'Antonio and Vitousek 1992). For example, cheat grass (*Bromus tectorum*), a highly flammable invasive annual found throughout western North America, recovers rapidly following fire and suppresses the growth of native species (Melgoza et al. 1990). The fire return interval of shrublands before invasions by cheat grass was 60–110 years (Whisenant 1990), but invaded sites burn every 3–5 years, and the latter are 500 times more likely to burn than sites without cheat grass cover (Stewart and Hull 1949). In some cases, invasive plants can introduce fire to a system with little or no fire history, which can cause high mortality of native species (Young and Evans 1978; Callison et al. 1985; Brown and Minnich 1986) and a major physiognomic shift from a shrub-dominated system to dominance by fire-adapted grasses (Walker and Smith 1997; Brooks and Chambers 2011). In addition to invasive grasses, there are numerous cases of invasive woody plant species that increase fire frequency and/or intensity (Mandle et al. 2011). Increased fire events and intensity after invasions can reduce live-tree C stocks through increased large tree mortality, and thus reduce potential C sequestration of invaded forests (Peltzer et al. 2010).

Production of allelochemicals is commonly associated with increased flammability, and thus allelopathic invasive plants commonly promote fire. For example, vegetation fires in invaded communities are often driven by highly allelopathic invasive plants with moderate to high flammability, such as *Ulex europaeus*, *Pinus radiata*, *Eucalyptus globulus*, and *Acacia melanoxylon* (Souto et al. 1994; Pyrke and Marsden-Smedley 2005; Wyse et al. 2018). These species release several inhibitory secondary metabolites that strongly reduce plant germination and growth (Souto et al. 1994; Pardo-Muras et al. 2018; Hozawa and Nawata 2020; López-Rodríguez et al. 2022). The high flammability of *U. europaeus* and *P. radiata*, in particular, has been associated with fire-promoting allelochemicals, such as essential oils, monoterpenes, and sesquiterpenes (Guerrero et al. 2021, 2022). *Ulex europaeus* also emits additional volatile organic compounds (Pardo-Muras et al. 2018), which further increases the flammability of invaded plant communities (Guerrero et al. 2021).

16.2.4 Secondary Invasions

Primary invasions by non-native plants can change the structure of vegetation and soil properties, and thus create novel conditions that can facilitate secondary invasions by other non-native species (Gioria et al. 2011). For example, the high production of litter by cheat grass increases water availability that promotes seed germination of several exotic species in desert shrublands (Evans and Young 1970; Ashton et al. 2016). In Asia and Africa, Siam weed (*Chromolaena odorata*) is a highly invasive species that provides feeding niches for the agricultural pest *Zonocerus variegatus* (painted grasshopper) and increases its survival and reproductive success. In this case, secondary compounds implicated as allelochemicals specifically serve as non-nutritional resources for agricultural pest species. *Zonocerus variegatus* obtains pyrrolizidine alkaloids (PAs) from Siam weed and stores the PAs for protection against predation; the PAs are particularly important to protect their diapausing eggs from predators or parasitoids (e.g., larvae of Mylabris beetles) (Boppré et al. 1992). *Lantana camara*, another aggressive invasive neotropical shrub, provides habitat for the stream-dwelling tsetse fly (*Glossina* spp.) in Africa, which increases the incidence of sleeping sickness in humans, as well as in domesticated and wild animals (Greathead 1968).

Once a non-native invasive species has become established in its introduced environment, eradication is generally nearly impossible, and the only option is to control the spread of the invader and prevent further ecological damage (Mack et al. 2000). The most widely used methods to slow the spread of invasive species are chemical, mechanical, and biological control, but each management approach has its caveats, as discussed earlier. In some cases, invasive plants not only displace native species from their habitat but also cause major ecological changes that have long-term negative effects on the ecosystem. Therefore, mitigation strategies should optimally aim to reduce the abundance of invasive plants and attempt to improve site conditions in favor of native species. Biochar represents a new potential tool for this purpose. The next section will discuss the different mechanisms through which biochar can reduce the impacts of plant invasions on native species and simultaneously mitigate climate change.

16.3 Potential of Biochar to Mitigate Climate Change and Allelopathic Effects

16.3.1 Biochar: A Recalcitrant Form of Carbon

Biochar is a carbon-rich product obtained from biomass sources through pyrolysis or thermal decomposition in a low-oxygen environment at approx. 350 °C to 700 °C (Glaser et al. 2001; Singh et al. 2015). The high recalcitrance of biochar, with a carbon half-life of 500–50,000 years (Spokas et al. 2010), is the main mechanism for increased C sequestration and climate change mitigation (Woolf et al. 2010; Lehmann et al. 2021). Biochar is also considered to be a critical component of carbon-negative energy systems due to its capacity for C sequestration (Smith 2016; Werner et al. 2022). In natural systems, pyrogenic C generated by wildfires is an important soil C sink (Krull et al. 2008; Jauss et al. 2015). Natural chars can increase soil fertility and further enhance C sequestration through increased tree growth (Mao et al. 2012; Gale and Thomas 2021). Fire suppression has greatly reduced pyrogenic C inputs in forest soils in the last 100 years (Steel et al. 2015). In cases where prescribed burns are not possible, biochar additions to forest soils can partially "emulate" the ecological role of fires and increase C sequestration (Wardle et al. 1998; Thomas and Gale 2015).

16.3.2 Biochar Use As a Soil Conditioner and Fertilizer

Conventional chemical fertilizers, as well as organic sources of nutrients, can enhance soil denitrification and emissions of nitrous oxide (Ding et al. 2016). Biochar is a climate-forward renewable alternative to fertilizers because it can improve soil fertility without releasing greenhouse gases (GHGs). Although biochar can directly provide some mineral nutrients inherited from feedstocks (Gezahegn et al. 2019), it mainly acts as a soil conditioner by increasing nutrient retention in soils (Randolph et al. 2017; Zhang et al. 2017). Biochar generally decreases available soil N by increasing N immobilization and decreasing N mineralization (Bruun et al. 2012; Dempster et al. 2012; Clough et al. 2013). However, it can act as a carrier for N fertilizers and mineralized N, and thus can enhance crop yields by increasing N use efficiency (Kammann et al. 2011; Khajavi-Shojaei et al. 2020; Sashidhar et al. 2020). Low soil pH limits plant growth and utilization of many essential nutrients by plants (Black 1993; Chintala et al. 2012a); biochar also acts as a liming agent and increases nutrient uptake by plants (Chintala et al. 2012b; Gezahegn et al. 2019). Biochar additions to agricultural soils have been found to increase the yield of many common commercial crops, such as rice, wheat, maize, and soybean (Jeffery et al. 2015). The use of biochar as soil amendment in agriculture can thus mitigate climate change by reducing fertilizer inputs, nutrient leaching, soil GHG emissions, and soil degradation.

16.3.3 Biochar Addition for Carbon Sequestration in Forest Systems

In addition to the sequestration of C present in biochar (Somerville et al. 2020), biochar addition can potentially increase C sequestration in forest systems through enhanced tree growth. Trees generally show positive growth responses to biochar, with a meta-analysis finding an average 41% increase in biomass (Thomas and Gale 2015). However, conifers show lower growth responses to biochar than angiosperms (Thomas and Gale 2015), likely due to adaptations to less productive environments and acidic soils, and lower rates of nutrient uptake (Coomes et al. 2005; Lusk 2011). Several mechanisms may account for enhanced tree growth in response to biochar, including increased nutrient availability, increased water supply related to increases in soil water retention, increased soil pH on acid soils (Pluchon et al. 2014; Thomas and Gale 2015), and sorption of growth-inhibiting substances such as allelochemicals (Wardle et al. 1998; Sujeeun and Thomas 2022). Nutrient availability is reduced in acidic soils and chars can increase soil pH, and thus increase concentrations of P, N, Mg, and Ca in acidic forest soils (Sackett et al. 2015; Bieser and Thomas 2019; Zhou et al. 2020). The highly porous structure and surface charge characteristics of biochar also contribute to increased soil water and nutrient retention (Atkinson et al. 2010; Biederman and Harpole 2013).

16.3.4 Mitigation of Allelopathic Effects in Invasive Plants

Biochar has mostly received attention as a soil amendment in agricultural systems; however, there has been recent interest in using biochar for soil restoration due to its ability to immobilize growth-inhibitors, such as heavy metals, salts, and phenolic compounds (Gundale and DeLuca 2006; Thomas et al. 2013; Sujeeun and Thomas 2017). Some forms of pyrogenic carbon, such as biochar and naturally occurring charcoal, and non-pyrogenic carbon such as activated carbon, have sorptive properties that can reduce the bioavailability of allelochemicals (Mahall and Callaway 1992; Wardle et al. 1998; Sujeeun and Thomas 2017). Activated carbon has often been used in allelopathy studies due to its high affinity for organic compounds but is distinct from biochar (and natural forms of PyC) in terms of its physicochemical properties (Hale et al. 2011). Biochar is a more relevant alternative to activated carbon for reducing the bioavailability of allelochemicals in the soil because of its low cost and its similarity to natural forms of PyC. Biochar has been found to reduce the inhibitory effects of several allelopathic species, including highly invasive plants, through the sorption or breakdown of allelopathic compounds in leaf extracts and crop residues (Mahall and Callaway 1992; Rogovska et al. 2012; Sujeeun and Thomas 2017). Black walnut (*Juglans nigra*) is the best-known case of allelopathy among tree species (Gabriel 1975; Jose and Gillespie 1998), and juglone is the main compound responsible for the strong allelopathic effects of black walnut (Willis 2000; Scott and Sullivan 2007). Biochar additions to soils containing juglone or black walnut litter can dramatically increase tree growth compared to allelopathic soils without biochar (Sujeeun and Thomas 2023). Biochar also has the potential to alleviate the inhibitory effects of highly invasive allelopathic species, such as strawberry guava and Norway maple (*Acer platanoides*), with large growth responses in tropical and temperate native tree species (Sujeeun and Thomas 2022; Sujeeun and Thomas 2023). Biochar can increase the productivity of forest systems, with particularly beneficial effects in cases where invasive plants have reduced the productivity of a site.

The main mechanism for reduced allelopathic effects after biochar addition is almost certainly sorption of allelochemicals (Hall et al. 2014; Sujeeun and Thomas 2017; Bieser et al. 2022). Biochar has the capacity to sorb a wide variety of polyaromatic hydrocarbons with chemical structures similar to common allelochemicals (Zhou et al. 2022). Biochars contain micropores that can enhance sorption of phenolic compounds (Zhao et al. 2020) and reduce their availability for plant uptake. Biochar addition was found to increase sorption of toxic organic compounds in soils with a higher clay content (Askeland et al. 2020), possibly because clay particles bind to biochar and enhance its capacity for ion exchange (Yao et al. 2014). After sorption of allelochemicals, biochar can potentially promote their microbial breakdown into non-inhibitory compounds. For example, biochar provides a substrate for *Pseudomonas* bacteria and promotes the biodegradation of allelochemicals (Yang et al. 2016; Zhao et al. 2020). *Pseudomonas* bacteria can use juglone as their only source of carbon and rapidly degrade juglone in soils (Schmidt 1988).

16.3.5 Mitigation of Soil GHG Emissions

The biogenic greenhouse gases (GHGs) such as carbon dioxide (CO_2), methane (CH_4), and nitrous oxide (N_2O) are the most important contributors to radiative forcing in the atmosphere and some soils can be a large source of these emissions (Khalil 1999; Oertel et al. 2016). Soil respiration, the main pathway for the release of soil organic carbon into the atmosphere, is the primary mechanism of carbon loss from terrestrial systems (Peng et al. 2008; Xu and Shang 2016). Biochar amendments to soils can affect GHG emissions by changing soil chemical, physical, or microbiological properties and associated processes (van Zwieten et al. 2009, 2010; Jones et al. 2011; Singh et al. 2019).

Biochar addition can have positive, negative, or negligible priming effects on soil CO_2 efflux (Spokas et al. 2009; Smith et al. 2010; Scheer et al. 2011; Zimmerman et al. 2011). Positive priming effects of biochar are associated with increased soil labile organic C that can promote organic matter decomposition to CO_2 emission (Smith et al. 2010; Wang et al. 2014). Soil CO_2 emissions increase with higher soil temperatures due to increased microbial decomposition of organic matter (Ray et al. 2020). Thus, greater CO_2 emissions in response to biochar amendments can also be due to lower surface albedo, which increases soil temperature (Usowicz et al. 2016). However, negative priming effects can also occur because biochar has the ability to (i) adsorb labile C, reducing its availability to microorganisms (Jones et al. 2011), and (ii) adsorb enzymes involved in the decomposition of soil organic matter (Woolf and Lehmann 2012). Lower CO_2 emissions from biochar-amended forest soils have also been attributed to sorption of CO_2 to biochar particles (Kasozi et al. 2010; Liang et al. 2010), and changes to soil properties, such as water content, porosity, aggregation, pH, CEC (Liang et al. 2010; Jones et al. 2011), and soil microbial communities (Mitchell et al. 2015; Liu et al. 2016; Wang et al. 2016).

Compared to unamended soils, biochar additions to rice paddies, wetland fields, and upland soils suppressed CH_4 emission by 7% on average, due to a lower ratio of methanogens to methanotrophs (Lyu et al. 2022). Biochar can increase CH_4 uptake and reduce soil CH_4 emissions by two main mechanisms. First, biochar amendments increase soil pH which promotes the growth of methanotrophs (Inubushi et al. 2005; Jeffery et al. 2016). Second, biochar additions decrease soil bulk density and increase soil porosity and aeration, which promotes CH_4 oxidation and uptake activity by soil bacteria (Brassard et al. 2016). While biochar effects on CH_4 flux vary among systems (Jeffery et al. 2016), there is more consistent evidence that biochar generally reduces N_2O emissions, with meta-analyses suggesting an average reduction of ~30–60% (Cayuela et al. 2014; Lyu et al. 2022). The reduction in N_2O emissions has been found to be positively correlated to the application rate of biochar (Borchard et al. 2019; Lyu et al. 2022). Potential mechanisms for reduced N_2O emissions by biochar are increased soil aeration (Yanai et al. 2007; van Zwieten et al. 2010), sorption of NH_4^+ or NO_3^- (Singh et al. 2010; van Zwieten et al. 2010) and reduced microbial ammonium nitrification through increased production of the microbial inhibitor, ethylene (Spokas et al. 2010).

Plant invasions can amplify the effects of climate change by contributing to GHG emissions (Tong et al. 2012; Qiu 2015). Although invasive plants can increase carbon sequestration due to their higher growth rates compared to native plants, they can also enhance GHG emissions, with the potential to accelerate global warming (Zhang et al. 2014; Qiu 2015). A recent meta-analysis found that methane emissions from invaded wetlands were almost double compared to native wetlands (Bezabih Beyene et al. 2022), possibly due to CH_4 transport via the aerenchyma system in invasive plants, and greater production of litter that promotes methanogenic activity in the soil (Duke et al. 2015; Bansal et al. 2020; Granse et al. 2022). Plant invasions also increased N_2O emissions from grasslands by about 77% due to higher turnover and availability of soil N (Bezabih Beyene et al. 2022). Invasive plant litter can increase soil CO_2 emissions due to a higher decomposition rate than native litter, which increases soil respiration (Zhang et al. 2014). Biochar can potentially mitigate GHG emissions from invasive plants by reducing their spread and by promoting the growth of native species, which can further suppress invasive species. Biochar additions might also immobilize N from invasive plant litter and reduce soil available N, and thus decrease N_2O emissions from invaded systems (Cui et al. 2017).

16.4 Biochar Use for Control of Allelopathic Invasive Species: A Case Study on the Effects on Strawberry Guava

Biochar has been found to alleviate the allelopathic effects of several dominant invasive species globally, including strawberry guava, lemongrass, yellow sweet-grass, Norway maple, garlic mustard, and red river gum (Sujeeun and Thomas 2017; Alshahrani and Suansa 2020; Bieser et al. 2022; Sujeeun and Thomas 2023). Activated carbon (which when derived from nutshell or wood feedstocks is essentially a physically or chemically treated form of biochar) reduced the growth of the invasive plant *Centaurea diffusa* in the presence of native species after controlling for spatial root niche partitioning, suggesting that the competitive advantage of *C. diffusa* is, at least in part, mediated by allelopathic effects (Callaway and Aschehoug 2000).

Strawberry guava (*Psidium cattleianum*) is a highly invasive tree species that threatens diverse forest ecosystems on tropical islands, with particularly devastating effects on the native species in Mauritius. The indigenous forests of Mauritius only exist as isolated fragments, and the vast majority of these are dominated by non-native flora, of which strawberry guava is by far the most common (Page and D'Argent 1997). In the lowland wet forests of Mauritius, native plant performance is severely suppressed by high densities of strawberry guava (Zimmerman et al. 2008; Monty et al. 2013). Mauritius is part of the globally significant Madagascar and Indian Ocean Islands biodiversity hotspot (Myers et al. 2000; Florens et al. 2012) and is home to ~700 native angiosperm species, of which ~40% are endemic to the island. More than 80% of the endemic flora is threatened by invasions of non-native species such as strawberry guava (Baider et al. 2010). Several competitive traits contribute to strawberry guava's success, including high reproductive capacity, copious fruiting, clonal growth, resprouting ability, and adaptation to a wide range of light conditions (Huenneke and Vitousek 1990; Schumacher et al. 2008). However, studies also suggest that allelopathy can contribute greatly to its successful invasion and dominance (Virah-Sawmy et al. 2009; Patel 2012).

In a field trial established in Mauritius' largest national park in 2016 (Fig. 16.2), Sujeeun and Thomas (2022) investigated, in the first study of its kind, the effects of biochar additions on native tree species in areas invaded by strawberry guava. Biochar was found to significantly reduce the inhibitory effects of strawberry guava and suppressed its regeneration in native forest communities. The most likely mechanism for strawberry guava suppression by biochar is through increased growth and density of native species that can reduce its regeneration by competitive interactions. Biochar more than doubled the growth of native species and large positive responses on tree growth were still present 30 months after biochar

Fig. 16.2 Treatment plots at the study site in Mauritius: non-weeded control, weeded, and biochar plots (left to right)

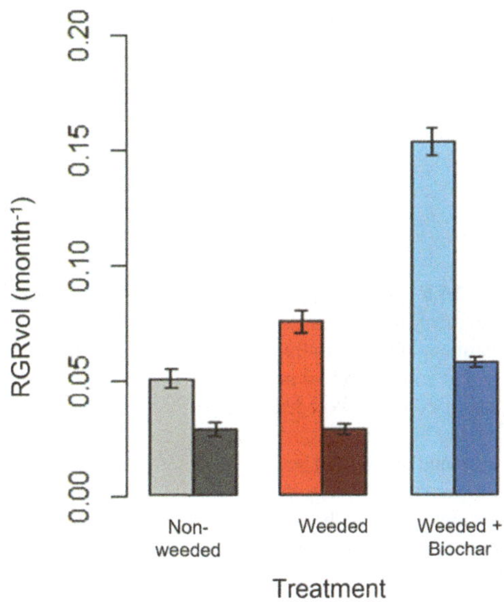

Fig. 16.3 Relative volume growth rate (RGRvol) of native trees examined after 6 months (lighter colors) and 30 (darker colors) months. "Weeded + Biochar" treatment is a combined average of RGRvol in treatment plots with biochar dosages of 25 and 50 t/ha

application (Fig. 16.3) (Sujeeun and Thomas 2022). Tree diversity in biochar plots was five times greater than non-weeded plots and almost doubled compared to weeded plots without biochar (Fig. 16.4). Increased growth of native species is consistent with biochar sorption of allelochemicals, but may also involve increased nutrient availability, increased soil pH, and increased soil water retention (Atkinson et al. 2010). In addition, biochar resulted in a dramatic reduction in the regeneration of strawberry guava compared to the control plots (Fig. 16.5) (Sujeeun and Thomas 2022). The combined effect of biochar sorption of allelochemicals in addition to enhanced resource availability likely increased the competitive ability of native species (Sujeeun and Thomas 2022) and may potentially allow them to reclaim dominance in the long term.

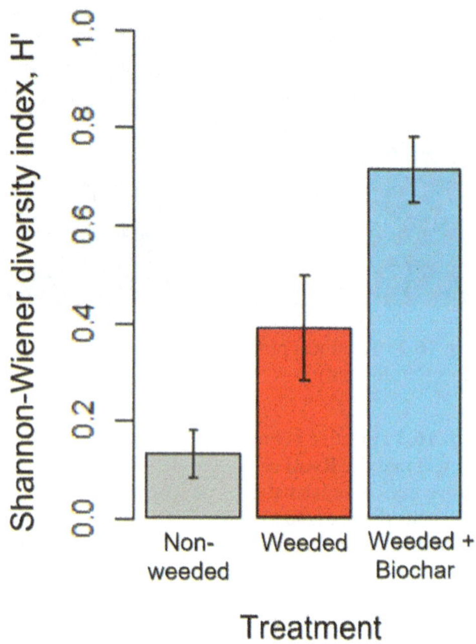

Fig. 16.4 Shannon diversity by treatment after 30 months. "Weeded + Biochar" treatment is a combined average of diversity in treatment plots with biochar dosages of 25 and 50 t/ha

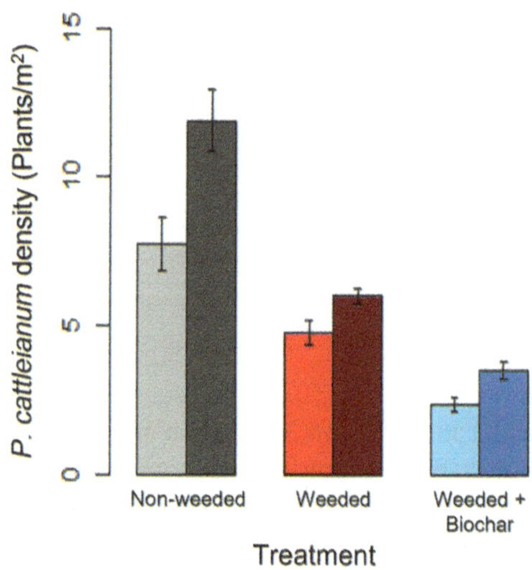

Fig. 16.5 Density of strawberry guava stems by treatment after 6 months (lighter colors) and 30 months (darker colors). "Weeded + Biochar" treatment is a combined average of strawberry guava density in treatment plots with biochar dosages of 25 and 50 t/ha

Biochar has been found to sorb some of the allelochemicals produced by strawberry guava, such as phenolic compounds and α-pinene (Wardle et al. 1998; Hale et al. 2015). In laboratory experiments, biochar was also shown to directly reduce the inhibitory effects of strawberry guava leaf extracts, consistent with sorption of allelochemicals (Sujeeun and Thomas 2017). The large negative responses of strawberry guava to biochar addition indicate that sorption of these allelochemicals was likely an important mechanism for reducing the regeneration of the invasive species (Sujeeun and Thomas 2022). Biochar can also suppress the growth and spread of strawberry guava by affecting soil nutrient availability because invasive plants typically benefit from nutrient additions, particularly of N, and N fertilization is often more beneficial to invasive species than to native species (Witkowski 1991; Mangla et al. 2011; Gioria and Osborne 2014). Strawberry guava performance is strongly affected by soil N levels; leaf N uptake and flower production increased rapidly after N fertilization (Normand and Habib 2001). Although biochar can increase soil nutrient availability, in many cases, it has been found to decrease available soil N (Clough et al. 2013). In this study, biochar additions to strawberry-guava-invaded communities reduced soil N levels and likely contributed to the suppression of strawberry guava regeneration (Sujeeun and Thomas 2022). The main allelochemical, β-caryophyllene, found in the leaves of strawberry guava, provides protection from herbivore damage by attracting natural enemies of the herbivores (Köllner et al. 2008; Chiriboga et al. 2018). Therefore, sorption of allelochemicals by biochar might make strawberry guava more vulnerable to attacks by pathogens and herbivores by weakening its defense system.

16.5 Conclusion and Future Perspectives

Climate change and invasive species are two of the most serious anthropogenic environmental issues, both of which require long-term and cost-effective solutions. Biochar appears to be a promising tool to combat both climate change and invasive plants, including highly allelopathic species with severe impacts on native ecosystems. Biochar can be used to mitigate climate change through increased carbon sequestration. Biochar additions can increase soil fertility, increase crop yields, reduce fertilizer use and leaching, increase tree growth and carbon uptake, and reduce soil GHG emissions. Biochar can also suppress invasive allelopathic plants through sorption and microbial breakdown of allelochemicals, which reduces the competitive advantage of the invasive species over native species. In addition, N immobilization by biochar reduces the growth of invasive species that thrive at high soil N levels. Biochar can potentially further decrease the regeneration of invasive species by increasing the growth and density of native species.

Biochar is an important new tool for mitigating the negative effects of climate change and invasive species, but several factors should be taken into consideration to maximize its potential benefits. The physicochemical properties of biochar are affected by the feedstock type and particle size, pyrolysis temperature, and heating rates. To date, biochar has only been tested with a handful of invasive plants. Invasive species that are strongly allelopathic and have particularly serious impacts on biodiversity, such as *Imperata cylindrica*, *Lantana camara*, and *Chromolaena odorata*, should be targeted in future trials. Further research is necessary to examine the properties and dosages of biochar that will optimize carbon sequestration and sorption of allelochemicals.

References

Abu-Romman S, Ammari T (2015) Allelopathic effect of 'Arundo donax,' a mediterranean invasive grass. Plant Omics 8:287–291

Ahmed R, Hoque ATM, Hossain MK (2008) Allelopathic effects of *Leucaena leucocephala* leaf litter on some forest and agricultural crops grown in nursery. J For Res 19:298–302

Alshahrani TS, Suansa NI (2020) Application of biochar to alleviate effects of allelopathic chemicals on seed germination and seedling growth. Bioresources 15:382–400

Annighöfer P, Schall P, Kawaletz H et al (2012) Vegetative growth response of black cherry (*Prunus serotina*) to different mechanical control methods in a biosphere reserve. Can J For Res 42:2037–2051

Ashton IW, Symstad AJ, Davis CJ et al (2016) Preserving prairies: understanding temporal and spatial patterns of invasive annual bromes in the Northern Great Plains. Ecosphere 7:e01438

Askeland M, Clarke BO, Cheema SA et al (2020) Biochar sorption of PFOS, PFOA, PFHxS and PFHxA in two soils with contrasting texture. Chemosphere 249:126072

Atkinson CJ, Fitzgerald JD, Hipps NA (2010) Potential mechanisms for achieving agricultural benefits from biochar application to temperate soils: a review. Plant Soil 337:1–18

Baider C, Florens FV, Baret S et al. (2010) Status of plant conservation in oceanic islands of the Western Indian Ocean. In: Proceedings of the 4th global botanic gardens congress, Dublin, Ireland, 13–18 June 2010

Bansal S, Johnson OF, Meier J et al (2020) Vegetation affects timing and location of wetland methane emissions. J Geophysical Res: Biogeosci. 125:e2020JG005777

Batten KM, Scow KM, Davies KF et al (2006) Two invasive plants alter soil microbial community composition in serpentine grasslands. Biol Invasions 8:217–230

Beesley L, Moreno-Jimenez E, Fellet G, Melo L, Sizmur T (2015) Biochar and heavy metals. In: Lehmann J, Joseph S (eds) Biochar for environmental management, 2nd edn. Routledge, New York, pp 563–594

Bezabih Beyene B, Li J, Yuan J et al (2022) Non-native plant invasion can accelerate global climate change by increasing wetland methane and terrestrial nitrous oxide emissions. Glob Change Biol 28:5453–5468

Biederman LA, Harpole WS (2013) Biochar and its effects on plant productivity and nutrient cycling: a meta-analysis. GCB Bioenergy 5:202–214

Bieser JM, Thomas SC (2019) Biochar and high-carbon wood ash effects on soil and vegetation in a boreal clearcut. Can J For Res 49:1124–1134

Bieser JM, Al-Zayat M, Murtada J et al (2022) Biochar mitigation of allelopathic effects in three invasive plants: evidence from seed germination trials. Can J Soil Sci 102:213–224

Black AC (1993) Soil fertility evaluation and control. Lewis Publishers, Boca Raton, FL

Boppré M, Biller A, Fischer OW et al (1992) The non-nutritional relationship of *Zonocerus* (Orthoptera) to *Chromolaena* (Asteraceae). In: Proceedings of the 8th International Symposium on Insect-Plant Relationships. Springer, Dordrecht, pp 89–90

Borchard N, Schirrmann M, Cayuela ML et al (2019) Biochar, soil and land-use interactions that reduce nitrate leaching and N_2O emissions: a meta-analysis. Sci Total Environ 651:2354–2364

Brassard P, Godbout S, Raghavan V (2016) Soil biochar amendment as a climate change mitigation tool: key parameters and mechanisms involved. J Environ Manag 181:484–497

Brooks ML, Chambers JC (2011) Resistance to invasion and resilience to fire in desert shrublands of North America. Rangel Ecol Manag 64:431–438

Brooks ML, D'antonio CM, Richardson DM et al (2004) Effects of invasive alien plants on fire regimes. BioScience 54:677–688

Brown KA, Gurevitch J (2004) Long-term impacts of logging on forest diversity in Madagascar. Proc Natl Am Sci 101:6045–6049

Brown DE, Minnich RA (1986) Fire and changes in creosote bush scrub of the western Sonoran Desert, California. Am Midl Nat 116:411–422

Broz AK, Manter DK, Vivanco JM (2007) Soil fungal abundance and diversity: another victim of the invasive plant *Centaurea maculosa*. ISME J 1:763–765

Bruun ES, Ambus P, Egsgaard H et al (2012) Effects of slow and fast pyrolysis biochar on soil C and N dynamics. Soil Biol Biochem 46:73–79

Callaway RM, Aschehoug ET (2000) Invasive plants versus their new and old neighbors: A mechanism for exotic invasion. Science 290:521–523

Callaway RM, Hierro JL (2006) Resistance and susceptibility of plant communities to invasion: revisiting Rabotnov's ideas about community homeostasis. In: Reigosa MJ, Pedrol N, González L (eds) Allelopathy: a physiological process with ecological implications. Springer, Netherlands, pp 395–414

Callison J, Brotherson JD, Bowns JE (1985) The effects of fire on the blackbrush (*Coleogyne ramosissima*) community of southwestern Utah. Rangeland Ecol Manag 38:535–538

Cappuccino N, Arnason JT (2006) Novel chemistry of invasive exotic plants. Biol Lett 2:189–193

Cayuela ML, Van Zwieten L, Singh BP, Jeffery S, Roig A, Sánchez-Monedero MA (2014) Biochar's role in mitigating soil nitrous oxide emissions: a review and meta-analysis. Ag Ecosys Env 191:5–16

Chen BM, Liao HX, Chen WB et al (2017) Role of allelopathy in plant invasion and control of invasive plants. Allelopathy J 41:155–166

Chintala R, McDonald LM, Bryan WB (2012a) Effect of soil water and nutrients on productivity of Kentucky bluegrass system in acidic soils. J Plant Nutr 35:288–303

Chintala R, McDonald LM, Bryan WB (2012b) Optimization of water potential and nutrient levels for Kentucky bluegrass-white clover mixture on acidic soils. Biotechnol Agron Soc Environ 16: 167–177

Chiriboga M, Guo H, Campos-Herrera R (2018) Root-colonizing bacteria enhance the levels of (E)-β-caryophyllene produced by maize roots in response to rootworm feeding. Oecologia 187: 459–468

Cipollini D (2016) A review of garlic mustard (*Alliaria petiolata*, Brassicaceae) as an allelopathic plant. J Torrey Bot Soc 143:339–348

Clewley GD, Eschen R, Shaw RH et al (2012) The effectiveness of classical biological control of invasive plants. J Appl Ecol 49:1287–1295

Clough TJ, Condron LM, Kammann C et al (2013) A review of biochar and soil nitrogen dynamics. Agronomy 3:275–293

Coomes DA, Allen RB, Bentley WA et al (2005) The hare, the tortoise and the crocodile: the ecology of angiosperm dominance, conifer persistence and fern filtering. J Ecol 93:918–935

Cronk QCB, Fuller JL (1995) Plant invaders: the threat to natural ecosystems. Routledge, London

Crooks JA (2002) Characterizing ecosystem-level consequences of biological invasions: the role of ecosystem engineers. Oikos 97:153–166

Cui J, Ge T, Kuzyakov Y et al (2017) Interactions between biochar and litter priming: A three-source ^{14}C and $\delta^{13}C$ partitioning study. Soil Biol Biochem 104:49–58

D'Antonio CM, Vitousek PM (1992) Biological invasions by exotic grasses, the grass/fire cycle, and global change. Annu Rev Ecol Syst 23:63–87

Dempster DN, Jones DL, Murphy DV (2012) Clay and biochar amendments decreased inorganic but not dissolved organic nitrogen leaching in soil. Soil Res 50:216–221

Di Tomaso JM (2000) Invasive weeds in rangelands: species, impacts, and management. Weed Sci 48:255–265

Ding Y, Liu Y, Liu S et al (2016) Biochar to improve soil fertility. A review. Agron Sustain Dev 36: 1–18

DiStefano JF (1982) The role of allelopathy in the invasion patterns of *Melaleuca quinquenervia* in southern Florida. Master's thesis,. University of Florida, Gainesville, FL

Dostál P, Müllerová J, Pyšek P et al (2013) The impact of an invasive plant changes over time. Ecol Lett 16:1277–1284

Duenas MA, Ruffhead HJ, Wakefield NH et al (2018) The role played by invasive species in interactions with endangered and threatened species in the United States: a systematic review. Biodivers Conserv 27:3171–3183

Duke ST, Francoeur SN, Judd KE (2015) Effects of *Phragmites australis* invasion on carbon dynamics in a freshwater marsh. Wetlands 35:311–321

Ehrenfeld JG (2010) Ecosystem consequences of biological invasions. Annu Rev Ecol Evol Syst 41:59–80

Evans RA, Young J (1970) Plant litter and establishment of alien annual weed species in rangeland communities. Weed Sci 18:697–703

Fatunbi AO, Dube S, Yakubu MT et al (2009) Allelopathic potential of *Acacia mearnsii* De Wild. World Appl Sci J 7:1488–1493

Florens FV, Baider C, Martin GM et al (2012) Surviving 370 years of human impact: what remains of tree diversity and structure of the lowland wet forests of oceanic island Mauritius? Biodivers Conserv 21:2139–2167

Foster JB, Magdoff F (1998) Liebig, Marx, and the depletion of soil fertility: relevance for today's agriculture. Mon Rev 50:32

Gabriel WJ (1975) Allelopathic effects of black walnut on white birches. J Forest 73:234–237

Gaertner M, Breeyen AD, Hui C et al (2009) Impacts of alien plant invasions on species richness in Mediterranean-type ecosystems: a metaanalysis. Progr Phys Geogr 33:319–338

Gale NV, Thomas SC (2021) Spatial heterogeneity in soil pyrogenic carbon mediates tree growth and physiology following wildfire. J Ecol 109:1479–1490

Gámiz B, López-Cabeza R, Velarde P et al (2021) Biochar changes the bioavailability and bioefficacy of the allelochemical coumarin in agricultural soils. Pest Manag Sci 77:834–843

Gezahegn S, Sain M, Thomas SC (2019) Variation in feedstock wood chemistry strongly influences biochar liming potential. Soil Syst 3:26

Gioria M, Osborne BA (2014) Resource competition in plant invasions: Emerging patterns and research needs. Front Plant Sci 5:501

Gioria M, Dieterich B, Osborne B (2011) Battle of the giants: primary and secondary invasions by large herbaceous species. In: Biology and Environment: Proceedings of the Royal Irish Academy, Dublin, December 2011

Glaser B, Haumaier L, Guggenberger G et al (2001) The 'Terra Preta' phenomenon: a model for sustainable agriculture in the humid tropics. Naturwissenschaften 88:37–41

Granse D, Titschack J, Ainouche M et al (2022) Subsurface aeration of tidal wetland soils: Root-system structure and aerenchyma connectivity in *Spartina* (Poaceae). Sci Total Environ 802:149771

Greathead DJ (1968) Biological control of *Lantana*—a review and discussion of recent developments in East Africa. Internat J Pest Manage 14:167–175

Grove S, Haubensak KA, Parker IM (2012) Direct and indirect effects of allelopathy in the soil legacy of an exotic plant invasion. Plant Ecol 213:1869–1882

Guerrero F, Hernández C, Toledo M et al (2021) Leaf thermal and chemical properties as natural drivers of plant flammability of native and exotic tree species of the Valparaíso region, Chile. Int J Env Res Pub He 18:7191

Guerrero F, Carmona C, Hernández C et al (2022) Drivers of flammability of *Eucalyptus globulus* Labill leaves: terpenes, essential oils, and moisture content. Forests 13:908

Gundale MJ, DeLuca TH (2006) Temperature and source material influence ecological attributes of ponderosa pine and Douglas-fir charcoal. Forest Ecol Manag 231:86–93

Gurevitch J, Padilla DK (2004) Are invasive species a major cause of extinctions? Trends Ecol Evol 19:470–474

Hagan DL, Jose S, Lin CH (2013) Allelopathic exudates of cogongrass (*Imperata cylindrica*): Implications for the performance of native pine savanna plant species in the Southeastern US. J Chem Ecol 39:312–322

Hale SE, Hanley K, Lehmann J et al (2011) Effects of chemical, biological, and physical aging as well as soil addition on the sorption of pyrene to activated carbon and biochar. Environ Sci Technol 45:10445–10453

Hale SE, Endo S, Arp HPH et al (2015) Sorption of the monoterpenes α-pinene and limonene to carbonaceous geosorbents including biochar. Chemosphere 119:881–888

Hall KE, Calderon MJ, Spokas KA et al (2014) Phenolic acid sorption to biochars from mixtures of feedstock materials. Water Air Soil Poll 225:1–9

Hejda M, Pyšek P, Jarošík V (2009) Impact of invasive plants on the species richness, diversity and composition of invaded communities. J Ecol 97:393–403

Henderson S, Dawson TP, Whittaker RJ (2006) Progress in invasive plants research. Prog Phys Geogr 30:25–46

Hernández-Aro M, Hernández-Pérez R, Guillén-Sánchez D et al (2016) Allelopathic influence of residues from *Sphagneticola trilobata* on weeds and crops. Planta Daninha 34:81–90

Hozawa M, Nawata E (2020) The interaction between leaf allelopathy and symbiosis with Rhizobium of *Ulex europaeus* on Hawaii Island. Plan Theory 9:226

Hu G, Zhang Z (2013) Allelopathic effects of *Chromolaena odorata* on native and non-native invasive herbs. J Food Agric Environ 11:878–882

Huenneke LF, Vitousek PM (1990) Seedling and clonal recruitment of the invasive tree *Psidium cattleianum*: Implications for management of native Hawaiian forests. Biol Conserv 53:199–211

Inderjit, Keating KI (1999) Allelopathy: principles, procedures, processes, and promises for biological control. In: Sparks DL (ed) Advances in agronomy, vol 67. Academic Press, New York, pp 141–231

Inubushi K, Otake S, Furukawa Y et al (2005) Factors influencing methane emission from peat soils: comparison of tropical and temperate wetlands. Nutr Cycl Agroecosyst 71:93–99

Jauss V, Johnson M, Krull et al (2015) Pyrogenic carbon controls across a soil catena in the Pacific Northwest. Catena 124:53–59

Jeffery S, Abalos D, Spokas KA et al (2015) Biochar effects on crop yield. In: Lehmann J, Joseph S (eds) Biochar for environmental management, 2nd edn. Routledge, New York, pp 333–358

Jeffery S, Verheijen FG, Kammann C et al (2016) Biochar effects on methane emissions from soils: a meta-analysis. Soil Biol Biochem 101:251–258

Jones DL, Murphy DV, Khalid M et al (2011) Short-term biochar-induced increase in soil CO2 release is both biotically and abiotically mediated. Soil Biol Biochem 43:1723–1731

Jose S, Gillespie AR (1998) Allelopathy in black walnut (*Juglans nigra* L.) alley cropping. I. Spatio-temporal variation in soil juglone in a black walnut–corn (*Zea mays* L.) alley cropping system in the midwestern USA. Plant Soil 203:191–197

Kalisz S, Kivlin SN, Bialic-Murphy L (2021) Allelopathy is pervasive in invasive plants. Biol Invasions 23:367–371

Kammann CI, Linsel S, Gößling JW et al (2011) Influence of biochar on drought tolerance of *Chenopodium quinoa* Willd and on soil–plant relations. Plant Soil 345:195–210

Karim SM (2017) Inhibiting effects of siam weed (*Chromolaena odorata* (L.) King & Robinson) on seed germination and seedling growth of four crops. Bangladesh J Bot 46:473–480

Kasozi GN, Zimmerman AR, Nkedi-kizza P et al (2010) Catechol and humic acid sorption onto a range of laboratory-produced black carbons (biochars). Environ Sci Technol 44:6189–6195

Kato-Noguchi H (2022) Allelopathy and allelochemicals of *Imperata cylindrica* as an invasive plant species. Plan Theory 11:2551

Kato-Noguchi H, Kurniadie D (2021) Allelopathy of *Lantana camara* as an invasive plant. Plan Theory 10:1028

Khajavi-Shojaei S, Moezzi A, Norouzi Masir M et al (2020) Synthesis modified biochar-based slow-release nitrogen fertilizer increases nitrogen use efficiency and corn (*Zea mays* L.) growth. Biomass Convers Biorefin 13:593–601

Khalil MAK (1999) Non-CO_2 greenhouse gases in the atmosphere. Annu Rev Environ Resour 24: 645

Köllner TG, Held M, Lenk C et al (2008) maize (E)-β-caryophyllene synthase implicated in indirect defense responses against herbivores is not expressed in most American maize varieties. Plant Cell 20:482–494

Koocheki A, Lalegani B, Hosseini SA (2013) Ecological consequences of allelopathy. In: Cheema ZA, Farooq M, Wahid A (eds) Allelopathy. Springer, Berlin, Heidelberg, pp 23–38

Koodkaew I, Senaphan C, Sengseang N et al (2018) Characterization of phytochemical profile and phytotoxic activity of *Mimosa pigra* L. Agric Nat Resour 52:162–168

Krull E, Lehmann J, Skjemstad J et al (2008) The global extent of black C in soils: is it everywhere. In: Schröder H (ed) Grasslands: ecology, management and restoration. Nova Science Publishers, Inc., New York, pp 13–17

Lavergne C, Rameau JC, Figier J (1999) The invasive woody weed *Ligustrum robustum* subsp. walkeri threatens native forests on La Réunion. Biol Invasions 1:377–392

Lehmann J, Joseph S (2015) Biochar for environmental management: Science, technology and implementation. Routledge, London

Lehmann J, Cowie A, Masiello CA, Kammann C et al (2021) Biochar in climate change mitigation. Nat Geosci 14:883–892

Liang BQ, Lehmann J, Sohi SP et al (2010) Black carbon affects the cycling of non-black carbon in soil. Org Geochem 41:206–213

Liao C, Peng R, Luo Y et al (2008) Altered ecosystem carbon and nitrogen cycles by plant invasion: a meta-analysis. New Phytol 177:706–714

Liu S, Zhang Y, Zong Y et al (2016) Response of soil carbon dioxide fluxes, soil organic carbon and microbial biomass carbon to biochar amendment: a meta-analysis. GCB Bioenergy 8:392–406

Lodhi MAK (1978) Allelopathic effects of decaying litter of dominant trees and their associated soil in a lowland forest community. Am J Bot 65:340–344

López-Rodríguez A, Hernández M, Carrillo-Galvez A et al (2022) Phytotoxic activity of *Ulex europaeus*, an invasive plant on Chilean ecosystems: separation and identification of potential allelochemicals. Nat Prod Res 2022:1–7

Lusk CH (2011) Conifer-angiosperm interactions: physiological ecology and life history. In: Turner BL, Cernusak LA (eds) Ecology of tropical podocarps. Smithsonian Institution Press, Smithsonian Contributions to Botany, Washington, DC, pp 157–164

Lyu H, Zhang H, Chu M et al (2022) Biochar affects greenhouse gas emissions in various environments: a critical review. Land Degrad Dev 33:3327–3342

Mack RN, Simberloff D, Mark LW et al (2000) Biotic invasions: causes, epidemiology, global consequences, and control. Ecol Appl 10:689–710

Mahall BE, Callaway RM (1992) Root communication mechanisms and intracommunity distributions of two Mojave Desert shrubs. Ecology 73:2145–2151

Mandle L, Bufford JL, Schmidt IB, Daehler CC (2011) Woody exotic plant invasions and fire: reciprocal impacts and consequences for native ecosystems. Biol Invasions 13:1815–1827

Mangla S, Sheley RL, James JJ et al (2011) Intra and interspecific competition among invasive and native species during early stages of plant growth. Plant Ecol 212:531–542

Mao JD, Johnson RL, Lehmann J et al (2012) Abundant and stable char residues in soils: implications for soil fertility and carbon sequestration. Environ Sci Technol 46:9571–9576

Melgoza G, Nowak RS, Tausch RJ (1990) Soil water exploitation after fire: competition between *Bromus tectorum* (cheatgrass) and two native species. Oecologia 83:7–13

Mitchell PJ, Simpson AJ, Soong R et al (2015) Shifts in microbial community and water-extractable organic matter composition with biochar amendment in a temperate forest soil. Soil Biol Biochem 81:244–254

Monty MLF, Florens FBV, Baider C (2013) Invasive alien plants elicit reduced production of flowers and fruits in various native forest species on the tropical Island of Mauritius (Mascarenes, Indian Ocean). Trop Conserv Sci 6:35–49

Morgan EC, Overholt WA (2005) Potential allelopathic effects of Brazilian pepper (*Schinus terebinthifolius* Raddi, Anacardiaceae) aqueous extract on germination and growth of selected Florida native plants. J Torrey Bot Soc 132:11–15

Muller CH (1966) The role of chemical inhibition (allelopathy) in vegetational composition. Bull Torrey Bot Club 93:332–351
Myers N, Mittermeier RA, Mittermeier CG et al (2000) Biodiversity hotspots for conservation priorities. Nature 403:853–858
Nissanka SP, Mohotti KM, Wijetunga ASTB (2005) Alleopathic influences of *Pinus caribea* on vegetation regeneration and soil biodiversity. In: Fourth World Congress in Allelopathy. Charles Sturt University, Wagga Wagga, NSW, Australia
Normand F, Habib R (2001) Nitrogen fertilisation induces floriferous flush in strawberry guava (*Psidium cattleianum*). Agronomie 21:735–742
Nunez-Gonzalez N, Rodriguez J, Gonzalez L (2021) Managing the invasive plant *Carpobrotus edulis*: is mechanical control or specialized natural enemy more effective? J Environ Manag 298:113554
Oertel C, Matschullat J, Zurba K et al (2016) Greenhouse gas emissions from soils - a review. Geochemistry 76:327–352
Page W, D'Argent G (1997) A vegetation survey of Mauritius. Mauritian Wildlife Foundation, Mauritius, Port Louis
Pardo-Muras M, Puig CG, López-Nogueira A et al (2018) On the bioherbicide potential of *Ulex europaeus* and *Cytisus scoparius*: profiles of volatile organic compounds and their phytotoxic effects. PLoS One 13:e0205997
Patel S (2012) Exotic tropical plant *Psidium cattleianum*: a review on prospects and threats. Rev Environ Sci Biotechnol 11:243–248
Pejchar L, Mooney HA (2009) Invasive species, ecosystem services and human well-being. Trends Ecol Evol 24:497–504
Peltzer DA, Allen RB, Lovett GM et al (2010) Effects of biological invasions on forest carbon sequestration. Glob Change Biol 16:732–746
Peng YY, Thomas SC, Tian DL (2008) Forest management and soil respiration: implications for carbon sequestration. Environ Rev 16:93–111
Pimentel D, Lach L, Zuniga R, Morrison D (2000) Environmental and economic costs of nonindigenous species in the United States. Bioscience 50:53–65
Pluchon N, Gundale MJ, Nilsson MC et al (2014) Stimulation of boreal tree seedling growth by wood-derived charcoal: effects of charcoal properties, seedling species and soil fertility. Funct Ecol 28:766–775
Powell KI, Chase JM, Knight TM (2011) A synthesis of plant invasion effects on biodiversity across spatial scales. Am J Bot 98:539–548
Pyrke AF, Marsden-Smedley JB (2005) Fire-attributes categories, fire sensitivity, and flammability of Tasmanian vegetation communities. Tasforests 16:35–46
Pyšek P, Jarošík V, Hulme PE et al (2012) A global assessment of invasive plant impacts on resident species, communities and ecosystems: the interaction of impact measures, invading species' traits and environment. Glob Change Biol 18:1725–1737
Pyšek P, Pergl J, Essl F et al (2017) Naturalized alien flora of the world: species diversity, taxonomic and phylogenetic patterns, geographic distribution and global hotspots of plant invasion. Preslia 89:203–274
Qambrani NA, Rahman MM, Won S et al (2017) Biochar properties and eco-friendly applications for climate change mitigation, waste management, and wastewater treatment: a review. Renew Sust Energ Rev 79:255–273
Qasem JR (1995) The allelopathic effect of three *Amaranthus* spp. (pigweeds) on wheat (*Triticum durum*). Weed Res 35:41–49
Qi SS, Dai ZC, Zhai DL et al (2014) Curvilinear effects of invasive plants on plant diversity: plant community invaded by *Sphagneticola trilobata*. PLoS One 9:e113964
Qiu J (2015) A global synthesis of the effects of biological invasions on greenhouse gas emissions. Glob Ecol Biogeogr 24:1351–1362
Quijas S, Jackson LE, Maass M et al (2012) Plant diversity and generation of ecosystem services at the landscape scale: expert knowledge assessment. J Appl Ecol 49:929–940

Randolph P, Bansode RR, Hassan OA et al (2017) Effect of biochars produced from solid organic municipal waste on soil quality parameters. J Environ Manag 192:271–280

Rashid MH, Asaeda T, Uddin MN (2010) The allelopathic potential of kudzu (*Pueraria montana*). Weed Sci 58:47–55

Ray RL, Griffin RW, Fares A et al (2020) Soil CO_2 emission in response to organic amendments, temperature, and rainfall. Sci Rep 10:1–14

Reigosa MJ, Sánchez-Moreiras A, González L (1999) Ecophysiological approaches to allelopathy. Crit Rev Plant Sci 18:577–608

Reigosa MJ, Pedrol N, Sánchez-Moreiras A et al (2002) Stress and allelopathy. In: Reigosa MJ, Pedrol N (eds) Allelopathy from molecules to ecosystems. Science Publishers, Enfield, pp 231–256

Reinhart KO, Callaway RM (2006) Soil biota and invasive plants. New Phytol 170:445–457

Rice EL (1984) Allelopathy. Academic Press, New York

Ridenour WM, Callaway RM (2001) The relative importance of allelopathy in interference: the effects of an invasive weed on a native bunchgrass. Oecologia 126:444–450

Rogovska N, Laird D, Cruse RM et al (2012) Germination tests for assessing biochar quality. J Environ Qual 41:1014–1022

Sackett TE, Basiliko N, Noyce GL et al (2015) Soil and greenhouse gas responses to biochar additions in a temperate hardwood forest. GCB Bioenergy 7:1062–1074

Sashidhar P, Kochar M, Singh B et al (2020) Biochar for delivery of agri-inputs: current status and future perspectives. Sci Total Environ 703:134892

Scheer C, Grace PR, Rowlings DW, Kimber S, Van Zwieten L (2011) Effect of biochar amendment on the soil-atmosphere exchange of greenhouse gases from an intensive subtropical pasture in northern New South Wales, Australia. Plant Soil 345:47–58

Schmidt SK (1988) Degradation of juglone by soil bacteria. J Chem Ecol 14:1561–1571

Schumacher E, Kueffer C, Tobler M et al (2008) Influence of drought and shade on seedling growth of native and invasive trees in the Seychelles. Biotropica 40:543–549

Scott R, Sullivan WC (2007) A review of suitable companion crops for black walnut. Agrofor Syst 71:185–193

Seastedt TR (2015) Biological control of invasive plant species: a reassessment for the Anthropocene. New Phytol 205:490–502

Siemens DH, Garner SH, Mitchell-Olds T et al (2002) Cost of defense in the context of plant competition: *Brassica rapa* may grow and defend. Ecology 83:505–517

Singh BP, Hatton BJ, Singh B et al (2010) Influence of biochars on nitrous oxide emission and nitrogen leaching from two contrasting soils. J Environ Qual 39:1224–1235

Singh R, Babu JN, Kumar R et al (2015) Multifaceted application of crop residue biochar as a tool for sustainable agriculture: an ecological perspective. Ecol Eng 77:324–347

Singh R, Srivastava P, Singh P et al (2019) Impact of rice-husk ash on the soil biophysical and agronomic parameters of wheat crop under a dry tropical ecosystem. Ecol Indic 105:505–515

Sinkkonen A (2006) Ecological relationships and allelopathy. In: Reigosa MJ, Pedrol N, González L (eds) Allelopathy: a physiological process with ecological implications. Springer, Netherlands, pp 373–393

Smith P (2016) Soil carbon sequestration and biochar as negative emission technologies. Glob Chang Biol 22:1315–1324

Smith RG, Maxwell BD, Menalled FD et al (2006) Lessons from agriculture may improve the management of invasive plants in wildland systems. Front Ecol Environ 4:428–434

Smith JL, Collins HP, Bailey VL (2010) The effect of young biochar on soil respiration. Soil Biol Biochem 42:2345–2347

Šoln K, Žnidaršič N, Dolenc Koce J (2022) Root growth inhibition and ultrastructural changes in radish root tips after treatment with aqueous extracts of *Fallopia japonica* and *F.*× *bohemica* rhizomes. Protoplasma 259:343–355

Somerville PD, Farrell C, May PB et al (2020) Biochar and compost equally improve urban soil physical and biological properties and tree growth, with no added benefit in combination. Sci Total Environ 706:135736

Souto XC, Gonzales L, Reigosa MJ (1994) Comparative analysis of allelopathic effects produced by four forestry species during decomposition process in their soils in Galicia (NW Spain). J Chem Ecol 20:3005–3015

Spokas KA, Koskinen WC, Baker JM, Reicosky DC (2009) Impacts of woodchip biochar additions on greenhouse gas production and sorption/degradation of two herbicides in a Minnesota soil. Chemosphere 77:574–581

Spokas KA, Baker JM, Reicosky DC (2010) Ethylene: Potential key for biochar amendment impacts. Plant Soil 333:443–452

Steel ZL, Safford HD, Viers JH (2015) The fire frequency-severity relationship and the legacy of fire suppression in California forests. Ecosphere 6:1–23

Steenhagen DA, Zimdahl RL (1979) Allelopathy of leafy spurge (*Euphorbia esula*). Weed Sci 27: 1–3

Stewart G, Hull AC (1949) Cheat-grass (*Bromus tectorum*) L. – an ecological intruder in southern Idaho. Ecology 30:58–74

Stricker KB, Hagan D, Flory SL (2015) Improving methods to evaluate the impacts of plant invasions: lessons from 40 years of research. AoB Plants 7:plv028

Sujeeun L, Thomas SC (2017) Potential of biochar to mitigate allelopathic effects in tropical Island invasive plants: evidence from seed germination trials. Trop Conserv Sci 10: 1940082917697264

Sujeeun L, Thomas SC (2022) Biochar rescues native trees in the biodiversity hotspot of Mauritius. Forests 13:277

Sujeeun L, Thomas SC (2023) Biochar mitigates allelopathic effects in temperate trees. Ecol Appl 33:e2832

Thelen GC, Vivanco JM, Newingham B et al (2005) Insect herbivory stimulates allelopathic exudation by an invasive plant and the suppression of natives. Ecol Lett 8:209–217

Thomas SC, Gale N (2015) Biochar and forest restoration: a review and meta-analysis of tree growth responses. New For 46:931–946

Thomas SC, Frye S, Gale N et al (2013) Biochar mitigates negative effects of salt additions on two herbaceous plant species. J Environ Manag 129:62–68

Tong C, Wang WQ, Huang JF et al (2012) Invasive alien plants increase CH_4 emissions from a subtropical tidal estuarine wetland. Biogeochemistry 111:677–693

Usowicz B, Lipiec J, Łukowski M et al (2016) The effect of biochar application on thermal properties and albedo of loess soil under grassland and fallow. Soil Tillage Res 164:45–51

van Kleunen M, Dawson W, Maurel N (2015) Characteristics of successful alien plants. Mol Ecol 24:1954–1968

van Zwieten L, Singh B, Joseph S et al (2009) Biochar and emissions of Non-CO_2 greenhouse gases from soil. In: Lehmann J, Joseph S (eds) Biochar for environmental management, 2nd edn. Routledge, New York, pp 259–282

van Zwieten L, Kimber S, Morris S et al (2010) Influence of biochars on flux of N_2O and CO_2 from ferrosol. Soil Res 48:555–568

Vilà M, Weiner J (2004) Are invasive plant species better competitors than native plant species?–evidence from pair-wise experiments. Oikos 105:229–238

Vilà M, Tessier M, Suehs CM et al (2006) Local and regional assessment of the impacts of plant invaders on vegetation structure and soil properties of Mediterranean islands. J Biogeogr 33: 853–861

Vilà M, Basnou C, Pyšek P et al (2010) How well do we understand the impacts of alien species on ecosystem services? A pan-European, cross-taxa assessment. Front Ecol Environ 8:135–144

Virah-Sawmy M, Mauremootoo J, Marie D et al (2009) Rapid degradation of a Mauritian rainforest following 60 years of plant invasion. Oryx 43:599

Vivrette NJ, Muller CH (1977) Mechanism of invasion and dominance of coastal grassland by *Mesembranthemum crystallinum*. Ecol Monogr 47:301–318

Walker LR, Smith SD (1997) Impacts of invasive plants on community and ecosystem properties. In: Luken JO, Thieret JW (eds) Assessment and management of plant invasions. Springer, New York, pp 69–86

Wan JZ, Wang CJ (2018) Expansion risk of invasive plants in regions of high plant diversity: A global assessment using 36 species. Ecol Inform 46:8–18

Wang Z, Li Y, Chang S et al (2014) Contrasting effects of bamboo leaf and its biochar on soil CO_2 efflux and labile organic carbon in an intensively managed Chinese chestnut plantation. Biol Fertil Soils 50:1109–1119

Wang JY, Xiong ZQ, Kuzyakov Y (2016) Biochar stability in soil: meta analysis of decomposition and priming effects. GCB Bioenergy 8:512–523

Wardle DA, Zackrisson O, Nilsson MC (1998) The charcoal effect in boreal forests: mechanisms and ecological consequences. Oecologia 115:419–426

Webb SL, Pendergast TH, Dwyer ME (2001) Response of native and exotic maple seedling banks to removal of the exotic, invasive Norway maple (*Acer platanoides*). J Torrey Bot Soc 128:141–149

Weidenhamer JD, Callaway RM (2010) Direct and indirect effects of invasive plants on soil chemistry and ecosystem function. J Chem Ecol 36:59–69

Weidlich EWA, Flórido FG, Sorrini TB et al (2020) Controlling invasive plant species in ecological restoration: A global review. J Appl Ecol 57:1806–1817

Werner C, Lucht W, Gerten D, Kammann C (2022) Potential of land-neutral negative emissions through biochar sequestration. Earth's Future 10:e2021EF002583

Whisenant S (1990) Changing fire frequencies on Idaho's Snake River plains: ecological and management implications. In: Proceedings from the symposium on cheatgrass invasion, shrub die-off and other aspects of shrub biology and management, U.S. Department of Agriculture, Forest Service, Intermountain Research Station, Ogden, 5–7 April 1989

Willis RJ (2000) *Juglans* spp., juglone and allelopathy. Allelopathy J 7:1–55

Witkowski ETF (1991) Growth and competition between seedlings of *Protea repens* (L.) L. and the alien invasive, *Acacia saligna* (Labill.) Wendl. in relation to nutrient availability. Funct Ecol 5: 101–110

Woolf D, Lehmann J (2012) Modelling the long-term response to positive and negative priming of soil organic carbon by black carbon. Biogeochemistry 111:83–95

Woolf D, Amonette JE, Street-Perrott FA et al (2010) Sustainable biochar to mitigate global climate change. Nat Commun 1:1–9

Wu AP, Yu H, Gao SQ et al (2009) Differential belowground allelopathic effects of leaf and root of *Mikania micrantha*. Trees 23:11–17

Wyse SV, Perry GL, Curran TJ (2018) Shoot-level flammability of species mixtures is driven by the most flammable species: implications for vegetation-fire feedbacks favouring invasive species. Ecosystems 21:886–900

Xu M, Shang H (2016) Contribution of soil respiration to the global carbon equation. J Plant Physiol 203:16–28

Yanai Y, Toyota K, Okazaki M (2007) Effects of charcoal addition on N_2O emissions from soil resulting from rewetting air-dried soil in short-term laboratory experiments. Soil Sci Plant Nutr 53:181–188

Yang F, Zhao L, Gao B et al (2016) The interfacial behavior between biochar and soil minerals and its effect on biochar stability. Environ Sci Technol 50:2264–2271

Yao Y, Gao B, Fang J et al (2014) Characterization and environmental applications of clay–biochar composites. Chem Eng J 242:136–143

Young JA, Evans RA (1978) Population dynamics after wildfires in sagebrush grasslands. Rangeland Ecol Manag 31:283–289

Zhang L, Wang H, Zou J et al (2014) Non-native plant litter enhances soil carbon dioxide emissions in an invaded annual grassland. PLoS One 9:e92301

Zhang R, Zhang Y, Song L et al (2017) Biochar enhances nut quality of *Torreya grandis* and soil fertility under simulated nitrogen deposition. For Ecol Manag 391:321–329

Zhao L, Xiao D, Liu Y et al (2020) Biochar as simultaneous shelter, adsorbent, pH buffer, and substrate of *Pseudomonas citronellolis* to promote biodegradation of high concentrations of phenol in wastewater. Water Res 172:115494

Zhou C, Heal K, Tigabu M et al (2020) Biochar addition to forest plantation soil enhances phosphorus availability and soil bacterial community diversity. Forest Ecol Manag 455:117635

Zhou X, Shi L, Moghaddam TB et al (2022) Adsorption mechanism of polycyclic aromatic hydrocarbons using wood waste-derived biochar. J Hazard Mater 425:128003

Zimmerman N, Hughes RF, Cordell S et al (2008) Patterns of primary succession of native and introduced plants in lowland wet forests in eastern Hawai'i. Biotropica 40:277–284

Zimmerman AR, Gao B, Ahn MY (2011) Positive and negative carbon mineralization priming effects among a variety of biochar-amended soils. Soil Biol Biochem 43:1169–1179

Part V

Plant Invasion and Policy Interventions

An Action Plan to Prevent and Manage Alien Plant Invasions in India

17

Achyut Kumar Banerjee and K. V. Sankaran

Abstract

Biological invasions are a major threat to native biodiversity, ecosystem services, socio-economic status, and good quality of life worldwide. Human preferences for species for food and feed, ornamental purposes, forestry, soil improvement, and other uses contributed significantly to the introduction of alien species outside their native range. Increased trade and travel and climate and land-use changes are some of the major causes and promoters of invasion by alien species. The invasions are predicted to increase in the future if these pathways and drivers are left unmanaged. Against this background, it is suggested that biosecurity measures may be updated and implemented meticulously, and efficient control measures adopted to prevent the introduction and spread of invasive alien species across the globe, especially in emerging economies like India.

In this chapter, we evaluated the existing capabilities and know-how in India to deal with the introduction and spread of invasive alien plant species (IAPS) and found that the current biosecurity policies, infrastructure and management measures are inadequate to address the problem successfully. We also identified several roadblocks to effective implementation of management actions against IAPS, like lack of coordination in activities, conflicts of interest among stakeholders, lack of dedicated regulations, inadequate experience, and scarcity of resources. By taking clues from IAS regulations used successfully to mitigate IAPS threats elsewhere, we propose certain mandatory changes in policies to

A. K. Banerjee (✉)
State Key Laboratory of Biocontrol and Guangdong Provincial Key Laboratory of Plant Resources, School of Life Sciences, Sun Yat-sen University, Guangzhou, China
e-mail: banerjee@mail.sysu.edu.cn

K. V. Sankaran
Kerala Forest Research Institute, Peechi, Thrissur, India

S. Tripathi et al. (eds.), *Plant Invasions and Global Climate Change*,
https://doi.org/10.1007/978-981-99-5910-5_17

regulate international and domestic trading, prevent accidental introductions, and manage existent invasions in the country. We also recommend the formation of a National Invasive Species Strategy and Action Plan and a dedicated agency to implement the action plans. The agency should ideally have a decentralized system, and it may develop a coordinated and multisectoral network of stakeholders for efficient operation. The need to raise public awareness through media, education, and citizen science programs and strengthen national response capacities through capacity building, scientific research, knowledge sharing, and collaboration are also raised. In this era of globalisation and rapid climate and land-use changes, the suggestions highlighted in this chapter would greatly assist the country in bringing down impending invasions by alien species and mitigating damages from the existing invasions.

Keywords

Biosecurity measures · Developing economy · Invasive alien plant species · Management · Policy · Response capacity

Abbreviations

IAPS	Invasive Alien Plant Species
IAS	Invasive Alien Species
MoEF&CC	Ministry of Environment, Forests, and Climate Change
NISSAP	National Invasive Species Strategy and Action Plan

17.1 Introduction

The Indian subcontinent has been colonised multiple times in the past, starting with the Arabs, followed by the Mughals and the Europeans. Numerous alien species were introduced intentionally or inadvertently along with the colonisation (Kannan et al. 2013a). With the development of botanical gardens and research facilities across the country under British rule, the import of foreign plant species, particularly those with commercial, medicinal, or decorative values, reached its zenith (Soumya and Sajeev 2020). Some of these species eventually escaped cultivation or as garden plants to become invaders (i.e., the self-sustaining population in the wild with individuals dispersing, surviving, and reproducing across multiple habitats in a multitude of ecosystems (Blackburn et al. 2011)). These invasive alien plant species (IAPS) formed one of the major threats to native biodiversity, ecosystem services, man and animal health, and the economy (Rai 2022).

Recent studies show that emerging economies like India face the greatest threat of biological invasion due to increased globalisation of trade and climate and land-use changes (Seebens et al. 2017). Also, reports of negative impacts of IAPS on human livelihood mostly come from Southeast Asia (Shackleton et al. 2019). Effective and timely implementation of biosecurity and management measures are, therefore, crucial to prevent the entry and spread of invasive alien species (IAS).

India has long recognised the negative impacts of IAS, similar to other environmental issues. The country became a party to the Convention on Biological Diversity (CBD) in February 1994. It developed the first national-level policy and action plans for biodiversity conservation and sustainable use in 1999. India established a National Biodiversity Action Plan (NBAP) in 2008, in which regulation of the introduction of invasive alien species and their management was included as one of the 11 objectives. In line with the CBD's Strategic Plan for Biodiversity 2011–2020, the NBAP was updated, and 12 National Biodiversity Targets (NBTs) were identified. The updated NBAP and the NBTs were submitted as part of India's fifth National Report to the CBD as a separate document, namely the Addendum 2014 to NBAP 2008. The IAS management was identified as one of the 12 NBTs (NBT 4), and nine action points (APs: 59–67) were outlined for regulation of the introduction of IAS and the management of those species which have already invaded and spread across the country (https://wii.gov.in/images/images/documents/NBAP_Addendum_2014.pdf; accessed on 27 December 2022) (see Box 17.1).

Box 17.1: Actionable Points to Regulate The Introduction of Invasive Alien Species and their Management

In India's fifth National Report to the Convention on Biological Diversity (CBD), the IAS management was identified as one of the 12 National Biodiversity Targets (NBTs). The following nine action points (APs: 59–67) were outlined in the addendum for the regulation of the introduction of IAS and the management of those already present in the country.

- Develop a unified national system for regulation of all introductions and carrying out rigorous quarantine checks (AP: 59).
- Strengthen domestic quarantine measures to contain the spread of invasive species to neighbouring areas (AP: 60).
- Promote intersectoral linkages to check unintended introductions and contain and manage the spread of invasive alien species (AP: 61).
- Develop a national database on invasive alien species reported in India (AP: 62).
- Develop appropriate early warning and awareness systems in response to new sightings of invasive alien species (AP: 63).
- Provide priority funding to basic research on managing invasive species (AP: 64).
- Support capacity building for managing invasive alien species at different levels with priority on local area activities (AP: 65).

(continued)

Box 17.1 (continued)

- Promote restorative measures of degraded ecosystems using preferably locally adapted native species for this purpose (AP: 66).
- Promote regional cooperation in adoption of uniform quarantine measures and containment of invasive exotics (AP: 67).

In the first section of this chapter, we have provided an overview of the existing policies and capabilities of the country to (1) regulate the introduction and spread of the IAPS and (2) identify the IAPS that have invaded the country to date. In the second section, potential hurdles to effectively implementing the IAPS regulations have been identified, and a set of policy interventions based on probable stakeholder perceptions is proposed. Finally, a national strategy and action plan to overcome the impediments to managing IAS in the country and how to successfully implement the action points included in India's fifth National Report to the CBD have been outlined (Box 17.1).

17.2 Existing Policies and Expertise to Manage IAPS in India

17.2.1 Regulatory Policies to Prevent Introduction and Manage Spread of IAPS

A review of the policies and regulations to prevent invasions by alien species enacted by Govt. of India can be found in Khetarpal et al. (2017) (see Box 17.2 for a summary of these legislations). The regulation for importing plant materials into India is addressed under The Plant Quarantine (Regulation of Import into India) Order (PQ Order), 2003, and amendments made to it subsequently. There were 95 amendments to the PQ Order until December 2021 (http://extwprlegs1.fao.org/docs/pdf/ind193998.pdf). The Directorate of Plant Protection, Quarantine, and Storage (DPPQ&S) is responsible for implementing the PQ Order. Under this Order, biosecurity and phytosanitary import permits and quarantine clearances are undertaken following standard operating procedures. A pest risk assessment is conducted for importing plant or plant materials into the country, after which import permits are issued. The plants and planting materials are categorised under four main categories: prohibited (Schedule IV), restricted (Schedule V), permitted to import with additional declarations (Schedule VI), plant materials for consumption purposes (Schedule VII), and prohibited quarantine weeds (Schedule VIII, amended in 2019).

Box 17.2: Environmental Protection Policies that Have Provisions for Invasive Alien Plant Species (IAPS) Management in India

In India's report on the transnational policy network submitted to the Convention on Biological Diversity in 2011, ten legislations related to IAS have been listed:

- The Prevention and Control of Infectious and Contagious Disease in Animals Act, 2009.
- The Plant Quarantine (Regulation of Import into India) Order, 2003.
- The Destructive Insects and Pests Act, 1914 (and amendments).
- The Plants, Fruits, and Seeds (Regulation of Import into India) Order, 1989.
- Livestock Importation Act, 1898.
- Environment Protection Act, 1986.
- The Biological Diversity Act, 2002.
- Indian Forest Act, 1927.
- Wildlife (Protection) Act, 1972.
- Forest (Conservation) Act, 1980.

Out of these ten legislations, four have provisions for IAPS management, the scopes of which are mentioned below. The Indian Forest Act, 1927 and Forest (Conservation) Act, 1980 (last amended 1996) have no mention of invasive alien species; these are focused on addressing the deforestation issue and improving the livelihood of the dependent people.

1. Name of the Regulation: Destructive Insects and Pests Act, 1914
   Scopes:
   - Regulation of import of plants and plant materials, diseases, and insects likely to cause infection or infestation into India.
   - Regulation of transport and carriage of the same within the country.
   - Inspection, detention, disinfection, or destruction of the same (by the state governments) identified by the central government.
2. Name of the Regulation: Plants, Fruits, and Seeds (Regulation of Import into India), 1984 (Revised in 1989)
   Scopes: Regulation of import of seeds or planting materials of plant species considered as agricultural commodities and thus having economic values.
3. Name of the Regulation: Environment (Protection) Act, 1986
   Scopes: Prohibit or restrict the movement and handling of hazardous substances that can cause potential damage to the environment.
4. Name of the Regulation: Plant Quarantine (Regulation of Import into India) Order, 2003
   Scopes:

(continued)

Box 17.2 (continued)

- Pest risk assessment for all imports.
- Import restriction of packaging material unless it is treated.
- Regulating import of timber, wooden logs, germplasm, soil, seeds.
- Agricultural commodities categorised as:
  - Schedule IV: 14 plant taxa, which are prohibited entry into India from designated countries.
  - Schedule V: 17 plant taxa are identified for which restricted imports are permissible with recommendations of authorised institutions and additional declarations from the eligible authority.
  - Schedule VI: 571 plant taxa, which are permitted to be imported in India if additional declarations are included in the phytosanitary certificate and special conditions are met during import.
  - Schedule VII: 279 plant taxa for consumption purposes and has the same conditions for import as Schedule VI.
  - Schedule VIII has the list of quarantine plants (prohibited, restricted, and regulated) that includes 57 species (amended in 2019).

An **Agricultural Biosecurity Bill** has been submitted in 2013. It has provision for the establishment of an authority for the prevention, eradication, and control of pests, including IAPS, and diseases of plants and animals and unwanted organisms. This bill is yet to be approved.

Domestic legislation having provisions for IAPS management (arranged chronologically)

- The Destructive Insects and Pests Act, 1914.
- The Madras Agricultural Pests and Diseases Act, 1919.
- The Travancore Plant Pests and Diseases Act, 1919.
- The Coorg Agricultural Pests and Diseases Act, 1933.
- The Patiala Destructive Insects and Pests Act, 1943.
- The Bombay Agricultural Pests and Diseases Act, 1947.
- The Rewa State Agricultural Pests and Diseases Act, 1947.
- The East Punjab Agricultural Pests, Diseases and Noxious Weeds Act, 1949.
- The East Punjab Agricultural Pests, Diseases and Noxious Weeds Act—extended to Himachal Pradesh, 1949.
- The Assam Agricultural Pests and Diseases Act, 1950.

On the domestic front, under the Destructive Insects and Pests Act 1914, the State Governments are empowered to prevent (restrict transportation, inspection, detention, disinfection, or destruction) the spread of plant pests that are destructive to crops and forestry within the country. At the regional and state level, there are several policies that are focused primarily on agricultural pests and insects but also have provisions for the management of the IAS (Box 17.2). For example, The Assam Agricultural Pests and Diseases Act 1950 has provisions for preventing the spread of insect pests, plant diseases, and noxious weeds in the State of Assam.

Management of IAS in India is a multi-agency affair, governed by several Ministries and implemented through multiple short- and long-term management programs. In India's transnational policy network report submitted to the Convention on Biological Diversity in 2011 (https://www.cbd.int/invasive/doc/legislation/India.pdf), twelve Government agencies have been cited that deal with IAS issues. In general, the Directorate of Plant Protection, Quarantine & Storage (DPPQ&S), under the Ministry of Agriculture and Farmers Welfare, is responsible for managing the introduction of cultivated plants and biological control agents and has supported projects on eradicating weedy plants, pathogens, and pests. The MoEF&CC (Ministry of Environment, Forests, and Climate Change) focuses on forest invasive species and supports research programmes on managing these species.

17.2.2 Identification of IAPS

Creating a national database of alien flora is necessary to identify emerging invaders and manage the ones already established (Pagad et al. 2018). The scientific documentation of Indian alien flora started in the mid-twentieth century when (Chatterjee 1940) identified the endemic (61.5%) and non-endemic (38.5%) dicotyledonous plant species of the Indo-Burma geographic region. Several national [e.g., (Nayar 1977)] and regional [e.g., (Pandey and Parmar 1994)] inventories of alien flora have been published since then. Based on these reports, (Reddy 2008) identified 173 IAPS in India belonging to 117 genera and 44 families. The Alien Flora of India, published in 2012 (Khuroo et al. 2012), identified 1599 species as alien in India, of which 145 species were cited as invasives. The most recent inventory of Indian alien flora is the Indian version of the Global Register of Invasive and Introduced Species (hereafter, GRIIS), which has identified 266 IAPS of the 2082 alien plant species (Sankaran et al. 2020). In addition to these national inventories, the state- [e.g., (Singh et al. 2010)] and regional-level [e.g., (Wani et al. 2022)] checklists of IAPS have also been brought out, providing new information.

Several Government agencies have also taken initiatives to enlist the IAPS of India. The most notable of these are the lists produced by the ENVIS Centre on Floral Diversity and the National Biodiversity Authority (NBA). The ENVIS checklist contained 173 IAPS (Reddy 2008). The NBA adopted an impact-assessment-based approach and recognised 63 IAPS in India in 2018 (http://nbaindia.org/uploaded/pdf/Iaslist.pdf, accessed on 27 December 2022). However, neither of these checklists were updated after they were produced.

17.3 Impediments to Manage Invasive Alien Species in India

IAS threat to the economy and environment in India is looming large, and invasions continue unabated despite efforts by Government agencies to prevent and manage invasions. The major impediments which frustrate the management of IAS in India are discussed below.

17.3.1 Lack of Cooperation and Coordination

Regardless of the number of Ministries and Government agencies involved in managing IAS and research institutions working on invasive species, there are significant gaps in information on the distribution of these species and the negative impacts they cause in the country. Also, the pathways and the major and minor drivers of invasion are unclear for most species. The list of alien plant species in India provided in the GRIIS database can be considered comprehensive (pending updation), though views may differ on whether a few of the species included are invasive or naturalised. However, the lists of invasive microbes, vertebrates, invertebrates, and marine species are grossly incomplete. In short, India is yet to produce a comprehensive list of IAS pending systematic inventories nationwide and across ecosystems. Against this background, prioritising a species for management and choice of management actions are challenging. Moreover, there is a lack of cooperation and coordination in activities implemented by Government agencies to manage IAS. For example, the ENVIS and the NBA (both under the MoEF&CC) have produced separate lists of invasive alien plants instead of working together to prepare a single authentic list.

Of late, certain environmental initiatives (e.g., Green India Mission) have been promoting the planting of fast-growing alien tree species to sequester carbon (to combat climate change) or for higher timber value without assessing the invasion risks they may pose. Similarly, the Social Forestry Scheme introduced by MoEF&CC also promoted the planting of alien trees across the country without assessing the risk of invasion by these species. For example, the tree species for raising forest plantations by the Department of Forests, Government of Tamil Nadu, included several Australian *Acacia* species identified as IAPS in India (https://www.tntreepedia.com/, accessed 6 September 2022). In fact, planting of Australian Acacias under the social forestry scheme (e.g., *Acacia auriculiformis* Benth. [*Racosperma auriculiformis* (Benth.) Pedley], *A. mangium* Willd. and *A. mearnsii* De Wild) has been adopted by several states from the 1980s. This shows a lack of consultation and coordination between different Government departments.

17.3.2 Want of Dedicated Policies and Regulatory Bodies

Quarantine is the primary biosecurity measure to prevent the entry of new invasive alien species (IAS) into any country. However, the list of 57 quarantined plant species in India (Schedule VIII in the PQ Order) is far from adequate for restricting the transborder movement of invasive alien plant species into the country. Moreover, the lists of taxa included in Schedules IV–VII are species of important pests or diseases of crops and forestry species. They include only very few IAPS recognised by the Government and Research Institutions in the country. For example, out of 571 plant taxa listed in Schedule VI, only six species are recognised IAPS in India. Even these species (except *Macroptilium lathyroides* (L.) Urb. and *Ricinus communis* L.) can be imported if the plant materials are free from seeds of 57 quarantined plant species listed under Schedule VIII. One reason for the under-representation of IAPS in the prohibited, restricted, and quarantined lists could be that the primary focus of Government agencies such as the DPPQS (for bulk material) and Indian Council of Agricultural Research (ICAR) (for research material), the nodal agencies for regulating plant imports to India, is agricultural pests. The PQ Order also lists a few forestry species, but these are listed along with the agricultural pests, not separately (Gupta and Sankaran 2021). Several Government bodies, such as MoEF&CC and the Department of Biotechnology, can regulate the import of GMOs and transgenics; however, no dedicated agency exists for controlling the transborder movement of alien plant species. In addition, the existing biosecurity regulations of India provide no stipulation for checking the unintentional introduction of alien plant species through air, sea, and land ports.

17.3.3 Lack of Awareness and Competence

India has taken several steps to mitigate the damages due to IAS. However, there is a general lack of awareness of the magnitude of damages caused by IAS among all concerned (e.g., policymakers, farmers, foresters, agriculturists, and the common man). The lack of competence to distinguish between alien and native species often delays management actions. Also, most stakeholders have no idea who to approach for management advice should they locate an alien species on their premises or the land they manage.

Resource managers, scientists, foresters, and other stakeholders often lack knowledge and skills in employing modern tools and techniques for surveillance, detection, prevention, and management of IAS. Evidence-based decision-making is unknown or not practiced, and hence there is confusion on which method of management would be effective and when during the different phases of invasion—pre-entry, entry of the species, establishment, spread, and widespread. The 'Integrated Forest Management (IFM)' scheme to control and eradicate IAPS in forest areas introduced by the MoEF&CC in 2009 did not fully succeed due to either a lack of awareness about IAPS or the provisions of the scheme were contradictory to the Wildlife Act, 1972, which prohibited harvesting of any life form, including the IAPS, from the protected areas (Kannan et al. 2013b).

17.3.4 Paucity of Resources for Long-Term Management Actions

The paucity of resources for the long-term management of IAS and continued surveillance of ecosystems to check re-invasion is a major issue hindering the success of management programmes. Most often, allocations from the Government may be limited only to a one-time action, whatever the method used—mechanical/physical, chemical, or biological. Chances to secure funding from private agencies are limited unless the success of management is proven and sustained, which is not practical since mechanical and chemical methods are effective only in the short term. And the efficacy of biological control will be visible only in the long term. Allocation of funds from international agencies is normally for a short-term research/management program in anticipation of continued funding from the respective country governments to carry on with the programs initiated (Boy and Witt 2013).

17.3.5 Need for Surveillance and Monitoring for IAS

Periodic surveillance for invasive alien species in forests, agriculture, and other ecosystems is uncommon in the country unless there is an urgency to trace the distribution of an invasive alien species, which is widespread in the neighbouring countries, posing a threat to India. It is possible that the species has spread more extensively by the time its distribution is mapped, necessitating species-led or site-based management. This lack of surveillance and monitoring hinders early detection and rapid control or containment of a species.

17.3.6 Conflicts of Interest

Management of some IAPS is often hindered in rural economies in the developing world since local communities depend on these species for their livelihood (Nunez and Pauchard 2010). Fast growth, abundance, and hardiness are some of the beneficial traits of these IAPS. A good example is the IAPS *Prosopis juliflora* (Sw.) DC., which is used as fodder and for making charcoal in several parts of India (Patnaik et al. 2017). In the urban context, some IAPS are valued as a trading commodity, primarily for their horticultural value. A recent study has shown that nearly 50% of the IAPS in India have ornamental values, of which more than 75% are actively traded in the market illegally (Banerjee et al. 2021). *Lantana camara* L. and *Sphagneticola trilobata* (L.) Pruski are some of the most common examples of this trade. It, in turn, generates community dependence on the species and promotes their introduction and spread into new regions (Negi et al. 2019). Illegal import of seeds of alien species and their trade by private agencies is another major pathway of invasion in various habitats.

India lacks an exclusive national policy or legislation to prevent trading or use of IAS. Also, no legal framework makes a government agency fully responsible for any unintended entry and spread of an IAS or its impact. The Environment (Protection)

Act, 1986 and Rules 1989 have provisions for restricting the movement and handling of substances that can cause harm to the environment; however, the problem of IAS is not explicitly mentioned in that order. Notably, society may oppose any policy regulation to manage a beneficial IAPS. Hence, the economic importance of an IAS may be considered while developing policy regulations for management (e.g., to identify IAPS for priority management) (Sandilyan et al. 2018).

17.4 Suggestions for Improving the Policy Framework

It is evident from the foregoing facts there is good scope for improving the existing biosecurity regulations in the country to achieve effective prevention and management of IAPS. Previous assessments have also highlighted the need for an improvement in the current legislation for successful prevention and management of biological invasions in India [e.g., see (Gupta and Sankaran 2021)]. In this section of the chapter, we have attempted to discuss how the biosecurity infrastructure of India can be strengthened. We relied heavily on IAS-related policies adopted by some of the developed and developing countries to make these proposals. It is hoped that these would help resolve the major impediments to the implementation of management actions and settle any potential conflict of interest among stakeholders.

17.4.1 Policies to Prevent Introduction and Spread of IAPS

Under the Plant Quarantine Order, 2003, the list of species to be quarantined was last amended in 2021. However, it is still inadequate to meet the challenges posed by IAPS in the country and needs to be expanded. Two approaches can be adopted to restrict the inflow of IAPS:

- Blocklist approach (equivalent to the 'blacklist' approach but without discriminating based on human race and colour), adopted by European countries (Essl et al. 2011) and North America (Simberloff 2006), which prohibits the introduction of potential IAPS.
- Safelist approach (equivalent to the 'whitelist' approach but without discriminating based on human race and colour), adopted by Australia and New Zealand (Auld 2012), which prohibits the introduction of alien species until they are declared safe by experts.

For either of these approaches, a weed risk assessment protocol may be established to identify the invasiveness of the alien species and subsequent integration of it into the national policy framework. The risk assessment procedure should ideally:

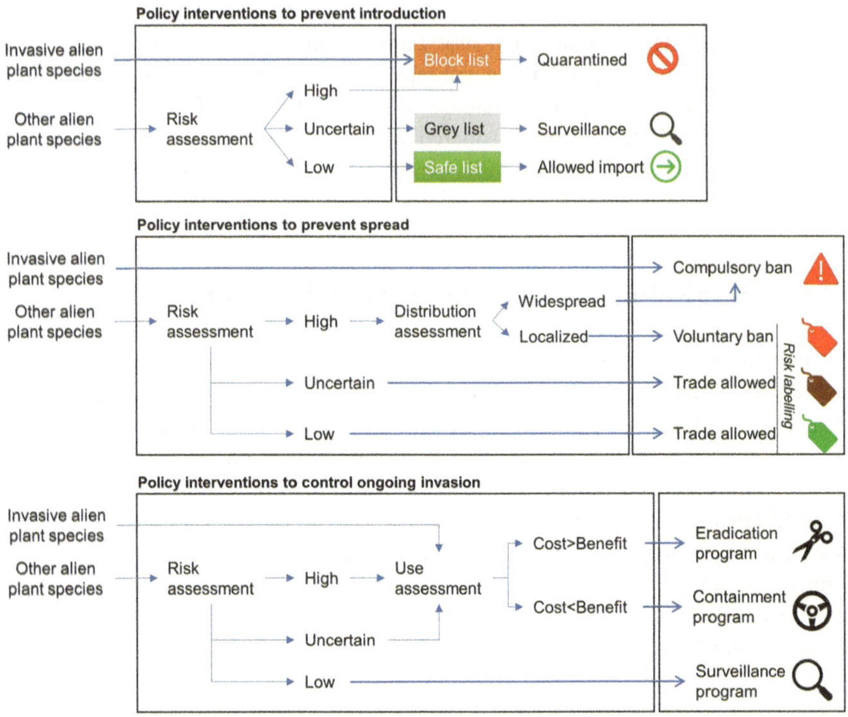

Fig. 17.1 Schematic representation of the proposed policy interventions to prevent the introduction and spread of alien plant species and control existing and ongoing invasions

- Follow a global standardised framework [e.g., (Leung et al. 2012)],
- Use consistent metrics for recording spatial abundance of the IAPS (Bradley et al. 2018)
- Include the socio-economic status of the IAPS [e.g., using the 'Socio-economic impact classification of alien taxa' (Bacher et al. 2018)]
- Include the cost-benefit analysis (CBA) as an essential decision-making component (Reyns et al. 2018)
- Should neither be too cautious nor too amenable in order to avoid omission and commission errors

Risk assessments of a large pool of alien species require sufficient resources. Also, if a consensus among stakeholders (experts, policymakers, and traders) on block listing criteria cannot be made, the solution could be:

- Based on expert judgement, the IAPS recognised as a threat to India are included in the blocklist (quarantine list) to restrict their entry into the country and ban their domestic trade (Fig. 17.1).

- The species identified as low risk may be introduced into the country, and trading is allowed in the domestic market (safelist).
- Species with uncertain risks may be included in the grey list. International and domestic trading of these species should be restricted until risk assessment is completed.

To expedite the implementation of the management actions and to avoid conflict of interest between stakeholders, especially for the block-listed species with economic uses and high trading values, the policy framework can consider the extent of spread and mode of dispersal of the species for decision-making (Fig. 17.1). The policymakers should prioritise legislating international and domestic trade-ban of the species with a localised distribution and having long-distance dispersal potential (Nathan et al. 2008). A voluntary sales ban with support from the industry can be imposed for species with widespread distribution. It is also essential that the legislative framework is dynamic. If an alien species is reported as invasive within the country or elsewhere, the focus of the policy measures should shift from preventing introduction to regulating domestic trade and managing existing populations.

Another strategy to restrict the introduction and spread of IAPS in the country could be shifting the consumers' focus from alien to native species. It can be achieved by:

- Imposing taxes on selling alien plants and incentivising the sale of native species, thus creating substantial price differences between native and alien species.
- Discouraging the traders and suppliers from selling alien plants and promoting traders to use government portals that deal exclusively with native species.
- Increasing citizen awareness by using colour-coded labels and mentioning the risks of the traded items (Fig. 17.1), thus helping them make an informed decision on purchase underpinning public opinion (Cordeiro et al. 2020).

17.4.2 Policies to Prevent Accidental Introductions

The following steps can be opted to prevent the unintentional movement of live organisms as contaminants of internationally traded commodities or as stowaways from transporting vessels and associated equipment.

- Importers may be required to declare that the imported stock is free from the block- and grey-listed species as a mandatory requirement for customs clearance. The importer may be made accountable for any unintended negative impact from the imported stock.
- Specific guidelines can be implemented for inspecting and cleaning vehicles and heavy machinery to make them free of IAPS propagules. These regulations should be specially targeted towards industries frequently affected by the IAPS, e.g., urban parks and recreational trails and public utilities (e.g., roadsides).

- The policy measures may also consider screening cargoes at the ports of entry to prevent the introduction of block- and grey-listed species.

17.4.3 Policies to Control Ongoing Invasions

Attempts at preventing the introduction and spread of an alien species should be followed by trials for eradication if prevention is unsuccessful. Eradication may be possible if the spread of the species is restricted to a small area. If eradication fails, containment of the species is the next option of management. If both eradication and containment of a species fail, the next option is to control it through species-based or site-based approaches involving mechanical, chemical, or biological methods. Site restoration with native species needs to be implemented wherever control measures are successful. The cost-effectiveness and stakeholder perspectives of eradication and containment may be considered before implementation (Fig. 17.1). In the cost-benefit analysis, both private (e.g., landowner) and social (e.g., public) perspectives should be considered. On the one hand, the economic impacts of IAPS measured in monetary terms characterise the private perspective of costs and benefits. The social perspective, on the other hand, considers both market and non-market (environmental) valuations. A suite of techniques is now available for estimating non-market prices in monetary terms [see Box 17.3 for an overview of these techniques (Hanley and Roberts 2019) and a hypothetical scenario of the decision-making process based on the cost-benefit analysis].

Box 17.3: Methods for Valuing Non-market Impacts

Environmental impacts have been estimated in monetary terms since the mid-1960s. Current available and widely used methodologies can be categorised into three types: stated preference method, revealed preference method, and production function method. All these methods are based on a standardised measure known as maximum willingness to pay (WTP). The WTP is a value people place on any environmental service and how much they are willing to pay to obtain it.

Stated preference methods (SPMs): Individuals are asked to make choices between different levels of environmental quality and the cost of provision. This method will show the value people place on, for example, funding an IAPS eradication program to protect native biodiversity. Two types of SPMs are available:

Contingent valuation: People are asked to vote whether they agree with the cost of provision (e.g., INR 500 to remove an IAPS from a waterbody for a year) for an environmental benefit (e.g., to maintain recreation opportunities).

(continued)

Box 17.3 (continued)

Choice modelling: People are asked to make choices between different packages of environmental benefits, from which the researcher can infer the economic value that people place on each of these benefits.

Revealed preference methods (RPMs): This method is based on actual behaviour rather than stated choices. The researcher infers the economic cost from consumer behaviours in markets related to non-market environmental goods. Two types of RPMs are available:

Travel cost models: In this method, the researcher estimates people's expenditures on outdoor recreation trips. In case of a change in the environmental quality of such a travelling destination (e.g., loss of tree covers due to an invasive pest), the economic losses are estimated in monetary terms.

The Hedonic pricing approach: This method examines the benefits of an IAS control where it affects the environmental benefit. For example, the spread of an IAPS can change the benefits of living at a lakeside location if the IAPS reduces the recreational opportunities in the lake.

Production function methods (PFMs): This method links IAS population changes to impacts on commercial crops and livestock or to human health.

A hypothetical scenario of the decision-making process based on the SPM choice modelling approach:

The decision to eradicate *Lantana camara* from an invaded landscape can be made through a choice modelling approach. People can make choices between benefits (e.g., increased household income) and losses (e.g., loss of native biodiversity, high management costs) incurred by the species. The findings can be integrated into the management decision, either in the form of implementing eradication measures or supporting public research programs to find alternate management techniques. From a landowner perspective, if the cost of managing a landscape invaded by a species (e.g., clearing a forest patch for uninterrupted movement of animals) overruns the benefit obtained from exploiting that species (e.g., by local industrial scale operation), the policies can focus on eradication programs and discouraging further use of the species.

17.5 Developing Infrastructure, Facilities, and Other Requirements

For effective management of IAS, India should develop and implement a National Invasive Species Strategy and Action Plan (NISSAP), as adopted by developed (e.g., North America; https://www.invasivespeciesinfo.gov/national-invasive-species-management-plan) and some developing economies (e.g., Tonga; https://leap.unep.org/countries/to/national-legislation/tonga-national-invasive-species-strategy-and-action-plan-2013). The Federal and State Governments may be made responsible for implementing the action plan to prevent, eradicate, control and site restoration. The government may identify IAS management as a priority item in its annual plans since economic and environmental impacts due to invasive species are huge.

Ideally, a single nodal agency should be identified or created to lead and implement NISSAP, as has been emphasised previously by several authors [e.g., (Khetarpal et al. 2017)]. The nodal agency may adopt a harmonised and integrated approach involving science and policy while implementing its actions. The action points that the nodal agency can prioritise are presented in Box 17.4. The major tasks of the agency are proposed to be—(a) timely and efficient enforcement of policies of NISSAP, (b) surveillance, prevention, eradication, or control of invasions, (c) increasing public awareness and promoting citizen participation in IAPS issues, (d) strengthening response capacities through research, training and knowledge sharing, and (e) ensuring stakeholder participation in all management programs. Here, we propose an outline of how the nodal agency is expected to function.

Box 17.4: Priority Action Points for the Nodal Agency

1. Create authentic baseline data on the distribution and spread of IAPS in the country.
2. Develop and implement a scientifically sound risk assessment scheme (in tune with internationally accepted methods) to prevent the introduction of potential IAPS and to distinguish between invasive and non-invasive alien species.
3. Oversee implementation of the updated biosecurity strategy to prevent invasion of alien species at the pre-border, border, and post-border.
4. Arrange periodic surveillance of IAPS in various ecosystems in collaboration with different departments (e.g., agriculture, forestry, National/state biodiversity boards) and local self-governments in different states.
5. Identify offices/officials who should be notified upon sighting a new alien species and who should be approached for identification and management advice.
6. Conduct and oversee early detection, rapid response, and containment plans wherever feasible.
7. Ensure stakeholder participation in all management activities. Local/indigenous knowledge, where available, may be incorporated into management plans.
8. Introduce the use of ‘decision-making frameworks’ to select targets and choose management options.
9. Identify which management plan (pathway management, eradication/containment/species-based or site-based management) may be effective in different contexts and for different species.
10. Ensure the availability of resources for basic research on IAS and to implement management actions. Funds need to be earmarked for an

(continued)

Box 17.4 (continued)

extended period of time to evaluate management responses and to check re-invasion.

11. Promote biological control since mechanical and chemical methods are only of short-term efficacy, labour intensive (mechanical/physical methods), or with non-target impacts (chemical control).
12. Avoid delay in approving, importing, and releasing of biocontrol agents—develop an effective framework at the Govt level to simplify procedures.
13. Restoration activities of sites with native species following the removal of all invasive species may be made mandatory.
14. Improve domestic quarantine regulations—employ qualified and trained staff at state borders.
15. Address how to address IAS in the climate /land-use change scenarios.
16. Address the existing confusion on terminologies related to IAS.
17. Address how to manage conflict species—harmful IAPS on which rural communities are dependent for their livelihood.

Necessary regulations for all the above actions may be included in the NISSAP.

17.5.1 Policy Formulation and Legal Enforcement

The discrepancies in the number of IAPS between government reports and scientific publications highlight that the agency should first create authentic baseline data on alien species and a scientifically informed and standardised risk assessment framework to identify alien species that are invasive (Action Point: 67, see Box 17.1). In addition, a legal framework should also be developed to address the IAS issue. Since pest risk assessment involves a high cost, the government may share the costs with any trading agencies wherever possible. Previous studies have shown that the costs of risk assessment borne by the industry have helped to reduce the number of IAS in New Zealand (Hulme et al. 2018). The practice may also increase the cost of IAPS in the (e.g., horticulture) market, shifting consumer preference towards native species. Therefore, the agency should emphasise corporate responsibility and voluntary codes of conduct for the industry, adhering to international norms and standards. While formulating the policy and legal frameworks, the agency should consider the interests of stakeholders and ensure public compliance. Once formulated, the agency should warrant and regularly audit the compliance of all actors to the legislation.

17.5.2 Networking and Information Sharing

The agency may ensure information sharing and maintain uniformity of management actions across different levels of operation, which can be achieved through the existing national network of the State Biodiversity Boards (SBBs) and the local Biodiversity Management Committees (BMCs). India shares borders with many South Asian countries, and trade volume is high through these open (e.g., Nepal) or porous (e.g., Bangladesh) borders. Therefore, the agency may create a collaborative network with the neighbouring nations for the exchange of information and the harmonisation of regulations in order to effectively regulate the import and export of IAPS. It will also be helpful to track imminent threats of IAPS from across the border and to share new techniques and tools to deal with IAPS.

17.5.3 Capacity Development

To effectively implement its regulatory policies, the agency may make use of its core resources to create a committed and qualified task force. Training divisions, much like the Centre for Biodiversity Policy and Law, may be developed to increase capacity and expand the nation's current biosecurity infrastructure. For example, customs and quarantine officials may be trained in the updated quarantine regulations and the use of sanitary and phytosanitary measures (in conformity with international standards) during export/import to prevent the accidental introduction of IAPS through trade and transport. Other stakeholders (scientists, agriculturists, forest managers, and plantation owners) may be trained on new tools and techniques to conduct surveillance (use of sensor networks, environmental DNA, remote sensing), detect and identify IAPS and manage them. The capacity-building programmes, such as those organised by the National Institute of Plant Health Management under the Ministry of Agriculture and Farmers Welfare, may be enhanced, and training programmes may be held more frequently. To handle the influx of information on IAS from various stakeholders and the dissemination of actions by the agency, an information exchange unit may be established in each state. Continuous surveillance of the trade network to trace the handling of IAPS also needs to be ensured. The agency may seek the help of non-governmental organisations for effective and long-term monitoring of the trade network.

17.5.4 Building Public Awareness

Effective management of IAS is possible only with the active participation of all stakeholders. The agency may, therefore, develop an effective communication strategy to increase public awareness (among common people, farmers, foresters, planters, and scientists to policymakers) on the damages due to IAS and ensure people's participation in planning and implementing management activities. To attract the wider attention of policymakers, officials, students, and the common man, the agency may publish articles on IAS in newspapers and magazines, prepare field guides in vernacular languages, exhibit posters on the damages due to IAS in air and seaports and offices of the forest, agriculture, and customs departments and schools and colleges to educate all concerned. Social media, interactive web portals, and mobile applications can entice people to participate voluntarily and raise much-needed knowledge of IAS among the general public. The stakeholders may be made aware of the existence of the agency and its legal framework.

Including information on invasive alien species in school and college curricula will encourage the younger generation to participate in IAS detection and management. Allotting short-term research projects to list IAS on school and college campuses will help engage students. Citizen science initiatives can be a valuable tool for real-time data collection on the IAPS and identifying potential IAPS like those that accidentally escaped cultivation.

17.5.5 Strengthening Scientific Research

The agency may identify research gaps and encourage scientists to conduct cutting-edge research on IAS detection and management. The scientists may also be involved in policy formulation on IAS prevention and management. A major area of research that can better inform the NISSAP and directly relate to policy development is nationwide inventories on IAPS with added information on introduction pathways, drivers, ecological characteristics, current distribution, and economic impacts. The information generated, irrespective of quantity, will be useful for framing policies and aiding future research [e.g., Indian Alien Flora Information Database (Pant et al. 2021)].

In short, information on pathways of introduction will help preparedness and prevention. The choice of management action will be greatly helped if we know where the invasive species in question is on the invasion continuum (introduction, establishment, spread, and widespread). This information will also help to identify imminent threats and enable proactive management measures. The lack of viable alternative options may impede management plans; therefore, the agency may promote research to identify native species with comparable consumer benefits as the IAPS. In summary, developing an information system on major IAPS (including information on probable pathways of invasion, drivers, species distribution, impacts, and management methods, if already known) would help formulate effective options for management.

17.6 Conclusions

There is no denying that India's rich biodiversity is at stake, and the economic and environmental damages due to invasive alien species are believed to be huge, though no reliable statistics are available. Therefore, it is suggested that the Government of India may attach greater importance and priority to preventing and managing biological invasions. The country has the potential to approach the issue proactively, but the existing biosecurity infrastructure is inadequate to successfully prevent and manage the problem and requires improvement. Also, there are several impediments to implementing management actions, such as lack of coordination between government agencies, absence of dedicated policies and regulatory bodies, inadequate capability and capacity to address the issue, scarcity of resources, and conflicts of interest among stakeholders. This chapter proposes certain simple measures to strengthen the existing policies to prevent future invasions and manage ongoing invasions in the country. It also emphasises the need for a dedicated strategy and action plan and the appointment of a single nodal agency to implement the policy regulations and action plans to manage IAS. Further, it stresses the need of networking and sharing information, creating awareness, capacity development, and strengthening research and includes a list of priority action points for the nodal agency to implement so as to overcome the current impediments to implementing management actions. The above suggestions, if applied scrupulously, would greatly assist the country in preventing impending invasions by alien species and mitigate damages from the existing invasions.

References

Auld B (2012) An overview of pre-border weed risk assessment and post- border weed risk management protocols. Plant Protect Quat 27:105–111

Bacher S, Blackburn TM, Essl F, Genovesi P, Heikkilä J, Jeschke JM et al (2018) Socio-economic impact classification of alien taxa (SEICAT). Methods Ecol Evol 9(1):159–168

Banerjee AK, Khuroo AA, Dehnen-Schmutz K, Pant V, Patwardhan C, Bhowmick AR et al (2021) An integrated policy framework and plan of action to prevent and control plant invasions in India. Environ Sci Pol 124:64–72. https://doi.org/10.1016/j.envsci.2021.06.003

Blackburn TM, Pyšek P, Bacher S, Carlton JT, Duncan RP, Jarošík V et al (2011) A proposed unified framework for biological invasions. Trends Ecol Evol 26(7):333–339. https://doi.org/10.1016/j.tree.2011.03.023

Boy G, Witt A (2013) Invasive alien plants and their management in Africa. CABI Africa, Nairobi

Bradley BA, Allen JM, O'Neill MW, Wallace RD, Bargeron CT, Richburg JA et al (2018) Invasive species risk assessments need more consistent spatial abundance data. Ecosphere 9(7):e02302. https://doi.org/10.1002/ecs2.2302

Chatterjee D (1940) Studies on the endemic flora of India and Burma. J Asiat Soc Bengal 5:19–67

Cordeiro B, Marchante H, Castro P, Marchante E (2020) Does public awareness about invasive plants pays off? An analysis of knowledge and perceptions of environmentally aware citizens in Portugal. Biol Invasions 22(7):2267–2281. https://doi.org/10.1007/s10530-020-02247-z

Essl F, Nehring S, Klingenstein F, Milasowszky N, Nowack C, Rabitsch W (2011) Review of risk assessment systems of IAS in Europe and introducing the German–Austrian Black List Information System (GABLIS). J Nat Conserv 19(6):339–350

Gupta K, Sankaran KV (2021) Forest biosecurity systems and processes: an Indian perspective (policy and practice reviews). Front For Global Change 4:699950. https://doi.org/10.3389/ffgc.2021.699950

Hanley N, Roberts M (2019) The economic benefits of invasive species management. People Nat 1(2):124–137. https://doi.org/10.1002/pan3.31

Hulme PE, Brundu G, Carboni M, Dehnen-Schmutz K, Dullinger S, Early R et al (2018) Integrating invasive species policies across ornamental horticulture supply chains to prevent plant invasions. J Appl Ecol 55(1):92–98. https://doi.org/10.1111/1365-2664.12953

Kannan R, Shackleton CM, Shaanker RU (2013a) Reconstructing the history of introduction and spread of the invasive species, lantana, at three spatial scales in India. Biol Invasions 15(6): 1287–1302

Kannan R, Shackleton CM, Shaanker RU (2013b) Playing with the forest: invasive alien plants, policy and protected areas in India. Curr Sci 104(9):1159–1165
Khetarpal R, Gupta K, Dev U, Dashora K (2017) Prevention and related measures for invasive alien plants in India: policy framework and other initiatives. In: Ellison CA, Sankaran KV, Murphy ST (eds) Invasive alien plants: impacts on development and options for management. CABI, Wallingford
Khuroo AA, Reshi ZA, Malik AH, Weber E, Rashid I, Dar G (2012) Alien Flora of India: taxonomic composition, invasion status and biogeographic affiliations. Biol Invasions 14(1): 99–113. https://doi.org/10.1007/s10530-011-9981-2
Leung B, Roura-Pascual N, Bacher S, Heikkilä J, Brotons L, Burgman MA et al (2012) TEASIng apart alien species risk assessments: a framework for best practices. Ecol Lett 15(12):1475–1493
Nathan R, Schurr FM, Spiegel O, Steinitz O, Trakhtenbrot A, Tsoar A (2008) Mechanisms of long-distance seed dispersal. Trends Ecol Evol 23(11):638–647. https://doi.org/10.1016/j.tree.2008.08.003
Nayar MP (1977) Changing patterns of the Indian flora. Bull Bot Surv India 19:145–155
Negi GCS, Sharma S, Vishvakarma SCR, Samant SS, Maikhuri RK, Prasad RC et al (2019) Ecology and use of Lantana camara in India (journal article). Bot Rev 85(2):109–130. https://doi.org/10.1007/s12229-019-09209-8
Nunez MA, Pauchard A (2010) Biological invasions in developing and developed countries: does one model fit all? Biol Invasions 12(4):707–714
Pagad S, Genovesi P, Carnevali L, Schigel D, McGeoch MA (2018) Introducing the global register of introduced and invasive species. Sci Data 5(1):170202. https://doi.org/10.1038/sdata.2017.202
Pandey R, Parmar PJ (1994) The exotic flora of Rajasthan. J Econ Taxon Bot 18:105–136
Pant V, Patwardhan C, Patil K, Bhowmick AR, Mukherjee A, Banerjee AK (2021) ILORA: a database of alien vascular Flora of India. Ecol Solut Evid 2(4):e312105. https://doi.org/10.1002/2688-8319.12105
Patnaik P, Abbasi T, Abbasi SA (2017) Prosopis (Prosopis juliflora): blessing and bane. Trop Ecol 58(3):455–483
Rai PK (2022) Environmental degradation by invasive alien plants in the anthropocene: challenges and prospects for sustainable restoration. Anthr Sci 1(1):5–28. https://doi.org/10.1007/s44177-021-00004-y
Reddy CS (2008) Catalogue of invasive alien flora of India. Life Sci J 5(2):84–89
Reyns N, Casaer J, De Smet L, Devos K, Huysentruyt F, Robertson PA et al (2018) Cost-benefit analysis for invasive species control: the case of greater Canada goose Branta canadensis in Flanders (northern Belgium). PeerJ 6:e4283. https://doi.org/10.7717/peerj.4283
Sandilyan S, Meenakumari B, Babu CR (2018) Guidelines for prioritization of invasive alien plants of India for management. National Biodiversity Authority, Chennai
Sankaran KV, Khuroo A, Raghavan R, Molur S, Kumar B, Wong LJ, et al (2020). Global register of introduced and invasive species - India. Version 1.3. GBIF. https://doi.org/10.15468/uvnf8m
Seebens H, Blackburn TM, Dyer EE, Genovesi P, Hulme PE, Jeschke JM et al (2017) No saturation in the accumulation of alien species worldwide. Nat Commun 8:14435
Shackleton RT, Shackleton CM, Kull CA (2019) The role of invasive alien species in shaping local livelihoods and human well-being: a review. J Environ Manag 229:145–157. https://doi.org/10.1016/j.jenvman.2018.05.007
Simberloff D (2006) Risk assessments, blacklists, and white lists for introduced species: are predictions good enough to be useful? Agric Resour Econ Rev 35:1–10
Singh K, Shukla AN, Singh J (2010) State-level inventory of invasive alien plants, their source regions and use potential. Curr Sci 99(1):107–114

Soumya R, Sajeev TV (2020) Introduction of Lantana camara L. and Chromolaena odorata (L.) king and robins to India and their spread to Kerala: political ecology perspective. Trop Ecol 61(3):387–399. https://doi.org/10.1007/s42965-020-00095-5

Wani SA, Ahmad R, Gulzar R, Rashid I, Malik AH, Rashid I et al (2022) Diversity, distribution and drivers of alien flora in the Indian Himalayan region. Global Ecol Conserv 38:e02246. https://doi.org/10.1016/j.gecco.2022.e02246

The manufacturer's authorised representative in the EU is Springer Nature Customer Service Centre GmbH, Europaplatz 3, 69115 Heidelberg, Germany. If you have any concerns regarding our products, please contact ProductSafety@springernature.com

Printed and bound by CPI Group (UK) Ltd, Croydon, CR0 4YY
30/09/2026
02225322-0016